Fourth International Conference on Information Technology and Digital Applications (ICITDA 2019)

IOP Conference Series: Materials Science and Engineering
Volume 803

Yogyakarta, Indonesia
15 November 2019

Editors:

Taufiq Hidayat
Moh. Idris
Fayruz Rahma

ISBN: 978-1-7138-3508-0
ISSN: 1757-8981

Printed from e-media with permission by:

Curran Associates, Inc.
57 Morehouse Lane
Red Hook, NY 12571

Some format issues inherent in the e-media version may also appear in this print version.

Institute of Physics
Dirac House, Temple Back
Bristol BS1 6BE UK

Phone: 44 1 17 929 7481
Fax: 44 1 17 920 0979

techtracking@iop.org

Additional copies of this publication are available from:

Curran Associates, Inc.
57 Morehouse Lane
Red Hook, NY 12571 USA
Phone: 845-758-0400
Fax: 845-758-2633
Email: curran@proceedings.com
Web: www.proceedings.com

TABLE OF CONTENTS

APPLICATION

HARD SCIENCE AND ENGINEERING

SOFTWARE ENGINEERING

IOP Conf. Series: Materials Science and Engineering **803** (2020) 011001 doi:10.1088/1757-899X/803/1/011001

PREFACE

Welcome to International Conference on Information Technology and Digital Applications 2019 (ICITDA 2019).

ICITDA is intended to become an international forum for exchanging of ideas, knowledge, and experience on the latest development in the field of information technology and digital applications among researchers from academia, engineers, and practitioners from industries. This conference is also expected to enable opportunities for collaboration among the participants to advance the theory and practice in the field of information technology.

We also expect that the ICITDA will become a forum for collaboration among universities from around the globe. In fact, we worked together with some universities to host this conference in previous years, for instance Rajamangala University of Technology Thanyaburi (Thailand), Youngsan University (South Korea) and University of the Cordilleras (Philippines).

We thank The God for success in organizing ICITDA which is annual conference since 2016. Every year more than 50 papers were submitted to ICITDA. Around 50% were accepted and presented. The papers came from various continents e.g. America, Europe, Australia, and Asia.

This year, ICITDA received 115 submissions from 9 countries, and 56 papers were accepted in 3 tracks: applications, hard science and engineering, and software engineering. The authors of the papers came from institutions of 6 countries: Philippines, Indonesia, South Korea, New Zealand, Malaysia and China.

Last but not least, we wish you a lot of benefits from this conference.

Taufiq Hidayat, Chair

November 2019

IOP Conf. Series: Materials Science and Engineering **803** (2020) 011001 doi:10.1088/1757-899X/803/1/011001

CHAIRS

Host

- Department of Informatics, Faculty of Industrial Technology, Universitas Islam Indonesia

Editors

- Taufiq Hidayat
- Moh. Idris
- Fayruz Rahma

Organizing Committee

- Taufiq Hidayat (Chairman of The Committee)
- Hendrik
- Ahmad M.R. Pratama
- Mohammad Idris
- Fayruz Rahma
- Hari Setiaji

Program Committee

- Th.P. van der Weide (Radboud University Nijmegen, Netherlands)
- Jefferson Tan (IBM Research, Australia)
- Agfianto Eko Putra (Universitas Gadjah Mada, Indonesia)
- Kridanto Suhendro (Institut Teknologi Bandung, Indonesia)
- Taufiq Asyhari (Cranfield University, United Kingdom)
- Arif Djunaidy (Institut Teknologi Sepuluh Nopember, Indonesia)
- Alvin Malicdem (Don Mariano Marcos Memorial State University, Philippines)
- Yudi Prayudi (Universitas Islam Indonesia, Indonesia)
- Sri Kusumadewi (Universitas Islam Indonesia, Indonesia)
- Zainudin Zukhri (Universitas Islam Indonesia, Indonesia)
- Hendra Setiawan (Universitas Islam Indonesia, Indonesia)
- Raden Teduh Dirgahayu (Universitas Islam Indonesia, Indonesia)
- Mukhammad Andri Setiawan (Universitas Islam Indonesia, Indonesia)

IOP Conf. Series: Materials Science and Engineering **803** (2020) 011001 doi:10.1088/1757-899X/803/1/011001

Keynote Speeches

EVENING OUT THE BOTTLENECKS FOR TODAY'S BLOCK CHAIN SYSTEMS

Prof. Hai Jin

Huazhong University of Science and Technology

Prof. Hai Jin is a Professor of Computer Science and Engineering at the Huazhong University of Science and Technology (HUST) in China. He is now the Dean of School of Computer Science and Technology at HUST.

He received his Ph.D. in computer engineering from HUST in 1994. In 1996, he was awarded German Academic Exchange Service (DAAD) fellowship for visiting the Technical University of Chemnitz in Germany.

He worked for the University of Hong Kong between 1998 and 2000 and participated in the HKU Cluster project. He worked as a visiting scholar at the University of Southern California between 1999 and 2000. He is the chief scientist of the largest grid computing project, ChinaGrid, in China.

Dr. Jin is a senior member of IEEE and member of ACM. He is the member of Grid Forum Steering Group (GFSG). His research interests include computer architecture, cluster computing and grid computing, virtualization technology, peer-to-peer computing, network storage, network security.

CAMERA PLACEMENT BASED ON RISK MAPS

Assoc. Prof. Vijanth S. Asirvadam

Universiti Teknologi PETRONAS

Dr. Vijanth S. Asirvadam is from an old mining city of Malaysia called Ipoh. He studied at University of Putra, Malaysia in Bachelor Science(Hon) majoring in Statistics and graduated in April 1997 before leaving to Queen's University Belfast to do his Masters where he received his Master's Science in Engineering Computation with a Distinction.

IOP Conf. Series: Materials Science and Engineering **803** (2020) 011001 doi:10.1088/1757-899X/803/1/011001

He later joined the Intelligent Systems and Control Research Group at Queen's University Belfast in November 1999 where he completed his Doctorate (Ph.D.) researching into Online and Constructive Neural Learning methods.

He took previous employments as a System Engineer and later as a Lecturer at the Multimedia University Malaysia. He was also a Senior Lecturer at Faculty of Engineering and Computer Technology, AIMST University. Since November 2006, he served as a Senior Lecturer and later as Associate Professor at the department of Electrical and Electronics Engineering, Universiti Teknologi PETRONAS.

His research interest in theory includes linear and nonlinear system identification and model validation. On application side he research into computing techniques in signal, image and video processing. Dr Vijanth is member of Institute of Electrical and Electronics Engineering (IEEE) and also a member of Institute of Engineering Technology (IET) . He has over 100 publications in proceedings and journals in the field of computational intelligence, control, signal and image processing.

IOP Conf. Series: Materials Science and Engineering **803** (2020) 011002 doi:10.1088/1757-899X/803/1/011002

Peer review statement

All papers published in this volume of *IOP Conference Series: Materials Science and Engineering* have been peer reviewed through processes administered by the proceedings Editors. Reviews were conducted by expert referees to the professional and scientific standards expected of a proceedings journal published by IOP Publishing.

IOP Conf. Series: Materials Science and Engineering **803** (2020) 012001 doi:10.1088/1757-899X/803/1/012001

Flipped Learning in Information Technology Courses: Effectiveness and Impact

Annissa Kurniasari, Hendrik

Department of Informatics, Islamic University of Indonesia, Yogyakarta

Email: anisakurniasari97@gmail.com, hendrik@uii.ac.id

Abstract. This study's purpose is to investigate the implication of the use of flipped learning on students' performance and perception. Participants are the second-year and third-year undergraduate students enrolled in the Information Technology courses. One hundred and twenty-five students were the participants associate with an experimental group and a control group. The experimental group consisted of the students of the ISD class A and BI courses. Meanwhile, the students of ISD class D and MIT were the control group. The experiments employed two instruments: (1) students' final score, and (2) questionnaire. This study used the independent sample t-test and two-proportion z-test to analyze the collected data. The result showed that the students' performance in the FL classes was superior to a traditional learning setting. Students' perception of both courses is higher than the experimental group. Besides, we observed that studying in an FL model had a positive impact on students' motivation. These findings suggest that the FL could be a promising way of enhancing students' learning outcomes.

1. Introduction

Flipped learning (FL) is one of the innovative learning models that have been developed by Bergmann and Sams in the early-mid 2000s. This approach refers to a learning model that reorders how students spent their time in and out of the classroom by bringing lectures outside the classroom to provide valuable learning time[1]. The traditional way of teaching making students listening to lectures in class while having them reading textbooks and working on assignments outside of the classroom. In an FL, presentations, recorded lectures,and online readingsare assigned to the students by the teachers. This approach allows the students to assume the ownership of learning by learning on their own time [2].

In the last two decades, the availability of the internet and computers in the learning process has strengthened their commitment to using the computer to improve the learning outcomes. This strategy is believed to improve the student's cognitive abilities during the class. Most of the early studies in the FL evaluated the implementation in non-technological subjects, such as foreign languages, algebra, mathematics, chemistry, and others [3]–[7]. It happened because the researcher believed that the teaching material is more theoretical than technological courses. The concept of material is easier to understand, provided by the availability of online material [7]. It provides positive results for non-technological subjects. Several studies on the effectiveness of FL also conducted on computer science subjects, including the introduction of programming (JAVA, XML, and advanced JAVA), software engineering, and computing I [8]. The ability to write and understand programs is an essential skill needed. The results showed that FL gave positive results in terms of students' performances and attitudes. The FL design also allows students to become more involved in the learning process. The students also realized that they could understand the content independently at home before the class began [9].

At the beginning of 2018, our department applies FL as a learning model for some subjects. This study provides a review of the FL implementation. Besides, this study examined whether the performance and perception of students in the learning process varied from the traditional model previously used by the department.Lastly, this research explores the perspective of students to regenerate the FL model, which improves the FL model's teaching and learning efficiency.The results of this study could help improve the potential implementation of the FL model. This study addressed two specific research questions: (1) Does FL affects students' performance?; (2) how effective are the learning method of FL and its effect on the level of students' perception?

The structure of this article consists of: background on the introduction section and review of relevant literature on related works section, followed with a description of research methodology, result, discussion, and conclusion.

2. Related Works

A flipped learning (FL) is a part of a blended learning environment, in which students learn instructional content, watch video lectures at home and do the homework in class. At the same time, the instructors also provide customized instruction and collaboration with other students rather than lecturing[10]. Bishop and Verleger (2013) explain flipped learning as one that consists of two parts: interactive group learning activities within the classroom and an individual direct computer-based instruction outside the classroom. First, FL is an interactive learning group activity that goes into the classroom. Second, the outside class session is directing by an individual computer-based instruction. By allowing this approach, the students become more involved and responsible for their learning process. The students also get immediate feedback from the teachers during class activities.

The FL model, which creates a more learner-centered environment, indicated an impressive learning achievement. Moreover, it reflected behavioral improvement and enhanced communication in the FL classroom between teachers and students [10]–[11]. The FL enables students to learn course content with online educational materials such as video, film, and voices outside of the classroom environment. Also, this model allocates time for active learning activities such as question-answer, discussion, and problem-solving. Students are responsible for their progress and success in learning in the FL model [7]. Using this learning model, they increase their knowledge, enhance their comprehension, and increase their responsibilities for each course content [2], [4], [12], [13].

Research shows two primary advantages of implementing the FL. First, the FL replaces lecture time in class for active, collaborative, and problem-based learning and helps to improve the concepts of learning without reducing the content [14], [15]. Second, the FL can motivate students to become self-learners and life-long learners [7], [16], [17]. The researchers agree that FL has a positive influence on students. Not only improve students learning outcomes but also students' attitude and participation.

In terms of computer science courses especially programming courses, the findings study found that the FL model has a great potential to improve students' programming capability in introductory programming as well as students' academic performance [10], [13], [16]. The study showed that the FL model is suitable for an introductory programming course, where students find programming difficult [16]. It is beneficial using the FL classroom supported by technology. The approach provides the student with comprehensive access to valuable resources and learning material [13] in a medium that they believe applicable and appropriate to their learning style.

However, it also has several issues in the implementation. For example, students often find it difficult to complete online assignments and set their priorities in self-study sessions. Another one is most studies used video-based learning as a media of FL [3], [18]–[20]. Several study results it is known that not all of the students like the FL model that uses video lectures from the instructor (video record) as a media for delivering the material [18]. According to the students, the video only contains instructors who explain the material verbally and monotonously without any additional animation or visualization, so students feel bored, and the learning process becomes unattractive.

3. Research Methodology

3.1. *Hypotheses*
Based on the above research questions, the authors formulated the following research hypotheses.
H_1: The experimental group will result in significantly higher students' learning performance when compared to the control group.
H_2: The experimental group will result in more extensive changes in levels of students' perception when compared to the control group

3.2. *Participants*
The participants of this study involved 98 students from two different Information System Development (ISD) classes, 50 students from the ISD class A as the experimental class, and 48 students from ISD class D as the control class. There also 27 students from the Business Intelligence (BI) class as the experimental group and Management of Information Technology (MIT) as the control group. In total, there are 125 students, and all of them have completed the survey questionnaires.

3.3. *Course design and procedures*
All of the courses (ISD, BI, and MIT) were six credits courses which taught twice a week. There are five ISD classes, but for this study, we only examined two classes, class A as the experimental group (used FL) and class D as a control group. Both have the same content material but the different instructors. Besides, we investigated BI class as an experimental group that applied the FL approach, and the MIT course became the control group. Students of the FL class learned the material by watching educational video lectures and reading other materials posted by Google Classroom. Later in the classroom session, students took part in face-to-face classroom activities such as group discussions and student presentations. The students were given several quiz questions as pre-class materials. In designing the online quiz questions, the instructor used the Google Form and Kahoot!. All classes have to complete five achievements.

In the FL model, students were assigned to watch or read the relevant lecture online before class. These lectures averaged 45 minutes in length that divided into 4 to 5 difference videos. It consisted of a PowerPoint presentation narrated by a lecturer. Students were encouraged to take notes and write down any questions they had. For viewing the entire term, all videos and materials are available online.

During the classroom session, the class began with approximately 10 to 15 minutes of quiz and 10 minutes of question and answer. For the next minutes, the students worked from the assignment on a problem that set in pairs or groups so that they can discuss it. While the pairs or groups worked on their problem sets, the lecturer, availed to students who had any problems or questions. This work was not collected offline, but the students are required to upload it online on Google Classroom.

3.4. *Instruments*

3.4.1. Students' performance. To measure the students' performances, we compare the final score of students. The final score is the accumulation of students' learning outcomes in exams, quizzes, assignments, mid and final test.

3.4.2. Students' perception. At the end of the study, students were given a survey to measure students' perceptions (**Table 1**). The students' perception in this study included students' background information about the participants, such as subject, class level, and their opinion about the learning method that used. Next, students were asked to rate their abilities to complete a particular task successfully and their level of understanding on the course content using a 4-point Likert scale (Strongly Disagree = 1, Disagree = 2, Agree = 3, Strongly Agree = 4). The students also were asked the learning media used in class. Then, the students choose the learning styles that work on them. At the end of the survey, the open-ended questions asked the students to give their viewson the learning process and their suggestions on how the

FL can be improved. A total of 63 in the experimental group and 71 in the control group, valid responses were received.

Table 1. Students' perception section questionnaire

Sub-section	Questions
Students' ability to complete the task	I was excited to study the course content before the class begins. I pay attention to the learning content during the learning process. I ask the teacher when having difficulties. I only study the course content material in-class sessions (face-to-face). I prefer to be silent when I do not understand the course content material. I only study for quizzes and exams. I enjoyed learning using discussion. I become silent when discussion sessions in groups. I want to learn to explore knowledge. I try to find references from other sources when I have difficulty in understanding the material. I enjoy exploring the material provided by teachers.
Students' capability to understand the course content	I know the course content that will be taught by teachers. I understand the relevant examples given by lecturers from the concepts taught. I understand the current issues in the field being taught. I can apply the subject matter in daily life. I can analyze the course content. I can apply the course content more broadly. I can do the quizzes and test that given well.

3.5. Data analysis

The Statistical Package for Social Sciences (SPSS) has been used to analyze the data. The students' performance was summarized using descriptive analysis by present the mean (M) and standard deviation (SD). Independent samples t-test was applied to examine the students' performance both in the experimental group and control group. We put forward a significance level of α= < 0.05.

Responses to the closed-ended questions in the survey were analyzed using a two-proportion z-test. The test was used to determine if students' perception of their ability to complete a specific task differed across experimental and control groups. Two proportion z-test test was also used to determine if students' perception of their capability of understanding of the course content differed between the experimental and control group. A significance level was set as $\alpha = 0.05$.

4. Result

4.1. Students' performance

We evaluate the students' performance in ISD, BI, and MIT courses by employing an independent samples t-test to compare the students' final scores. The mean scores of the students in the experimental and the control group showed significant differences. Overall, the analysis showed that the experimental group's mean score ($M = 79.873, SD = 4.559$; $M = 77.096, SD = 13.845$) was higher than the control group ($M = 72.807, SD = 10.363$; $M = 69.593, SD = 12.451$) at the 0.05 level of significance. For ISD course $t = 4.338$, $t > t_\alpha$, while BI-MIT course has $t = 2.094$, $t > t_\alpha$. These results confirmed the hypothesis (H_1) that the experimental group result is significantly higher on students' learning performance when compared to the control group in information technology courses. It means that FL makes students performed academically better than traditional learning. Table *2* presents the summary of the results.

Table 2. Independent samples t-test result of students' performance on experimental and control group

	EG (n=77)		CG (n=75)		*t-value*
	M	*SD*	*M*	*SD*	
ISD	79.873	4.559	72.807	10.363	4.338
BI-MIT	77.096	13.845	69.593	12.451	2.094

Note: EG (experimental group); CG (control group); ISD (information technology development); BI-MIT (Business Intelligence-Management of Information Technology); M (mean); SD (standard deviation).

We also determined the students' grade distribution based on their final scores. In this section, we compared the students' final scores on ISD, BI, and MIT courses between experimental (n=77)and control group (n=75). From **Figure 1** below, we can see whetherthe experimental group or control group has varying values. Based on its courses, most of the students in the experimental group on the ISD course has A/B and B- grade with a total of 10 students (20%). Meanwhile, in the control group, the highest percentage of students was found on the A/B grade of 35 students (72%).

Different things happen in BI and MIT courses. The course showed that most grades got by the control group was A with 40 percent of students. However, the students who got grade F (failed) almost as high as the A. On the contrary, in the experimental group, the grades more distributed. The highest percentage of the students got B+. Nevertheless, less than 15 percent of students who got A.

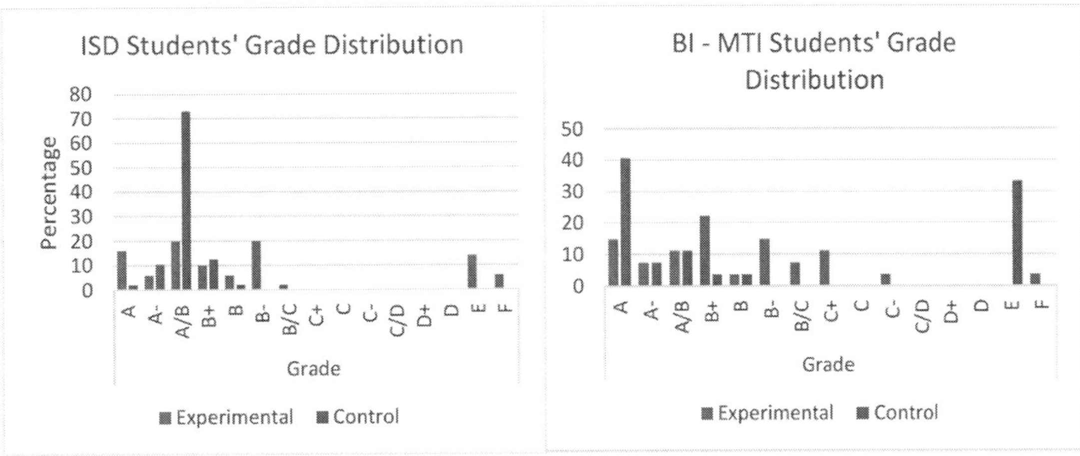

Figure 1. Students' grade distribution based on courses

4.2. Students' perception

Generally, the students had a positive response toward the FL model. They stated that they liked the FL because they can fulfil their curiosity and effectively understand the course content. Some of the course content is delivered using videos. Students stated that using the videos facilitated their understanding because the videos can be access anytime and anywhere. They also agreed that the FL is making them more active in discussion and learning session. The quizzes that held in the class were also fun. The teacher used multiple platforms and helped them understand the content more.

The students' perception was focused on students' confidence in their abilities to complete a particular task and their capability of understanding the course content material. When asked about their confidence to complete the task, 45 participants (71%) agreed that they exited to learn and understand the course content before the class begins. Students stated that they could prepare themselves before the class begin and ask the teacher for further. Up to 31 participants (49%) will explore the course content as a reference. They become more active in participating in every class session of FL class than in the traditional class. In the closed-ended question on the course content understanding section, up to 50 participants (77%) agree that they can understand the material and analyze each course and example. Furthermore, 44 participants (68%) state that they can do de quizzes, assignments, and tests easily.

We determine the FL's impact in improving students' confidence to complete the task and their capability to understand the course content by using a two-proportion z-test to compare them. Twenty-seven percent of the experimental group, perceive that their confidence to complete the task is "Very High" ($z = 0.995,\ z > z_\alpha$). However, only 20 percent of participants in the control group sense the same condition. The results show that the students' confidence in the experimental group to complete the task was higher compared with the control group. Afterward, we also used a two-proportion z-test for analyzing students' capability to understand the course content. As shown in Table 3., 11 participants categorized as "Very High" on students' capability to understand the course content ($z = 0.823,\ z > z_\alpha$). Nevertheless, only 7 participants categorized as "Very High" on the control group. The results show that the students' capability to understand the course content was higher compared with the control group.

These results confirmed the second hypothesis (H_2) that students' perception of the experimental group was higher when compared to the control group in information technology courses.

Table 3. Students' perception categorized from experimental and control group

Students' perception	**Experimental group (n=63)**				**Control group (n=71)**			
	Frequency (%)							
	Very High	**High**	**Low**	**Very Low**	**Very High**	**High**	**Low**	**Very Low**
Students' ability to complete the task	27	68	5	0	20	74	6	0
Students' capability to understand the course content	11	68	21	0	7	70	23	0

5. Discussion

The results of this study show that the students in the FL increased students' performance and perception of information technology courses. Regarding students' performance, scores of the students who used the FL model were higher than the students who used the traditional learning model. This result cannot be separated from the fact that the students learn the course content before the class begins. They learn the content material and looking for a similar reference for their understanding. The students were also actively asking the questions to the instructor in the classroom session. They receive immediate feedback from the instructor and have much interaction with their friends and instructor during a lesson. The students came to the class more prepared and it increased the student-student and the student-instructor interaction[10], [16], [18], [21], [22]. In terms of the students' performance, [8], [10], [15], [22] showed the improvement. However, the other findings have shown that the FL model has a less positive impact on students' performance[3], [13], [23]. The differences could be related to the various materials, the classroom environments, and the processes used to apply and support the FL model. The types of the materials used and the surrounding learning environment also can affect students' performance.

Based on the open-ended questions section, student response positively toward the FL model. Students stated that FL supported their learning process when it comes to practicing section. They loved to involve and remain to be motivated in every learning process. It is easier for them to apply the concept that they have learned before. Students had access to learning resources at their own pace whenever they needed it. This approach ensures that students can study and understand the course content on their own. This study confirms the other studies, such as[2], [13], [21]. The students also stated that the FL change their old learning habits, and they enjoyed it because they could teach themselves before class and discuss with others in the class, looking for answers to question actively. Unfortunately, some students find that discussion is hard if the class environment does not support it. It might be easier for them to know what will they learn over the next six months.

Some of earlier research studies showed that the FL model was more suitable for upper educational level [23], [24], as the FL model may be hard and challenging for students who have not developed strong study skills [13], [24]. Also, the FL model may be more suitable for a theoretical course. It happens because students learn and understand the theoretical part easier [7], [13]. There is a lack of earlier research on the applicability of the FL model and suggest future studies should investigate this research gap. This study reports the success of implementing the FL in technology course that more practical.

6. Conclusion

This study found some promising significant impact of implementing the FL model on information technology courses. The improvement of students' performance is proof that the FL model design appears to make a difference in the educational field. By using the FL model, students can spend enough time watching video lectures, preparing themselves well before attending the class (face-to-face) session in which they are actively involved in solving problems based on the questions given. In terms of students' perception, the FL model results in more significant changes in the students' abilities to complete a specific task and their capability of understanding the course content as compared to the traditional learning model. The study showed that the FL model also suitable for technology courses that have more practicing sections.

The FL effectively improves the students' performance and changes the students' perception in the learning process. However, the limitations of this study should be pointed out. Further research in the form of a further study and more practical courses may be useful in clarifying the other aspects that impact the student.

References

[1] Polytechnic N 2014 Flipped classroom for differentiated learning *Yap Tat Kwong and Pang Nai Kiat School of Information Technology, Nanyang Polytechnic, Singapore* PANG_Nai_Kiat@nyp.edu.sg Issue September
[2] O'Flaherty J and Phillips C 2015 The use of flipped classrooms in higher education: A scoping review *Internet High. Educ.* vol **25** pp 85–95
[3] Chilingaryan K and Zvereva E 2017 Methodology of Flipped Classroom as a Learning Technology in Foreign Language Teaching *Procedia - Soc. Behav. Sci.* vol **237** June 2016 pp 1500–4
[4] Baepler P, Walker J D and Driessen M 2014 It's not about seat time: Blending, flipping, and efficiency in active learning classrooms *Comput. Educ.* vol **78** pp 227–36
[5] Kumar K, Chang C and Chang C 2017 International Forum of Educational Technology & Society The Impact of the Flipped Classroom on Mathematics Concept Learning in High School *International Forum of Educational Technology & Society Linked references are available on JSTOR* vol **19** no 3 pp 134–42
[6] Gilboy M B, Heinerichs S and Pazzaglia G 2015 Enhancing student engagement using the flipped classroom *J. Nutr. Educ. Behav.* vol **47** no 1 pp 109–14
[7] Evseeva A and Solozhenko A 2015 Use of Flipped Classroom Technology in Language Learning *Procedia - Soc. Behav. Sci.* vol **206** November pp 205–9
[8] Wang F H 2017 An exploration of online behaviour engagement and achievement in flipped classroom supported by learning management system *Comput. Educ.* vol **114** pp 79–91
[9] Giannakos M, Krogstie J and Chrisochoides N 2014 Reviewing the Flipped Classroom Research: Reflections for Computer Science Education *ACM Ref. Format* vol **7** pp 23–9
[10] Yilmaz R 2017 Exploring the role of e-learning readiness on student satisfaction and motivation in flipped classroom *Comput. Human Behav.* vol **70** pp 251–60
[11] Mohamed H and Lamia M 2018 Implementing flipped classroom that used an intelligent tutoring

system into learning process *Comput. Educ.* vol **124** December 2017 pp 62–76
[12] Chao C Y, Chen Y T and Chuang K Y 2015 Exploring students' learning attitude and achievement in flipped learning supported computer aided design curriculum: A study in high school engineering education *Comput. Appl. Eng. Educ.* vol **23** no 4 pp 514–26
[13] Tanner M and Scott E 2015 A Flipped Classroom Approach to Teaching Systems Analysis, Design and Implementation *J. Inf. Technol. Educ.* vol **14** pp 219–41
[14] Mok H 2014 Teaching tip: The flipped classroom *J. Inf. Syst. Educ.* vol **25** no 1 pp 7–11
[15] Lo C K, Lie C W and Hew K F 2018 Applying 'First Principles of Instruction' as a design theory of the flipped classroom: Findings from a collective study of four secondary school subjects *Comput. Educ.* vol **118** no July 2017 pp 150–65
[16] D'Souza M J and Rodrigues P 2015 Investigating the effectiveness of the flipped classroom in an introductory programming course *New Educ. Rev.* vol **40** no 2 pp 129–39
[17] Kong S C 2014 Developing information literacy and critical thinking skills through domain knowledge learning in digital classrooms: An experience of practicing flipped classroom strategy *Comput. Educ.* vol **78** pp 160–73
[18] Awidi I T and Paynter M 2019 The impact of a flipped classroom approach on student learning experience *Comput. Educ.* vol **128** September 2018 pp 269–83
[19] Long T, Logan J and Waugh M 2016 Students' Perceptions of the Value of Using Videos as a Pre-class Learning Experience in the Flipped Classroom *TechTrends* vol **60** no 3 pp 245–52
[20] Caviglia-Harris J 2016 Flipping the undergraduate economics classroom: Using online videos to enhance teaching and learning *South. Econ. J.* vol **83** no 1 pp 321–31
[21] Thai N T T, De Wever B and Valcke M 2017 The impact of a flipped classroom design on learning performance in higher education: Looking for the best 'blend' of lectures and guiding questions with feedback *Comput. Educ.* vol **107** pp 113–26
[22] Zainuddin Z 2018 Students' learning performance and perceived motivation in gamified flipped-class instruction *Comput. Educ.* vol **126** July pp 75–88
[23] McNally B, *et al* 2017 Flipped classroom experiences: student preferences and flip strategy in a higher education context *High. Educ.* vol **73** no 2 pp 281–98
[24] Akçayır G and Akçayır M 2018 The flipped classroom: A review of its advantages and challenges *Comput. Educ.* vol **126** August pp 334–45

A Web-Enabled GIS Platform for Agricultural Land Use

Gemma G. Acedo

Catanduanes State University, Virac, Catanduanes Philippines

Email: ggacedo@gmail.com

Abstract. One of the important and leading sources of livelihood in Catanduanes is agriculture. Agriculture has been an avenue to help improve the Philippine economy. To further boost the agriculture industry, the researcher aimed to design and develop a web-enabled GIS platform for agricultural land use that contains a database socio-demographic profile and geographic physical information details of agricultural lands of the province. Data collected are subjected to analysis of an approach in the development and implementation of the study and set as bases for government implementation and a possible policy that assist farmers and local communities. The features of the system help the local leaders to strategize activities and easily facilitate the identification, monitoring, forecasting, visualization, and analysis of situations on Agricultural land use. The acceptability rating of the system developed is 4.72 highly acceptable according to system users. The web-enabled GIS serves as an initiative to improve agricultural land use system by integrating GIS in terms of locating, mapping, profiling, and land management in the province. It is expected through this endeavour that an increase in harvest rate will benefit farmers and local communities. Utilization is expected to help combat climate change and preserve the environment and its ecosystem.

1. Introduction

The province of Catanduanes faces the Pacific Ocean to its east where a vast majority of typhoons usually hit the Philippines. With these calamities that frequent the island there have been many occasions that food supply becomes scarce due to delay in deliveries caused by disruptions in the sea and air transportation and damage to road networks.

Catanduanes is rich in natural resources which became the center for main development and growth of the province. The study utilizes Geographic Information System to assess agricultural land. The developing urbanization of Catanduanes is taking steps to lessen the accessibility of agricultural profitable lands in return for other, non-horticultural, settlement ventures. The growing urbanization of Catanduanes is threatening to diminish the availability of agriculturally productive lands in exchange for non-agricultural, settlement projects.Likewise, climate change has a major effect on water resources, crops, land usability, and health issues. Basically GIS is used in the identification of dynamics to the vulnerability of lands scheme and programs. It is used to statistically plan and answer problems in the utilization and maximization of land use which is one of the priority areas of the government.

With the use of Geographical Information System, it was found out that agricultural lands are affected by the increase in population and urbanization of the province to convert exchange for non-agricultural and settlement projects.

Dhungeldesigned and developed a web mapping application to manage large spatial data in an open source development in a three-tiered cloud architecture. It supports the need of spatial cloud computing to come up with the different domains of physical science. A wide variety of global change using Geospatial technology assists to monitor land use and understand the spatial and temporal distribution of land use [2]. The goal of this study is to design and develop a web enabled GIS mapping agricultural lands in the Province. The communality of Geospatial technology in the study is utilized to monitor land use in either spatial or manual survey in distributing land for mapping [4]. Acedostressed the importance of GIS to the relative needs of mapping and identifying the different kinds of agricultural land in light of an arrangement of inclinations, criteria and markers for the Province. The web enabled GIS fosters engagement in open and participatory governance through the use of information technology. This will improve open access to information and the government can enhance its effort to safeguard agricultural lands, preserve and combat natural and human calamities including climate change [5]. Agriculture is still the top priority of the government in addressing poverty, food safety and security of the people.

Geographical Information System is a computer assisted system that provides spatially referred data, GIS mapping is an important tool for managing agricultural area to attain and apply the correct information into an accurate planning environment [6]. Through this tool it can visualize and forecast by analysing the data acquired on how to manage and control resources to improve the current system in acquiring and generate a forecasted analysis on the outcome of the result.

With ICT support, the participatory governance of ICT and the government initiated an approach by sharing ideas to farmers and giving local communities priority as participants [7]. Orienting and educating farmers and local communities on the importance of land use and agriculture, help alleviate food security. Thus, the main objective of the study is to design and develop web enabled GIS platform for agricultural lands use. This system contains a database profile of agricultural lands of the province with specific location, latitude and longitude and specific area. Data collected are subjected to analysis for a better approach in the development and implementation of the study. The processed data sets as bases for government implementation and a possible policy that will assist farmers and the local communities.

2. Methods

The researcher used an experimental research design technique as the research relies on how the respondents assess and evaluate the web-enabled GIS platform for agricultural land use. The researcher used survey questionnaires which were of great help in identifying the perceptions of the system users in the web-enabled GIS platform for agricultural land use[8].The study used the Agile Iterative Development Model (shown in Figure 1), a combination of incremental and iterative process model that focus on client's fulfilment and process adaptability by immediate delivery of working applicationsystem [9]. This suits the web enabled GIS platform for agricultural land use perfectly considering the purpose of the research which is easily facilitate the identification, monitoring, forecasting, visualize and analyse situations on Agricultural land use.Scrum in agile programming can be applied to different software projects to manage anything that consists of team with different deliverables. Thus, it was applied in this GIS web enabled project. The web enabled GIS have seen the biggest benefit where there are so many tools and techniques developed for a conventional project which supports agile methods. The scrum method of agile programming creates an epic with system architecture and database design – geodatabase design where eachmethods contains user stories.

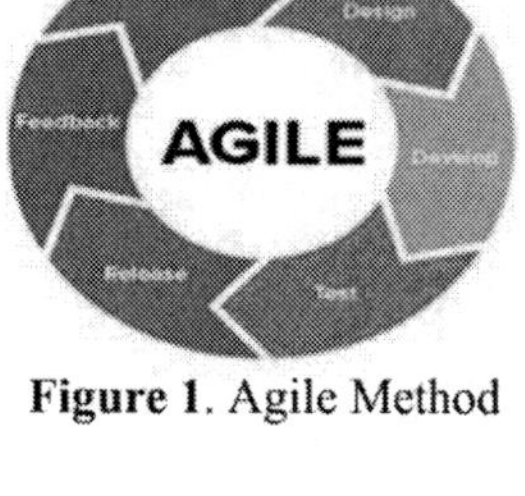

Figure 1. Agile Method

IOP Conf. Series: Materials Science and Engineering **803** (2020) 012002 doi:10.1088/1757-899X/803/1/012002

All the steps are done in every sprint thus defying the traditional waterfall approach.

2.1. Planning and Requirements Analysis

This is done during Sprint. This activity was done to set the objectives of the proposed system was identified based on the goal of the study. Gantt chart was created that set as the timeline in managing the activities of the study. Requirements which were identified as user stories and were evaluated based on priorities. These were collected from different government agencies and personnel. The Provincial Agriculture, Municipal Agriculture, Register of Deeds, Department of Agriculture, Provincial Planning Office, Municipal Planning and Development Office as concerned government agencies in agriculture, Secondary data and relevant concepts and studies from online resources.

2.2. Design

A technical solution design was built based on the user stories in the sprint backlog. User interfaces, data models, reports, and other things are designed for development.

2.3. Development

All tasks under the sprint backlog which were designed in full detail were converted into a running prototype. During the development different iterations were conducted with acceptable results in parallel with the users' acceptance. Identification of bugs and other issues were fixed after the iteration. Every iteration was tested to determine the full functionality as identified in sprint.

2.4. Testing

Testing was done during each sprint. The test cases for each sprint are based on the acceptance criteria provided by the client. Test cases are based on the agreement during sprint planning of the team depending on the timeline of the sprint. Agile Scrum methodology made possible changes for the research easier to include in later sprints.

3. Results and Discussions

This section highlights the result of the study conducted. This shows the features of the system with the architectural model in the deployment of the study and the result of technical user acceptance in terms of its functionality and usability

Table 1. Solution Details

#	API	Source	Target	Interface Type	Interface Description	New/Reuse/ Enhance
1.	Login as User	Web/App	API Gateway	Internal	Process Request	Reuse
2.	Validate User	Web/App	Web-GIS Server	Response	SOAP/REST	Reuse
3.	View Simulated Land Cover GIS	Web/App	Web-GIS Server	Internal	Process Request	Reuse
4.	View Data Visualization	Web/App	Web-GIS Server	Internal	Process Request	Reuse

Table 2. User Stories

Business Requirements/User Stories	*Computing Facility*	*Acceptance Criteria*
Municipal Agriculturist Should be able to receive any necessary documents	*Web tool*	• *Can update records of land owners* • *Generate thematic analysis, surveys, forecast and view visualization of land* • *Predict base on the key in documents*
1) As a Provincial Agriculturist officer - Should be the main administrator of the system -	*Web tool/ web accessible to mobile*	• *Validate users of the system* • *Receive any necessary information* • *View visualization* • *Generate analysis report*
2) As a Farmer - Should be able to manage any other account	*Web tool/ web accessible to mobile*	• *Login to validate Account* • *View and Print necessary Information needed*

3.1. Architectural Framework

As a web enabled system, the system is display-configured to tablets and mobile phones. As a security standard, access is given to rightful user roles only. Log in procedures are also embedded in the system. The system uses PHP and MySQL coordinated with a GIS software, ArcGIS Web mapping APIs. PHP and MYSQL are both open source software while ArcGIS is a licensed or proprietary software [10].

With these emerging innovations, the system gives a cutting-edge method for creating data in content, geospatial and mixed media designs. The framework in this context will screen and oversee agricultural lands practices and management. Using ArcGIS mapping algorithms for thematic analysis was conducted to graphically visualize the specific land area and automatically goes to Arc-GIS map database forwarded to the web-GIS website dedicated online facility.

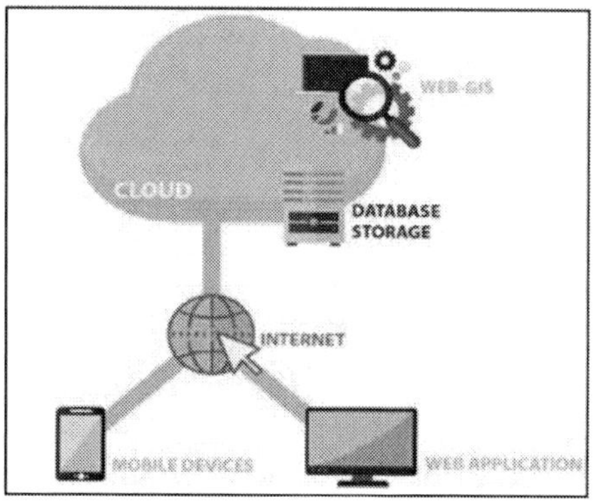

Figure 2. Architectural Framework

3.2. Features

The web enabled GIS provides a platform for all users to view the simulated Land Cover mapping as shown in Figure 3. They can zoom the picture to see the actual area they want to view. This can be available upon log-in as shown in figure 4. Log in features are embedded which are validated by the administrator. This can be the Municipal Agriculturist. Upon validation data visualization are displayed in graphical forms shown in Figure 5. Thematic analysis on forecasting and projecting crop adoption with regards to geographic physical information and land use pattern are also viewed. A user guide was developed to assist and help the users in navigating the web enabled GIS.

Figure 3. Log in features

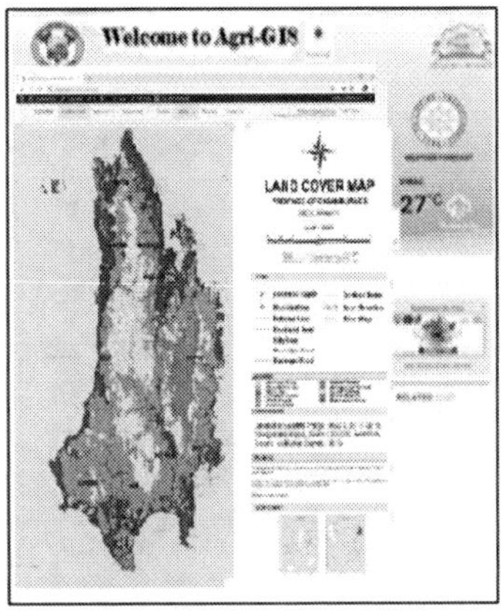

Figure 4. Land Cover Map

3.3. User Acceptability

The evaluation of usability of the system follows the software quality model ISO 9126 characteristics and sub characteristics (ISO 991) which contains the three (3) characteristics namely Functionality, Usability, Efficiency. Testing was conducted before deployment.The system was evaluated by 30 respondents and undergo two types of evaluation. The first evaluation was conducted by five (5) IT experts who primarily evaluated the system flow of the Information System but are not users of the system. The second evaluators are the main users of the systems composed of (11) Municipal Agriculturist, one (1) Provincial Agriculturist and thirteen (13) farmers. The Likert scale in Table 3 was used for the interpretation of qualitative (QL) and quantitative (QN) in interpreting the weighted mean, the following scale was used:

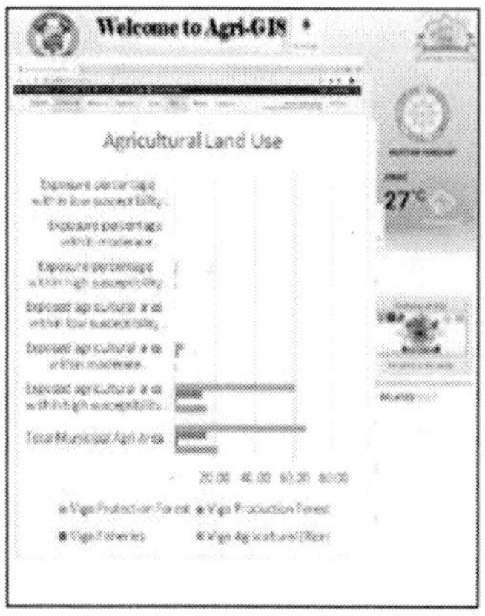

Figure 5. Data Visualization

Table 3. Five-PointLikert Scale

Numerical	Rating	Descriptive Rating
4.51 - 5.00	5	Strongly Agree
3.51 - 4.50	4	Highly Agree
2.51 - 3.50	3	Slightly Agree
1.51 - 2.50	2	Acceptable
1.50 and below	1	Not Acceptable

Table 4. Web-GIS Functionality Evaluation

Functionality	Weighted Mean	Qn	QI
1. The software has suitable but appropriate set of functions in accordance to its objective.	4.68	5	Strongly Agree
2. It is interoperability with defined sets of computing environment.	4.75	5	Strongly Agree
3. The software provides accurate results.	4.78	5	Strongly Agree
4. The software addresses the defined set of needs.	4.37	4	Strongly Agree
5. It is capable of presenting unauthorized access whether accidental or deliberate to program or data.	4.56	4	Highly Agree
AVERAGE WEIGHTED MEAN:	**4.60**	**5**	**Strongly Agree**

3.3.1. Functionality. The system was tested for functionality based on the five indicators such as suitability, accurateness, interoperability, compliance and security. The result shows in terms of functionality the average weighted mean is 4.68 with QI of "Strongly Agree". Hence, the system shows good functionality in terms of suitability, accuracy, interoperability, compliance and security with an average weighted mean of 4.60.

3.3.2. Usability. This suggests that the respondents "Strongly Agree" that the system is usable. This makes it easier to implement without much effort on the training of new users. A heuristic evaluation of each different usability methods was checked and evaluated to generate good results. Respondents strongly agree that the system is easy for the users to recognize embedded operations. Finally, it shows that in terms of usability of the system the average weighted mean is 4.79 that has a qualitative rating of "Strongly Agree".

Table 5. Web-GIS Usability Evaluation

Usability	Weighted Mean	Qn	Ql
1. It is easy for the users to recognize the embedded operations.	4.75	5	Strongly Agree
2. The instructions given in the system are self-explanatory	4.82	5	Strongly Agree
3. The software is easy to navigate	4.79	5	Strongly Agree
AVERAGE WEIGHTED MEAN:	**4.79**	**5**	**Strongly Agree**

Table 6. Web-GIS Efficiency Evaluation

Efficiency	Weighted Mean	Qn	Ql
1. It has acceptable responsive processing time and throughput rates.	4.79	5	Strongly Agree
2. It consistently uses enough computing resources (memory system) for all of its functions.	4.72	5	Strongly Agree
AVERAGE WEIGHTED MEAN:	**4.76**	**5**	**Strongly Agree**

Legend: Qn: Quantitative Rating
Ql: Qualitative Rating

Table 7. Summary of Results

Indicators	Weighted Mean	Qn	Ql
Functionality	4.60	5	Strongly Agree
Usability	4.79	5	Strongly Agree
Efficiency	4.76	5	Strongly Agree
TOTAL WEIGHTED MEAN:	**4.72**	**5**	**Strongly Agree**

3.3.3. Efficiency. The implication of this result of this study is that the time behaviour and the resource behaviour of the system makes it more useful to process collaboration and to plan, adapt to the needs of the users. Having an average weighted mean of 4.76 with a qualitative rating of "Strong agree" the user of the system has a clear understanding of what process implies and how users will handle the system tasks that make it easy for them to use.

4. Conclusions

Based on the result, the presence of spatial data generated using ubiquitous computing generated in web enabled GIS stresses consistent and flexible features. The system serves as initiative to improve agricultural land use in the province by means of locating, mapping, and profiling land management of agricultural lands. With the high acceptability rating of 4.72 by the respondents stated on the result it is recommended for utilization and be presented to the provincial Government of Catanduanes for implementation. These will help the government of Catanduanes strategize activities to easily facilitate the identification, monitoring, forecasting, visualize and analyse situations on Agricultural land use. Thus,enable the province help benefit farmers, local communities and will eventually help combat climate change and preserve the environment and its ecosystem. For the scientific contribution, the

IOP Conf. Series: Materials Science and Engineering **803** (2020) 012002 doi:10.1088/1757-899X/803/1/012002

study contributes on the enhancement of biodiversity assessment in spatial information on temporal monitoring of agricultural land use system and determine the crop and vegetation status for global change studies.

References

[1] Dhungel R 2012 *Web Mapping and Application Towards a Cloud: Enabling a WebGIS Prototype in an Open Source Environment* Thesis (Illinois: Southern Illinois University Carbondale)

[2] Roy P, Roy A, Joshi P, Kale M, Srivastava V, Srivastava S, Dwevidi R, Joshi C, Behera M, Meiyappan P and Sharma Y 2015 Development of decadal (1985–1995–2005) land use and land cover database for India *Remote Sensing* **7(3)** pp 2401-30

[3] Acedo G G and Palaoag T D 2018 GIS Based Mangrove Location Mapping and Information System in Catanduanes *Proc. of the 2018 2nd Int. Conf. on Algorithms Computing and Systems* pp 89-93

[4] Cooperative G I and Collins F 1988 The unique qualities of a geographic information system: a commentary *Photogrammetric Engineering and Remote Sensing* **54(11)** pp 1547-9

[5] Setyowati H A, Dwinugroho M P, Murti B S, Yulianto A, Ajiwihanto N E, Hadinata J and Sanjiwana A K 2018 ESDM One Map Indonesia: Opportunities and Challenges to Support One Map Policy based on Applied Web-GIS *IOP Conf. Series: Earth and Environmental Science* Vol **165** No 1 p 012021

[6] Van den Hoven J 2007 ICT and value sensitive design *The information society: Innovation, legitimacy, ethics and democracy in honor of Professor Jacques Berleur SJ* pp 67-72 (Boston, MA: Springer)

[7] Dunn C E 2007 Participatory GIS—a people's GIS? *Progress in human geography* Oct **31(5)** pp 616-37

[8] Weiner D, Harris T M and Craig W J 2002 Community participation and geographic information systems *Community Participation and Geographical Information Systems* Apr 4 pp 29-42 (Florida: CRC Press)

[9] Welling L and Thomson L 2003 *PHP and MySQL Web development* (Indiana: Sams Publishing)

[10] MySQL 2013 *Manual Documentation Mysql 5.7 reference manual* [online] https://dev. mysql. com/doc/refman/5.7/en/[Janauary 15, 2018]

Efficacy of virtual reality in reducing fear of public speaking: A systematic review

M M Daniels[1], T Palaoag[2], M Daniels[3]

[1]Ifugao State University, Ifugao, Philippines
[2]University of the Cordilleras, Baguio City, Philippines
[3]Department of Social Welfare and Development

Email: mylemelodydaniels@gmail.com, tpalaoag@gmail.com, my13dan@gmail.com

Abstract. Public speaking is no doubt one of the most feared mode of oral communication. Such anxiety is often treated through cognitive behavior therapy (CBT). An arising alternative to this treatment is virtual reality exposure therapy (VRET) where user confronts feared stimuli in a controlled 3D environment designed for such purpose. Sense of presence in this medium elicits similar feeling when standing in front of a real audience. Because of this, several studies were conducted to test its effect on individuals suffering from PSA. This paper attempts to look into the undertakings of studies within the last 10 years with the objective of gaining enlightenment on the efficacy of virtual reality in reducing public speaking anxiety. From Google Scholar search engine and online libraries PLOS ONE and ScienceDirect, 14 were short-listed for final review based on the set inclusion/exclusion criteria. Findings of these studies reveal that VRET can be as effective as CBT, and more advantageous in terms of cost-efficiency and manageability.The use of virtual reality is then highly recommended as a tool for treating PSA. It is worth noting that the length of time spent in this therapy is a great factor in its efficacy.

1. Introduction

Public speaking is one of the most feared mode of oral communication. Survey from decades ago reveals one third of the general population experience anxiety when speaking to a larger audience, and a third of this group report some clinically significant distress or impairment due to their public speaking anxiety (PSA)[1].In an analytic study with 1135 undergraduates aged 17-58, 63.9% reported fear of public speaking, and 89.3% would like their undergraduate program to include classes to improve public speaking [2]. This condition is treated with cognitive behavioural therapy (CBT) that usually includes vivo exposure therapy. The procedure includes allowing the patient to physically encounter the feared factor until patient begins to understand that nothing fearful results from such encounter[3]. The long-term effectiveness of such treatment is evident even from earlier studies[4].

Virtual reality is well-known for its highly immersive digital 3D environment. The realistic graphics in VR gives a real sense of presence that arouses genuine emotions from user, causing them to respond as if it were real. Recreating situations and managing stimuli in this controlled environment is more cost-effective than collecting and arranging physical objects for the vivo procedure[5]. This is why the use of VR as an exposure medium became a point of interest for authors in psychology, leading them to explore this alternative treatment known as virtual reality exposure therapy (VRET). Studiesdemonstrated that the use of virtual reality systems in the therapeutic setting yields a promising outcome [6][7]. This is true across different phobias.In terms of attribution, a recently conducted meta-analytic examination indicated that virtual reality exposure and in vivo exposure therapy show similar rates of attrition[8].

While VRET is generally found useful for treatingdifferent kinds of phobia, the focus of this study is on the efficacy of VR in reducing public speaking anxiety. There is a rising demand for a good public speaking skill in the business industry, especially so that in business, marketing, sales pitches and advertisement would require such skill. One of the greatest hindrance to a persuasive speech is simply the fear of speaking in front of an audience. In vivo therapy for this scenario could be difficult to manage since it would involve a good number of people to serve as an audience. Studies were conducted to test the effect of VRET to public speaking with varying results. This paper digs into the undertakings of studies within the last 10 years with the goal ofgaining enlightenment on the efficacy of virtual reality in reducing public speaking anxiety through a systematic review guided by Khan's Five Steps in Conducting a Systematic Review[9].Speakers who wish to overcome this fear can gain insight from thispaper and evaluate whether resorting to this treatment is worth the investment.

2. Methodology

Online library PLOS ONE, ScienceDirect, and search engine Google Scholar was used to collect papers for this review. For PLOS ONE, a search for the keywords "virtual reality for public speaking", with specified filters (Article type: journal article; Where my keywords appear: title; Publication Date: January 2009 to September 2019), yielded one result. Similar search settings in ScienceDirect resulted to a list of 8 papers, while in Google Scholar, 34results. A quick scan through the list showed a couple ofduplicate studies and three were written in foreign language. First screening process was then performed to removethese duplicate papers and those written in foreign language. Second screening processselected only those in line with the context of this review, and with consideration of the exclusion criteria (type of research: systematic review, commentary, developmental). Final list of papers were summarized according to methods and outcomes.

3. Findings

Out of 24 studies filtered from the first screening process, 10 were excluded. Two were out of context such that one tested the effect of stereoscopic viewing on presence, anxiety and cyber-sickness in VR, and the other is on the effect of level presencebut not comparing it to a control group. Two did not measure PSA, and the rest of the excluded paperswere about the design and development of VR software or framework for public speaking. Test procedures to prove their usability or efficacy were not discussed in the manuscript. The final list of reviewed papers are listed in Table 1.

3.1. Methods

The total number of participants in all the reviewed papers is 630. There were 54% students, 34% are of recorded public anxiety level and the rest (12%) are randomly selected,of ages 18 through 60. Among the diverse scales utilized in measuring public speaking anxiety level were Liebowitz Social Anxiety Scale (LSAS) [10][11][12], Self-Statements during Public Speaking (SSPS) [13], Attitude Towards Public Speaking (ATPS)[14], Personal Report of Confidence as a Speaker (PRCS) [13][15], and Personal Report of Public Speaking Anxiety (PRPSA) [16], Personal Report of Communication Apprehension (PRCA)[17], and Public Speaking Anxiety Scale (PSAS)[18]. A study also utilized heart beat rate and skin conductance meter [14] for objectivity. Fifty seven percent (57%) of these papers compared VRET to either a cognitive behaviour therapy (CBT) treatment [10][19][11] or a control group [20][14][13][21] – those who did not receive any treatment at all.

3.2. Outcomes

Virtual reality is able to elicit public speaking distressand exposure to it is found to cause distress habituation [15]. This makes it a useful tool in coping with fear [14][12][22][23] as compared to mere visualization [17] or imagination technique [20]. These reports come from studies whose participants engaged with VR once, yet with promising outcome. There is one study that reported no improvement on PSA after immersing in VR [15]. Authors of such study highlighted, however, that participants only had one session in VR. They were instructed to complete a 5-minute mock public speaking task to a virtual audience while wearing 3D glasses. The control group delivered their speech to the researcher.

Table 1. Final list of reviewed papers

No	Title	Year	Participants				VR Sessions	PSA Scale	Reference
			VR	CBT/CT	Control	Total			
1	Virtual Reality Cognitive Behavior Therapy for Public Speaking Anxiety: A Randomized Clinical Trial	2009	28	30	30	88	7	LSAS; SSPS	[10]
2	The effectiveness of 3-D video virtual reality for the treatment of fear of public speaking	2010	20			20	6	SPSS	[24]
3	Virtual reality exposure versus Cognitive Restructuring for Treatment of Public Speaking Anxiety: A Pilot Study	2011	20	28	30	78	7	LSAS	[11]
4	Virtual Reality Therapy: A Means of Reducing Public Speaking Anxiety	2011	40	40	40	120	1	PRCA	[17]
5	The use of doppelgangers in virtual reality to treat public speaking anxiety: a gender comparison	2014	21		20	41	1	PRCA	[20]
6	Virtual reality therapy: an effective treatment for the fear of public speaking	2015	7		7	14	5	ATPS	[14]
7	The effect of virtual reality exposure on fear of public speaking using cloud-based software	2016	65		33	98	6	PRCS	[13]
8	Beat the Fear of Public Speaking: Mobile 360° Video Virtual Reality Exposure Training in Home Environment Reduces Public Speaking Anxiety	2017	35			35	4		[25]
9	Overcoming public speaking anxiety of software engineers using virtual reality exposure therapy	2017	6			6	1	LSAS	[12]
10	Therapist-led and at-home one-session Virtual Reality exposure therapy for public speaking anxiety using consumer hardware and software, with online maintenance: A randomized controlled trial	2017	25		25	50	4	PSAS	[18]
11	Public speaking anxiety decreases within repeated virtual reality training sessions	2018	19			19	1	PRCS-2; Heartrate	[15]
12	Behavioural therapy and virtual reality exposure for public speaking anxiety	2019	6			6	6	SSPS	[22]
13	A pilot of acceptance and commitment therapy for public speaking anxiety delivered with group videoconferencing and virtual reality	2019	15			15	6	Not indicated	[23]
14	The effects of immersion and presence in a virtual reality public speaking task	2017	20		20	40	1	PRPSA	[16]

After completing the task participants then completed the PRPSA. Findings revealed no improvement to PSA after participating in such task and increased immersion did not significantly reduce fear of public speaking within one session. Nevertheless, all the other reviewed papers came to a unanimous conclusion that VRET was successful in reducing PSA among participants, even equal to CBT[11][19], and decrement was strongest in participants with initially high speaking anxiety baseline levels[25].Additionally, Lister's study reported an increase in confidence and positive self-statements [13]. Finally, the level of presence in VR did not have significant effect on the PSA level of participants. Simple consumer VR hardware [18] and VR 3D video displayed using a standard CRT television[24]successfully triggered fear responseand was used for the treatment.

3.3. Extracted Pre- and Post-treatment PSA levels

In this section, the mean PSA level of participants in the reviewed papers are presented. The attempt to gather all participants'mean PSA levels on pre- and post-treatmentwas only successful in five papers shown in Table 2. This is because three of the other papers were abstract-only and the rest with full paperdid not indicate numeric data. Instead, theypresented data in graphs without specified value.

To highlight the changes on the PSA level of participants, the difference between pre- and post-treatment was computed. The percentage value of decrease is also presented. The percentage of decrease in the PSA level of participants ranged from 7.8% up to 54.7%, with mean value of 29.2 %. Those who had the highest record of pre-treatment PSA level (65)[12] showed a decrease of 46.2%; while those who had the least (14.55) [24] only had 15.7% decrease. These data support the findings of Stupar-Rutenfrans'study where VRET is found more effective when aimed at reducing high-state anxiety levels[25].

Table 2. Extracted PSA levels before and after treatment

Title	PSA Level		Difference	Percentage of decrease
	Pre-treatment	Post-treatment		
Virtual Reality Cognitive Behavior Therapy for Public Speaking Anxiety: A Randomized Clinical Trial	19.39	8.79	10.6	54.7
The effectiveness of 3-D video virtual reality for the treatment of fear of public speaking	14.55	12.27	2.28	15.7
Virtual reality exposure versus Cognitive Restructuring for Treatment of Public Speaking Anxiety: A Pilot Study	30.8	28.4	2.4	7.8
Virtual Reality Therapy: A Means of Reducing Public Speaking Anxiety	16.7	13.1	3.6	21.6
Overcoming public speaking anxiety of software engineers using virtual reality exposure therapy	65	35	30	46.2
			Mean =	**29.2**

3.4. Limitations of the reviewed studies

While the findings of the reviewed papers yielded a strong evidence to support the efficacy of VR in reducing PSA, some limitations in the undertakings were found. First, majority of the measurement tool utilized was self-report. Only 1 out of 14 studies utilized technical measures (heart rate and skin conductance) in addition to self-report scale for testing the effect of VRET to fear of public speaking. Self-report can be biased in this aspect since subjects are expecting to improve in their anxiety level after the treatment. Second, there is lack of comparison between VRET and control group. Out of the 14studies, only 8comparedVRET to a comparison group. Findings would have been more robust if a control groups were included for comparison.

4. Conclusion

Having positive reports from 13 out of 14 reviewed studies, and with the support of T-test analysis, this paper finds VR as an effective tool in reducing PSA. The tool is then recommended to speakers as an alternative treatment for PSA since it is portable, manageability, cost-efficient, and more convenient to use anywhere at any time. A close look at the single study that reported no significant effect of VRET to PSA leads this paper to suggest that the efficacy of virtual reality as a tool for reducing PSA is highly dependent on the length of time user is exposed to such medium. While a specific number of sessions cannot be recommended due to lack of studies in this aspect, users are advised to undergo VRET as needed for habituation. Findings of this review call for authors to further their research and exploreother features of VR that could complement its ability to elicit real-life emotions (through sense of presence) and make it more useful not only in reducing PSA but also in enhancing public speaking skills.

References#

[1] Stein M, Walker J and Forde D 1996 Public-speaking fears in a community sample Prevalence, impact on functioning, and diagnostic classification *Archives of General Psychiatry* vol **53** no 2 pp 169-74

[2] Marinho A C F, de Medeiros A M, Gama A C C and Teixeira L C 2017 Fear of Public Speaking: Perception of College Students and Correlates *J. of Voice* vol **31** no 1 pp 127e7-127e11

[3] The Center for Treatment of Anxiety & Mood Disorders 2019 *The Center for Treatment of Anxiety & Mood Disorders* 2017 [Online] Available: https://centerforanxietydisorders.com/treatment-programs/in-vivo-exposure-therapy/ [Accessed 18 September 2019]

[4] Mersch P P 1995 The treatment of social phobia: The differential effectiveness of exposure in vivo and an integration of exposure in vivo, rational emotive therapy and social skills training *Behaviour Research and Therapy* vol **33** no 3 pp 259-69

[5] Bouchard S, Dumoulin S, Robillard G, Guitard T, Klinger É, Forget H, Loranger C and Roucaut F X 2017 Virtual reality compared with in vivo exposure in the treatment of social anxiety

disorder: A three-arm randomised controlled trial *The British J. of Psychiatry* vol **210** no 4 pp 276-83

[6] Valmaggia L, Latif L, Kempton M and Rus-Calafell M 2016 Virtual reality in the psychological treatment for mental health problems: A systematic review of recent evidence *Psychiatry Tres* vol **236** pp 189-95

[7] Carl E, Stein A, Levihn-Coon A, Pogue J, Rothbaum B, Emmelkamp P and Asmundson G 2019 Virtual reality exposure therapy for anxiety and related disorders: A meta-analysis of randomized controlled trials *Anxiety Disord* vol **61** pp 27-36

[8] Benbow A A and Anderson P L 2019 A meta-analytic examination of attrition in virtual reality exposure therapy for anxiety disorders *J. of Anxiety Disorders* vol **61** pp 18-26

[9] Khan K, Kunz R, Kleijnen J and Antes G 2003 Five steps to conducting a systematic review *J. of the Royal Society of Medicine* pp 118-21

[10] Wallach H, Safir M and Bar-Zvi M 2009 Virtual Reality Cognitive Behavior Therapy for Public Speaking Anxiety: A Randomized Clinical Trial *Behavior Modification* **33** pp 314-38 DOI: 10.1177/0145445509331926.

[11] Wallach H, Safir M and Bar-Zvi M 2011 Virtual reality exposure versus Cognitive Restructuring for Treatment of Public Speaking Anxiety: A Pilot Study *Isr. J. Psychiatry Reelat. Sci.* vol **48** no 2

[12] Nazligul M D, Yilmaz M, Gulec U, Gozcu M A, O'Conner R and Clarke P 2017 Overcoming public speaking anxiety of software engineers using virtual reality exposure therapy *European Conf. on Software Process Improvement* vol **748** pp 191-202

[13] Lister H 2016 *The effect of virtual reality exposure on fear of public speaking using cloud-based software* Thesis University of New Brunswick: Heather Lister

[14] North M M, North S M and Coble J R 2015 Virtual reality therapy: An effective treatment for the fear of public speaking *Int. J. of Virtual Reality* vol **03** no 3 pp 1-6

[15] Takac M, Collett J, Blom K J, Conduit R and Rehm I 2019 Public speaking anxiety decreases within repeated virtual reality training sessions *PLoS ONE* vol **14** no 5

[16] Wilsdon L and Fullwood C 2017 The effect of immersion and presence in a virtual reality public speaking task *Annual Review of CyberTherapy and Telemedicine* url: http://hdl.handle.net/2436/622249

[17] Heuett B L and Heuett K B 2011 Virtual Reality Therapy: A Means of Reducing Public Speaking Anxiety *Int. J. of Humanities and Social Science* vol **1** no 16

[18] Lindner P, Miloff A, Fagernas S and Andersen J 2017 Therapist-led and at-home one-session Virtual Reality exposure therapy for public speaking anxiety using consumer hardware and software, with online maintenance: A randomized controlled trial*9th Swedish Congress on internet interventions* pp 19-20

[19] Safir M P, Wallach H S and Bar-Zvi M 2011 Virtual Reality Cognitive-Behaviour Therapy for Public Speaking Anxiety: One-Year Follow-Up *Behavior Modification* **33(3)** pp 314-38

[20] Aymerich-Franch L and Baileson J 2014 The use of doppelgangers in virtual reality to treat public speaking anxiety: a gender comparison *Proc. of the Int. Society for Presence Research* pp 173-86

[21] Niebuhr O and Michalsky J 2018 Virtual reality simulations as new tool for practicing presentations and refining public-speaking skills *Proc. of the 9th Int. Conf. on Speech Prosody 2018* pp 309-13

[22] Zacarin M R, Borloti E and Hardy V B 2019 Behavioral therapy and virtual reality exposure for public speaking anxiety *Trends in Psychology* vol **27** no 2 pp 2-14

[23] Yuen E, Goetter E, Stasio M, Ash P, Mansour B, McNally E, Sanchez M, Hobar E, Forte S,

Zulaica K and Watkins J 2019 A pilot of acceptance and commitment therapy for public speaking anxiety delivered with group videoconferencing and virtual reality *J. of Contextual Behavioral Science*vol **12** pp 47-54

[24] Lister H, Piercey D and Joordens C 2010 The effectiveness of 3-D video virtual reality for the treatment of fear of public speaking *J. of Cyber Therapy & Rehabilitation* vol **3** no 4 pp 375-81

[25] Stupar-Rutenfrans S, Ketelaars L and van Gisbergen M 2017 Beat the Fear of Public Speaking: Mobile 360° Video Virtual Reality Exposure Training in Home Environment Reduces Public Speaking Anxiety *Cyberpsychology, Behavior, and Social Networking* vol **20** no 10

Deep Residual U-Net Based Lung Image Segmentation for Lung Disease Detection

E Mique, Jr[1] and A Malicdem[2]

[1]Faculty, College of Information Technology, Don Mariano Marcos Memorial State University
[2]Dean, College of Information Technology, Don Mariano Marcos Memorial State University

Email: emique@dmmmsu.edu.ph

Abstract. The World Health Organization (WHO) estimated that by the year 2030, lung disorders such as Chronic Obstructive Pulmonary Disease (COPD) would be one of the leading cause of death all over the world. Consequently, accurate and timely detection of lung diseases may prevent further death. It is therefore vital that the early detection may lead to treatment and prevention of mortality among patients. However, there are only a minimum number of experts or well-trained radiologists reading Chest X-Ray (CXR) that delays the timely diagnosis of lung diseases. In order to aid the radiologist in reading CXR images, a computer-aided tool is proposed. Before the processing of images, it needs to be segmented to make it easier for the machine to understand. This study is focused on developing a model that will segment the lung from CXR images. Using Residual U-Net (ResUnet) architecture based semantic segmentation, the researchers were able to develop and train a model using a set of 562 CXR images and lung mask images, 70% of the images were used as training data and 30% as test data. The model was trained with 40 epochs and a batch size of 16. Dice coefficient was used to assess the similarity of the segmented result and the ground truth mask. The developed model has achieved a Dice coefficient of 0.9860. The developed model can then be used in classifying lung diseases by focusing on the segmented image rather than focusing on the entire CXR image.

1. Introduction

Lung diseases are disorders affecting the lungs which include asthma, chronic obstructive pulmonary disease (COPD), tuberculosis, influenza, lung cancer, pneumonia and other breathing problems. Worldwide population are affected by huge number of related lung diseases[1] and respiratory tract infections are the top cause of death and disability[2]. World health organization (WHO) estimates that by 2030, lung diseases such as COPD will be one of the top causes of mortality. To prevent further death, accurate and timely diagnosis of pulmonary disease is needed [1]. Hence, to treat the disease, it is essential that at its early stage the disease is detected[3]. Often, chest X-Ray is the first procedure patients undergo if doctors suspect lung disease. This is due to the fact that it is economical, an effective diagnostic tool and because of its noninvasive characteristics in showing pathological modifications. Chest X-Ray(CXR) gives an overall orientation as an underlying symptomatic examination and is particularly valuable in the conclusion of pneumonia, malignant growth and COPD [4]. By analyzing CXR image, radiologists can diagnose many lung related diseases[5]. Skilled radiologists use CXR to recognize diseases, for example tuberculosis, pneumonia, interstitial lung malady, and cancer [6].

Classifying CXR irregularities is considered as a dreary task for radiologists[5]. Many studies have been conducted in the domain of Artificial Intelligence to aid radiologists in reading CXR Images. In the study conducted by Ramalho et. al, a new approach was proposed for classification of lung illnesses. CXR images were segmented using Adaptive Crisp Active Contours Models(ACACM) and a novel method for lung illness identification using feature extraction for segmented lung structure[1]. In the study of Abiyev and Ma'aitah, they demonstrated that classifying chest pathologies in CXR utilizing conventional and deep learning strategy is viable. The researchers presented Convolutional neural networks (CNN) in diagnosing chest illnesses[5]. Another study conducted by Kumar et. al, focuses on finding nodules, early symptoms of cancer diseases, appearing in patient's lungs. The researchers utilized an altered watershed segmentation strategy to segregate the lung in an x-ray image[7].

For better classification of lung diseases, CXR images must be first segmented. The area of interest must be first separated from the entire image. Through segmentation, it aims to divide the image in a series of region based on the characteristics of the image that are almost constant in each of the region[8]. Lung segmentation is an important step in developing a Computer-aided Diagnosis (CAD) system for the diagnosis of lung illnesses in radiographs[9]. The purpose of image segmentation is to extract the area or region of interest (ROI) in an image may it be automated or semi-automated procedure. Its goal in medical field is to extricate quantitative facts like morphometric data with regard to an organ of interest. There are two related tasks to be considered in segmentation problem, one is object recognition and the second is delineation. In the first task, the location of the object on the image is determined. The object's composition is drawn in the object delineation task[10].

Many research works have been undertaken on the use of image segmentation especially in the medical field. Norouzi et al. made a comparison of the different algorithm used in medical image segmentation. In their study, they categorized image segmentation into four: clustering method, region-based method, classifier method, and hybrid method. The pros and cons of each method were discussed. Each algorithm of the four methods was discussed for the examination of grey-level images. It was further stressed that image segmentation procedures can be chosen based on different parameters[11]. In the study of Christ et. al., cascading fully convolutional neural networks (CFCNs) and dense 3D conditional random fields (CRFs) were used in automatically segmenting liver in CT images. In their study, a 2-fold cross-validation on 3DIRCAD was used in training CFCN models. Based on their findings, CFCN-based semantic liver segmentation attained a dice score of more than 94% for liver with calculation under 100 s per volume[12].

Deep learning is also gaining popularity in image segmentation. It has been the primary option for images segmentation, medical image segmentation in particular. Deep learning-based image segmentation has been set up as a vigorous instrument in image segmentation and has been generally used to isolate homogeneous areas as the first and basic part of diagnosis and treatment pipeline[13]. Alom et. al. proposed models for medical image segmentation: A Recurrent Convolutional Neural Network (RCNN) and Recurrent Residual Convolutional Neural Network (RRCNN), both are based on U-Net architecture. The proposed models were tried on three different standard datasets. Compared to other models, the outcomes show predominant performance on segmentation tasks [14]. In the study of Vesal, Ravikumar and Maier, they proposed a deep learning framework to segment organs-at-risk (OAR) in thoracic CT images, explicitly for the trachea, heart, aorta and esophagus. They employed in the bottleneck of U-net network an expanded convolutions and collected residual connections. The model attained a final Dice score of 91.57% on 20 not yet seen samples[15]. Numerous researches have also been conducted in segmenting the lungs from CXR images like the study conducted by Kalinovsky and Kovalev, which presented the result of the first exploratory stage of research and development on segmentation of lungs in X-rays using deep learning methods and encoder-decoder convolutional neural network(EDCNN). The result obtained in this study concludes that EDCNN networks may be considered as an encouraging tool for lung segmentation in large-scale projects[16].

IOP Conf. Series: Materials Science and Engineering **803** (2020) 012004 doi:10.1088/1757-899X/803/1/012004

In the study of Heo et. al, U-Net was utilized to segment the lung from CXR. The segmented images were then used for the detection of tuberculosis. In training the U-Net, the researchers used 140 images with lung mask. The trained U-Net attained a mean Dice Coefficient of .9621 for 60 validation sets[17]. Considering the very small number of images in training the U-Net model, it was able to achieve high result. If the number of images can still be increased, the accuracy may still be improved.

This prompted the researchers to conduct a study that will create a model for lung segmentation based on Deep Residual U-net architecture to be used for lung disease detection. Specifically, the study aims to:

a. identify CXR images and its equivalent lung mask to be used in training the model.

b. Develop a model based on Deep Residual U-Net Architecture to segment the lungs from CXR image for lung disease detection.

2. Methodology

2.1. Data Collection
The dataset used in training the model was downloaded from the National Institute of Health(NIH) particularly Shenzhen hospital X-ray set. The dataset was collected by Shenzhen hospital in Shenzhen, Guangdong province, China. It contains a mixture of normal and abnormal x-rays. Another dataset was obtained from Kaggle, a manually segmented lung masks dataset for Shenzhen X-ray set which were manually created by teachers and students of Computer Engineering Department of National Technical University of Ukraine "Igor Sikorsky Kyiv Polytechnic Institute".

2.2. Deep Residual U-Net Architecture
In developing the model for lung image segmentation, the researcher utilized a deep learning technique U-Net with residual connections. U-Net Architecture was created for Bio-Medical Image Segmentation by Olaf Ronneberger et. al.[18]. U-Net can still be enhanced through residual unit instead of plain unit. Through residual connection, it will maximize the performance and the capability of the network to learn. Figure 1 shows the structure of the deep residual U-net (ResUnet) Architecture. ResUnet combines both the strength of U-Net and residual neural network. The residual

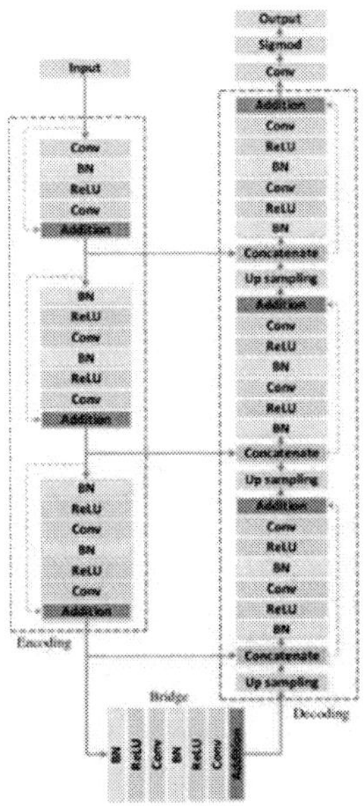

Figure 1. Deep Residual U-Net Architecture

will allow easier network training and the skip connection in the residual unit will make the propagation of information easier without degradation. ResUnet is comprised of 3 main parts, the encoding, bridge and decoding. In the encoding part, the image which serves as an input is encoded into a denser representation. The decoding part on the other hand, recovers the depictions to a pixel-wise categorization. The bridge joins the encoding and decoding paths. These parts are created with residual units consisting of two convolution blocks with a size of 3 x 3 and an identity mapping which joins the input and the output of the unit. The convolution block has a branch normalization layer (BN) and it also contains both a Rectified linear unit activation (ReLu) and a convolutional layer. The size of the feature map is reduced by half using a stride of 2 in the convolution block instead of utilizing pooling operation in each of the residual unit in the encoding part to downsample the feature map size. Prior to the residual unit in the decoding part, there is an up-sampling of maps from lower part and joining of maps from the related encoding part. A sigmod activation layer and a convolution layer with a size of 1 x 1 is utilized to project the multi-channel maps in the target segmentation next to the final part of decoding path[19].

U-Net based with residual unit semantic segmentation was preferred by the researchers because it works with very few training samples and provides better performance for segmentation tasks [11]. Moreover, U-Net was basically created for biomedical image segmentation which is suited for this study.

2.3. Training the Model

The model in segmenting lung from CXR image was developed and trained using Python, Tensorflow, and Keras. Keras is a high-level neural network API which can run on top of Tensorflow. The use of Keras allows the researchers to do fast experiments without writing large number of codes.

The training dataset consists of the original CXR images and manually segmented mask images. The dataset consists of 562 sets of mask and unmask CXR images which were divided into training and validation data. 70% of the images were used for training and 30% were used for validation. The images were resized and were trained in batches to facilitate faster training without the need for additional computing resources.

2.4. Dice Coefficient

To assess the similarity of the manually mask CXR images and the result, dice coefficient was used. Dice coefficient is a measure of intersection between objects. The measure ranges from 0 to 1 where a dice coefficient of 1 means perfect and complete overlap[20]. It can be calculated using the formula:

$$\mathrm{DC} = \frac{2TP}{2TP + FP + FN} = \frac{2|X \cap Y|}{|X| + |Y|}$$

where TP are the true positives,
FP are false positives, and
FN as false negatives

The number of true positives are the numbers that the model finds or correctly identify while false positives are the negative that the model classify as positive.

3. Results and Discussions

The developed model, which was based on U-Net architecture were trained using 562 sets of CXR images and manually mask images from Shenzhen No. 3 hospital. The images were resized and trained with a batch size of 16 at 40 epochs or the times the whole dataset is passed through the neural network both forward and backward path.

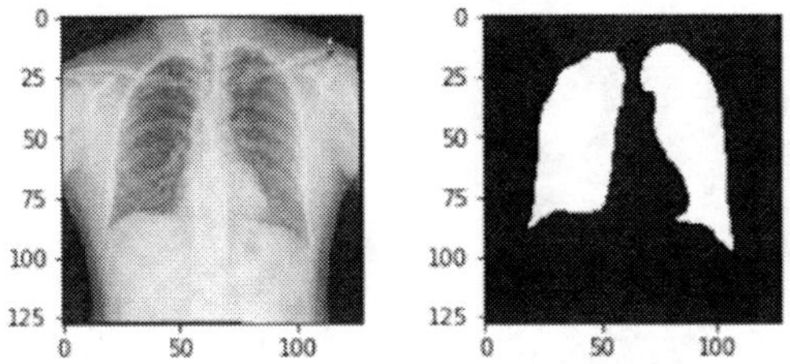

Figure 2. CXR Image Sample

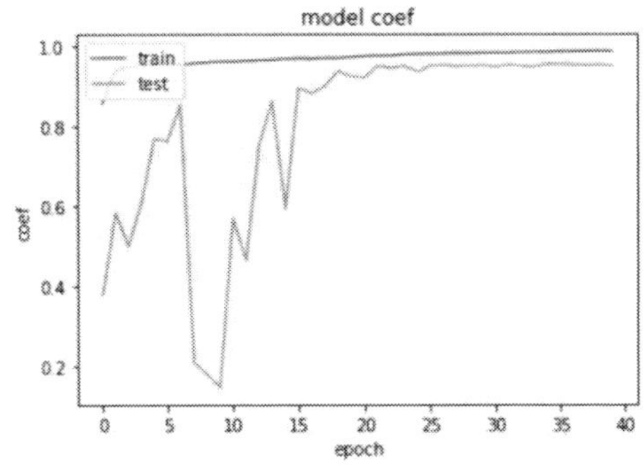

Figure 3. Training and Validation Dice Coefficent

Figure 2 shows a sample dataset image. The image on the left is a sample CXR image from Shenzhen Hospital and the image on the right is the manually mask CXR image prepared by teachers and students of National Technical University of Ukraine "Igor Sikorsky Kyiv Polytechnic Institute". Figure 3 shows the dice similarity coefficient graph of the model. The blue line indicates training dice

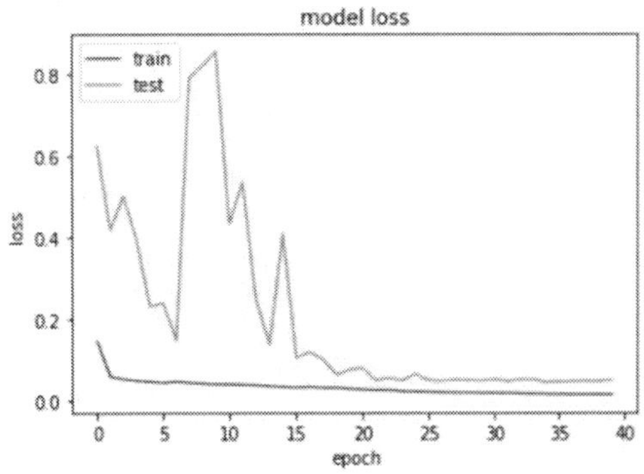

Figure 4. Training and Validation Loss Graph

coefficient graph while the orange line indicates the validation dice coefficient graph. The trained model was able to achieve a training dice coefficient of 0.9860 and validation dice coefficient of

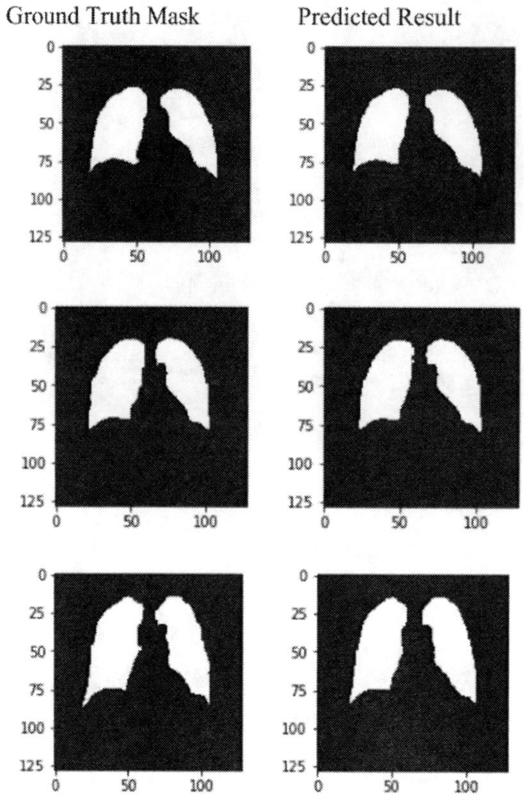

Figure 5. Sample Prediction Results

0.9496. The result shows that the comparison between the ground truth mask and segmented result are almost identical. Moreover, the validation and accuracy loss are very low as seen in Figure 4. At around 30 epochs, validation and training loss stabilized which indicates that trained model is neither overfit nor underfit. A model is said to be overfit when the train set's performance continuously improves while the validation set's performance improves at a certain point and then begins to go down. Underfit model on the other hand, is when the training dataset performs well while validation dataset performs poorly. A good fit is when both training and validation dataset perform well and it stabilized at some point[21]. Based on the graph as seen in figure 3, the model exhibited a good fit model. This indicates further that the model could predict unseen CXR images with high prediction accuracy and with very little margin of error.

Figure 5 shows sample prediction results of the developed model. The images on the left are the ground truth mask CXR images while the images on the right are the predicted lung mask of the CXR images. The ground truth and the predicted lung mask are almost identical which shows that the prediction accuracy of the model is very high.

4. Conclusion

In this paper, the researchers were able to create a model that can segment the lungs from CXR images based on Deep Residual U-Net architecture. 562 CXR images with manually mask lungs were used in training the model. 70% of the model were used as train data and the remaining 30% was used as testing data. The trained model was able to achieve a validation dice coefficient of 0.9496 and a loss of 0.0504. Therefore, the trained model can segment the lungs from CXR images with high accuracy.

The model can be used in segmenting the lungs from CXR image before predicting lung diseases as CXR includes structures other than the lungs, such as the heart and spine. These structures are not important or might hinder in detecting lung diseases[8]. Through the developed model, the classification and detection of lung diseases can be much faster as it only focuses on the lung structure rather than the entire CXR image. The findings in the study will also be useful for future researchers who will be conducting studies in the field of biomedical imaging.

References

[1] Ramalho G L B, Rebouças F P, Medeiros F N S and Cortez P C 2014 Lung disease detection using feature extraction and extreme learning machine *Revista Brasileira de Engenharia Biomédica* **30(3)** pp 207- 14

[2] Bağcı U, Bray M, Caban J, Yao J and Mollura DJ 2012 Computer-assisted detection of infectious lung diseases: a review *Comput Med Imaging Graph* 2012 Jan **36(1)** pp 72-84 doi: 10.1016/j.compmedimag.2011.06.002. Epub 2011 Jul 1

[3] Singh S, Singh Y and Vijay R 2016 Image Processing: A Better means for diagnosis of lung cancer *Int. J. of Eng. Applied Sci. and Technol.* Vol **1** Issue 3 ISSN No. 2455-2143 pp 121-3

[4] Wielpütz M O, Heußel C P, Herth F J and Kauczor H U 2014 Radiological diagnosis in lung disease: factoring treatment options into the choice of diagnostic modality *Dtsch Arztebl Int.* Mar **14111(11)** pp181-7 doi: 10.3238/arztebl.2014.0181

[5] Abiyev R and Ma'aitah M K S 2018 Deep convolutional neural networks for chest diseases detection *J. of Healthcare Eng. Volume 2018* Article ID 4168538 https://doi.org/10.1155/2018/4168538

[6] Qin C, Yao D, Shi Y and Song Z 2018 Computer-aided detection in chest radiography based on artificial intelligence: a survey *BioMedical Engineering OnLine* 201817:113. https://doi.org/10.1186/s12938-018-0544-y

[7] Kumar V, Garg K and Kumar R 2012 Pulmonary nodules diagnosis from x-ray imaging using image processing *Conf.: Int. Conf. on Recent Trends of Computer Technol. in Academia*

[8] Stolojescu-Crisan C and Holban S 2014 An Interactive X-Ray Image Segmentation Technique for Bone Extraction *Int. Work-Conf. on Bioinformatics and Biomedical Eng. (IWBBIO)* pp 1164-71

[9] Pattrapisetwong P and Chiracharit W 2016 Automatic lung segmentation in chest radiographs using shadow filter and local thresholding *IEEE Conf. on Computational Intelligence in Bioinformatics and Computational Biology (CIBCB)* Chiang Mai pp 1-6 doi: 10.1109/CIBCB.2016.7758113

[10] Mansoor A, Bagci U, Foster B, Xu Z, Papadakis G Z, Folio L R and Mollura D J 2015 Segmentation and image analysis of abnormal lungs at CT: current approaches, challenges, and future trends *Radiographics : a review publication of the Radiological Society of North America Inc* **35(4)** pp 1056–76. doi:10.1148/rg.2015140232

[11] Norouzi A, Rahim M, Altameem A, Saba T & Ehsani Rad A, Rehman A and Uddin M 2014 Medical image segmentation methods, algorithms, and applications *IETE Technical Review* **31** pp 199-213 10.1080/02564602.2014.906861

[12] Christ PF *et al* 2016 Automatic liver and lesion segmentation in CT using cascaded fully convolutional neural networks and 3D conditional random fields *Medical Image Computing and Computer-Assisted Intervention – MICCAI 2016 Lecture Notes in Computer Science* vol **9901** ed S Ourselin, L Joskowicz, *et al* (Cham: Springer)

[13] Hesamian MH, Jia W, He X *et al* 2019 Deep learning techniques for medical image segmentation: achievements and challenges *J. of Digital Imaging.* https://doi.org/10.1007/s10278-019-00227-x

[14] Alom Md Z, Yakopcic C, Hasan M, Taha T, Asari, V 2019 Recurrent residual U-Net for medical image segmentation *J. of Medical Imaging* **6(1)** doi 10.1117/1.JMI.6.1.014006

[15] Vesal S, Ravikumar N and Maier A K 2019 A 2D dilated residual U-Net for multi-organ

segmentation in thoracic CT *The Technical Writer's Handbook* ed M Young (Mill Valley, CA: University Science) *ArXiv* abs/1905.07710

[16] Kalinovsky A and Kovalev VA 2016 Lung image segmentation using deep learning methods and convolutional neural networks *XIII Int. Conf. on Pattern Recognition and Information Processing*

[17] Heo S, Kim Y, Yun S, Lim S, Kim J, Nam C, Park E, Jung I and Yoon J 2019 Deep learning algorithms with demographic information help to detect tuberculosis in chest radiographs in annual workers' health examination data *Int. J. of Environmental Research and Public Health* **16(2)** 10.3390/ijerph16020250

[18] Ronneberger O, Fischer P and Brox T 2105 U-Net: Convolutional networks for biomedical image segmentation *Medical Image Computing and Computer-Assisted Intervention – MICCAI 2015 Lecture Notes in Computer Science* vol **9351** ed N Navab, J. Hornegger, *et al* (Cham: Springer)

[19] Zhang Z, Liu Q and Wang Y 2018 Road extraction by deep residual U-Net *IEEE Geoscience and Remote Sensing Letters* **15** pp 749-53

[20] Jordan J 2019 *An overview of semantic image segmentation* [Online] Available:https://www.jeremyjordan.me/semantic-segmentation/ [Accessed 19 July, 2019]

[21] Brownlee J 2019 *How to diagnose overfitting and underfitting of LSTM models* [Online] Available:https://machinelearningmastery.com/diagnose-overfitting-underfitting-lstm-models/ [Accessed 19 July, 2019]

A Procurement Data Standard Tool for State Universities and Colleges

Floradel S. Relucio & Josephine S. Dela Cruz

Bicol University Polangui Campus, Polangui, Albay, Philippines; University of the Cordilleras, Baguio City, Philippines

dimplerelucio@yahoo.com; delacruzpen@gmail.com

Abstract. The study presents the development of a procurement data standard tool to promote transparency for compliance monitoring on the procurement process in all government agencies. Extreme programming was utilized as a methodology during the development of the system. Procurement Data Standard was validated and evaluated using the USE Questionnaire to check the usability of the data standard tool that focuses on three factors namely: usefulness, ease of use and satisfaction and get its equivalent based on the respective percentage, for usefulness is 4.12 (30%); ease of use is 4.25 (40%) and satisfaction is 3.96 (30%). Based on the result the overall weighted mean got a total of 4.11 and was interpreted as Very Satisfactory, it clearly indicates that the developed standard tool is easy to use and useful. This study will aid government agencies to promote transparency and accountability in Procurement Activities of the Government.

1. Introduction

Among many activities of the government, public procurement takes a very important share because of each its monetary side and its contribution to decision-making method. In general, the goal of procurement is to make purchasing simpler and economical to produce the most effective value for money. In this respect, organizations can use some techniques like performance measures, quality control and standardized procedures to make their procurement activities efficient and cost-effective. However, procurement within the government is complicated and needs some adjustment to associated processes [1].

Procurement is essential to organizations and its strategies have progressed toward becoming part of a business achievement. It supports proficiency and aggressiveness and to understand these, it is indispensable to give emphasis about the vital components that influence performance of the procurement function, activities of procurement usually suffer from neglect, poor coordination, absence of open competition and transparency. Corruption may likewise happen in the various processes of procurement, inflexible and bureaucratic systems of procurement add to contract delays, expanded expenses and absence of fair competition, all of which affect the procurement process and performance negatively. [2].

This will add on the advancement of technology when it comes to procurement process in the government sector. The planning and development process may also be a strong basis for other researchers conducting similar studies. And most importantly, the inclusion of evaluation which is can be used for assessment of continuous improvement necessary on scientific method.

As consequence of transparency, e-Procurement improves public administration further by fighting against corruption. Through the improved accessibility of all parties to the public procurement information and electronic work of all transactions, equal treatment within the public sector business opportunities can be achieved and also the chance of detection of illegal transactions can be enlarged [3].

As a summary, a good and efficient procurement process provides public institutions to achieve additional comprehensive image of their overall procurement activities, initiate aggregate purchasing with others and improve relationships with the vendors. In addition, public institutions can reduce the maverick shopping which is outlined as the purchasing of goods and services that do not meet the specified standards or do not seem to be equipped from the approved vendors.

It is now widely acknowledged that one of the key tenets of achieving sensible governance is transparency in the way public affairs are run. This is more so in the public financial management system and in particular in public procurement since it is one area of Government that is most at risk of corruption [4].

Open Contracting Data Standard (OCDS) is a core product of the Open Contracting Partnership (OCP), developed by the World Wide Web Foundation through a project supported by the Omidyar Network and the Word Bank – sets out to transform the way governments publish contracting data and to shine a light on how trillions of public money are spent. OCDS is an open data standard for publication of structured information on all stages of a contracting process: from planning to implementation. It was designed with a focus on public procurement of goods, works and services, but can be extended for use in other contexts. The standard lay out key documents and data that should be published at each stage of a contracting process, giving governments a common format and set of practical tools that will allow them to publish contracting data consistently and openly. It enables disclosure of data and documents at all stages of the contracting process by defining a common data model.

The Bicol UnversityPolangui Campus (BUPC) as a State University and as mandated by Republic Act No. 9184, created the local Bids and Awards Committee (BAC) to handle all the Procurement Process. However, upon the initial data gathering of the researcher, there were instances that some documents of transactions are not reported and it is hard to access. There are times that it is difficult in acquiring information about the procurement process and lack of document transparency.

Along with this, the main aim of this study is to develop a data standard tool on the procurement transactions that would be beneficial as necessary to monitor the procurement activities of Bicol University Polangui Campus (BUPC), as mandated by the law to promote transparency, available to the public with its detailed view and fast generation of necessary reports. Thus, this will hasten the verification of the procurement documents and will check the completeness of the data.

2. Methodology

The study used Extreme Programming as an agile software development framework that aims to produce higher quality software, and higher quality of life for the development team. It allows flexibility within the modelling process. XP is the most specific of the agile frameworks regarding appropriate engineering practices for software development. The methodology consists of five phases. The planning, designing, coding, testing and lastly listening phase [5].

2.1. Planning

For this phase, the researchers set the goals and scope of the entire project and certain iterative cycles. Specification documents was prepared as well as determining what customer wants and how the researchers will get into more resource consumable phase. Further, the researchers have a better understanding on the requirements needed for the system to be developed in relation to how the institution conducts their day-to-day transactions.

2.2. Designing

As the researchers determine the requirements of the developed data standard tool and define the main features of the data standard tool in this phase, it also drafted the blueprint of the developed system at the same time. The developed data standard tool features include uploading of rendered services, flagging system, and community feedback.

2.3. Coding
In this phase, the source code is developed in relation to the specification and design features of the system. The researchers created also the database schema based on the data requirements necessary to producetheneededinformation.API'sandscriptshasalsobeenincludedtofurther enhance the content and functionality of the system.

2.4. Testing
The researchers conducted unit testing before the deployment to ensure that there are no errors and the designs are working based on their intended functions. Acceptance testing through the use of evaluation tool has been done also to conform that the functions of the system have met the specification requirements set by the users.

2.5. Listening Phase
This is the phase where feedbacks from the intended users is collected which is essential for future iterations of the system. Along with acceptance testing, valuable comments and suggestions have been received.Thesefeedbackshavebeenconsideredandshallbeusedforthesystem'senhancement.

3. Results and Discussion
This portion presented the results and analysis of findings made upon conducting the study. It includes the main features of the system, developed Procurement Data Standard Tool and Deployment Diagram.

3.1. Main Features of the Data Standard Tool
There are three major features involved in the tool. The posting of procurement information is initiated by the procuring entity. This entity will post the tender of the procuring agency to inform the contractors, suppliers, and/or service provider about the possible opportunity of procurement. The data includes reference number, procuring entity, title, area of delivery, solicitation number, trade agreement, procurement mode, classification, category, approved budget for the contract, delivery period, contact information, associated components, bid supplements, document request list, date published, closing date and time, and description [6].

As for uploading rendered services, the contractors, suppliers, and/or service provider will upload the necessary documents, generally the class A and B documents, as well as the technical and financial documents to see how many procurements have been awarded to them, and to check if they have complied standard in procurement process. Once the document has been uploaded, the procuring agency that received the service can rate and provide feedback about the experience they had which will be verified by the system validator.

Another feature is the flagging, the procuring entity can flag a contractor, supplier, or service provider that had a bad experience before, with low rating, and/or have negative feedbacks. This particular contractor, supplier, or service provider will not be seen on respective webform. Also, the system validator can flag a contractor, supplier, or service provider if discrepancies have been found in the documents uploaded, and have lots of verified bad services or negative feedbacks.

And lastly, the threading, For the community, the public can start a thread by creating a topic and others can comment and reply back on the specific topic and the public can post comments on a particular procurement information or on the feedback provided by the procuring entity. The comments may include further information, inquiries, or additional feedbacks.

3.2. Procurement Data Standard Tool
Figures 1, 2 and 3 shows the screenshots of the developed data standard tool. It shows the sample web forms, community feedback and procurement reports. The developed standard tool aimed to provide a

powerful tool for BUPC to track contracting processes and to enables users and partners to analyze and share data.

Figure 1 presents the sample webform screenshot of the developed data standard tool. The form allows the user to input data used for the procurement process such as the procurement entity, tenders, planning, status, award, contract duration and so on.

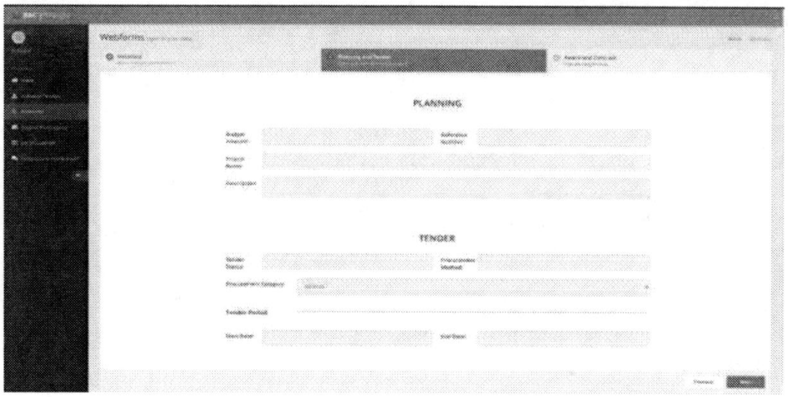

Figure 1. Sample Webform of the Developed Data Standard Tool

Community Feedback of the developed data standard tool that was shown in figure 2 has a two-way monitoring platform that will be displayed for public viewing. The data will be visualized using Graphs to make sense of the data being displayed to the public. The public can create a thread about the data. They can flag the data if they see anomalies about it and they can only report the same data once. This would increase public participation in the procurement process and will promote transparency in the transactions of the government in the procurement of projects.

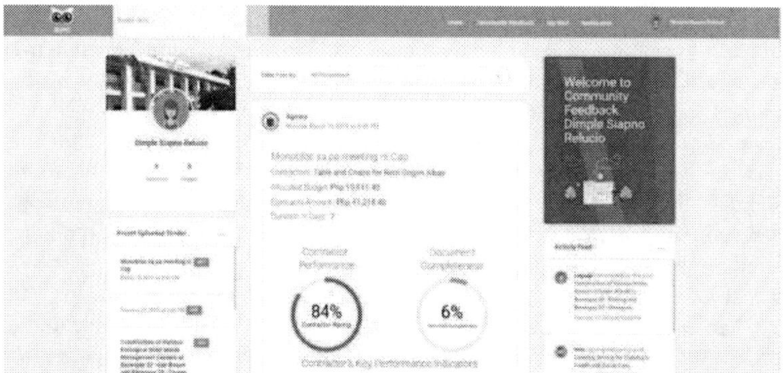

Figure 2. Community Feedback Page of the Developed Data Standard Tool

For flagging of data, it will checked or monitor if the data has found errors or anomalies, then the system would flag the data. The public can also flag the data if they have probable reasons, for example, the project was delayed, unfinished or has substandard quality.

Figure 3 describes the generation of necessary reports regarding all the data uploaded to the system. Most of the reports are also visualized to using graphics and etc. to be more understandable. The developed standard tool would provide the compliance checking of the documents of the tendered items. It checks if the documents uploaded by the BAC complies with the law and the open contracting data standard. It would display the thoughts and comments of the public and would use a data visualization to show the necessary reports or important data of the procurement process to the public.

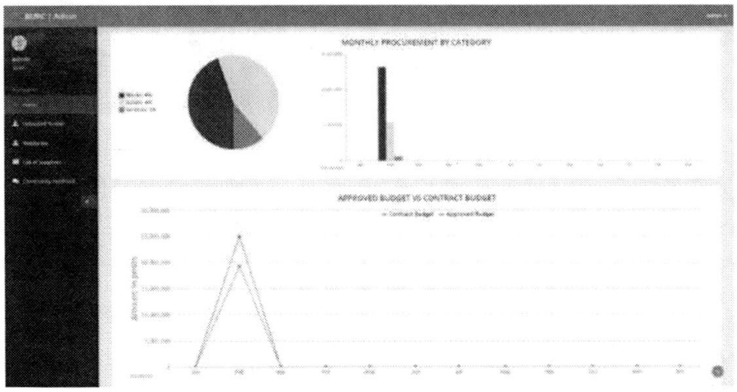

Figure 3. Procurement Reports

3.3. Deployment Diagram

The deployment diagram as shown in figure 4 shows the configuration of run time processing nodes and the components of the developed system. It also shows the physical component architecture of the developed system as deployment or distribution of software artifacts to deployment targets.

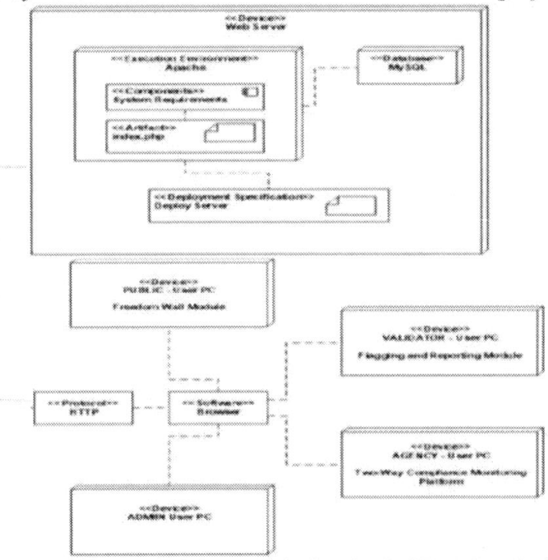

Figure 4. Deployment Diagram of the Deveopled Data Standard Tool

3.4. User's Evaluation

Shown in Table 1 is the Expert Respondent Level of Evaluation and Acceptability of the Procurement Data Standard Tool for BUPC. The study adopts the USE Questionnaire[7] in assessing the developed standard tool focusing on three factors namely usefulness, satisfaction and ease of use. Respondents for this study were fifty five (55) composed of all members of the BUPC BAC Personnel, BUPC end users like Department Chairs, Coordinators, office heads, selected faculty members, civil society groups and suppliers. In summary, the weighted mean of usefulness is 4.12 which means that the developed system are effective, productive and useful; in ease of use is 4.25 which is excellent to use, user friendly and flexible; and satisfaction is 3.96 which interpreted as very satisfactory. The overall weighted mean got a total of 4.11,theevaluationresultwasinterpretedas“very satisfactory”but it clearly indicatesthat the developed data standard tool is easy to use and useful.

Table 1. Evaluation Result of the Procurement Data Standard for BUPC

Indicator	Weighted Mean	Interpretation
Usefulness	4.12	Very Satisfactory
Satisfaction	3.96	Very Satisfactory
Ease of use	4.25	Excellent
Total	**4.11**	**Very Satisfactory**

Numerical Scale: 4.21-5.00 – Excellent; 3.41-4.20 – Very Satisfactory; 2.61-3.40 – Satisfactory; 1.81-2.60 – Fair; 1.00-1.80 - Poor

In order to interpret the result of the survey, researchers used the descriptive interpretation of the usability result as shown in Table 1 to verify the level of usability in the area of usefulness, ease of use, and satisfaction. The descriptive interpretation was derived from the compared scores collected in other usability studies to contextualize the performance of the developed data standard tool relative to other systems [8].

4. Conclusion

The developed data standard tool focused mainly on uploading of rendered services, flagging system, and community feedback. The result of the usability in the area of usefulness is 4.12 which is very satisfactory, for satisfaction is 3.96 which interpreted as very satisfactory and ease of use has weighted mean of 4.25 which is excellent, it clearly makes the usability performance and user-friendliness of the developed data standard tool more reliable and error-free. With an overall result of the evaluation rated an average of 4.11, which is very satisfactory. Therefore, in line with derived findings, the developed procurement data standard tool can help all the government agencies in promoting transparency in procurement process, monitor the procurement activities to possibly eliminate mismanagement and corruption, and made it easier to produce reports.

References

[1] Hinson C 1999 *Welcome to Public Procurement National Institute of Government Purchasing*

[2] Gätjen J 2014 *Open Contracting-what is it and how good is it?* thesis Retrieved from https://essay.utwente.nl/65272/

[3] OECD 2003 *Retrieved from Transparency in government procurement: the benefit of efficient governance and orientation for achieving it* http://www.oecd.org/LongAbstract?0,3425,en_2469_3727_2510174,00&&enUSS_01BC.html

[4] Open Contracting 2014 *Open Contracting Data Standard: Documentation* Retrieved from http://standard.open-contracting.org

[5] Agile Allieance 2019 *Extreme Programming* Retrieved from: https://www.agilealliance.org/glossary/xp/

[6] *Philippine Government Electronic Procurement System* Retrieved from https://www.philgeps.gov.ph

[7] Lund A M 2001 *Measuring Usability with USE Questionnaire STC Usability SIG Newsletter* Retrieved from https://garyperlman.com/quest/quest.cgi?form=USE

[8] Bangor A, Kortum P and Miller J 2009 Determining What Individual SUS Scores Mean: Adding an Adjective Rating Scale *J. of Usability Studies* Vol **4** Issue 3 pp 114-23 Retrieved from http://uxpajournal.org/determining-what-individual-sus-scores-mean-adding-an-adjective-rating-scale/

Web-based Knowledge Management System for Camarines Norte State College

D E Maligat Jr.[1], J O Torio[2], R T Bigueras[3], M C Arispe[4] and T D Palaoag[5]

[1,2,3]Camarines Norte State College, Daet, Camarines Norte, Philippines 4600
[4]Bicol University Polangui Campus
[5]University of Cordilleras

Abstract. One of the issues of the academic institution is knowledge management. How the academe disseminate information over the academic community on real time and the availability of knowledge anytime and anywhere is challenging task in every institution. The objective of this study is to develop Knowledge Management System for Camarines Norte State College that is capable of knowledge sharing, creating and knowledge repository. The developed KM system is designed specifically to cater the requirements of the academic institutions especially Camarines Norte State College in terms of knowledge management. A usability test was conducted to find out how far the academic community can interact efficiently and effectively and how far the system application provides user guidance, provide information, and provide consistent interaction. The pooled mean of the overall usability test was 4.55with adjectival rating of SA or strongly agree, which means that the system has the functionality needed by the users, easy to use and the system is very useful.Moreover, KMS in Higher Education Institution (HEI) could help in the growth of learner centered knowledge and action learning, growth in work-related learning, movement from closed to open knowledge systems and extensive development in computer-based communication technologies.

1. Introduction

The new frontier towards industrial economy is the knowledge or information-based system where this wealth of knowledge is an intellectual asset that if given proper management would offer greater benefits and competitive advantage to an institution. Knowledge management (KM) can be described as: the organized process of generating, utilizing and sharing individual, group and team knowledge utilizing technology, values, strategy and individual in enhancing innovation, efficiency, completion times and organizational performance [1].

Knowledge Management Systems (KMS) is not easy to develop and deploy, there are several challenges on the manner. It includes IT infrastructure to form new knowledge, store it, share and diffuse it, and to apply it for effective actions [2]. Moreover, knowledge management once applied to Higher Education Institutions (HEI's) poses a great deal of challenges particularly from the stakeholders. Given the growing needs of HEI's as they are now more aware of the clear advantage of having KMS. However, there is limited researches in this field while there is no evidence of an effective framework that will answer HEI's needs for better KMS [3].

Since most KMS available in the market was Industry Based KMS or Enterprise KMS. Industry operations is different from academic operations and every academic institutions has its own differences in their operation. With this, it is better to build customized academic KMS based on their needs and operations, moreover, commercially KMS is very expenses that most academic institutions can't afford to purchase.

Being one of the member of SUC's that aims for quality education and services as it enhances management efficiency, fiscal autonomy, and institutional competitiveness. Therefore, the Knowledge Management System in Camarines Norte State College (CNSC) is expected to benefit the organization through utilization of its intellectual assets and foster competitive edge and a capacity for learning. However, at present there are no HEI's in the Bicol Region that is equipped with a KMS deployed in their school. As a primer educational institution that will implement a fully working KMS in a Higher Educational Institution (HEI) this would establish a common benchmark that other institutions in Camarines Norte and possibly the Bicol region could adopt. Hence, this is the challenge and motivation that the researcher is trying to address in this study.The objective of this studyis to develop Knowledge Management System for Camarines Norte State College that is capable of knowledge sharing, creating and knowledge repository.

2. Literature Review

The literature survey was done to show the contributions of theories and concepts to the improvement of this study.

2.1. Knowledge Management System and its Features

KM is an approach to accomplishing key targets by visualizing, sharing, and utilizing intangible assets of an organization and its partners [4] [5]. In this regard, one of its core process and technique is data mining and knowledge production which is one of the important steps in knowledge discovery and a subfield of knowledge management [6] [7].

The concept of KMS is not new but an important emerging technology that would enable better management towards better decision for an organization. Although, it found worldwide recognition as an important strategic tool, its integration into academic organization becomes lagged and it have much lesser collaboration with its counterpart which is the industry [8] [9] [10]. Even if there are disagreements in several studies regarding knowledge and knowledge management [11] [12] [13], but still most authors concur on the bases of KMS and its practice as key to competitiveness.

As such, knowledge imparting is imperative to the success of KM practices in all organizations including universities. An effective KMS sharing is basicto organization to benefit from the knowledge its organizations have produced [14] [15]. However, even the higher administration and various stakeholders in universities might have heard KM but they do not have a clear idea how to initiate and implement it even the richness of literature for the past three decades have given various points in its development and implementation [16]. Information sharing is imperative to the success of KM practices and even though there are some elements affecting the preparedness to share information which can be classified as organizational, individual, and technology factors; this can be answered through driving academics to involve in knowledge sharing activity [14].

In fact, KM can also support higher education in learning circle and social interaction which can be used by teachers and students to improve teaching-learning process [17].It is essential for KMS technology and systems to bond the gap of knowledge formation, creation, giving, sharing, or application. KMS in Higher Education Institution (HEI) could facilitate in the development of learner centered knowledge and action learning, progress in work-related learning, undertaking from closed to open knowledge systems and widespread development in computer-based communication technologies [18].

2.2. Knowledge Management System Development

Knowledge management system development involves different approaches and techniques. One of such, is that the KMS architecture based on four layers which includes the application layer, technology layer, infrastructure layer and repository layer. Its functions also required internet, extranet, and intranet infrastructure using client and server computing [19] [20]. Another is taking considerations of an organization framework through proper observation and survey to provide effective periphery participation within the system. Taken in consideration the design of KMS to facilitate movement of information and decisions from the center towards the periphery. It is highly encouraged to have a two-way interaction between center and periphery for effective utilization of KMS where character of periphery participation is generally directed at knowledge discovery and application, instead of than creation and sharing [21].

In terms of framework, there's a very important factor to be considered such as setting up a schema as [22] have done which was utilized as a beginning point of repetitive process until getting the expected and desired final and common KM framework. Several schematic categories were identified such as knowledge and knowledge management artefacts, knowledge management frameworks and models, knowledge management systems, knowledge management ecosystem, influencing factors, and knowledge management reference disciplines and research method. It is necessary to have a proper infrastructure to primarily reengineer knowledge foundation and delivery established on principles of KM and organizational learning [23].

A different but relevant framework was presented in the study of [24] that might be used as a guide in developing and implementing KMS but this is often supported a Socio-Technical Systems based on Pan and Scarborough view. It comprises of Infrastructure which refers to the technology, Info structure which is also known as organizational structure and Info culture or organizational culture which is believed to give clear understanding on KMS. Knowledge management systems can improve firms' agility. Their framework supports the positive impact of knowledge management systems in the absorptive capacity through better collaboration and decision-oriented management system promoting dynamic capabilities [2].

Moreover, in developing a university knowledge management system essential points are to consider such as cooperation, tacit knowledge, knowledge continuity, power relationships, and social knowledge management. Cooperation is important to enhance KM performance while a multi-dimensional view of knowledge encompassing the tacit is better for capturing complex phenomena of knowledge in an organization [25]. There is a distinct difference among tacit knowledge and the explicit knowledge types and determining the use of separate knowledge modules using their own knowledge management strategy depending on the knowledge type of the stored information was necessary to support KM initiative successfully [26].

Also, holding essential operational knowledge should be incorporated into KMS framework while considering that power and its consequences have both negative and positive impacts on the system where KM is expected to be seriously influenced by the potential of social/media software through interactivity, social processes to promote organizational knowledge. Tacit knowledge is never easy and its collection through interviews and videos has not succeeded hence, a story-telling approach might be useful on the process [27]. Therefore, it is essential to be able to realize the whole concept and process before the actual development and such can be done using roadmap or even knowledge models [28].

One of the challenges within the development of an effective KMS is the financial issue that may arise from building it towards its completion since there are many things to consider such as its value, internal business process, growth and learning, as well as the infrastructure that is involved. Moreover, KM is not widely adopted in industries due to a lack of an effective approach [29]. On the case of its development so far, there many KMS "silver bullet" that is significant for the successful application of KMSs in organizations [30].

2.3. Knowledge Management System Implementation

The implementation of KMS have brought up different views and ideas based on its needs and purpose. Several implementation means have been covered by different literatures and one of the major concern in its deployment is to have an effective resolution of cultural and organizational issues which is consistent to information system management literature, advocating organizational and behavioural change management as critical success factors in the implementation of information systems [11]. Even its increased usage forms virtual communities, expert localization and establishment of knowledge taxonomies, knowledge transfer and sharing processes, incubation and

Mentorship, collaborative software development and their role in creating entrepreneurship initiatives and knowledge economies [31].

However, in its implementation there are several challenges that could be used as motivations and even best practices that could be made to make key success in small and medium sized organizations [32], [33]. Some of the challenges in its implementation were economic and information managementtechnologysecurity, and top management support and strategy, acknowledgement, user's enthusiasm and values and project management, management transformation, preparation and training. Four of these determined the success of its implementation such as the financial or economic and technology and information security, top management support strategy, information technology management, and acceptance [34]. Another research finds sustainability as one of the major challenged in KMS implementation because stakeholders have various expectations and firms need to decide how they intend to address these which becomes a great pressure in its effective implementation [35].

3. Methodology

3.1. Methods of Research

The descriptive survey method of research was used in the study. The questionnaire-checklist supplemented by informal interview and evaluative observation was utilized in gathering the needed data.

3.2. Software Development Methodology

The researcher used the Rational Unified Process (RUP) methodology since the present study requires documentation and information processing that RUP can handle and because it is supported by tools, visual modelling processes as well as documentation guides and testing hence, it became an important component in the development and deployment of the proposed system entitled "Web-based Knowledge Management System for Camarines Norte State College". The RUP Methodology is shown in Figure 1 which has nine (9) disciplines that use repetitive approach for establishing projects in terms of workflow and stages, each involving of one or more iterations. Every project iteration cycle initiates with a plan outlining what were completed and done with an assessment of whether objectives have been met.

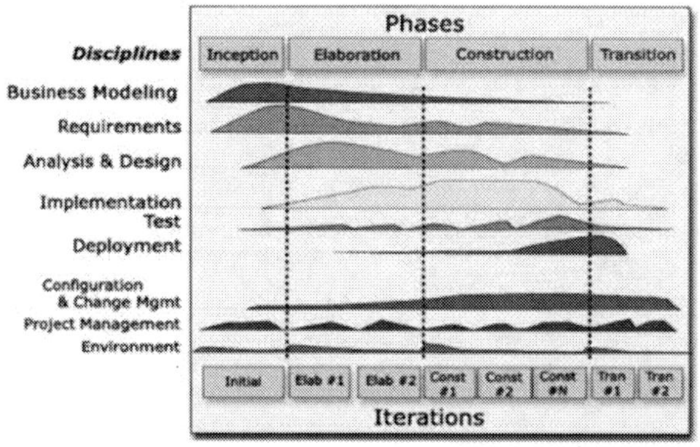

Figure 1. The RUP Methodology

4. Results and Discussions

The system design and development led to the creation of the full-packaged Knowledge Management System for the Camarines Norte State College using KM Architectural framework for academe as a guide in the development of the system. USE tool evaluation was conducted to determine the usability of the system's developed.

4.1. Knowledge Management System framework
The development of the KMS system was based on the academic KMS architectural framework concept developed by the researcher. It was customized based on the Business Intelligence concept and Enterprise Architecture of Tucker and Debrosse. The framework modifies the organizational architecture layers in to only three, *people*, *technology and application* as one, and *process*. This framework was also served a conceptual framework of the study and served as a guide in the development of the knowledge management system.

Figure 2 shows the three (3) organizational architecture layers as: people, technology and application, and process.

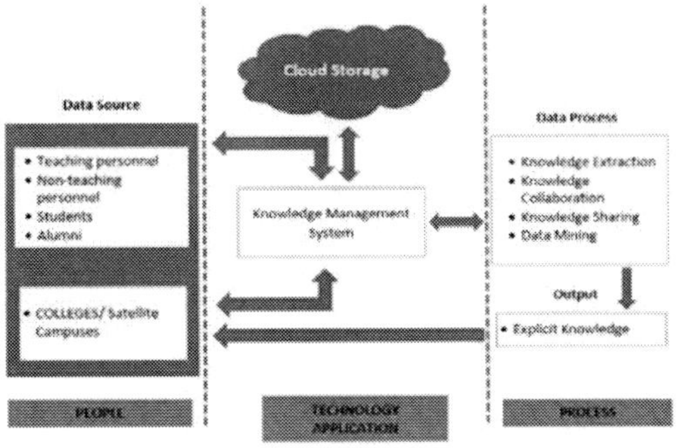

Figure2. KM Architectural framework for academe

4.2. Use case diagram of the system
Figure 3, shows the different actors that will use the Knowledge Management System for CNSC. Figure depicts the interaction on the system and it also shows the different level of users who will interact with the system. As shown in diagram Administrator can access all the features of the system and the Dean/Directors can create, rename and delete folders. They can also upload and delete files. On the level of faculty users, they can only upload and download files on the system.

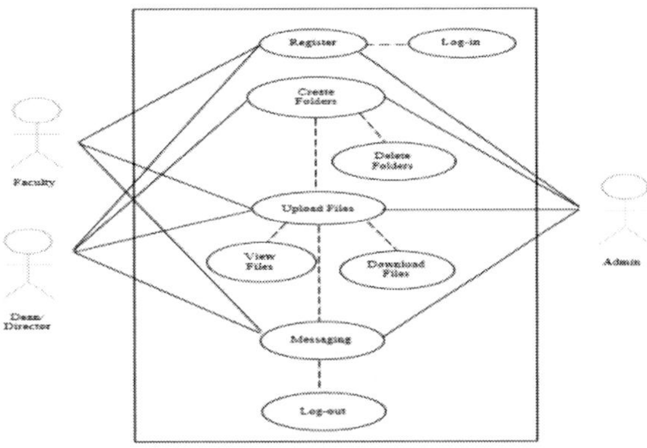

Figure 3. Use case diagram of the system

4.3. CNSC Knowledge Management System features
There are three main features of the developed system KMS: digital databank, knowledge creation and sharing and knowledge collaboration.

4.3.1. KMS digital databank. The system has database for storing digital knowledge/information that are useful for academic use, such as research, data gathering for accreditation, documentations and other academic uses.

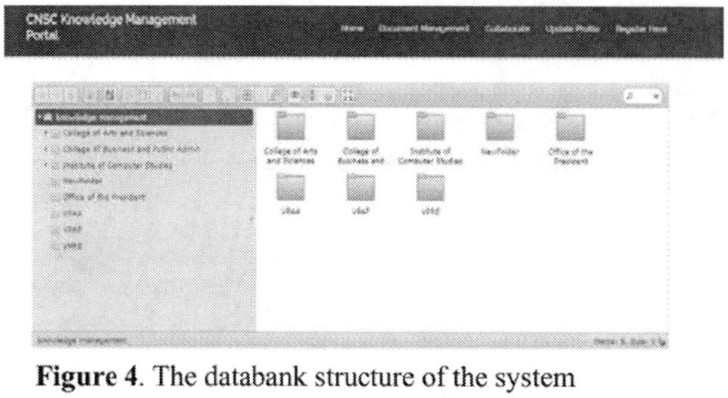

Figure 4. The databank structure of the system

Figure 4 shows the interface of the documents that are stored on the CNSC Knowledge management system. All documents that stored by different stakeholders has its own directory for convenience and easy access of the users. Moreover, availability of information on the system is 24/7.

4.3.2. KMS knowledge creation and sharing. The system allows the users to create and share documents. These features let the users to store and retrieve documents/information from the CNSC KMS. All documents that store by the users shall be available to the academic community for the purpose of academic use. Figure 5 (a) shows the download interface of the system while figure 5 (b) shows the upload interface of the system.

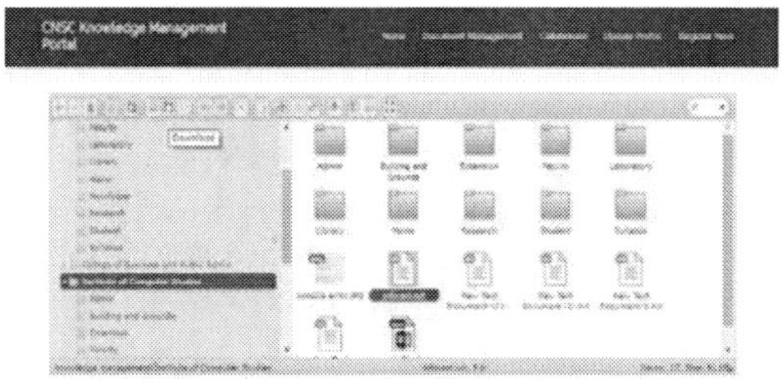

Figure5. (a) Document download interface of the system

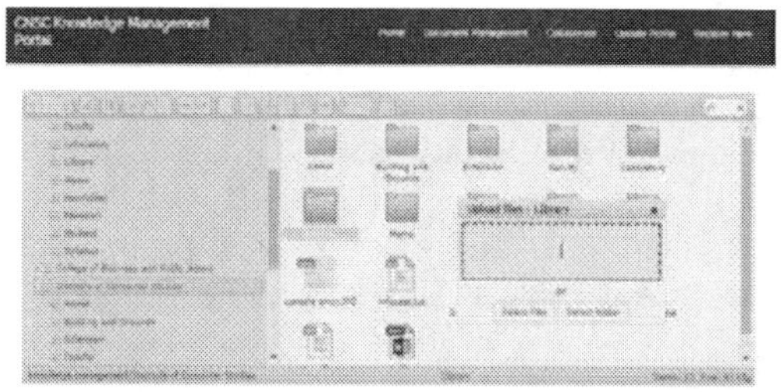

Figure5.(b) Document upload interface of the system

4.3.3. KMS knowledge collaboration. As shown in figure 6, the system has its collaboration features, the users can be allowed to modify, edit and update the document. Only documents subject for revisions can be modify or edited. Moreover, all login users can communicate to each other through a chat interface of the system.

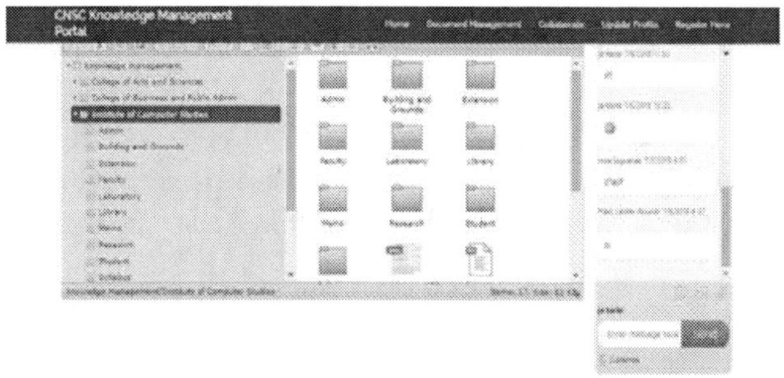

Figure 6. Collaboration interface of system

4.4. System Evaluation using USE tool

Upon completion of the system, the researchers conducted a test to determine the usability of the system. The researcher presented to the stakeholders, particularly to the faculty, deans/directors and the administrator of Camarines Norte State College in Daet, Camarines Norte.

This USE tool evaluation wishes to find out how the stakeholders be able to interact efficiently and effectively and also how the system application affords user guidance, deliver information, and keep consistent interaction [36]. The process in facilitating a USE tool evaluation is to use a questionnaire, one of which is the USE questionnaire. The USE questionnaire used 30 questions to evaluate the system application from 4 aspects: usefulness, satisfaction,ease of use and ease of learning.

Table 1. Overall Usability of the CNSC KMS

Criteria	Weighted Mean	Adjectival Rating
Usefulness	4.80	Strongly Agree
Satisfaction	4.40	Agree
Ease of use	4.60	Strongly Agree
Ease of learning	4.40	Agree
Pooled Mean	**4.55**	**Strongly Agree**

Legend(5-point Likert Rating Scale): Strongly Agree(4.50-5.00); Agree(3.50-4.49); Neutral(2.50-3.49); Disagree(1.50-2.49); Strongly Disagree(1.0-1.49)

Table 1 shows that he target users is satisfied with regards to the usability of the system. Based on the rating, the users strongly agrees with the pooled mean of 4.55 which means that the system has the functionality needed by the users, easy to use and the system is very useful.

5. Conclusion

The researchers were able to determine one of the major problems in academic operations especially in Camarines Norte State College. This problem is focused knowledge management. The need to have knowledge management system with Digital Databank, Knowledge Creation and Sharing, and Knowledge Collaboration features will cater to the ever growing cases of knowledge management issues in the academic community.

The new system is designed specifically to cater the needs of the academic institutions especially Camarines Norte State College in terms of knowledge management. With the CNSC KMS helps in managing information and making sure that knowledge is available anytime, anywhere and 24/7 operational.

As a future work area, in order to increase the flexibility of the software, the researchers would like to include comprehensive data warehouse data analytics built with intelligent decision-support system for Camarines Norte State College administrations and personnel.

References

[1] Oluikpe P, Sohail M and Odhiambo F 2011 The role of knowledge management in development projects *J. of Information & Knowledge Management* Dec **10(04)** pp 315-26

[2] Marhraoui M A and Manouar A El 2017 Towards a new framework linking knowledge management systems and organizational agility: an empirical study *Int. J. of Computer Science & Information Technology IJCSIT* **91** pp 21–36 https://doi.org/10.5121/ijcsit.2017.9103

[3] Miguel J, Nunes B and Arif M 2017 Knowledge Management Practices in Higher Education Institutions: a Systematic Literature Review *IFLA WLIC 2017* pp 1–16

[4] Gerami M 2010 Knowledge Management *Int. J. of Computer Science and Information Security* **72** pp 234–8

[5] Mitre-Hernandez H, Mora-Soto A, Lopez-Portillo H and Lara-Alvarez C 2014 Strategies for

fostering Knowledge Management Programs in Public Organizations *The 16th European Conf. on Knowledge Management*

[6] Odumuyiwa V and David A 2009 Collaborative knowledge creation and management in information retrieval *Int. Symp. on Collaboration Technologies and Systems* pp 1–12 USA

[7] Silwattananusarn T and Tuamsuk K 2012 Data Mining and Its Applications for Knowledge Management: A Literature Review from 2007 to 2012 *Int. J. of Data Mining & Knowledge Management Process IJDKP* **25** pp 13–24

[8] Grossman M 2007 The emerging academic discipline of knowledge management Management *J. of Information Systems Education* **18** Retrieved from http://vc.bridgew.edu/management_fac/11

[9] Iqbal J and Mahmood Y 2012 Reviewing Knowledge Management Literature Interdisciplinary *J. of Contemporary Research in Business* **4** 6 October pp 1005–26 Retrieved from http://www.ijcrb.webs.com

[10] Sureephong P, Chakpitak N, Ouzrout Y and Bouras A 2008 An Ontology-based Knowledge Management System for Industry Clusters *Int. Conf. on Advanced Design and Manufacture (ICADAM 2008)* pp 1–10

[11] Stenberg E and Vu-Thi X 2011 *A Literature Review of the field of Knowledge Management Systems* Dissertation (Halmstad: Halmstad University)

[12] Xu J, Sankaran G, Sankaran S and Clarke D 2008 Knowledge Management in Twenty-first Century: Literature Review and Future Research Directions Academic Global Publications *Int. Technology Management Review* **12** pp 16–24 Retrieved from http://www.academicglobalpublications.com/itmr/

[13] Xue C T 2017 A Literature Review on Knowledge Management in Organizations Research *Business and Management* **41** https://doi.org/10.5296/rbm.v4i1.10786

[14] Cheng M, Ho J S and Lau P M 2007 Knowledge Sharing in Academic Institutions: a Study of Multimedia University Malaysia *Electronic J. of Knowledge Management* **73** pp 313–324

[15] Varblane U and Al-Obaidi Z 2003 *General Manager's Role in Balancing Subsidiary between Internal Competition and Knowledge Sharing*

[16] Agarwal N K and Marouf L N 2014 Initiating Knowledge Management in Colleges and Universities: A template *Int. J. of Knowledge Content Development & Technology* **42** pp 67–95

[17] Michel C and Prévôt P 2009 Knowledge management concepts for training by project: An observation of the case of project management education *Int. Conf. on Knowledge Management and Information Sharing KMIS 2009* pp 1–6 (Madeira Portugal)

[18] Ramakrishnan K and Yasin N M 2012 Knowledge Management System and Higher Education Institutions In 2012 *Int. Conf. on Information and Network Technology ICINT 2012* Vol **37** pp 67–71 (Singapore)

[19] Abdullah R, Selamat M H, Sahibudin S and Alias R A 2005 A Framework For Management System Implementation In Collaborative Environment For Higher Learning Institution *J. of Knowledge Management Practice* **March** Retrieved from www.tlainc.com/articl83.ht Knowledge Management tools Retrived from http://www.knowledge-management-tools.net/three-km-models.html

[20] Bourdon I and Kimble C 2007 An Analysis of Key Factors for the Success of the Communal Management of Knowledge University of York *ArXiv abs/0804.2844*

[21] Dube R 2017 *An Analysis of the Knowledge Management Systems in an Organization with a Center-Periphery Knowledge Management Landscape* March

[22] Fteimi N 2015 Analyzing the Literature on Knowledge Management Frameworks: Towards a Normative Knowledge Management Classification Schema *Twenty-Third European Conf. on Information Systems ECIS 2015* Münster Germany: Association for Information Systems AIS Electronic Library (AISeL) Retrieved from http://aisel.aisnet.org/ecis2015_cr

[23] Piccoli G, Ahmad R and Ives B 2000 Knowledge management in academia: A proposed

framework *Information Technology and Management* 1 pp 229–45
[24] Assegaff S, Razak A and Hussin C 2015 Review of Knowledge Management Systems As Socio Technical System *ArXiv abs/1212.0387*
[25] Kalkan V D 2017 Understanding Knowledge Management in Academic Units: A Framework for Theory and Research *European J. of Business and Social Sciences* **512** 1–14 Retrieved from http://www.ejbss.com/recent.aspx-/
[26] Mian D, Kajewski S and Perera N 2014 *A Framework for an Innovative Knowledge Management System in Knowledge-based Firms* (Brisbane: Queensland University of Technology)
[27] Shaw N 2018 *A Knowledge Management System KMS Using a Storytelling-based Approach to Collect Tacit Knowledge* 1033
[28] Ahmad H 2015 Conceptual Knowledge Relationship Model of Enterprise Architecture and Top Management Roles In Australasian *Conf. on Information Systems* pp 1–13 Adelaide
[29] Xu J, Houssin R, Caillaud E and Gardoni M 2010 *Fostering continuous innovation in design with an integrated knowledge management approach* pp 1–17 (Strasbourg: Universite de Strasbourg)
[30] Whitfield J F 2008 *An Empirical Investigation of Factors Influencing Knowledge Management System Success Air Force Institute of Technology Air University*
[31] Mwangi K E, Thuku L X and Kangethe J P *Software Development Industry in East Africa: Knowledge Management Perspective and Value Proposition* 1 AJBUMA Publishing 26–37 2012
[32] Aiken D 2016 *Best Practices for the Implementation of a Knowledge Management System in Small and Medium Enterprises* Vol **1277** Oregon
[33] Elden W 2000 *Knowledge Management for Enterprises* ACM January 1–22
[34] Akhavan P, Hosnavi R and Adalati S 2010 Essential Issues in Knowledge Management System Implementation: Lessons from Iranian IT-Based Companies In *11th Int. Conf. on Social Responsibility Professional Ethics and Management* Ankara Turkey
[35] Hinz A 2012 *Knowledge Management as a Requirement for successful Sustainability Management University of St Gallen*
[36] Aelani K 2012 Pengukuran usability sistem menggunakan USE questionnaire *Seminar Nasional Aplikasi Teknologi Informasi*

IOP Conf. Series: Materials Science and Engineering **803** (2020) 012007 doi:10.1088/1757-899X/803/1/012007

A Gamified Approach on Learning Logic Gates to Improve Student's Engagement

J O Torio[1], R T Bigueras[2], D E Maligat, Jr.[3], M A Arispe[4], and J S Dela Cruz[5]

Camarines Norte State College[1,2,3], Bicol University Polangui Campus[4] and University of the Cordilleras[5]

E-mail: jocelynt77@gmail.com[1], rosemariebigueras@gmail.com[2], dmj234@yahoo.com[3], cham.arispe@gmail.com[4]anddelacruzpen@gmail.com[5]

Abstract. Gamification is the application of game design elements and game mechanics in a non-game context. This concept has been successfully used in several fields such as businesses and education. As such, the gamification e-learning approach could be used in education, particularly in a digital logic subject area, as a tool to improve student engagement. However, few research and design guidelines exist regarding a gamified learning approach for digital logic. The objective of this paper is to propose a complete framework for the development of a gamified learning application for digital logic gates that will improve student's engagement. We first investigate the design, the principles, and parameters used to measure student engagement from a variety of related work and identify which design, principles and parameters that shall be applicable and significant in the development of a gamified learning application that will drive an engaged student. This framework shall provide the game developers a guide and shall help standardize the development process of a gamified learning approach in a digital logic subject area in the future.

1. Introduction

In education, gamification has been used to make learning more interesting, engaging, empowering, and motivating for most of the students or learners. According to Suresh Kumar DN, games helped in releasing happy hormones such as dopamine and serotonin and these have been proven through several implementations [1]. Unsurprisingly, education is one of the most common applications arenas of gamification with the aim to promote entertaining, challenging and motivating experience for learners through the use of levels, feedback and engaging challenges. Hence, it is pragmatic that the retention rate and the subject matter learned at school or at work were better applied because of the gamified learning approach.

Reference [2] emphasized that gamification is applying game mechanics to non-game settings which have as of late turned into a drifting point over a wide scope of industries and has been introduced as a possibly disruptive vigor in the field of education. It is based on the principle that it can encourage motivation and engagement and thus contribute to the learning development and when it is appropriately utilized, it has the ability to engage, inform, and educate [3]. Reference [4] agreed that gamification in e-learning stages appears to can possibly to increment and improve student motivation, yet it is important to accomplish that impact, and a major exertion of effort is required in the plan, design, and implementation of the experience for it to be completely motivating and engaging for learners.

Currently, the digital logic application in the field of education here in the Philippines seems to call the need for utilizing the game as an instructional tool. As a matter of fact, Oxales O. stated that gamification is another revolutionary pedagogical trend in education that enable instructors or professors devise teaching methods and strategies that will better instill creativity and critical thinking among learners [5]. Also, Department of Education Secretary Leonor M. Briones said that ICT can be a game-changer both inside the classroom and the agency's bureaucracy where it is primarily a tool which helps them in making decisions, in understanding the world and in finding solutions. Although this shows an inspiring scenario for the adoption of gamification as an instructional tool, there is still little research in the measurement of student's engagement in a gamified learning approach.

Gamification as a learning tool is an alternate approach of using game thinking, game elements, and aesthetics in a different learning situation. However, most of the current gamified applications depend on providing other forms of rewards as an appreciation for the achievement of the player. These strategies use common game elements like badges, points, level, leaderboards, etc. In order to be effective on a long-term basis, gamification must be more than just incorporating these kinds of an element to a non-game context to act on the intrinsic motivation, engagement, performance and learning experience and which can also be useful in instruction [6] and [7].

The adoption of gamified application in the education arena can help improve the student's motivation when it comes to learning and performance. Even before, teachers and instructors have been applying the game-like elements approach in their teaching methods through implementing strategies in the class by giving points or scores or rewards to those students who have completed the homework or activities, these scores are badges or known as final grades of the students [8].

Despite the efforts to adopt gamification in the field of education, it remains interesting when it comes to fully engaged learners to particular lessons or activities [9]. Core topics of every subject offered in a curriculum should be given emphasis so that learners should learn the entire subject with ease, motivated and engaged. Digital logic is the core topic of digital design in which most of our current life technologies and innovations deal with the idea of digital design. Thus, it plays an important role in today's everyday life. Gamified digital logic gates would be the best tool for learning since traditional materials are declining, thus it will improve motivation for learners as well as adopting a novel tool for teaching and learning strategy [10].

Considering this, the paper focuses on determining the main principles and game design elements that are employed in the gamification and the parameters used to measure student engagement on how these parameters can be used to increase student engagement. More specifically, this paper presents the framework on the development of a gamified approach on learning digital logic gates that will help improve student engagement.

2. Development of Gamification Framework

With all the benefits of a gamified learning approach in education, this paper presents a gamification framework for logic gates to improve student's engagement. This will provide guidance and suggestion to mobile game developers and researchers in gamifying some educational applications.

2.1. Principles of Gamification

Based on our review of related literature, we identified five (5) main principles of gamification: (i) Goal orientation; (ii) Achievement, (iii) Reinforcement, (iv) Competition, and (v) Fun Orientation. These main principles were identified as important and significant in the development of a gamified learning approach. These principles were derived from the Mechanics, Dynamics, Emotion (MDE) framework of gamification principles of K. Robson et.al. [11] which is shown in figure 1. The MDE framework served as the backbone of the five (5) main principles to be more specific with what to consider on the development.

2.1.1. Mechanics. Mechanics are *the* boundaries of the situation to be gamified as what the designers have set. This includes setting up of stories, rules, and game progression. It is important for a gamified learning tool to be structured and organized with various levels of goals [12].

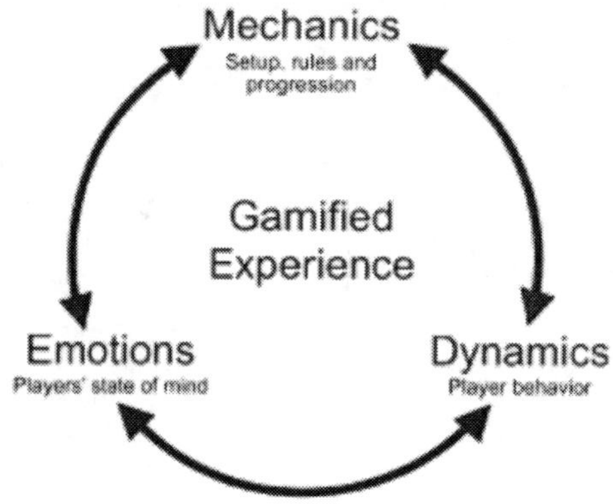

Figure 1. MDE Framework for Gamification Principles (K.Robson, et.al, 2015).

2.1.2. Dynamics. Dynamics are the different types of how players behave while experiencing the game. If mechanics are the ones that are set by the developer, dynamics are generated by how players track the mechanics with strategic action, that is set by the designer [12].

2.1.3. Emotions. Emotions are the result of how players track the mechanics and then later produce dynamics. These are mental affective states and reactions experienced by individual players when playing the gamified learning tool [13]. Hence, mix emotions or even negative emotions such as disappointment when a reward is not achieved are also included in this principle.

2.2. Design Elements for Gamification

In the development of every game application, an outline of the set of design mechanisms should be prepared before starting with the development. However, most of the current gamification literature focuses on the general effectiveness of game design elements in encouraging particular behavioural results [14]. Reference [15] conducted a literature review that have commonly identified gamification elements which are: leaderboards, progress, challenges, ranking levels, points, and rewards while other gamification elements are unique to a particular literature such as task unlock, team play, time constraints, karma, storyline, long/short-term goals, social networking tools, rating, and virtual goods. The frequently used design elements are badges, leaderboards, progress, and challenges.

Reference [16] integrated game elements namely; points, levels, challenges/ achievement, leaderboard, story, and time constraint. The student earned points once a method is correctly applied. Levels serve as the metric of progress and guide to the game. Challenges or achievements are a motivation factor that will help students apply the gained knowledge from the lessons and a way to maximize the points earned in every level. A leaderboard is an effective way to trigger competition among students. A rule or story is a challenge to the student of reach of the goal. Time constraints add more challenges to students to finish the tasks in every level.

In this paper, we determined and selected the basic game design elements applicable specifically for a digital logic gamified application based on the literature review [15] and [16]. To avoid complications on the gamified application, six (6) elements that are identified to be significant in the development of a digital logic gamification are: (i) Story and Rules; (ii) Points or Scores; (iii) Levels; (iv) Challenges/ Quests; (v) Leaderboards; and (vi) Time Constraints.

2.2.1. Story and Rules. This design element will be the first thing to prepare upon the development of a gamified application. There should be a concrete story, rule, and goal to sustain the student's engagement in order to complete the game.

2.2.2. Points or Scores. This design element is the basic scoring representation in a game to show the progress of a player. In a gamified application, points or score could also be used to claim rewards or exchange of any item in order to make some advancement on the game.

2.2.3. Levels. This design element indicates the milestones of players on the completion of different tasks. Badges can be offered on this part which can be used on the next levels of the game. This element is significant in enhancing the goal of the game in order to show achievement.

2.2.4. Challenges/Quests. This design element is significant to keep the players stay focused, engaged and motivated on the game. This is an important element of a gamified learning application for it gives opportunities for players to apply what they have learned and soon be rewarded for completing the challenge or quest.

2.2.5. Leaderboards. This design element brings pride to the player where it shows the entire community who visits the application the leading scorers or highest pointers of the game. This will make learners stay focused, motivated and engaged.

2.2.6. Time Constraints. This element brings a challenge where a player was given a time limit to complete the quest or task in every level of the game.

2.3. Student's Engagement Parameters

The key to understanding student involvement in gamification is measuring engagement, yet there are different notions of student engagement. Some students engage in different aspects throughout the game. It is significant that the student's engagement parameters should be identified to boost and quantify the level of student engagement [17]. To quantify the student engagement in utilizing gamification, table 1 shows the user engagement metrics and described how a particular metric be triggered [18] during gameplay.

Table 1. Engagement Metrics.

Metric	Description
Dropout Rate	It is the number of items completed by the player divided by the number of items attempted (completed or not), for a user.
Error message display time	It is the time a player spends in reading and analysing the reason why the player's answer was incorrect.
Total time spent vs. the total number of questions attempted	It is the total time a player spends in playing divided the time a player spends for finished or completed items.
Tips appreciation	A symbol like thumbs up or down is an example of appreciation where a player would select whether a tip is helpful or not.
Proactive feedback	It is the part where a player could send feedbacks, comments or messages.
Bounce Rate	It is the time where a player would visit the application with no activity divided by the total number of visits.
Return visits	It is the number of times where player visits the digital logic gamified application who have previously visited the application.

2.4. Gamification Framework

The proposed gamification framework for the digital logic gate is illustrated in figure 2. The framework is composed of three (3) factors to consider in the development of digital logic gates gamified application which are principles of gamification, design elements for gamification, and student engagement parameters.

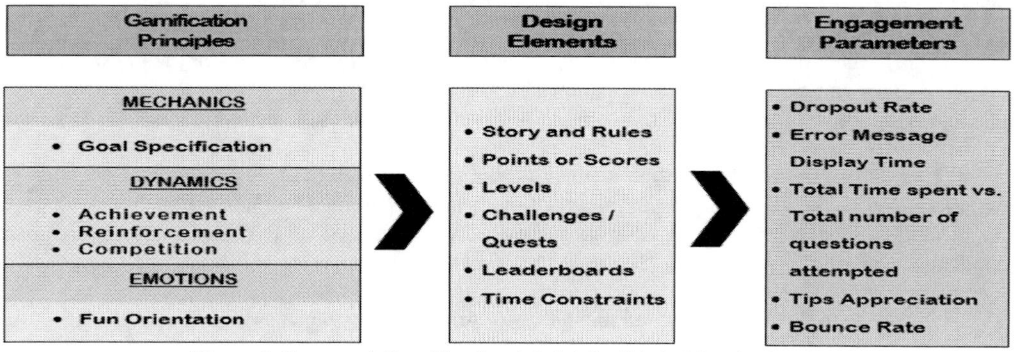

Figure2. Proposed Gamification Model for Digital Logic Gates.

2.4.1. Principles of Gamification. Principles of Gamification is the first factor to consider in the development of a gamified learning approach which is categorized into three (3); mechanics, dynamics, and emotions.

2.4.2. Mechanics. Mechanics are the restrictions of the situation to be gamified as what the designers have set which includes goal specification or goal orientation. It is important for a digital logic gates gamified learning tool to be structured and organized with various levels of goals. For example, to complete the final goal of the digital logic game, the player is presented with the preliminary goal of completing the tasks in every level such as finding the truth table, the Boolean expression, digital logic circuit, etc., in which each task is a prerequisite of the succeeding one. As the player advances with the tasks and levels, the challenge of the game increases and the player's engagement is sustained. A well-defined mechanics of the game also helps improve student's engagement and motivation.

2.4.3. Dynamics. Dynamics indicates the player's behaviour while experiencing the game and how players follow the mechanics with strategic action that is set by the designer. Achievement, reinforcement, and competition are some of the criteria of the dynamics. In digital logic gamified learning application, achievement can be recognized in the form of badges or awards like receiving electronic components that can be used in the next level to complete the circuit diagram and make it function or badges or awards that can be sold in the market. These principles will further enhance a player's motivation, engagement, performance and learning experience.

2.4.4. Emotions. Emotions are mental affective states and reactions experienced by individual players when experiencing the gamified learning tool. Since this is a gamification framework for an educational application, it is expected that fun-oriented and appealing emotions will be produced so that players will continue to play as they enjoy, and engage with the game while learning. In a digital logic gates gamified application, a fun-filled emotion will occur every time players will get badges or rewards like getting electronic components where the player could use in the next level and a negative emotion for a failed quest or challenges.

2.4.5. Design Elements for Gamification. From several reviews of literature, there are lots of design elements given and identified which will make the application a more complicated one. We have identified six (6) basic design elements that are most significant and appropriate for an educational gamified application like the digital logic gates gamified approach. The design elements are as follows:

a. *Story and Rules.* This design element will be the first thing to prepare upon the development of a digital logic gamified application. For example, a circuit map will be shown and a story will be stated while giving the rules so that players would know what to do and how to complete the circuit map;

b. *Points or Scores.* This is a significant design element for a digital logic gates gamified application to engage and challenge the players, points or score could also be used to claim rewards in order to make some advancement on the game. Players could use the points to buy electronic materials in the market or an exchange for a hint to make advancement;

c. *Levels.* This design element is common for all the games which make also significant in digital logic gates gamified application for it shows the milestones of players on the completion of different tasks. Badges can be offered on this part like earning electronic component which can be used on the next levels of the game;

d. *Challenges/Quests.* This design element is significant to keep digital logic players stay focused, engaged and motivated on the game. Strategies on giving challenges and quest during the classroom discussion about logic gates will be applied such as solving problems and designing or testing digital circuits;

e. *Leaderboards.* This design element brings pride to the player of digital logic gamified application where it shows the entire community who visits the application the status of the players and leading scorers or highest pointers of the game;

f. *Time Constraints.* This element brings a challenge to the digital logic gamified application where a player was given a time limit to complete the quest or task in every level of the game just like answering questions or solving problems in an allocated time limit.

Table 2. Engagement Parameters for Gamification.

Event	**Explanation**
Dropout Rate	This metric apprehends a player's engagement with a combined player's persistence and performance. A higher dropout rate will mean a less engaged and less motivated player.
Error message display time	When a player spends more time in understanding the error messages prompted in a game, the player will become less engaged. This data will help in improving message proper construction.
Total time spent vs. a total number of questions attempted	This metric captures the player's engagement when a lower score shows that a player is more engaged than with a higher score shows a less engaged player.
Tips appreciation	This metric helps increase a player's engagement. A helpful tip makes players encourage to stick and engage on the game.
Bounce Rate	This metric would determine the number of players who access the application to gauge the level of a player's interest. This would help developers to improve the digital logic gamified application.
Return visits	This metric indicates a player's engagement when a player keeps on returning with the application.

2.4.6. Student's Engagement Parameters. After we review and synthesize the related literature, metrics for engagement were identified. Table 2 shows how a particular event could be measured as an indicator of engagement on the digital logic gates gamified application.

These identified events to measure engagement parameters for gamification will be considered in the development of the application through including in the activity logs or event logs of the game.

3. Conclusions

With the widespread adoption of games and interactive technologies in the field of education, the learning and teaching experience has been through a constant transformation. The gamified learning approach has contributed and supplemented the traditional learning approach in the field of education in terms of empowering engagement, motivation and learning abilities. This paper sought to present a framework for the development of digital logic gamified application based on the analysis of gamification literature. The significant principles, design elements and engagement parameters that are applicable and appropriate for a digital logic gamified learning application were selected and identified so that the game developer could focus on a basic and important element on educational gamification during the game development phase. Also, this paper showed that the digital logic gamified learning approach is a focal instrument in student's engagement when appropriate design and principles are integrated into the development.

References

[1] Kumar S D N 2018 *8 Top eLearning Trends for 2019* Retrieved from https://elearningindustry.com October 16th

[2] Attali Y and Arieli-Attali M 2015 Gamification in assessment: Do points affect test performance? *Computers & Education* **83** pp 57-63

[3] Kapp K M 2012 *The gamification of learning and instruction: game-based methods and strategies for training and education* (New Jersey: John Wiley & Sons)

[4] DomíNguez A, Saenz-De-Navarrete J, De-Marcos L, FernáNdez-Sanz L, PagéS C and MartíNez-HerráIz J J 2013 Gamifying learning experiences: Practical implications and outcomes *Computers & Education* **63** pp 380-92

[5] Oxales 2018 O *Digital Transformation for Public Education. Manila Standard* Retrieved from http://manilastandard.net/opinion/columns/pens-es-by-fr-ranhilio-aquino/260736/digital-transformation-for-public-education.html March 12th

[6] Simões J, Redondo R and Vilas A 2013 A gamification framework to improve participation in social learning environments *Elearning Papers* **35**

[7] Kim B 2015 Designing Gamification in the Right Way *Library Technology Reports* **51(2)** 29-35

[8] Lee J J and Hammer J 2011 Gamification in education: What, how, why bother? *Academic exchange quarterly* **15(2)** 146

[9] Nah F F H, Telaprolu V R, Rallapalli S and Venkata P R 2013 Gamification of education using computer games *Int. Conf. on Human Interface and the Management of Information* pp 99-107 (Berlin, Heidelberg: Springer)

[10] Robal, T and Kalja A 2007 Applying e-environments in teaching the basics of digital logic *2007 IEEE Int. Conf. on Microelectronic Systems Education (MSE'07)* pp 41-42

[11] Robson K, Plangger K, Kietzmann J H, McCarthy I and Pitt L 2015 Is it all a game? Understanding the principles of gamification *Business Horizons* **58(4)** pp 411-20

[12] Nah F F H, Telaprolu V R, Rallapalli S and Venkata P R 2013 Gamification of education using computer games *Int. Conf. on Human Interface and the Management of Information* July pp 99-107 (Springer, Berlin: Heidelberg)

[13] Sweetser P and Wyeth P 2005 GameFlow: a model for evaluating player enjoyment in games *Computers in Entertainment (CIE)* **3(3)** p 3

[14] Seaborn K and Fels D I 2015 Gamification in theory and action: A survey *Int. J. of human-computer studies* **74** pp 14-31

[15] Khalil M, Wong J, de Koning B, Ebner M and Paas F 2018 Gamification in MOOCs: A Review of the State of the Art *2018 IEEE Global Engineering Education Conf. (EDUCON)* pp 1629-38

[16] Legaki N Z, Xi N, Hamari J and Assimakopoulos V 2019 Gamification of The Future: An Experiment on Gamifying Education of Forecasting *Proc. of the 52nd Hawaii Int. Conf. on System Sciences*

[17] Sheth S K, Bell J S and Kaiser G E 2012 Increasing student engagement in software engineering with gamification *Columbia University Computer Science Technical Reports*

[18] Atkins, A Wanick V and Wills G Metrics 2017 Feedback Cycle: measuring and improving user engagement in gamified eLearning systems *Int. J. of Serious Games* **4(4)** pp 3-19

Acknowledgments

We would like to express our gratitude to the faculty, students and administrators of Camarines Norte State College as a higher education institution, wherein, they guide the researchers in gathering information needed for the study and provide necessary inputs that help the researcher fully understand the course content, and requirements of the subject Logic Circuit and Switching Theory in creating a framework for a gamified learning approach particularly in digital logic gates. In addition, we would like to thank our adviser and fellow DIT students for their feedback, cooperation and of course friendship.

LogIO: An Adaptive Gamification Learning Approach on Digital Logic Gates

JO Torio[1], R T Bigueras[2], D E Maligat, Jr.[3], MA Arispe[4], and J S Dela Cruz[5]

[1,2,3]CamarinesNorte State College, [4]Bicol University Polangui Campus and [5]University of the Cordilleras

E-mail: [1]jocelynt77@gmail.com, [2]rosemariebigueras@gmail.com, [3]dmj234@yahoo.com, [4]cham.arispe@gmail.com, [5]delacruzpen@gmail.com

Abstract. The gamified e-learning approach has been widely used as a learning and teaching strategy, particularly in higher education. This strategy has been successfully used to engage, motivate and enhance learning performances. However, learners or students react differently to game designs and game mechanics depending on the subject area that a particular gamified application though there are several studies shows that gamification enhances motivation. In this paper, we designed and implemented a specific learning model for the gamified digital logic gates or LogIO. Thus, thirty (30) learners evaluated and we determined how do game design elements affect learners in terms of motivation, performance and learning experience. The main results of this study showed that learners agreed that the motivation level and usability of LogIO have a weighted mean of 4.56 and 4.49 respectively while performance and learning experience level has 81% during the utilization of the gamified mobile application. These results support the significance and relevance of adapting particular design elements for gamified digital logic gates to enhance motivation, performance and learning experience.

1. Introduction

Nowadays, the e-learning approach is a common tool in both teaching and learning experience particularly in higher education. Thus, increasing the effectiveness, motivation, performance and learning experience of students can be achieved through gamification. Gamification has become increasingly popular as an e-learning concept in higher education that needs to be properly integrated to have achieved a positive impact on both learning and teaching process which promotes satisfaction on motivation and engagement of learners [1] and [2]. There are several studies that give lots of game design elements and yet, it remains a challenge in determining which particular design element that actually affects learner's motivation, performance and a learning experience [3]. The proper use of game design principles is significant in leading game to motivation which then leads to learning and improved performance [4].

Teaching digital logic gate is a complex area where discussions and explanations should be accompanied with actual tools and electronic components which clearly demonstrates its operation. In the Philippines, the utilization of a gamified medium of instruction has revolutionized. As a matter of fact, the Secretary of Department of Education, Leonor M. Briones emphasized that ICT instructors or professors should devise teaching methods and strategies that will enhance creativity and motivation among learners [5] and it is time to adopt gamification strategy in education.

This paper focuses on determining how do game design elements affect learners in terms of motivation, performance and learning experience. More specifically, this paper presents an evaluation on the usability of a mobile application model of a gamified approach on learning digital logic gates in terms of usefulness, satisfaction and ease of useof the learning model.

2. Related Work

One of the focus of this study is to develop gamification with design elements for game mechanics and game dynamics specifically for learning digital logic gates. Strmečki et al. [6] presented the gamified design elements which correspond to the elements used in the research with a goal of investigating the effectiveness of gamification of an informatics online course against the conventional presentation of online learning content using text and illustrations. The result in their study indicated that an online gamification course related to computer graphics resulted in greater online course achievement. Similarly, Barata et al. [7] introduced gamification as a novel technique that applies game elements to non-game contexts to engage users and solve problems. Learning experience on a non-gamified and the gamified form of the course was compared and assessed their impact on student performance and satisfaction with a goal of introducing multiple game design elements into the course design, including experience points, levels, etc. The results in their study suggest that based on the number of lecture downloads and the number of online posts, the gamified course application will lead to greater student engagement in online course activities.

Meanwhile, Dominguez et al. [8] emphasized that gamification in e-learning platforms seems to likely increase student motivation, but it's not trivial to achieve that effect, and for it to be fully motivating for participants a big struggle is needed when it comes to the design and implementation of the experience. Their study aims to make a pragmatic indication in the field by designing, implementing and evaluating a gamified learning experience in tertiary education. The results of their study suggest that utilizing games in the field of education can be challenged on common beliefs about the benefits being obtained.

It is stated in the study of O'Donovan et al. [9] that tertiary education generally lacks an emphasis on engaging elements that help motivate active learning. The goal of their study is to improve interaction, understanding the topic, problem-solving skills, and engagement through gamification which have a great impact on the goal of our study. The results of their study show that gamification techniques used in their design significantly improved students' understanding and particularly their engagement.Furthermore, Osipov et al. [10] presented the new strategy to evaluate gamification tools in the e-learning approach which not only allows to deport quality training but presents an opportunity for statistical analysis of different parameters which are found in the log files in order to assess the effectiveness of the technical and pedagogical tool.

There are various studies on digital logic gates that are designed in different platforms such as web-based courseware, computer-aided application, simulators,etc. Mohandes et al. [11] stated that online courses in the electrical engineering curriculum can enhance both the teaching and learning processes since the digital logic circuit design course is the core and pre-requisite for a higher-level course in the areas of digital systems in electrical engineering course. The goal of this study is to develop a gamification application that will provide a convenient and self-paced e-learning tool for teaching and learning the fundamentals of Digital Logic Gates.

3. Design and Implementation

3.1. Design of Learning Model

The proposed gamification model for the digital logic gate is illustrated in Fig. 1. The model is composed of three (3) factors to consider in the development of digital logic gates gamified application which are principles of gamification, design elements for gamification, and student engagement parameters. This model was used as a guide in the development of the LogIO gamification for digital logic gates to enhance learners' motivation, performance and learning experience. Logic Gates runs with input and output and it is significant to name the gamified mobile game as logic gates with input and output or LogIO.

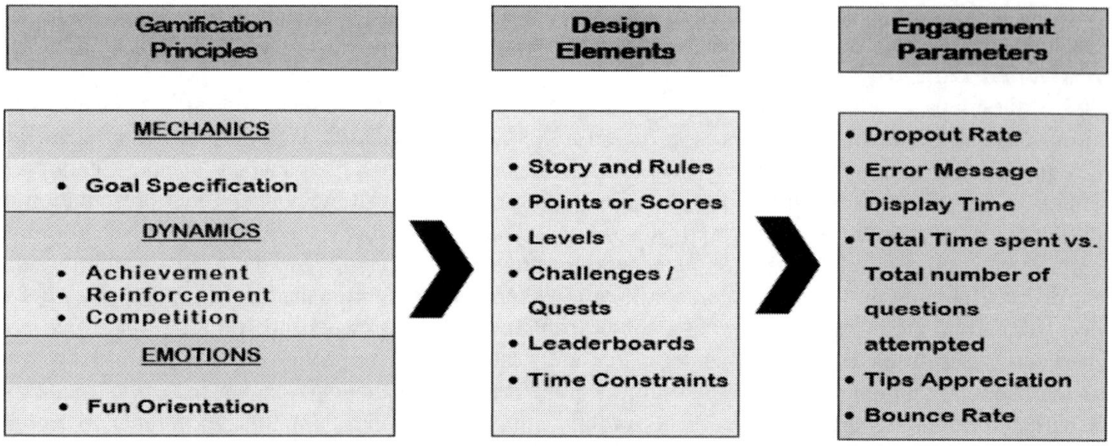

Figure 1. Proposed Gamification Model for Digital Logic Gates.

3.2. Implementation

LogIO was developed on top of Windows platform which supports 64-bit Android Emulator technology using an android studio. A minimum disk space requirement of 4GB (500MB for IDE + 1.5 GB for Android SDK and emulator system image). Also, a minimum screen resolution of 1280 x 800 is required. The content of the application is implemented in the English language as what is also used in one of the reference books of Moris M. Mano in the subject Logic Circuit and Switching Theory. The game starts with a quick review on binary systems which was included in the game for it is a pre-requisite topic for logic gates. Boolean Algebra and its properties, Boolean functions, logic operations, digital logic gates, and integrated circuits are the topics covered on the LogIO game [12].

3.3. Game Design Elements

Several game design elements were used and identified in numerous studies which make an application complicated and may not be relevant in the development and implementation for a gamified logic gate. This study identified the game design elements that are most significant and appropriate in thedevelopment of the gamified LogIO such as story and rules, points, levels, challenges, leaderboards, and time constraints.

As illustrated in figure 2a, the begin button will give the story and rules through a map that needs to be accomplished in sequence. The player cannot proceed with the next level as long as the current level is not yet accomplished where points will be earned at every level with different challenges. For example, a player could not play games on the integrated circuits (IC's) once the game on logic operations are not yet done, this will make the player hooked and motivated in the game. Figure 2b shows that once a challenge is completed, an item will be unlocked as a reward and can be used in the next level or challenge. In this feature, performance and learning experience will be enhanced as the player unlocked levels of different topics. Leaderboardsare also integrated to enhance a player's motivation as ranking and statistics are shown based on the gameplay. Time constraint is significant in the game design to show how a player performs and enhance his/her learning. Figure 2c shows the logs and game statistics every time a player completes a level to enhance a player's motivation.

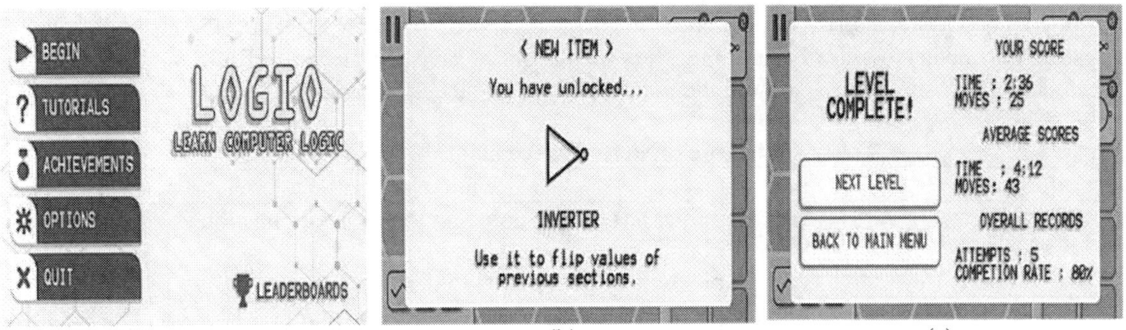

(a) (b) (c)

Figure 2. Samples screenshots for the design elements of LogIO gamified application.

4. Results and Discussion

The researchers administered an evaluation on the LogIO via convenience sampling in order to determine how do game elements affect the learners in terms of motivation, performance and learning experience. Simultaneously, thirty(30) BS Electrical Engineering students were randomly selected from different sections who have not yet taken the Logic Circuit and Switching Theory subject where digital logic gates are part of the course have evaluated the application. Different survey instruments were used to measure the motivation level, performance and learning experience level as well as the usability of the game.

4.1. Motivation

The researchers utilized the Instructional Materials Motivation Survey (IMMS) designed by Keller to measure and analyze the motivation level based on the four (4) components that affect motivation in the learning process such as attention, relevance, confidence, and satisfaction (ARCS)[13]. The IMMS survey was designed to examine the motivation levels whether the e-learning material is in lined with the motivation principles. This instrument has been used, applied and proved to be consistent and valid in examining learner's motivation for an e-learning application [14] and [15].

Table 1. Overall measure on motivation level.

Criteria	Weighted Mean	Description
Attention	4.60	Strongly Agree
Relevance	4.70	Strongly Agree
Confidence	4.43	Agree
Satisfaction	4.50	Strongly Agree
Pooled Mean	**4.56**	**Strongly Agree**

Legend(5-point Likert Rating Scale): Strongly Agree(4.50-5.00); Agree(3.50-4.49);Neutral(2.50-3.49); Disagree(1.50-2.49); Strongly Disagree(1.0-1.49).

Based on the responses of the students who responded on the survey after using the LogIO, the researchers tallied and computed for the weighted mean of a different parameters of each criterion as well as the general weighted mean. Table 1 shows that the attention factor of the gamified application has 4.60 weighted mean which means strongly agree responses. Confidence has 4.43 and satisfaction have 4.50 weighted mean and relevance has the highest weighted mean of 4.70. The general weighted mean is a strongly agree value of 4.56 which shows thatthe motivation level of LogIO is high and may be helpful in enhancing learners' motivation.

4.2. Performance and Learning Experience

To provide insights on the digital logic gates gamified application, mobile application reporting in google analytics (GA) will be utilized in this study to measure the learning experience of the learners.

Application reporting in google analytics is natively integrated with Firebase which is a google application developer platform, in order to provide reporting of distinct events. This will help the researcher understand clearly how the users or players behave during the utilization of the gamified mobile application. Table 2 showed the list of features that the application will log.

Table 2. Measure on Performance and Learning Experience.

Criteria	Description	Percentage
Earned Points/Rewards	When a player has earned points in the game application.	81%
Level Up	When a player levels-up in the game.	81%
Spend Points/Rewards	When a player has spent points in the game.	82%
Tutorial Begin	When a player selects and begins with tutorial.	82%
Tutorial Completed	When a player completes the tutorial in the game.	79%
Unlocked Achievement	When a player unlocks an achievement.	77%
Average percentage		**81%**

As players utilize the application, some criteriato measure the players' behavior in terms of performance and learning experience such as earned points, level-ups, spend points, tutorials begin and completed, and unlock achievements were recorded. Table 2 shows that the average percentage of frequency on performance and learning experience criteria is 81% with a base of zero (0). This means that a high percentage of utilizing the game with the integrated game design elements has also increased performance and learning experience level.

Figure 3 notably shows that the greatest percentage of LogIO players were interested in the game as the criterion for starting tutorials has 82% followed by both earned points and spend points. This simply shows that once a player keeps on accessing the tutorial page, the player is performing as well as learning as earned and spend points goes hand in hand and were also with a higher percentage of 81%.

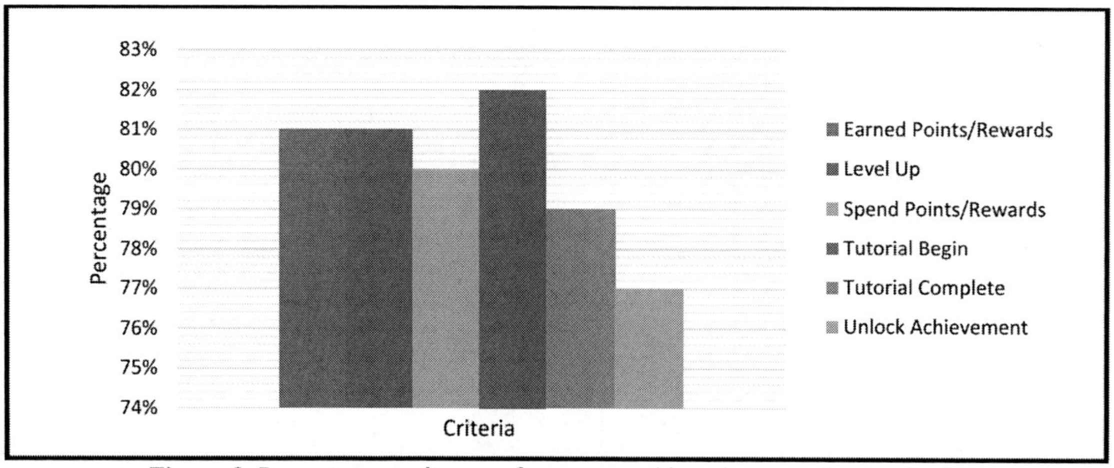

Figure 3. Percentage result on performance and learning experience criteria.

4.3. User's Evaluation

Shown in table 3 is the respondent level of evaluation to measure the usability of the LogIO gamified application for Bachelor of Science in Electrical Engineering of CamarinesNorte State College. The study adopts the USE questionnaire in measuring the usability of the application in terms of usefulness, satisfaction, and ease of use [16]. In summary, the weighted mean of usefulness is 4.52 which means that the application is effective, productive and useful, satisfaction has a weighted mean of 4.43 which is interpreted as respondents agreed to be satisfied in using the application while ease of use has a weighted mean of 4.53 which means that the application is user-friendly and flexible. The overall weighted mean has 4.49 in which respondents agreed for the usability of the game application.

Table 3. Overall usability of the game application.

Criteria	Weighted Mean	Description
LogIO as to Usefulness	4.52	Strongly Agree
LogIO as to Satisfaction	4.43	Agree
LogIO as to Ease of Use	4.53	Strongly Agree
Pooled Mean	**4.49**	**Agree**

Legend(5-point Likert Rating Scale): Strongly Agree(4.50-5.00); Agree(3.50-4.49);Neutral(2.50-3.49); Disagree(1.50-2.49); Strongly Disagree(1.0-1.49).

5. Conclusion

Gamification has been widely used in education for decades. This teaching and learning strategy have been successfully used in higher education in order to engage, motivate and enhance learning performances of learners. Thus, adapting game designs and game mechanics that are particularly appropriate and significant in gamified digital logic gates shall enhance motivation and engagement.This paper sought to design and implement a specific learning model for the gamified digital logic gates or LogIO and thirty (30) learners evaluated the gamified application and the researchers determined how do game design elements affect learners in terms of motivation, performance and learning experience using different survey tools. The results of this study emphasized that learners agreed that the motivation level and usability of LogIO have a weighted mean of 4.56 and 4.49 respectively while performance and learning experience level has 81% during the utilization of the gamified mobile application. These results shall be a basis to support the significance and relevance of adapting particular design elements in order to enhance motivation, performance and learning experience not only for a gamified digital logic gates topic but for other subject areas.

References

[1] Urh M, Vukovic G and Jereb E 2015 The model for introduction of gamification into e-learning in higher education *Procedia-Social and Behavioral Sciences* **197** pp 388-97

[2] Simões J, Redondo R D and Vilas A F 2013 A social gamification framework for a K-6 learning platform *Computers in Human Behavior* **29(2)** pp 345-53

[3] Deterding S, Sicart M, Nacke L, O'Hara K and Dixon D 2011 Gamification using game-design elements in non-gaming contexts *CHI'11 extended abstracts on human factors in computing systems* May 7th pp 2425-8

[4] Paras B 2005 Game, motivation, and effective learning: An integrated model for educational game design *DiGRA '05 – Proc. of the 2005 DiGRA Int. Conf.: Changing Views: Worlds in Play*

[5] Oxales O 2018 *Digital Transformation for Public Education Manila Standard* Retrieved from http://manilastandard.net/opinion/columns/pens-es-by-fr-ranhilio-aquino/260736/digital-transformation-for-public-education.html March 12th

[6] Strmečki D, Bernik A and Radošević D 2015 Gamification in e-Learning: introducing gamified design elements into e-learning systems *J. of Computer Science* Jan 1 **11(12)** 1108-17

[7] Barata G, Gama S, Jorge J and Gonçalves D 2013 Engaging engineering students with gamification *2013 5th Int. Conf. on Games and Virtual Worlds for Serious Applications (VS-GAMES)* Sep 11th pp 1-8

[8] DomíNguez A, Saenz-De-Navarrete J, De-Marcos L, FernáNdez-Sanz L, PagéS C and MartíNez-Herrálz JJ 2013 Gamifying learning experiences: Practical implications and outcomes *Computers & Education* Apr 1st **63** pp 380-92

[9] O'Donovan S, Gain J and Marais P 2013 A case study in the gamification of a university-level games development course *Proc. of the South African Institute for Computer Scientists and*

Information Technologists Conf. Oct 7th pp 242-51
[10] Osipov I V, Nikulchev E, Volinsky A A and Prasikova A Y 2015 Study of gamification effectiveness in online e-learning systems *Int. J. of advanced computer science and applications* Feb 1st **6(2)** pp 71-7
[11] Mohandes M, Dawoud M, Al Amoudi S and Hussain A A 2006 Online development of digital logic design course *2006 2nd Int. Conf. on Information & Communication Technologies* Apr 24th Vol **1** pp 42-7
[12] Mano M M and Kime C R 2001 *Logic and Computer Design Fundamentals* 2nd Edition Updated Chapter (New Jersey: Prentice Hall)
[13] Huang B and Hew K F 2016 Measuring learners' motivation level in massive open online courses *Int. J. of Information and Education Technology*
[14] Huang W H 2011 Evaluating learners' motivational and cognitive processing in an online game-based learning environment *Computers in Human Behavior* Mar 1st **27(2)** pp 694-704
[15] Pittenger A and Doering A 2010 Influence of motivational design on completion rates in online self-study pharmacy-content courses *Distance Education* Nov 1st **31(3)** pp 275-93
[16] Lund A M 2001 Measuring usability with the use questionnaire12 *Usability interface* **8(2)** pp 3-6

Acknowledgments
The researcherswould like to express our gratitude to the faculty, students and administrators of CamarinesNorte State College as a higher education institution, for providingnecessary inputs that helped the researchers fully understand the course content, and requirements of the subject Logic Circuit and Switching Theory in creating a framework for a gamified learning approach particularly in digital logic gates.

IOP Conf. Series: Materials Science and Engineering **803** (2020) 012009 doi:10.1088/1757-899X/803/1/012009

Web-based System for State Universities and Colleges Budget Office Management

Joni Neil Capucao[1], Thelma Palaoag[2] and Sean Sierra[3]

[1]College of Arts and Sciences, Partido State University, Camarines Sur, Philippines
[2]University of the Cordilleras, Benguet, Baguio, Philippines
[3]Partido State University, San Jose Campus, Camarines Sur, Philippines

E-mail: [1]jcapucao265@gmail.com, [2]tpalaoag@gmail.com, [3]sean.sierra@gmail.com

Abstract. The development of a prototype web-based system for budget offices of State Universities and Colleges is the primary goal of this study. In providing a vivid view of the existing budget offices' processes, graphic presentations of the budget process cycle, and flowchart of the budget execution process performed. As a result, under the old system, significant flaws in the transactions conducted surfaced. Considering this as a feedback for improvement in the budget process, as depicted in the swimlane and use case diagram, were proposed. The development of the proposed web-based budget management system was done using the model, view, and controller framework. In observing the software development life cycle, the researchers utilized the rapid application development methodology. In the systems' dry run, the faculty, the staff, and the students gave overall excellent (4.50, 4.62, and 4.67) ratings for functionality, usability, and efficiency. The respondents strongly recommended the adoption of the online system. The evaluation results manifested by *Accurateness in Functionality* (4.8, excellent), *Learnability in Usability* (4.8, excellent), and *Time Behaviour in Efficiency* (4.8, excellent) have all indicated that the system satisfied the requirements. After its full implementation, if some minor flaws may surface, adjustments could just be easily adapted.

1. Introduction

Budget preparation and utilization in State Universities and Colleges involves almost all the academic and administrative offices commonly called responsibility units in the finance department (campuses, colleges, institutes, departments, academic offices, and administrative offices). The bulk of transaction happens in the budget office during the consolidation of the proposed budget; monitoring of allotments against utilization; information dissemination, and reports generation. In Partido State University (PSU), budget preparation requires several days or even weeks to come up with a consolidated budget. PSU does not have an online system for budget preparation, budget synchronization, budget execution, and tracking of documents making it difficult for the administrators and personnel in monitoring into how well these offices is succeeding in meeting its goals.

Moreover, Higher Educational Institutions are mandated to follow the 9001:2015 International Organizations for Standardization, which specify requirements for a quality management system. Included in these guidelines is the prompt submission of required reports and proper logging of documents [1]. As practiced, logging and consolidating of documents is tiresome and often inadvertently neglected. The development of an online budget management system, which includes budget preparation, budget synchronization, budget allotment, and utilization management, is a professional committal in addressing such a gap by the researchers.

Previous works have shown evidence that the challenge can be addressed using an online budget management system. The system may be designed and developed using responsive website design [2] applicable to both mobile devices and desktop computers [3]. The use of barcodes can even be incorporated into the system for ease of entry [4]. Document Tracking System can be useful in administrative and academic institutions [5] for tracking documents and manage the movement of files from desk to desk and different colleges [6]. A mobile application can be used to track expenses incurred [7].

Along with this, applying budget management software will improve deductions, enhance accounts processing, and other vital aspects of every day work for budget teams. A study by Goasduff noted that integrating Business Process Management solutions (BPM) with financial workflow can boost annual ROI by 40 percent to 60 percent, and help minimize mistakes [8]. BPM software is essential because it can help PSU gain a competitive advantage in assisting university administrators to improve organizational performance, analyzing, designing, implementing, controlling, and modifying operational processes.

2. Methodology

Both descriptive and developmental methods were appropriately selected to support the discussion. The descriptive method is intended to give focus on the information processes describing the existing budget process flow, the problems encountered, and the processes that need to be enhanced or replaced in the budget office. The prototyping of the proposed system is through a developmental approach. Faculty, staff, and student responses during evaluation of the prototype suggest the system's functionality, usability, and efficiency. Statistical measures may further determine significant differences in responses

2.1. Investigating PSUExistingProcesses

Conducting a semi-structured interview with the office personnel of the document control office, finance office, colleges, office of the presidentand collecting sample forms used by different offices such as work instruction, procedure manual, organizational structure, the master list of logbooks, and observing the office procedures in processing documents may eventually benchmark in identifying current proceduresto assess the existing budget preparation, consolidation, and budget execution process in PSU. The expected information obtained can be made bases in creating the proposed Budget Management System.

2.2. Development of the Web-based System

The development of the Budget Management System focused on the four major PSU Budget Processes: budget preparation, budget consolidation, operational budget preparation, and budget execution. Designing the prototype system, Web Development Framework could be adopted. For the software development methodology, Rapid Application Development (RAD), shown in figure 2, can be objectively chosen by the researchers; as such, it could allow them to get immediate feedback from the end-users to enable necessary adjustments in developing the prototype.

2.2.1. Web Development Framework. A web development framework is a set of resources and tools for software developers to build and manage web applications, web services, and websites. CodeIgniter is one of the leading web development frameworks, developed by EllisLab, a popular web application framework. It is loosely implied on Model, View, and Controller (MVC) architecture, as shown in figure 1. CodeIgniter promises with exceptional performance, nearly zero-configuration, and no large-scale monolithic libraries [9]. In the MVC Framework, the Model component corresponds to all the data-related logic that the user is working. It can represent either the data that is transferring between the View and Controller components or any other business logic-related data. The View component caters to all the UI logic of the application. Controllers act as an interface between Model and View components to process all the business logic and incoming requests, manipulate data using the Model component, and interact with the Views to render the final output [10]. CodeIgniter MVC framework contains bootstrap classes that make the website dynamic or responsive. A responsive website can adapt to different resolutions and devices on which it is displayed, such as PCs, laptops, tablets, smartphones, and the like without undergoing degradation. The site should also recognize what device it is being displayed on and can automatically respond to the preferences of the user.

2.2.2. Rapid Application Development. For the software development methodology, Rapid Application Development (RAD), shown in figure 2, can be objectively chosen by the researchers, as such it could allow them to get immediate feedback from the end-users to enable necessary adjustments in developing the prototype [11].

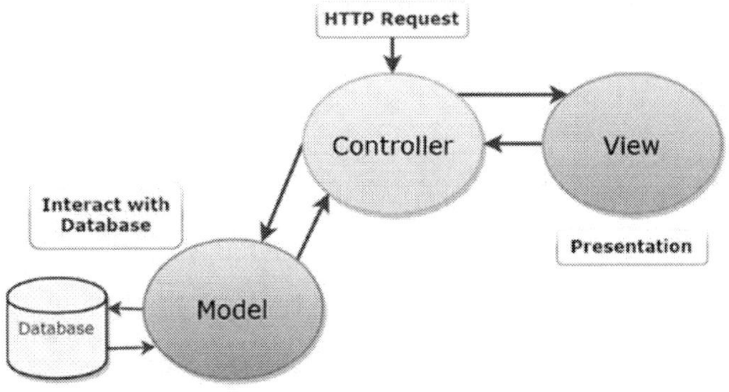

Figure 1. Model, View, and Controller Framework.

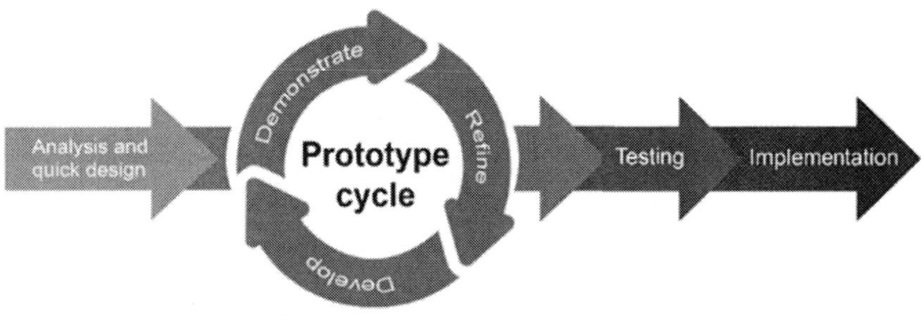

Figure 2. Rapid Application Development (RAD) Methodology.

3. Results and Discussion

3.1. Overview of PSU Budget Office Business Process

State Universities and Colleges (SUCs) are accustom to the manual system in budget management. Figure 3 shows the seven (7) phases in the budget process wherein the significant transaction happens during budget preparation, budget consolidation, preparation of operational budget, and budget allotment and expenditures.

In budget preparation, each unit members list down their priority needs. The unit head consolidates the same and submits the printed document to the budget office. Budget office personnel consolidate them into a university budget proposal. The University submits the budget proposal to the Department of Budget and Management (DBM) for inclusion in the National Expenditure Program. The DBMs' NEP becomes the basis of each responsibility center to create an operational budget. Once the responsibility units have created the operational budget, the budget officer consolidates them for submission to the Congress or Senate for deliberation, and once the president of the Philippines approved the budget, the Government Appropriation Act (GAA) becomes the guide of each agency in executing its expenditures. Budget Management in SUCs involves several sources of funds. This paper focuses more on the management of the General Fund and Income of the University since Trust Funds, Revolving Funds, and other funding sources already have an allotted expenditure.

From the GAA, unit members may submit their payments and claims for obligation in the budget office. The request for the obligation of allotment follows several steps as specified in figure 4. It states that the Budget Aid officer receives the request for funding or payment, checks for its completeness, and attach a checklist of documents necessary for the said request. Budget officer is responsible for the preparation of Obligation Request (OR) and Obligation Utilization Request (BUR), records the amount to the Registry of Allotment and Obligation (RAO) and forwards all papers to the chief budget officer for signature and sends back all the documents to the Budget Aid for submission to the Accounting Section for processing.

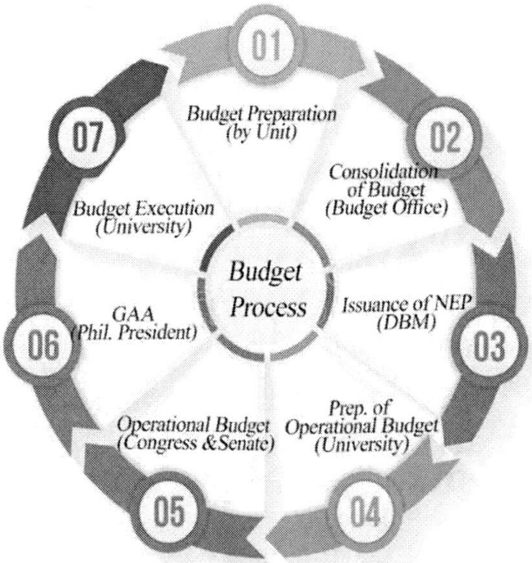

Figure 3. Existing Process in Budget Services

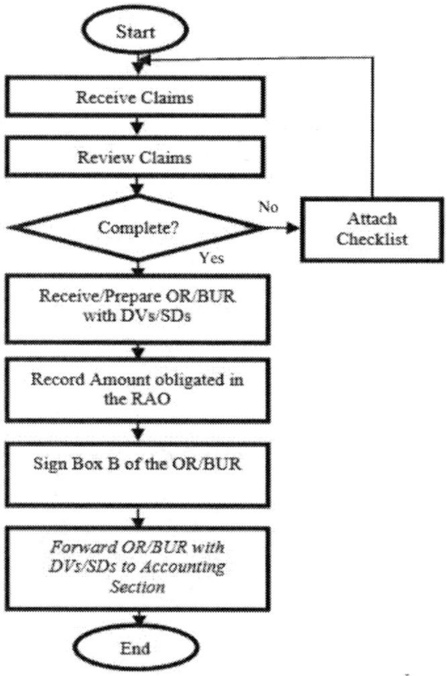

Figure 4. Flowchart of Existing Process in Budget Services

3.2. Prototype System Development

The prototype Budget Management System was developed using Rapid Application Development (RAD) methodology and guided by the design of the Swimlane and Use Case Diagram.

3.2.1. Rapid Application Development Methodology. RAD methodology as shown in figure 3 has four phases, namely; 1. Analysis and Quick Design, 2. Prototype Cycle, 3. Testing, and 4. Implementation.

Phase 1: Analysis and quick design. In this stage, the researchers conducted interviews with faculty (n=160), staff (n=77), and students (n=340) of PSU regarding the budget office flow of documents andthe accessibility issues of documents and report generation that is needed in the design of the proposed system.

Phase 2: Prototype cycle. In this phase, the researchers gathered documents from responsibility units such as Procedure Manuals (PM) and Work Instructions (WI) in determining the scope of the system; and identified the needs of the stakeholders in terms of budget preparation, collaboration of budget allocations, document management and monitoring of budget allotment against utilization. These requirements were analyzed, findings of the study were discussed, and the architecture design of the system was made, including the design of the database, components, and modules of the system.

Phase 3: Testing. For this phase, the researcher analyzed the requirements for the system and designed a solution to be implemented taking into consideration the requirements, constraints and all applicable standards and guidelines. The proposed system has been developed and unit testing have been done to ensure that the system works based on the design of the system and meet all the requirements specified in the specification documents.

Phase 4: Implementation. The last part of the RAD methodology was the implementation of the prototype system. The researchers conducted an initial implementation of the prototype delimiting its coverage to those offices with current basic facilities.

3.2.2. Process Flow Diagrams. The researchers determined the system's scope. It also identified the needs of the university in terms of budget management. The researchers analyzed the requirements for the system and provide solution design, taking into consideration the requirements, constraints, and all applicable standards and guidelines. One such change set up is the online consolidation of budget proposal making all transactions happen in a single "push button" action instead of the old system's process where all transactions are laboriously accomplished solely in the budget office. Presented below, in figure 5 and 6 are the processes involve in the proposed budget management system – swimlane diagram and use case diagram.

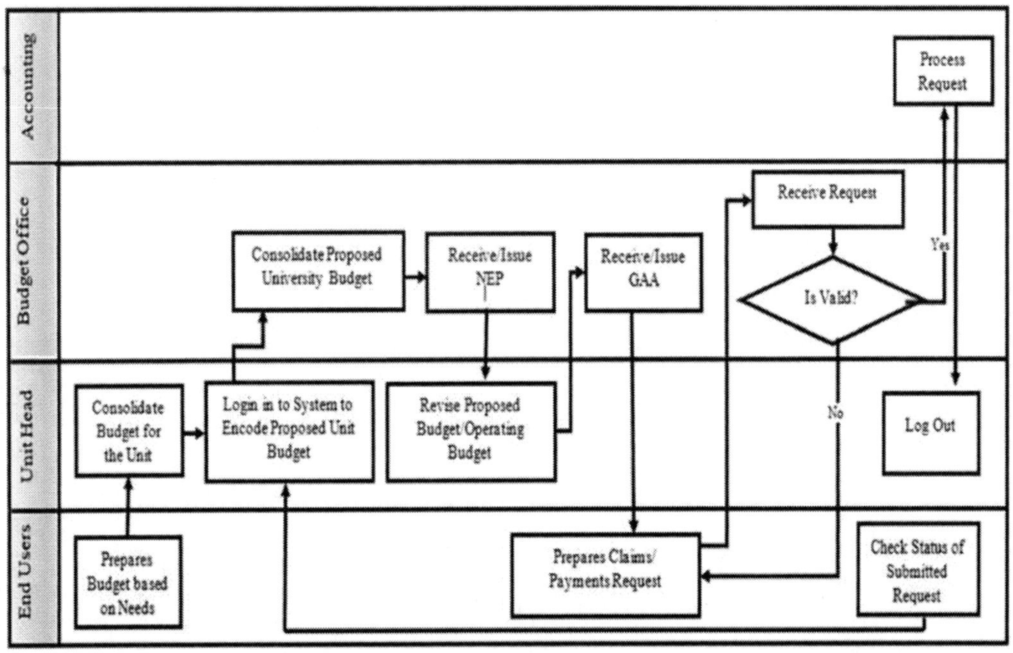

Figure 5. Proposed Budget Management Process using Swimlane Diagram

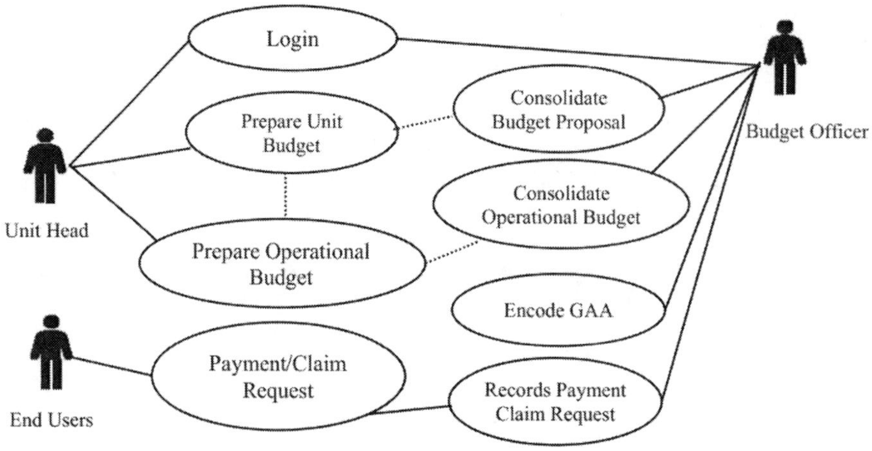

Figure 6. Use Case Diagram

3.2.3. User Interface. Figure 7 shows sample users' interfaces, indicating among other things that this prototype system can be accessed either through mobile phones or laptop/desktop computers by respective clients and other interested parties.

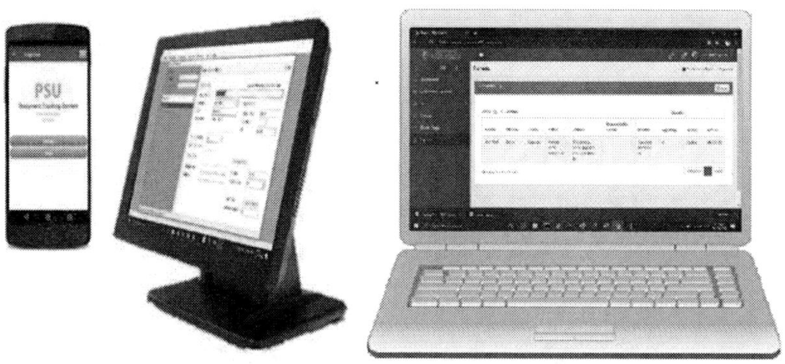

Figure 7. User Interface for Mobile Phone, Laptop, and Desktop Computer.

3.3. Evaluation of the Proposed Budget Management System

From the total target population of two thousand eight hundred sixty-seven (2,867), faculty members shared two hundred seventy-one (271), staff contributed ninety-six (96), and for the students two thousand five hundred (2,500).

The main respondents of the present study are the faculty (160), staff (77), and students (340) of Partido State University. From the universal population of 2,867 only 577 has been considered sample size which is within the 5 percent tolerable margin of sampling error. Though purposively (homogenous) selected, the intention is to provide higher *representative.* (see Table 1).

The evaluation result of the proposed system in terms of: (a) functionality; (b) usability; and (c) efficiency using the ISO 9126 Characteristic and Sub-characteristics (ISO 1991) [12]is shown in Table 2. All responses along functionality (4.50, excellent), usability (4.62, excellent) and efficiency (4.67, excellent) showed no significant difference at alpha .05, therefore suggesting that faculty, staff, and students have the same observations in favor of using the prototype budget management software.

Table 1. Distribution of Respondents

Faculty		Staff		Students	
Target Population	Sample Size	Target Population	Sample Size	Target Population	Sample Size
271	160	96	77	2500	340

Table 2. ISO 9126 Characteristic and sub-characteristics(ISO 1991)

Characteristic	Sub-characteristic	Explanation	Weighted Mean				Interpretation
			Faculty	Staff	Students	Total	
Functionality	Suitability	Can software perform the tasks required?	4.5	4.7	4.6	4.60	Excellent
	Accurateness	Is the result as expected?	4.7	4.8	4.9	4.80	Excellent
	Interoperability	Can the system interact with another system?	4.1	4.2	4.3	4.20	Very Good
	Security	Does the software prevent unauthorized access?	4.4	4.3	4.5	4.40	Very Good
Usability	Understandability	Does the user comprehend how to use the system easily?	4.3	4.4	4.3	4.33	Very Good
	Learnability	Can the user learn to use the system easily?	4.7	4.8	4.9	4.80	Excellent
	Operability	Can the user use the system without much effort?	4.7	4.7	4.8	4.73	Excellent
	Attractiveness	Does the interface look good?	4.5	4.7	4.6	4.60	Excellent
Efficiency	Time Behaviour	How quickly does the system respond?	4.8	4.8	4.8	4.80	Excellent
	Resource Utilization	Does the system utilize resources efficiently?	4.5	4.6	4.5	4.53	Excellent

H_o*Significant @.05 **Significant @.01

4. Conclusion

In this study, a prototype budget management systemwas presented using an MVC web development framework. The PSU budget office business processes considered and exhibited using a flowchart, swim lane diagram, and use case diagram. In the systems' dry run, the faculty, the staff, and the students gave an overall excellent (4.50, 4.62, and 4.67) ratings for functionality, usability, and efficiency. The respondents strongly recommend the adoption of the online system. The evaluation results manifested by the *Accurateness in Functionality* (4.8, excellent), *Learnability in Usability* (4.8, excellent), and *Time Behaviour in Efficiency* (4.8, excellent) have indicated that the system satisfied all the requirements. After its full implementation, if some minor flaws may surface, adjustments could just be easily adapted.

References

[1] ISO/TC176/SC2 ISO9001 2015 *Quality management systems requirements* Available from: https://www.iso.org/standard/62085.html.

[2] Giurgiu L and Gligorea I 2017 Responsive web design techniques *Int. Conf. Knowledge-Based Organization* DOI:10.1515/kbo-2017-0153

[3] Waseem B I and Abdelaziz H I 2016 Responsive web design techniques *Int. J. of Comp. App.* 0975–8887 ResearchGate pp 18-27

[4] Gao J Z, Prakash L and Jagatesan R 2007 Understanding 2D-barbode technology and applications in m-commerce design and implementation of a 2D barcode processing solution *31st Annual Int. Computer Software and Applicatioin Conf.*

[5] Ajoge N S and Hamisu I A 2015 Document e-tracker: a panacea for efficient internal document management system in kaduna polytechnic *JOSTE* **3(2)**

[6] Omoregbe N A, Azeta A A and Adewumi A 2014 Development of a file tracking system for tertiary institutions *Proc. of edulearn14 Conf.* pp 3963-70

[7] Thanapal M P, Patel Y, Lokesh R T P and Satheesh K J 2015 *Income and expense tracker* (Tamil Nadu, India: Vellore)

[8] Goasduff L 2016 *BPM is critical to business transformat ion success* Available from: https://www.gartner.com/smarterwithgartner/bpm-is-critical-to-business-transformation-success/.

[9] Gupta H 2018 *Top 10 Frameworks for Web Applications Retrieve* [Online] cited October 09th 2019 Available from:https:// www.geeksforgeeks.org/top-10-frameworks-for-web-applications

[10] Majeed A and Rauf I 2018 MVC Architecture: A Detailed Insight to the Modern Web Applications Development *Peer Review J. of Solar & Photoenergy Systems* **1(1)** p 1-7

[11] Kralev V S and Kraleva R S 2017 Methods and Tools for Rapid Application Development *Proc. of the III International Scientific and Practical Conf.: "Methodology of Modern Research" Dubai, UAE* Research Gate pp 21-4

[12] ISO/IEC 9126 1991 *Software engineering-product quality ISO* Retrieved from https://www.iso.org/standard/16722.html

IOP Conf. Series: Materials Science and Engineering **803** (2020) 012010 doi:10.1088/1757-899X/803/1/012010

Temperature and Humidity Monitoring System in Broiler Poultry Farm

Irving V Paputungan[1,3], Abidurrahman Al Faruq[2], Fitri Puspasari[2], Furqaan Al Hakim[2], Imam Fahrurrozi[2], Unan Y Oktiawati[2], Iing Mutakhiroh[4]

[1]Department of Informatics, Universitas Islam Indonesia
[2]Department of Electronics and Informatics Engineering, UniversitasGadjahMada Yogyakarta, Indonesia
[3]Department of Electrical and Electronics Engineering, University Teknologi PETRONAS, Malaysia
[4]Huazhong University of Science and Technology, Wuhan, China

E-mail: irving@uii.ac.id, unan_yusmaniar@ugm.ac.id

Abstract. Broiler chicken is a livestock that is characterized by fast growing. Temperature and humidity become the most important factors during their growth. Unmonitored and uncontrolled temperature and humidity within the cage can lead to reduced productivity and high rate of mortality. This paper presents a prototype that is able to monitor such factors automatically. A prototype of sensor using DHT11 and Arduino DUE microcontroller has been developed. Android-based app is deployed to support the prototype. A web-based application is also built to show the monitoring result comprehensively. The result shows that the sensor is able to continuously capture the current temperature and humidity of the chicken cage for some periods. After the implementation, the number of death chicks are decreasing from 5 to 4 chicks per day. Controlling both factors will be the future work.

1. Introduction

Poultry meat is the most preferred halal cheap protein source. As one of nations with the largest Muslim population in the world (85% of total population), data from Frost and Sullivan in 2017 shows that Indonesia poultry consumption is only 10.1 kg per capita compared with Malaysia (47.5 kg) and Thailand (16 kg) [2]. This situation makes poultry sector remain promising and needs further improvement to all stakeholders.

However, poultry farming, especially chickens, is not an easy task in Indonesia during dry season. High temperature imposes severe stress to chickens [3] [4] because they have difficulties to cast away the heat from their body. Chicken is a hot-blood animal that has no sweat-gland and most of their body is covered with feather. Such kind of stress leads to reduced performance and increase the mortality rate.

In this paper, we presents a prototype system that can monitor temperature and humidity level in chicken house/cage. Monitoring such parameters is expected able to reduce the mortality rate to avoid further productivity loss on broiler. The rest of this paper as follows. Section 2 provides the system design and development phase. Section 3 presents some of the captured data and the prototype. Lastly, summaries are provided in the section 5.

2. System Design and Development

Figure 1 depicts the architecture of the system. It contains the hardware and software that are used during development. DHT11 sensor is used to detect temperature and humidity [1]. DHT11 is a very low price and low power consumption sensor that has basic capability data logging, converting from analog to digital. It is good at 0-50°C temperature reading with ±2°C accuracy and 20-80% humidity readings with 5% accuracy. Such measurement is suitable enough for any chicken cage in Indonesia.

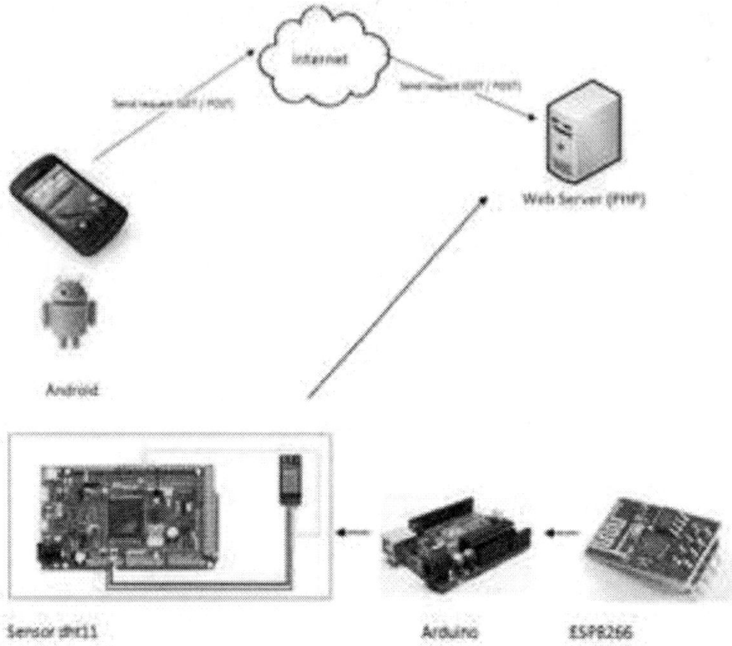

Figure 1. System Architecture.

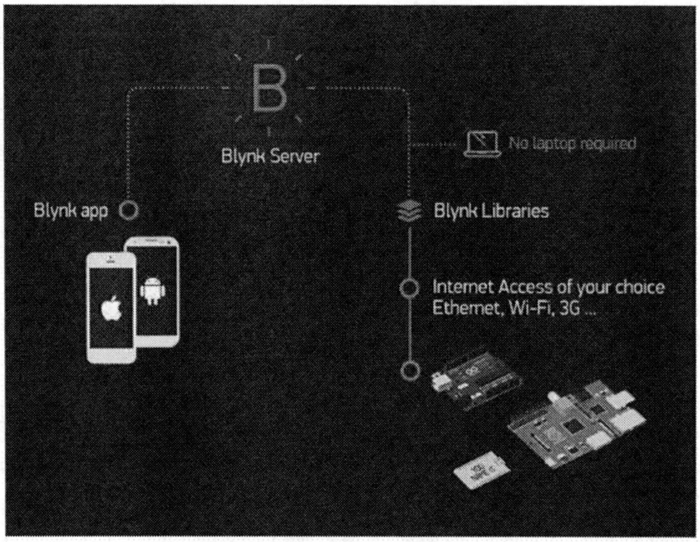

Figure 2. Blynk Architecture

Arduino DUE [5] [6] is used to capture the sensor data from DHT11. Arduino DUE is a powerful 84 MHz and 32-bit processor microcontroller. It has Direct Memory Access (DMA) feature that is able to support multi-tasking within the microcontroller. USB OTG capability is also a benefit of this

IOP Conf. Series: Materials Science and Engineering **803** (2020) 012010 doi:10.1088/1757-899X/803/1/012010

Arduino to enable Wi-Fi module connection. Low cost Wi-Fi module ESP8266 is added to the prototype to support internet connection. Using such connection, the data that has been processed in Arduino will be transmitted to the IoT platform.

Blynk app [7] is used as a platform to control and monitor the deployed device. This platform is also very useful to connect the device to the Blynk cloud server and analyse the telemetry data over Wi-Fi or any internet connection. It has several libraries to support data analysis. Another pros is Blynk cloud server is open-source and deployable in minutes. See Figure 2.

The sensor is then installed in the chicken cage with some broilers inside. The cage area is 80 x 11 m. Sensor is placed every 15 m in one side of the cage. The monitoring prototype is deployed and obtaining data from the sensor that is connected using Wi-Fi.

3. Result and Discussion

Figure 3 shows the temperature and humidity that are reported in Blynk app. This app makes user able to acknowledge the statusof the chicken cage from time to time and from anywhere. Figure 4 depicts the web-based monitoring system that has been deployed. Such website is able to show the status within the cage and will be used later during controlling the humidity and temperature.

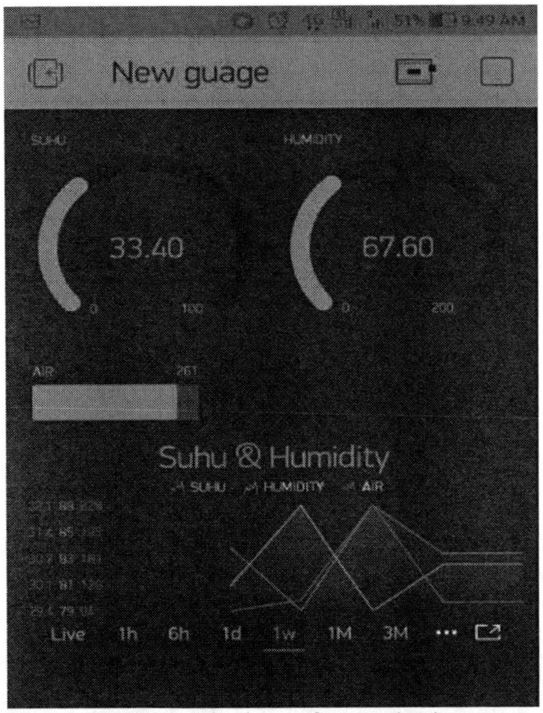

Figure 3. Blynk app for monitoring

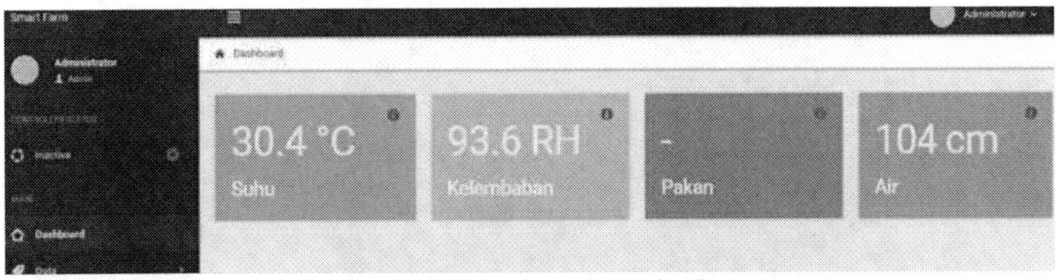

Figure 4. Web-based monitoring

IOP Conf. Series: Materials Science and Engineering **803** (2020) 012010 doi:10.1088/1757-899X/803/1/012010

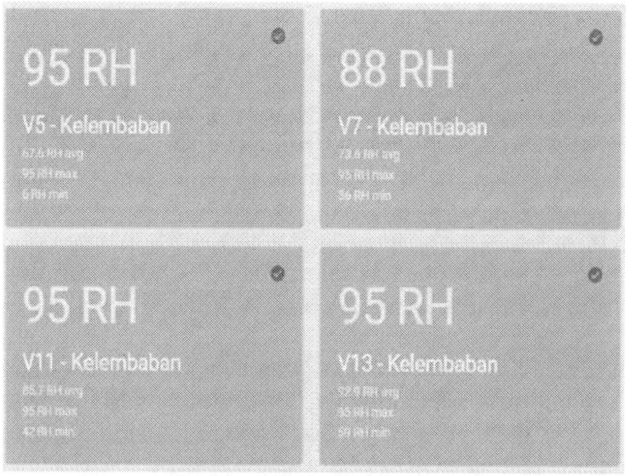

Figure 5. Humidity rate

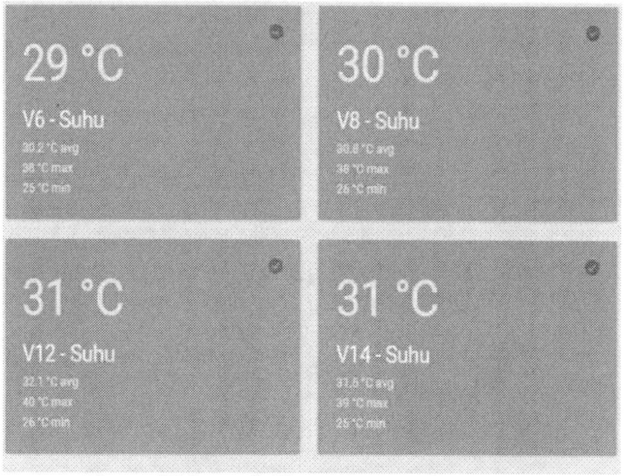

Figure 6. Temperature rate

Figure 5 and 6 are the screen captured from the website that shows some humidity and temperature level based on continuous monitoring. The chicken cage where the experiment is taken place is shown in Figure 7. Some chicks are now growing up well during monitoring.

Figure 7. Chick's house

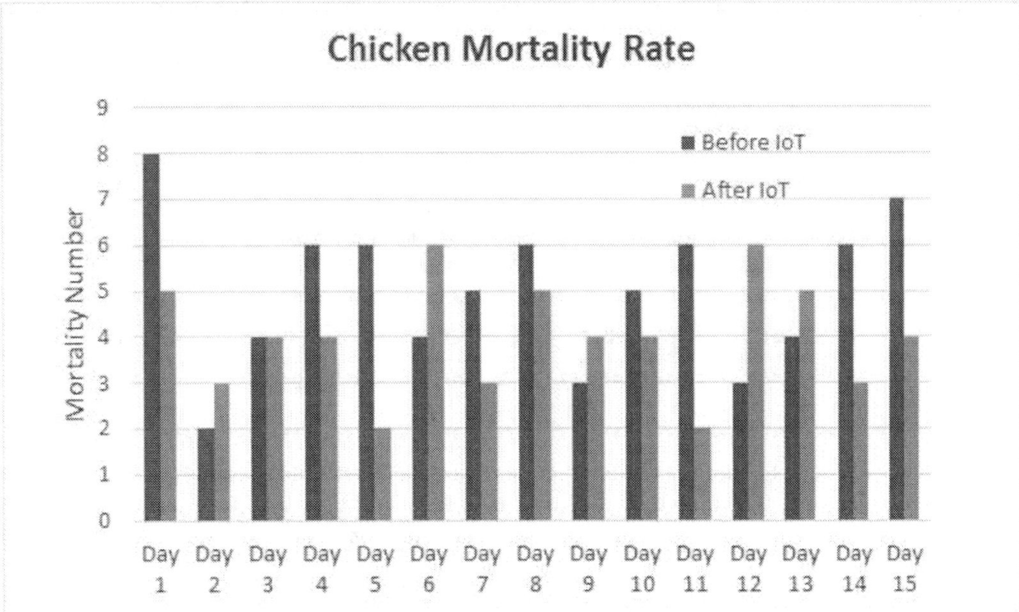

Figure 8. Mortality Rate Comparison for 15 days

Figure 8 depicts the mortality rate of the chicken during observation period (14 days). Chicks are difficult to adapt the environment within the age of 14 days [8]. It shows that, in average, 5 chicken died before the monitoring system implementation and 4 chicken after the implementation. It means that the system is able to support in reducing the number of chicken mortality rate.

During observation, the average humidity level during observation period is 95% and the average temperature is 28 °C. Such values are off from the comfort zone of any chicks. See Table 1 on the ideal value of temperature and humidity. Such situation makes chicks are prone to death after some days in the cage. See Table 2 on the temperature felt by the chicks based on humidity level.

Table 1. Ideal temperature and humidity for chicken

Broiler		
age (days)	temperature (°C)	humidity (%)
1	32-29	60-70
3	30-27	60-70
6	28-25	60-70
9	27-25	60-70
12	26-25	60-70
>=15	25-24	60-70

Table 2. How chicken feels the temperature

Effective Temperature felt by the chicken (°C)	Cage Humidity level (%)				
	40	50	60	70	80
	Cage Temperature (°C)				
30	36.0	33.2	30.8	29.2	27.0
28	33.7	31.2	28.9	27.3	26.0
27	32.5	29.9	27.7	26.0	25.0
26	31.3	28.6	26.7	25.0	24.0
25	30.2	27.8	25.7	24.0	23.0
24	29.0	26.8	24.8	23.0	22.0

4. Summary

The development of humidity and temperature monitoring equipment for poultry farm is a stepping stone toward Internet of Things (IoT) implementation. IoT is a new emerging area of using any devices that are connected to the internet. Such devices are used to collect some data to be sent to the server for further processing. The use of IoT in poultry farm is very beneficial in improving the broiler quality by monitoring and controlling the environment.

This paper has presented part of monitoring and controlling the humidity and temperature within a chicken cage. Monitoring has been done for 14 days while controlling will be performed in the next period after most of the chicken is replaced with newborn ones. Adding some lamps in the cage is suggested in order to keep the cage warm (higher temperature and lower humidity).

References

[1] Ada L 2019 *DHT11, DHT22, and AM2302 sensors Adafruit Learning System* Retrieved from https://learn.adafruit.com/dht/overview
[2] RHB 2018 Indonesia poultry sector update *RHB Report*
[3] Lara L J and Rostagno M H 2013 Impact of heat stress on poultry production *J. Animals* **3** pp 356-69
[4] Bhadauria P, Keshava, Mamgai P, Murai A and Jadoun Y S 2016 Management of heat stress in poultry production system *Technical Bulletin ICAR-Agricultural Technology Application Research Institute*
[5] Bitjoka L, Ndje M, Boum A T and Song-Manguelle J 2017 Implementation of quadratic dynamic matrix control on arduino due ARM cortex-M3 microcontroller board *J. of Engineering Technology* **6** no 2 pp 682-95
[6] Sunitha S 2017 Distance measurement using ultrasonic sensor and NodeMCU *Int. Research J. of Engineering and Technology* **4** no 6 pp 1794-7
[7] https://blynk.io/
[8] Ross Broiler Handbook 2018

Improving Inland Aquaculture: A WBIS Perspective

Alexy Gene Castillo[1], Josephine Dela Cruz[2]

[1]Mabini Colleges, Inc., [2]University of the Cordilleras

E-mail: [1]axc.mit@gmail.com, [2]delacruzpen@gmail.com

Abstract. Water pollution is the contamination of water bodies that is devastating to the aquatic organisms for it destroys aquatic life and reduces its reproduction. Due to polluted waters of inland aquatic farms in Bicol Region, Philippines the production of fishery resources and the ecosystem is duly affected. Thus, this study assessed the capability of the developedWeb-based InformationSystem (WBIS)using factors such as functionality, usability and efficiency garnering a grand mean of 4.65having an adjectival interpretation of highly capable. Hence, the WBIS (iARMIS: Inland Aquatic Resources Management Information System)could be an effectivedecision support information generatorthat imparts Web-GIS technology inbioremediation process to improve and sustain the aquaculture industry in Camarines Norte without compromising the environment.

1. Introduction

Aquaculture magnifies its importance as an industry that contributes to the nation's increase of Fish Production, Food Security and Economic Growth. With the mentioned fish production demand that delves with earnings and income [1], record shows that the environment is being compromised due to harmful farm practices [2] resultingwith polluted aquatic pond waters [3]. Water pollution is the contamination of water bodies [4] that disrate and destroys the reproduction of aquatic life [5].

Awareness on environmental preservation and consciousness on water quality [6] is very much required as it caters aquatic life its production [7]. Bioremediation is characterized as an applied bio-technology processes that uses beneficial micro-organisms to eradicate contaminating and polluting substances in soil and water [8]. The concept of this technology is to actively promote bacterial nutrient processing into forms suitable for uptake by aquatic species, while limiting the presence of pathogenic microorganisms in aqua cultural environments through competitive exclusion [9].

The papers of Candelaria and Corre [10], Hamdan and Mahmoud [11], significantly presented how bioremediation employing probiotics were extensively used as eco-friendly approach to improve aquatic species' disease resistance and survival, feed utilization, growth performance and immune status; while its vitality in managing the water qualityof fish farms is discussed in the studies ofTuan and Duc[12], and MartinezCruz et al [13].

On the other hand, a Web-based Information System (WBIS) is an organized facilitydeveloped employing database management principles to generate and disseminateinformation services using the internet[14].While GeographicInformation System (GIS) a computer assisted system that capable of using geo-referenced data for a range of spatial analyses and outputs[15].Essentially, these innovations were reported to establishtheirusesbeneficial tofish farming sector in terms of site location and facility mapping[16], and in the development and management of resources[17].

The study of Nieves [18] and Cruz-Lacierda [19] depicts the status and gaps of Bicol Philippines' inland and marine aquaculture in terms of its characteristics, structure, resources and management. They cited that fish diseases is caused by poor farming practices which negatively affects both production [20]and the eco-system [21]. Consequently, Castillo and Dela Cruz [22] describes that aquaculture sector in Camarines Norte (Lat14.16 °, Long 122.75 °) posed challenges and issues particularly in the monitoring of aquatic resources in the area, which compromiseslaws and orders such as RA 8550, RA 7160, RA 9003, RA 9275 and other related provisions that oversees both Philippine aquaculture and environmental safety. Hence, they proposed a model using Web-GIStechnology and bioremediation procedure (iARMS: Inland Aquatic Resource Management System) to address stated concerns.

Accentuating the absence and the demand of clearly monitoring aquaculture resources in Camarines Norte, the present study evaluated the capability of the developedWeb-Based Information System (WBIS)based from the aforementioned iARMS Model incatering the information management ofits inland aquatic resources thru geo-mapping and profiling the biophysical and agri-fisheries livelihoodcharacteristics, and improving the water qualityof fish farms by generating bioremediation prescription. These actions were taken to support the aims of the Bureau of Fisheries and Aquatic Resources (DA-BFAR) towards sustainable aquaculture and safe environment.

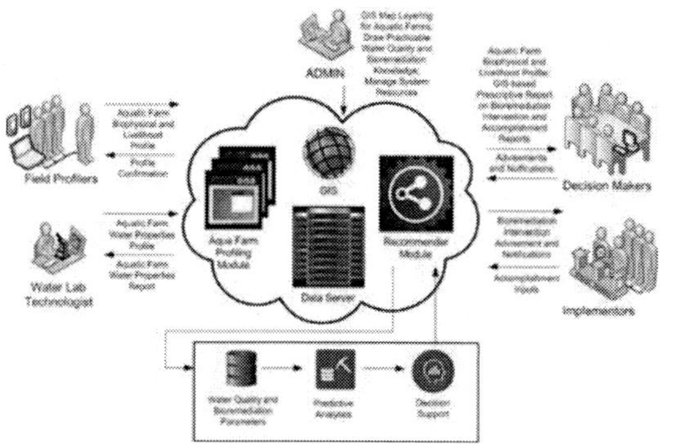

Figure 1. The iARMS Model

2. Methodology

The architectural framework of iARMS is illustrated by Figure 1. As conceptually described by the proponents, it depicts a decision-support (DSS) [23] Web-GIS [24] system design based from defined operational procedures to monitor and manage the information of inland aquatic resources of the province as to fish farms profiling, water quality management – bioremediation and production capacity.

One of the policies embodied in BFAR directives is to conduct monitoring and evaluation that affects conservation, operation and management of fishery resources. Thus, Field Profilers are sent off to characterize inland aquatic farms for its biophysical and agri-fisheries livelihood [25]production capacity profiling using the model's profiling module.Water samples are collected, labeled and forwarded to water quality laboratory for testing. Water Lab Technologist supplied the preliminary profile of the surveyed aquatic farms.

Using an online GIS software, these profiles are utilized by the System Administrator tocreatemaps which pin-pointsthe location of each fish farm geographically.As an online facility, the model provides dedicated URLs to board information contentof each farm in web-pages format.

In terms of bioremediation intervention, the Recommender Module examines the recorded water properties of the pond based from the stored practicable knowledge on aquaculture water qualitystandards, bioremediation procedures, as well aslocation-route planning,and generatesprescriptive reports for the Decision Makers and Implementers (Bioremediation Workforce) to take action.

With aquaculture and eco-system is at stake, the present study employed Rapid Application Development methodology [26]to systematically realise iARMS Model'sperceived technical specifications as a fullyfunctional WBIS in the shortest possible period of time. GPS Map Camera app[27], PHP [28] and MySQL [29], ArcGIS software (Online and Web Mapping APIs) [30] and Fiber Optic internet [31]were used in building-up thebackbone anddeveloping the information systemfacility. Water samples taken from identified aquatic farms was tested atCNWD-Primewater[32] laboratory to serve as the foundation of the model's knowledge management-basedRecommender Module for bioremediation intervention. The data gathered served as the reference for determining the water quality of the pond comparing to that of the standardized water quality criteria for aquaculture [33]. Prescription of probiotics application [34] [35] [36], route plans [15], including water qualityacceptability were grounded fromthe concepts of predictive analytics [37].

The capability of the developed WBIS' Profiling and Recommender Moduleswere assessed bysix (6) IT and Bioremediation experts,three (3) BFAR Aqua-culturists, and three(3) Fish farm operators using modified ISO 9126-1 [38] metric-based rating instrumentwith afive-point Likert scale ranges of 4.51-5.00: Highly Capable; 4.01-4.50: Capable; 3.51-4.50: Moderately Capable; 2.51-3.50: Less Capable; 1.50-Below: Non-Capable. Descriptive statistics using weighted mean is applied to verbally interpret the result.

Table 1. Biophysical and Agri-Fisheries Livelihood Profile of Identified Inland Aquatic Farms in 2nd District of Camarines Norte using iARMS Model

Aquatic Farm ID	Geographical Coordinates	Area (HA)	Culture Environment	Cultured Specie/s	Water Quality Mgt. Method	Years of Operation	FLA
AF 0-001	Lat. 14.0427 Long. 122.9319	0.08	Fresh Water Earth Pond	Tilapia	Antibiotics / Partial Water Freshening	2	None
AF 1-001	Lat. 14.0741 Long.122.5858	0.8	Brackish Earth Pond	Milkfish	Formalin / Partial Water Freshening	2	None
AF 2-001	Lat. 14.2227 Long. 122.9006	2.5	Brackish Earth Pond	Tilapia / Milkfish	Hydrogen Peroxide / Partial Water Freshening	45	None

3. Result and Discussion

Taking iARMS model into operation, Table 1 presents the profile of the surveyed inland aquatic farms in the locale, From these data inputs, parcelled maps that geographically visualize the locations of each aquatic farms were created using ArcGIS' selection and mapping algorithms [29], as a working component of the WBIS. While the other modulesprovidesdedicated and secured pages for system users tomanage the inland aquatic resources information, ArcGIS makes it available for the facilityto generate geo-referenced data relevant to its purposes. The index page of the developed WBIS (iARMIS: Inland Aquatic Resources Management Information System) is graphically presented by Figure 2.

The capability assessment of iARMIS was undertaken thru actual system testing (Figure 3). The summary of the said evaluation is presented in Table 2. It could be gleaned from the matrix that the assessors rated theWBIS, "Highly Capable" at all factors that describes its technical characteristics, user-friendliness, access and execution time, as to fish farm profiling and bioremediation recommendation operations. With a grand weighted mean of 4.65, it could then be stated that iARMIS isan onlineinformation system facility capable to manage inland aquatic resources information.

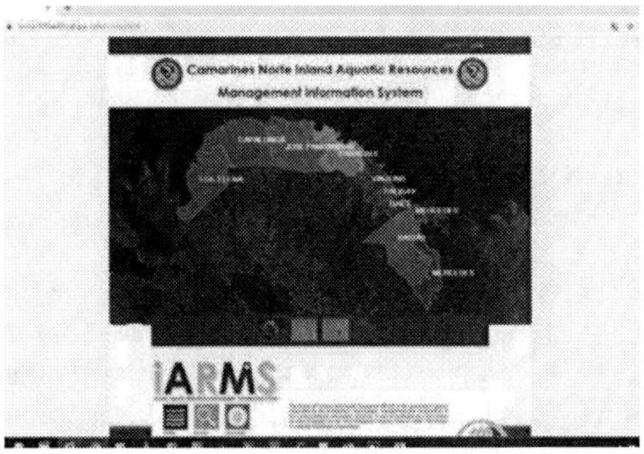

Figure 2. Index page of iARMIS WBIS

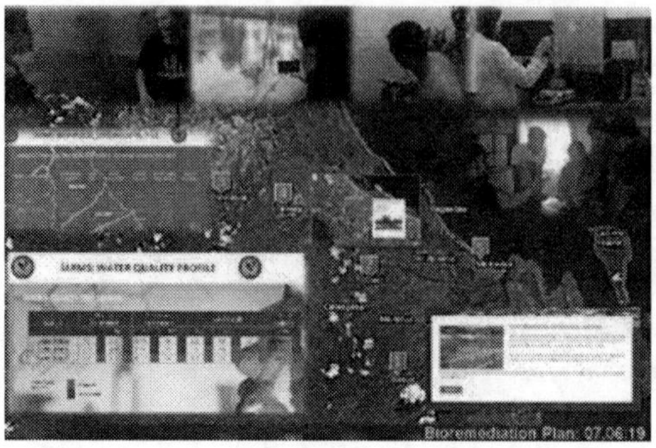

Figure 3. iARMIS Operating Environment

Table 2. The iARMIS Capability Evaluation Summary.

Capability Factors	Weighted Mean	Verbal Interpretation
Functionality	4.68	Highly Capable
Usability	4.52	Highly Capable
Efficiency	4.75	Highly Capable
Grand Weighted Mean	4.65	Highly Capable

Table 3. Water quality acceptabilitytest of AF-1001 using iARMIS.

Acceptable Range of Water Quality Parameters for Aquaculture [32]		Results Prior to Bioremediation		Results After Bioremediation Process	
pH	6.5~8.5	4.8	UA	6.6	A
DO (mg/L)	5.0	6.9	UA	5.0	A
P (mg/L)	0.20	0.44	UA	0.19	A
Hg (μg/L)	<= 2.0	2.2	UA	1.7	A
Pb (μg/L)	<= 50	53	UA	42	A
Cd (μg/L)	<= 10	6	UA	4	A
Total Coliform (MPN)	<= 5000	6558	UA	4327	A

*UA – Unacceptable
* A – Acceptable

Drawn from thewater quality test ledger of iARMIS profiler and recommender module, Table 3 presents the water quality test results forAF (Aquatic Farm) 1-001. Based from thelab test resultsprior to bioremediation, the system automatically evaluates [36] each data input against the water quality criteria and standards, in which are interpreted to be "Unacceptable." After the bioremediation process was conducted the water quality of the pond was improved, hence "Acceptable."

The results obtained was determined due to the bioremediation prescription [36] of iARMIS which comprises preparatory measures, dosages of liquid-probiotics to be applied and its itinerary [33] [34] [35] and location-route plan, with the area and geographical coordinates of the aquatic farm as reference. This notably displays the potential of theWBIS for bioremediation intervention [6] [9] [12] [13].

4. Conclusion

The present study, assessed the capability of the developed WBIS that caters the information management of inland aquatic resources subject for bioremediation in Camarines Norte. The evaluation garnered an overall "Highly Capable" rating in terms of functionality, usability and efficiency, and found to have a potential to be an effective decision support tool for bioremediation intervention. With theUnited Nationsgoalsof improving all forms of agriculture while preserving the natural resources for better life, iARMIS could serve as an information generator that will assist the Bureau of Fisheries and Aquatic Resources to perform their functions as the agency in the Philippines that oversees aquaculture, in line with the national development plan in attaining the Sustainable Development Goals (UN SDGs).Thus, the adaption of iARMIS as a WBIS infrastructure is recommended to disclose its technical capabilities, and respond withoperational and other relatedissues that could be encountered during its implementation.

Acknowledgment
The authors acknowledges their families and affiliations (Mabini Colleges, Inc. and University of the Cordilleras) for the untiring support and inspiration along way. They would also like to extend their gratitude to Min. JuliusarNemis, Dr. Thelma Palaoag, Prof. Sylvia Salvan, and the CCS GISP-Access Team for their valuable means and insights. To God Almighty be all the glory.

References
[1] Béné C, Arthur R, Norbury H, Allison E, Beveridge M, Bush S and Thilsted S 2016 Contribution of fisheries and aquaculture to food security and poverty reduction: assessing the current evidence *World Development* **79** pp 177-96
[2] Bureau of Fisheries and Aquatic Resources-PHILMINAQ Project 2007 *Managing aquaculture and its impact a guidebook for local governments* (Quezon City Philippines: Bureau of Fisheries and Aquatic Resources) p vii
[3] Musyoka S 2016 *Concept of microbial bioremediation in aquaculture wastes: a review* Retrieved from https://www.researchgate.net/publication/310425025
[4] Bender M 2019 *The effects of water pollution on ecosystems - Ecosystem Management & Associates* Retrieved from http://ecosystem-management.net/effects-of-water-pollution-on-ecosystems/
[5] Bhargava R 2017 *Environmental pollutants and their bioremediation approaches* Retrieved from https://www.crcpress.com/Environmental-Pollutants-and-their-Bioremediation-Approaches/Bharagava/p/book/9781138628892#googlePreviewContainer
[6] Mannan M, Islam M, Rimi R, Suravi and Meghla N 2012 Impact of water quality on fish growth and production in semi-intensively managed aquaculture farm *Bangladesh J. Of Environmental Science* **23** 1561-9206 pp 110-1
[7] Bureau of Fisheries and Aquatic Resources-PHILMINAQ Project 2007 *Managing aquaculture and its impact a guidebook for local governments* (Quezon City Philippines: Bureau of Fisheries and Aquatic Resources) pp 12-3
[8] Divya M, Aanand S, Srinivasan A and Ahilan B 2015 Bioremediation–an eco-friendly tool for effluent treatment: a review *Int. J. of Applied Research* **1(12)** p 530
[9] Panigrahi A, Mohapatra A and Panigrahi A 2005 *Bioremediation an Environmental Friendly Approach For sustainable Aquaculture* Retrieved from https://pdfs.semanticscholar.org/e60e/09ec3949169295cba00977fd5d969087827c.pdf
[10] Candelaria P and Corre V 2012 Probiotics as control for luminescent vibrosis mud crab (scylia serrate forsskal, 1175) larval rearing *Proc. of the 3rd Bicol Fisheries Research, Development & Extension (RDE) Symp.* pp 54-66
[11] Hamdan A, El Sayed A and Mahmoud M 2016 Effects of a novel marine probiotic, on growth performance and immune response of nile tilapia (oreochromisniloticus) *J. of Applied Microbiology* **120(4)** 1061-73
[12] Tuan T, Duc P and Hatai K 2013 Overview of the use of probiotics in aquaculture *Int. J. of Research in Fisheries and Aquaculture* **3(3)** pp 89-97
[13] Martinez Cruz P, Ibáñez A, Monroy Hermosillo O and RamírezSaad H 2012 Use of probiotics in aquaculture *ISRN microbiology 2012*

[14] Suplicy F, Vianna L, Rupp G, Novaes A, Garbossa L, de Souza R and dos Santos A 2017 Planning and management for sustainable coastal aquaculture development in Santa Catarina State, south Brazil *Reviews in Aquaculture* **9(2)** pp 107-24
[15] Meaden, J and Aguilar-Manjarrez 2013 Advances in geographic information systems and remote sensing for fisheries and aquaculture *FAO Fisheries and Aquaculture Technical Paper* **552(8I)** pp 25-43 49-78 81-82
[16] McLeod I, Pantus F and Preston N 2002 The use of a geographical information system for land based aquaculture planning *Aquaculture Research* **33(4)** pp 241-50
[17] Falconer L, Telfer T, Pham K and Ross L 2018 *GIS Technologies for Sustainable Aquaculture* Retrieved from https://dspace.stir.ac.uk/handle/1893/25366#.XY7zphYzaM8
[18] Nieves P 2012 Status of aquaculture industry in Bicol *Proc. of the 3rd Bicol Fisheries Research, Development & Extension (RDE) Symp.* pp 18-20
[19] Cruz-Lacierda E, De la Peña L and Lumanlan-Mayo S 2000 The use of chemicals in aquaculture in the Philippines *Use of Chemicals in Aquaculture in Asia: Proc. of the Meeting on the Use of Chemicals in Aquaculture in Asia 20-22 May 1996* pp 155-6
[20] Philippine Statistics Authority 2017 *Fisheries Situation Report* **14** pp 5-6
[21] Lazaro R 2018 *Fish kill hits Bulacan ponds* Retrieved from https://www.philstar.com/nation/2018/05/08/1813072/fish-kill-hits-bulacan-ponds
[22] Castillo A and Dela Cruz J 2019 Improving Aquaculture: A Web-GIS Initiative *3rd Global Conf. on Computing and Media Technology 17-18 July 2019* (KL Malaysia)
[23] Wang R, Chen D, and Fu Z 2006 AWQEE-DSS: A decision support system for aquaculture water quality evaluation and early-warning *2006 IEEE Int. Conf. on Computational Intelligence and Security* **2** pp 959-62
[24] Gobakis K, Mavrigiannaki A and Kalaitzakis K 2017 Design and development of a web based GIS platform for zero energy settlements monitoring *Energy Procedia* **134** pp 48-60
[25] Bhandari B 2003 Participatory Rural Appraisal (PRA): Module 4 *Institute for Global Environmental Strategies (IGES)* Retrieved from https:// www.iges.org
[26] Tang X, Shen Y and Shen J 2012 An arcobjects-based approach to rapid application development of GIS *2012 Int. Conf. on Automation and Logistics* pp 411-4
[27] Softonic 2019 *GPS Map Camera* Retrieved from https://gps-map-camera-google-map.en.softonic.com/android
[28] Gkatzoflias D, MelliosG and Samaras Z 2013 Development of a web GIS application for emissions inventory spatial allocation based on open source software tools *Computers & Geosciences* **52** pp 21-33
[29] MySQL 2019 *MySQL 5.7 Reference Manual 11.5.3 Supported spatial data formats* Retrieved from https://dev.mysql.com/doc/refman/5.7/en/gis-data-formats.html
[30] Wagner J 2015 *How to use APIs to build GIS, mapping and location applications* Retrieved from https://www.programmableweb.com/news/how-to-use-apis-to-build-gis-mapping-and-location-applications/analysis/2015/04/13
[31] Hayes J and Sheckler P 2019 *Fiber optic networks* Retrieved from https://www1.freewebs.com/masanton/Bab%20III.pdf
[32] *CNWD-Primewater infrastructure corporation* Retrieved from http://cnwd.gov.ph/contact.html
[33] White P, Cromey C, Palerud Rune, Hernandez J, Rosario W, Regpala R and Lopez Nelson 2013 *Mitigating aquaculture impact in the Philippines PHILMINAQ Outputs and Recommendations* Retrieved from https://www.researchgate.net/publication/259089454
[34] Jahangiri L and Esteban M 2018 Administration of probiotics in the water in finfish aquaculture systems: a review *Fishes* **3(3)** Retrieved from https://www.mdpi.com/2410-3888/3/3/33
[35] Boyd C and Tucker C 2012 Pond aquaculture water quality management *Springer Science & Business Media* pp 525-8
[36] *PVS PROCON-PS (SPL)* Retrieved from https://www.indiamart.com/proddetail/procon-ps-spl-6489624388.html

[37] Williamson B 2016 Digital education governance: data visualization, predictive analytics, and 'real-time'policy instruments *J. of Education Policy* **31(2)** pp 123-41
[38] Jankovic D and Milidragovic R 2012 Selecting the optimal opensource GIS software for local authorities by combining the ISO 9126 standard and AHP approach *2012 Proc. of the 35th Int. convention MIPRO* pp 1661-65

IOP Conf. Series: Materials Science and Engineering **803** (2020) 012012 doi:10.1088/1757-899X/803/1/012012

Blood Glucose Prediction Model for Type 1 Diabetes based on Extreme Gradient Boosting

Ganjar Alfian[1], Muhammad Syafrudin[2], Jongtae Rhee[2], Muhammad Anshari[3], M. Mustakim[4], Imam Fahrurrozi[5]

[1] Nano Information Technology Academy, Dongguk University, Seoul, Korea
[2] Department of Industrial and Systems Engineering, Dongguk University, Seoul, Korea
[3] School of Business & Economics, Universiti Brunei Darussalam, Brunei Darussalam
[4] Teknik Informatika, Universitas Islam Negeri Sunan Kalijaga, Indonesia
[5] Departemen Teknik Elektro dan Informatika, Sekolah Vokasi, Universitas Gadjah Mada, Indonesia

E-mail: ganjar@dongguk.edu, udin@dongguk.edu, jtrhee@dongguk.edu, anshari.ali@ubd.edu.bn, taqiem235@yahoo.com, imam.fahrurrozi@ugm.ac.id

Abstract. Predicting future blood glucose (BG) level for diabetic patients will help them to avoid critical conditions in the future. This study proposed Extreme Gradient Boosting (XGBoost), an ensemble learning model to predict the future blood glucose value of diabetic patients. The clinical dataset of Type 1 Diabetes (T1D) patients was utilized and the prediction models were generated to predict future BG of 30 and 60 minutes ahead of time. The prediction models have been tested tofive children who develop T1D and showed that BG prediction model based on XGBoost outperformed other models, with average of Root Mean Square Error (RMSE) are 23.219 mg/dL and 35.800 mg/dL for prediction horizon (PH) 30 and 60 minutes respectively. In addition, the result showed that by utilizing statistical-based features as additional attributes, most of the performance of predictions model were increased.

1. Introduction

Diabetes is chronic disease when the body cannotprovideenough insulin (T1D, Type 1 diabetes) or the body cannot effectively use its produced insulin (T2D, Type 2 diabetes) [1]. It is expected for the patient to manage their BG level in the normal value between70–180 mg/dL. High blood glucose or hyperglycemia is the condition where BG > 180 mg/dL and can affect several complications such as cardiovascular disease and retinopathy. Other condition that occurs to the diabetic patients is low blood sugar or hypoglycemia (BG < 70 mg/dL) and can cause death [2]. To monitor blood glucose level, the diabetes patient can utilize continuous glucose monitoring (CGM) device. CGM deviceisadvanced wearable medical devices that can monitor and record glucose levels concentration every 1–5 min and has shown improvement for diabetes management [3,4].

Several studies have demonstrated BG prediction model, so that future value of BG can be obtained earlier. Blood glucose prediction models have been presented in previous studies by considering glucose level as single input. Perez-Gandia et al [5] proposed neural network (NN) to predict future blood glucose by using the last 20 minutes of BG level as input. Ben Ali et al [6] proposed neural network-basedmodelto predict BG level of Type 1 Diabetes (T1D) using only CGM data as inputs.

Finally, Hamdi et al [7] proposed prediction model by utilizing vector regression and differential evolution algorithms to forecast blood glucose. The proposed method was testedon CGM data and showed that the proposed model achieved high prediction accuracy.

For the case of internet of things (IoT) sensor, the previous studies have shown that statistical based features can be utilized as input for prediction model. Park et al [8] proposed fault detection system from IoT sensor in the manufacturing. The statistical features were extracted from sensor data to be utilized as input for prediction model. Finally, Alfian et al [9] utilized statistical features from the received signal strength (RSS) of radio frequency identification (RFID) data as input for prediction model. The prediction model showed significant result on predicting the movement of RFID tag.

Previous studies showed that the blood glucose can be utilized as single input for blood glucose prediction model with high accuracy. In addition, the statistical features also showed significant result on improving the performance of prediction model for IoT sensor. Therefore, in present study the prediction model by considering blood glucose as input and statistical features as additional attributes are presented. The proposed prediction model based on XGBoost will be utilized to predict future blood glucose level of 30 and 60 minutes ahead of time.

2. Methodology

The history of blood glucose of diabetic patients needs to be gathered by CGM device, so that forecasting from this time-series data could be executed. For predicting future blood glucose, multi-step forecasting can be utilized where the task is predicting next Hvalues $[y_{N+1}, \dots, y_{N+H}]$ of a historical time series $[y_1, \dots, y_N]$ composed of N observations and $H > 1$ denotes the forecasting horizon [10]. In multi-step forecasting $(t + h)$, given $n$previous values and N total data, the training set can be derived by creating the input data matrix$X$ and output vector$Y$.

Figure 1 shows the illustration of sliding window approach to convert time-series data into input matrix X and output vector Y. In this scenario, the time series dataset was split into several windows with the size $n = 6$ as previousdata for learning and is used to predict the blood glucose for the next $h = 6$. In our study, the blood glucose dataset was measured for every 5 minutes, therefore our prediction model (as presented in Figure 1) used the last 30 minutes of historical data to predict the next 30 minutes of blood glucose. In Figure 1, the windows (presented in red lines) were collected and presented as input matrix X, while the collection of next 30 minutes value (presented in blue lines) were used as output vector Y.

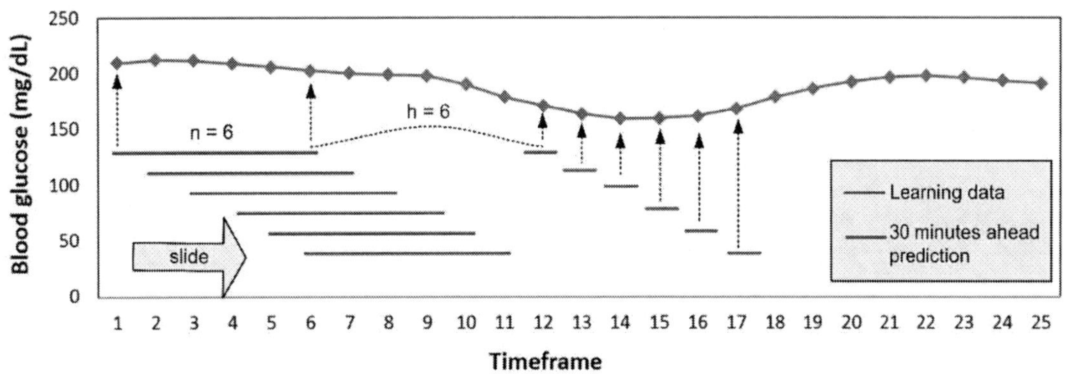

Figure 1. Sliding window approach for timeseries blood glucose data.

We propose statistical features as attributes addition for the $n$previous values which is generated by sliding window approach. Most of the previous studies, only considered $n$historical data as the input. Table 1 shows the proposed eight relevant statistical features extracted from each window (historical data) to improve forecasting performance.

Table 1. Statistical attributes extracted from window.

Attribute name	Description
Min	Minimum value of BG in a window.
Max	Maximum value of BG in a window.
Mean	Average value of BG in a window.
Std	Standard deviation value of BG in a window.
Diff	Difference between highest and lowest BG value in a window.
Median	Middle value of BG in a window.
Kurtosis	Describing tails of BG distribution differ from a normal distribution.
skew	Distribution asymmetry of BG in a window.

Next step, the additional statistical features are combined with $n$previous values generated by sliding window approach. Finally, given $n$previous values or window size, $h$forecasting horizon, 8 (eight) number of statistical features and N total data, the input X can be derived by creating the $[(N-n-h+1)\times(n+8))]$

$$X=\begin{bmatrix} y_1 & \cdots & y_5 & y_n & min\,(y_1,\dots,y_n) & \cdots & skew\,(y_1,\dots,y_n) \\ \vdots & \vdots & \vdots & \vdots & \vdots & \vdots & \vdots \\ y_{N-n-h} & \cdots & y_{N-h-2} & y_{N-h-1} & \cdots & \cdots & skew\,(y_{N-n-h},\dots,y_{N-h-1}) \\ y_{N-n-h+1} & \cdots & y_{N-h-1} & y_{N-h} & \cdots & \cdots & skew\,(y_{N-n-h+1},\dots,y_{N-h}) \end{bmatrix} \quad (1)$$

and the $[(N-n-h+1)\times 1]$ output vector Y.

$$Y=\begin{bmatrix} y_{n+h} \\ \vdots \\ y_{N-1} \\ y_N \end{bmatrix} \quad (2)$$

In order to investigate how the proposed prediction model can predict future blood glucose, the dataset from DirecNet was utilized for this study [11].The subjects are children, age 3 to less than 7 and 12 to 17 with diabetes type 1. The subjects were asked to use the Guardian-RT (i.e., continuous glucose monitoring device) to record their blood glucose for approximately 6 days. In this study, we selected blood glucose measurement from 5 patients, and the detail dataset distribution can be seen in Table 2.

Table 2. Distribution of BG dataset from type 1 diabetes patients.

Patient ID	#Data Point	Blood Glucose Level			
		Min	Max	Mean	STD
1	1496	41.17	400.0	200.351	68.353
15	2145	66.37	353.0	179.361	58.381
4	2212	53.74	303.72	151.202	52.796
21	2076	53.25	400.0	158.012	72.418
22	2173	40.0	232.33	121.561	37.766

In this study, we utilized XGBoost model to predict future blood glucose level of type 1 diabetes patient. XGBoost is decision-tree-based ensemble model that utilizes gradient boosting [12]. The objective function contains training loss and regularization. The objective function (loss function and regularization) can be presented as follows.

$$\mathcal{L}(\phi) = \sum_i l\left(\hat{y}_i, y_i\right) + \sum_k \Omega(f_k) \quad (3)$$
$$\text{where } \Omega(f) = \gamma T + \frac{1}{2}\lambda\|w\|^2$$

Here, l is a differentiable loss function that shows the difference between the target y_i and the prediction $\hat{y}_i$. The regularized term Ω penalizes the complexity of the model and T represents the total number of leaves in tree. Each $f_k$represents tree q and weight of leaf w. The term γ corresponds to the threshold and pre-pruning is performed while optimizing to limit the growth of the tree and λ is used to smooth the final learned weights to prevent overfitting.

The prediction models were implemented in Python V3.6.6 and Scikit-learn V0.19.1 [13]. To simplify implementation of machine learning models, we used default parameters provided by Scikit-learn. Datasets were split into two parts: approximately 50:50 for training and testing. To present the performance of prediction models, the correlation coefficient (r), RMSE and MAPE will be used. The prediction model fits the data when the RMSE and MAPE generates small values. The modelcorrectly fits the data when the r generates higher value. The detailed formula of r, RMSE and MAPE for a dataset of size n, predicted value $\hat{y}$andactual value y are presented in Table 3.

Table 3. The performance metrics for the forecasting models.

Performance Metric	Formula
Root Mean Square Error (RMSE)	$\sqrt{\frac{1}{n}\sum_{i=1}^{n}(y_i - \hat{y}_i)^2}$
Pearson correlation coefficient (r)	$\frac{\sum_{i=1}^{n}(y_i - \bar{y}_i)(\hat{y}_i - \bar{\hat{y}}_i)}{\sqrt{\sum_{i=1}^{n}(y_i - \bar{y}_i)^2}\sqrt{\sum_{i=1}^{n}(\hat{y}_i - \bar{\hat{y}}_i)^2}}$
Mean Absolute Percentage Error (MAPE)	$100\% \times \frac{1}{n}\sum_{i=1}^{n}\frac{\lvert y_i - \hat{y}_i\rvert}{y_i}$

3. Result and Discussion

Table 4 showed the performance of prediction models to predict the future blood glucose for different prediction horizon (PH). Multi-Layer Perceptron (MLP), K-Nearest Neighbour (KNN), Support Vector Regression (SVR), Random Forest (RF), Decision Tree (DT) and AdaBoost have been compared with proposed XGBoost model for predicting multi-step ahead of blood glucose. In this experiment, window size or previous values n = 6 (last 30 minutes) has been utilized for machine learning models to predict future blood glucose (BG). Each prediction model was generated for each patient and the performance average was computed over 5 models (generated from 5 patients). The statistical features were utilized by proposed XGBoost model as additional attributes (see formula 1 and 2), while other models only utilized last 30 minutes values generated by sliding window approach (Figure 1). The result showed that as increasing the prediction horizon, the prediction models generated higher RMSE and MAPE, and lower r. The proposed XGBoost model showed highest performance by generating lowest RMSE and MAPE as compared to other prediction models. The RMSE of proposed XGBoost are 23.219 mg/dL and 35.800 mg/dL for prediction horizon (PH) 30 and 60 minutes respectively, while the MAPE of proposed model are 10.896 and 17.681 for PH 30 and 60 minutes respectively. Furthermore, the proposed model also generated highest r (correlation coefficient) compared to other models, they are 0.900 and 0.750 for PH 30 and 60 minutes respectively.

Table 4. Performance of blood glucose prediction models.

Method	PH = 30 min			PH = 60 min		
	RMSE	r	MAPE	RMSE	r	MAPE
MLP	30.997	0.854	13.683	42.277	0.705	19.046
SVR	31.839	0.855	17.138	40.553	0.707	22.665
KNN	26.439	0.872	12.562	41.302	0.698	20.291
DT	33.788	0.824	16.446	52.638	0.580	25.967
RF	26.791	0.873	13.000	41.115	0.689	20.269
AdaBoost	27.986	0.864	15.314	39.868	0.710	22.919
Proposed model	23.219	0.900	10.896	35.800	0.750	17.681

In addition, the impact of feature type on the performance of prediction models are presented in Figure 2. The result showed that by combining sliding window approach and statistical features (StatFeature) provided lower RMSE as compared to prediction models with sliding window-based attributes, except for SVR and DT for PH 30 minutes (Figure 2a), and SVR and AdaBoost for PH 60 minutes (Figure 2b). By integrating sliding window approach and statistical features as attributes for prediction models, the average of RMSE was reduced as much as 0.668 and 1.688 for PH 30 and 60 minutes respectively as compared to prediction models with sliding window-based attributes only.

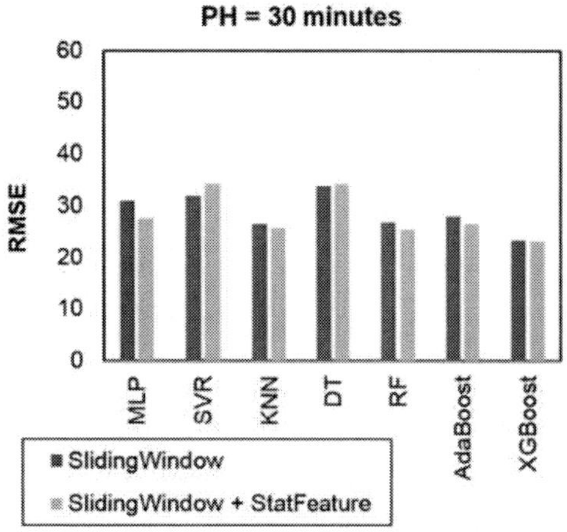

(a)

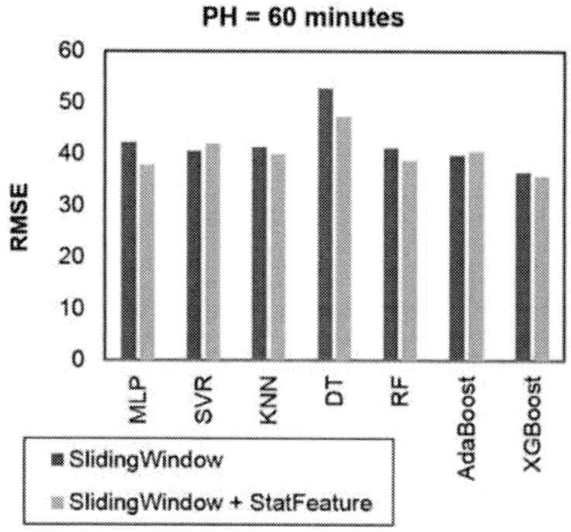

(b)

Figure 2. Impact of feature type on the model performance for PH (a) 30 minutes and (b) 60 minutes.

4. Conclusion and Future Works

In this study, the proposed XGBoost model was utilized to predict future blood glucose of 30 and 60 minutes ahead of time. The proposed model has been tested to the dataset consist of 5 (five) children who develop T1D. The proposed XGBoost model showed highest performance by generating lowest RMSE and MAPE as compared to other prediction models. The RMSE of proposed XGBoost are 23.219 mg/dL and 35.800 mg/dL for prediction horizon (PH) 30 and 60 minutes respectively, while the MAPE of proposed model are 10.896 and 17.681 for PH 30 and 60 minutes respectively. Furthermore, the proposed model also generated highest r (correlation coefficient) compared to other models, they are 0.900 and 0.750 for PH 30 and 60 minutes respectively. In addition, by integrating sliding window approach and statistical features as attributes for prediction models, the average of RMSE was reduced as much as 0.668 and 1.688 for PH 30 and 60 minutes respectively as compared to prediction models with sliding window-based attributes only.

The dataset in this study was limited to small set of type 1 diabetes for children, so it is difficult to generalize the robust prediction model to be applied for different purposes. The other type of datasets such as that collected from other CGM sensor or glucose meter need to be utilized in the near future. Furthermore, extending the comparison with other prediction models could be presented in the near future.

Acknowledgments
This research was financially supported by the Ministry of Trade, Industry and Energy (MOTIE) and the Korea Institute for Advancement of Technology (KIAT) through the International Cooperative R&D program (Grant number N0002301).

References

[1] World Health Organization 1999 *Definition, Diagnosis, and Classification of Diabetes Mellitus and Its Complications. Part 1: Diagnosis and Classification of Diabetes Mellitus* https://apps.who.int/iris/handle/10665/66040 retrieved at September 1st 2019
[2] American Diabetes Association 2018 Introduction:Standards of medical care in diabetes—2018 *Diabetes Care* **41** pp S1–S2 DOI:10.2337/dc18-Sint01
[3] Bode B W and Battelino T 2010 Continuous glucose monitoring *Int. J. Clin. Pract. Suppl.* **64** pp 11–5 DOI: 10.1111/j.1742-1241.2009.02272.x
[4] Torres I, Baena M G, Cayon M, Ortego-Rojo J and Aguilar-Diosdado M 2010 Use of sensors in the treatment and follow-up of patients with diabetes mellitus *Sensors***10** pp 7404–20 DOI:10.3390/s100807404
[5] Pérez-Gandía C, Facchinetti A, Sparacino G, Cobelli C, Gómez E J, Rigla M, de Leiva A and Hernando M E 2010 Artificial Neural Network Algorithm for Online Glucose Prediction from Continuous Glucose Monitoring *Diabetes Technol. Ther.* **12(1)** pp 81-8 DOI: 10.1089/dia.2009.0076
[6] Ben Ali J, Hamdi T, Fnaiech N, Di Costanzo V, Fnaiech F and Ginoux J 2018 Continuous blood glucose level prediction of Type 1 Diabetes based on Artificial Neural Network *Biocybernetics and Biomedical Engineering* **38(4)** pp 828-40 DOI:10.1016/j.bbe.2018.06.005
[7] Hamdi T, Ben Ali J, Di Costanzo V, Fnaiech F, Moreau E and Ginoux J 2018 Accurate prediction of continuous blood glucose based on support vector regression and differential evolution algorithm *Biocybernetics and Biomedical Engineering* **38(2)** pp 362-72 DOI:10.1016/j.bbe.2018.02.005
[8] Park D, Kim S, An Y and Jung J-Y 2018 LiReD: A Light-Weight Real-Time Fault Detection System for Edge Computing Using LSTM Recurrent Neural Networks *Sensors* **18(7)** p 2110 DOI:10.3390/s18072110
[9] Alfian G, Syafrudin M, Yoon B and Rhee J 2019 False Positive RFID Detection Using Classification Models *Applied Sciences* **9(6)** p 1154 DOI:10.3390/app9061154
[10] Ben Taieb S, Bontempi G, Atiya A F and Sorjamaa A 2012 A review and comparison of strategies for multi-step ahead time series forecasting based on the NN5 forecasting competition *Expert Systems with Applications* **39(8)** pp 7067–83 DOI:10.1016/j.eswa.2012.01.039
[11] *Evaluation of Counter-regulatory Hormone Responses during Hypoglycemia and the Accuracy of Continuous Glucose Monitors in Children with T1DM* https://public.jaeb.org/direcnet/stdy/167 retrieved at September 1st 2019
[12] Chen T and Guestrin C 2016 XGBoost: A Scalable Tree Boosting System *Proc. of the 22nd ACM SIGKDD Int. Conf. on Knowledge Discovery and Data Mining - KDD '16* pp 785–94

DOI:10.1145/2939672.2939785
[13] Pedregosa F *et al* 2011 Scikit-learn: Machine Learning in Python *The J. of Machine Learning Research* **12** pp 2825-30

Development of Material Recovery Trading System: An Innovative Tool for Community Waste Management Support

JamaicaM. Dioquino[1], Anna RoseV. Cac[2], Dionisio R. Tandingan Jr.[3]

University of the Cordilleras, Governor Pack Road, Baguio City, Philippines

E-mail: [1]jamaicadioquino@yahoo.com, ph[2]annarosecac@gmail.com, [3]dion_tand@yahoo.com

Abstract. Waste management is a perennial problem in rural and urban areas. Many initiatives and projects were implemented by local government units, either improving their current manual system or adopting ready to use technology platforms, to help motivate residents in their proper disposal of wastes most especially recyclable materials. This study was intended to develop an online material recovery trading system (MRTS) that shall serve as an innovative tool to support the community waste management project of barangay Bued, Calasiao, Pangasinan, Philippines in retrieving reusable materials from its residents. The web app that serves as tool for trading adopted the use of points system to provide incentives to residents based on their accumulated points such as exchange of goods or supporting a cause or advocacy. The app was tested and evaluated by the stakeholders using the SUS tool. The app's usability scored an average of 80.40 which means that the developed system is easy to use and has well-integrated functions. Further, the app is being considered by barangay officials to be implemented to support the community waste management effort in recovering recyclable materials from its residents.

1. Introduction

The era of rapid urbanization that has been the norm for most cities in developing countries, like the Philippines, has led to a number of problems for local government, especially the management of solid waste. Increases in the city populations have further deteriorated the situation [1][2]. It has now become a well-known fact that the majority of the municipalities in the developing countries are grappling with the problems of high volumes of waste.

The worsening condition of solid waste create environmental and health problems [3]. Garbage is considered the third pollution inextricably interlocked with the air and water pollution creating environmental hazard. Improper disposal of solid wastes contribute to air, soil, and water pollution.Solid waste clogs drains, creates stagnant water for insect breeding, and causes floods during rainy seasons especially in urban areas. Insect and rodents vectors can also spread diseases such as cholera and dengue fever due to improper waste disposal.

Solid waste, in general, consist of domestic wastes from households, both biodegradable and non-biodegradable[4]. These include paper, plastic, scrap metals, organic and non-organic matter, glass, and rubber materials. Based on the increasing dilemma of waste in the Philippines, there are municipalities which are initiating projects to help their community in properly disposing solid wastes.

In Cebu City, Philippines, its local government officials had opened a unique barter store with the purposeof reducing its plastic wastes and helping poor families in the city by providing some of their daily needs [5]. This plastic barter store, located at BalaySamaritano along Mabini Street, Cebu City, accepts different plastic wastes in exchange for goods or various donated items. Community residents can buy bag of rice, cloth, or school supplies in exchange for four kilos of plastic wastes. Or, they can buy toys, crayons, or slippers, in exchange for two kilos of plastics.

In Cauayan City in northeastern Philippines' Isabela province, the city officials had started their own plastic barter scheme called "Basuramo, Papalitan ng Bigas D2" (Trash for Rice) [6]. Aside from encouraging the residents to trade in their plastic wastes, the project also aims to teach locals' proper waste segregation. Two kilograms of plastic bottles or three kilograms of assorted plastic or one kilo of plastic wrappers can be exchanged for a kilo of rice.

Similarly in Brgy. Talon, Cavite, Philippines, the Sangguniang Kabataan (youth council) supported by Cavite Governor Jesus Crispin Remulla had the waste reduction project, dubbed as "Plastic palitBigas" (plastic in exchange for rice) [7]. These trashes include "chichirya" (snack food) and candy wrappers, empty shampoo, ketchup, soy sauce, vinegar sachets, among other plastic containers and receptacles. This only limits with a kilo of clean plastics in exchange for a kilo of rice.

With the emergence of application development cloud platform such as Google Firebase, one can easily develop and implement web-based app [8] that can support community waste management projects. Firebase supports system that can store huge amount of real-time data and offers various services like user authentication, backend for real-time database, and dynamic web hosting services across multiple platforms such as web browsers for iOS and Android apps [9]. These gathered data could further be utilized with its Firebase analytics for decision-making and ad-hoc resolution.

Based on the problems and available development tool stated above, the researchers aim to (1)develop an online trading system as an innovative tool to support community waste management in recovering reusable materials from the residents of Barangay Bued,Calasiao,Pangasinan, Philippines, (2) design the features of the online trading system, and (3) determine the usability of the proposed system.

2. Review of Related Literature

Barter or trade is the exchange of goods, property or services for other goods, without using money. It is considered as the worlds' oldest form on exchange. Traditionally, it involves direct exchange of goods and/or services between two or more trading partners. Barter trade has, however, changed a lot over the years with newer forms characterized by variations of conditions and agreements [10].

In 2011, many local government units in the Philippines are applying a unified system of waste management which includes reducing, recycling, composting and re-using wastes [11]. Several laws and ordinances were passed to address the issue of Solid Waste Management (SWM). Republic Act No. 9003 or as Ecological Solid Waste Management was signed by Former President Gloria Arroyo on 2001. Ordered by RA 9003, Cebu City formed the City Solid Waste Management Board (CMSWMB) that shall organize, present and execute projects for an ensured and clean management of solid wastes in areas under its command.The current status of SWM in Cebu City are as follows: collects not lower than 350 tons of garbage every day, implementation of "No Segregation, No Collection" Policy but lack of enforcement in segregating in most barangays [12].

Gawad Kalinga Pinagsama Villages in Western BicutanTaguig and Brgy. UP Campus Quezon City had an awareness raising with the collection and recycling of biodegrable and non-biodegradable waste. The said areas were given a segregation frame with a pale for food waste and hooks for bags for cans, bottles, plastic cups, and plastic film.The community has established a recyclable storage facility (RSF) in their premises, and each household takes turn in transferring the recyclables from the unit courtyard to the RSF every week. The recyclables waste were white paper, newspaper/colored paper, cartons, waste paper, tin cans and glass.

Information technology has further enabled the use of trade-shares in barter. In electronic commerce, cash has all but been entirely eliminated. Standardized data systems that allow cashless exchange has been developed and will continue to be further developed [13]. The barter tradeas compared to monetary transactions is associated with high risk. One of the risks is the threat ofmismanagement and fraud. If proper management practices are not in place, accounting for products received in barter trade may be difficult or not properly managed. Good management practices are essential to effectively take advantage of the opportunities and deal with the potential problems commonly associated with barter trade [14][15].

Our study is distinct from these past researches because it uses recyclable materials as trade for points that will be used to trade in to buy basic needs or used in buying school supplies that can be donated to charities specifically in the Day Care Center of Brgy. Bued, Calasiao, Pangasinan, Philippines.

3. Methods

The researchers employed the Agile'sFeature-Driven Development (FDD) methodology [16] in developing the proposed app. The researchers conducted series of interviews with the barangay captain of the local community in order to determine the projects and plans of the leadership in terms of their material recovery and waste management. Several factors affecting their project were identified including improper waste segregation, burning of garbage, and an increase in recyclable wastes that are supposed to be sent to the material recovery facility located in the same barangay. The proponents then evaluated the current situation in order to delimit the most significant factors of the project and identify what can improve their current manual system of material recovery management.

The researchers conducted researches and literature reviews to identify possible solutions to the problem of waste disposal, reuse, and recycling which was identified as factors affecting the waste management of the community. No projects are implemented yet by the community in order to strengthen their advocacy in making waste materials more useful. Thus, the creation of a trading system as a tool to help in material recovery and reusability was aimed by the study.

The Bootstrap front-end framework and the Google Firebase cloud services were utilized to developthe proposed app. Data used were provided by the barangay captain from their most current census in the community. The use of points system for every trade-in of reusable and recyclable waste materials was implemented in order to motivate people to donate waste materials through the use of incentives generated from their collected points.

3.1. System Architectural Design

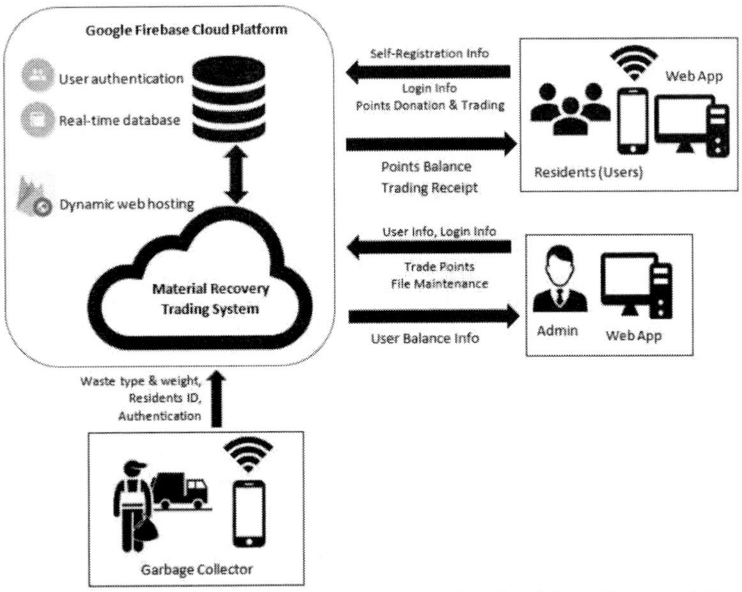

The system's architectural design shows the processes involved in using the Material Recovery Trading system. The Google Firebase cloud platform was utilized to host the proposed web app with its user authentication and real-time database services. Residents can participate in trading waste materials by having a self-registration using the app. With the resident's information uploaded into the cloud through Firebase, user can login to the app by providing user name and password. Trade points could be earned by the users through trading the waste materials to the garbage collector. The garbage collector inputs necessary wastes information such as the type of wastes and weight to the proposed app to determine the equivalent points earned by the resident.The accumulated trade points of the user can either be donated to the charity through his/her app or trade in for available basic needs or goods in the trading store. The Admin account that handles the trading store can manage the file maintenance for point's equivalent per kilogram of wastes and for maintaining the equivalent points to buy the basic needs or goods. This account is also responsible in creating the garbage collector accounts.

3.2. Features

Table 1 shows the list of features in the web app for each user account. Features that will be available for access would depend on the type of account logged in to the application. The garbage collectors for example can access the web app through their mobile phone and be redirected to the page specified for them to use for converting trade-in waste items into points. The resident account, on the other hand, can access specified pages connected to trading points to items through web and mobile. The admin account can access the web app and its database to check and provide the requested items by the resident.

Table 1. Features of the Material Recovery Trading System

Resident account	Garbage collector account	Admin account
Self-Registration	Data entry oftrading wastes materials	File Maintenance for waste material types and item trade-in
Points Balance Inquiry		Trade Points to basic needs orgoods
Points Donation and Trading		Trade points to cash
		Creation of garbage collector accounts

4. Software design

The Material Recovery Trading System (MRTS) web app serves as a tool for trading process of recyclable and reusable materials. Garbage collector accepts waste materials from the residents and exchange them into points equivalent to the material's type and weight. On the other hand, users can use these points to trade-in to their preferred options from the given choices from the app. The web app, which can be accessed through mobile and desktop, is available for the three (3) user accounts namely the resident, garbage collector and the admin accounts. The garbage collector inputs the type and the generated weight of the waste materials collected from a residents. The app automatically computes for the equivalent points that can be added to the resident's accumulated points which can be inquired by the residents through the app.

Table 2 shows the list of equivalent points that can be garnered for every kilogram of waste material traded with garbage collectors. Points are subject to change by the Administrator based on the changes in the economical values of materials.

Table 2. Points garnered per kilogram of waste material type

Waste Type	Points per Kilogram
White Plastic	30
Misc. Plastic	20
Paper	15
Kitchen Waste	10
Cardboard	15
Glass	20
Electronic Waste	15

The garbage collector and the resident accounts can only view and edit certain parts of the system. Figure 1 shows the user interface of the web app accessed through mobile for the garbage collector. This is where he/shecan input the weight and waste material type traded by the residents. This will then be converted into points saved into the database.

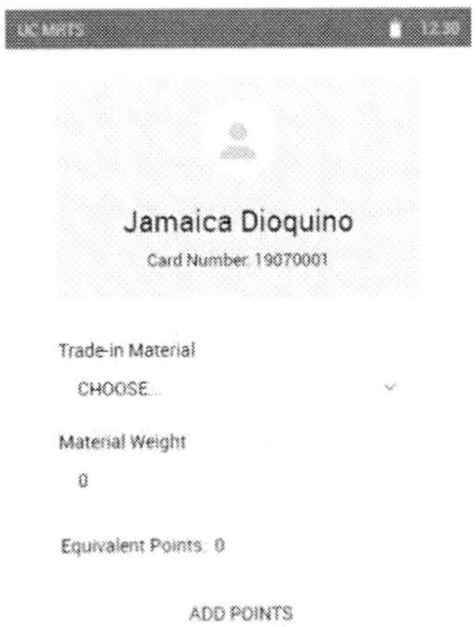

Figure 1. Garbage collector's user interface

The web-based app can be accessed by the residents through their personal computer, mobile phones or through a machine located in the community trade-instore. This will be used by the community members in trading their points. Users can choose either to exchange points into goods, donate the points to a chosen charitable institution, or trade the points to an equivalent cash amount. After choosing an action and filling out forms, the app will generate a receipt that can be presented to the admin to go through with the trading process.

Figure 2 shows the Main Menu page of the web app. Resident users can choose an action whether to trade-in, donate, change pin, or log-out of the application. Upon choosing trade-in, the user will be redirected to the item trade-in page as shown in Figure 3 where they can choose an item to trade-in with their accumulated points. Chosen items can be stored on the Cart (Figure 4) where they can edit and finalize their choices.

Choosing Donate on the main menu will redirect the user into the donation page (Figure 5) where users can donate the points and have it computed for its equivalent cash to be donated to their chosen institution or group within the community. The points of the users could also be traded into cash equivalents if they choose to as shown in Figure 6.

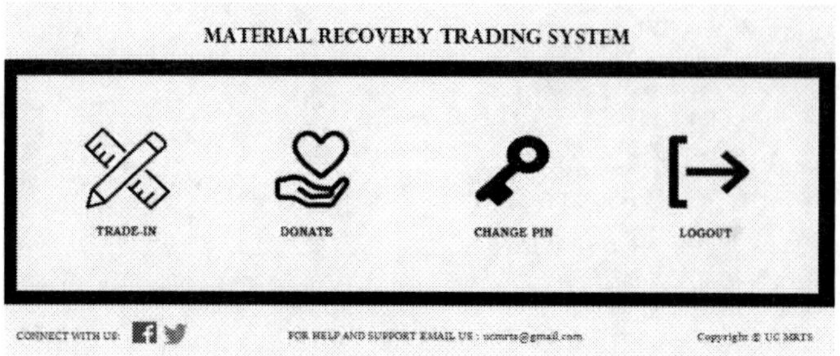

Figure 2. Trade-inmain menu

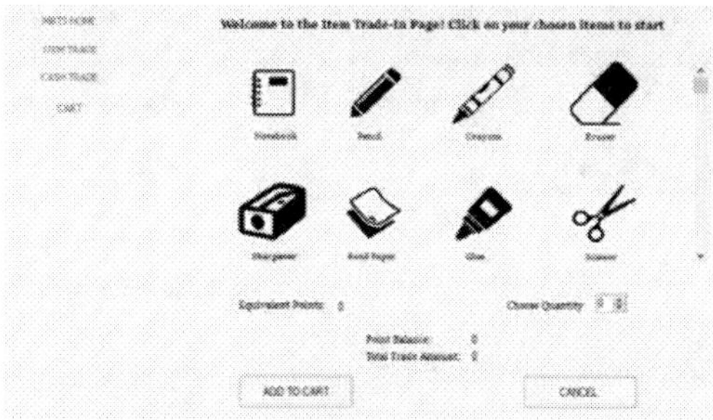

Figure 3. Item trade-in page.

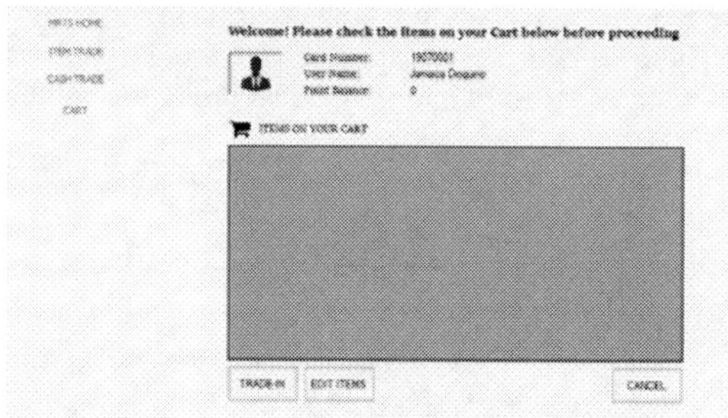

Figure 4. The trade-in Cart.

Figure 5. Donation page.

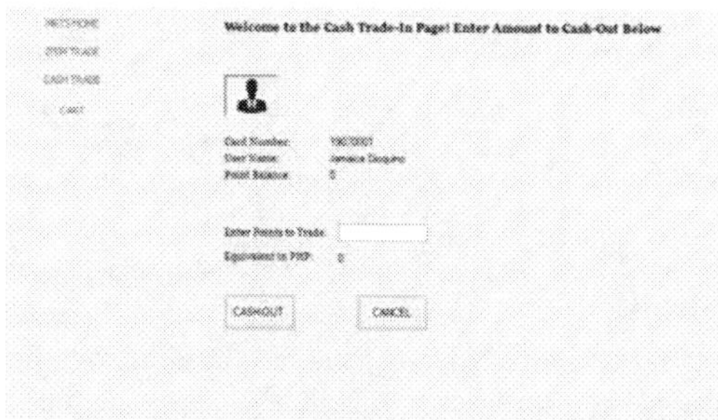

Figure 6. Cash Trade-in page.

5. Findings

The researchers utilized the system usability scale (SUS) questionnaire to evaluate the usability and success of the proposed app. Fifty five (55)respondents from the community including 8 barangay officials, 5 garbage collectors, and 42 residents took part in the survey. After answering the questionnaires, their responses were then tallied and computed based on the SUS scoring formula. The proposedapp got a score of 80.40which indicates that the developed app is easy to use and has well-integrated functions.

The use of an online trading system was also found to be an effective way to encourage residents of the community to trade-in their plastic or recyclable wastes instead of burning them. An interview was conducted with the residents who have used the system and respondents showed a positive impact of the study as they say it encourages them to trade more to gain more.A resident indicated "the system not only encouraged me and my family to segregate and trade-in our garbage but it also made an impact on our way of keeping our household clean for the sake of our health, specially our kids".

The accessibility of the web application through mobile devices also made it easier for the users to gain points and trade from their own household.Findings also showed the potential of online trading systems in the innovation of waste management and recovery endeavor of the community. Further, the app is believed to support the community waste management effort of the barangay in recovering recyclable materials from the residents.

6. Conclusion

The utilization of innovative tools such as ready to use IT tools could beofbig help in the implementation of community projects especially those aiming for a greener community. Through this, technological advancements can now be done without risking our environment, proving that green technology does exist. Furthermore, the system has made a big impact on the community by encouraging residents to engage in solid waste management and overall cleaning of the community.

7. Recommendation

The researchers recommend the establishment of a one-stop trade-in store wherein the basic commodities or goods listed in the web-based application can be claimed by the residents.

The adaptation of the study as part of the smart community is also recommended to ensure a greener technological advancement of communities in the near future. Furthermore, the researchers recommend conducting studies as a continuation to further enhancement of the proposed system.

References

[1] Aggarwal P, *et al* 2009 *Interactive environmental education book VIII* (New Delhi: Pitambar Publishing Co) p 86 ISBN 8120913736

[2] Maxwell J C 1892 *A Treatise on Electricity and Magnetism* 3rd ed vol 2 (Oxford: Clarendon) pp 68–73

[3] Alam P and Ahmade K 2013 Impact of solid waste on health and the environment *Special Issue of Int. J. of Sustainable Development and Green Economics (IJSDGE)* ISSN No: 2315-4721 **V-2** pp I-1 – I-2

[4] Jambeck J R, Geyer R and Wilcox C 2015 Plastic waste inputs from land into the ocean *Science* **347** (6223) pp 768-71

[5] Lachica I 2019 *Barter store opens in Cebu City to accept plastic garbage from street dwellers in exchange for goods* Retrieved from https://cebudailynews.inquirer.net/239902/barter-store-opens-in-cebu-city-to-accept-plastic-garbage-from-street-dwellers-in-exchange-for-goods#ixzz64IHJzB91 on June 18th 2019 at 02:15 pm

[6] Visaya V Jr 2019 *Cauayan City residents exchange plastic waste for rice* Retrieved fromhttps://www.pna.gov.ph/articles/1073033 on June 21 2019 at 9:40 pm

[7] *Cavite town's 'plastic waste for rice' project gains ground* Retrieved from https://www.pna.gov.ph/articles/1054066 on November 16th 2018 at 1:49pm

[8] Li W J, Yen C, Lin Y S and Tung S C 2018 JustIoT Internet of Things based on the Firebase real-time database *2018 IEEE Int. Conf. on Smart Manufacturing, Industrial & Logistics Engineering (SMILE)* Hsinchu,Taiwan

[9] Chatterjee N, Chakraborty S, Decosta A and Nath A 2018 Real-time Communication Application Based on Android Using Google Firebase *Int. J. of Advance Research in Computer Science and Management Studies* Volume **6** Issue 4 April 2018

[10] Frikken K and Opyrchal L 2008 PBS: Private bartering systems *Lect. Notes Comput. Sci.* **5143** pp 113-27

[11] Zurbrugg C 2002 Urban solid waste management in low-income countries of Asia how to cope

with the garbage crisis *Presented for: Scientific Committee on Problems of the Environment (SCOPE) Urban Solid Waste Management Review Session Durban South Africa* pp 1-13

[12] DOST 2010 *Challenges on Municipal Solid Waste (MSW) Management in Cebu* Retrieved from http://pcieerd.dost.gov.ph/images/downloads/presentation_materials/gmi_04242012/07%20challenges%20on%20msw_artajo.pdf

[13] Deloitte Consulting 2000 *The unstoppable economic tidal wave: How the Internet continues to shatter the world's fundamental business paradigms* www.dc.com

[14] Egan C and Shipley D 1996 Strategic orientations towards countertrade opportunities in Emerging markets *Int. Mark. Rev.* **13(4)** pp 102-20

[15] Ference M 2009 *Fight Economic Woes By Trading Your Way To New Business promotional products association magazine* Available from: http://www.ppbmag.com/Article.aspx?id=4519

[16] Kumar G and Bhatia P K 2012 Impact of Agile Methodology on Software Development Process *Int. J. of Computer Technology and Electronics Engineering (IJCTEE)* Volume **2** Issue 4 August 2012

IOP Conf. Series: Materials Science and Engineering **803** (2020) 012014 doi:10.1088/1757-899X/803/1/012014

Development of an Augmented Reality Mobile Application to Enhance the Pedagogical Approach in Teaching History

A Elivera[1] and T Palaoag[2]

[1]Eastern Samar State University, Guiuan, Eastern Samar, Philippines
[2] University of the Cordilleras, Baguio City, Philippines

E-mail:[1]aidaf.ae95@gmail.com,[2]tpalaoag@gmail.com

Abstract. The rapid advancement of technology nowadays opens new opportunities for schools and universities worldwide to improve the quality of teaching and learning experience using Augmented Reality can help make classes more interactive and allow learners to focus more on practice instead of just theories. The researcher identified the problem that pedagogy has not been learner-cantered in most approaches in teaching and that teachers need to adapt to the new method of teaching. The goal of the study is to develop an Augmented Reality (AR) mobile application to enhance the pedagogical approach in teaching. This will use a marker-based method for transmitting virtual objects into the realworld. Blippar was used in the development of AR. This resulted in the form of instructional material and technology that can be used in teaching. During the testing of the material Students' interest was aroused and it even makes them more eager to learn the subject. End-users and experts also deem that the AR mobile application as usable and acceptable with a rating of 78% using the System Usability Scale (SUS). This is an excellent contribution to the educational domain through augmented reality and improving the pedagogical approach in teaching.

1. Introduction

History aims for students to understand the present existing social, political, religious, and economic conditions of people. The educational activity of history helps the students to explain the present, to examine it and to trace its course. To make it full of life, there is a need to embrace a technology that could save the lesson lively using Augmented Reality.

Augmented Reality (AR) is one of the most recent advances that offer another way to educate [1] an innovation that gives the environment of the existing real-world by utilizing sensors, sound, video, illustrations, pictures, Global Positioning System (GPS) and content.

Considering the traditional method of teaching and learning history subject, nowadays,it does not give meaning to learn of younger generations who are much more inclined to use mobile phones and gadgets. Instead of depriving these students in using their mobile phones, why not use it in the teaching-learning process using augmented reality. Educators need to maintain abreast of change and embrace curriculum design and pedagogical approach using AR technology. Educators should recognize that the learning process is about creativity and collaboration. Teachers' goal is to bring students interested in a field of study. For a generation that's been produced on shared technologies bringing AR into the classroom and curriculum can also help encourage active engagement and contribute to students' retention [2]. AR content provides students who are visual learners gain a fuller intellect and knowledge in History [2]. Educational activity and learning theory subjects such as history is a major worry because it has stories, dialogues, images, and the bulk of the contents are paragraphs lots of paragraphs. This is likely the reason history is one of the less favored subjects by the students. For teachers, letting students understand history lessons is a challenge.

The usage of AR in education has become more popular [3].The study [4] focuses on answering difficulty in reading human body shape. In paper [7], teaching tourism and history get a new way of delivering the lessons through the emerging of new technologies. Composition [8] is a work that creates instructional media in Indonesia through augmented reality. In paper [9], and AR was used to discover information about historical graffiti in the Temple of Debod. Some other field of AR [10] is disrupting the present with remnants of history. The majority results of the different studies using augmented reality in education arouse the interest of the students to take the example [2], [5], [6]. Most research papers prove that Augmented Reality improves the caliber of instruction and learning experience [1], [5], [6], [7], [9], [14], [15].

The researcher identified the problem that pedagogy has not been learner-centered; students are less interested in learning history.

The end of the work is to develop an Augmented Reality (AR) mobile application to enhance the pedagogical approach in teaching history and would satisfy the interest of the students in learning history.

2. Methodology

The Instructional System Design is the systematic approach to the Analysis, Design, Development, Implementation, and Evaluation of learning materials and actions. The instructional design aims for a learner-centered rather than the traditional teacher-focused access to teaching so that efficient learning can take place [11]. This intends that every element of the command is governed by the learning outcomes, which have been decided after a thoroughgoing analysis of the learners' needs. The educational content using Augmented Reality (AR) technology was produced according to the ADDIE model [12]. This is an iterative instructional design process, where the results of the formative evaluation of each phase may lead the instructional designer back to any previous phase.

2.1. Analysis

The definition of the problem is the source of the analysis phase, followed by identifying the root of the problem and determining possible answers. The result of this phase will be the input in the development phase.

2.2. Design

Educational content is included in the conception stage, and it should incorporate several media elementsto apply to diverse learning modes. Designing augmented reality application [13] that will be practiced, should be thoroughly thought of and meticulously planned. The user should be the priority, what the purpose and outcomes are identified in designing and developing these tools. The study of [14] discusses the iterations of creating a mobile augmented reality (MAR) application for the tourism context. To indicate who will utilize the AR technology, a use case will be made. And to fully understand the flow, an activity diagram will be produced. The author proposed a system model for creating a mobile application.

Using the Blippar to develop the AR is the idea, Blippar's working principle that is a marker-based [15]. This entails that the marker is needed.

2.3. Development

Methods and processes employed by the researcher in developing the AR technology are discussed in this phase.Names and thumbnails of historical sub-topics should appear when scanning the marker image using the Blippar application that is installed on a mobile device. Methods for developing AR applications applyto different pedagogical approaches that are compatible with the "Active learning" framework: constructivist learning, situated learning, game-based learning, and inquiry-based [15].The development of AR comprises of creation of multimedia elements that would be mixed during the execution. Some papers [1], [4], the researcher uses marker-based AR on Android smartphones.

2.4. Implementation
After all, the multimedia elements have been created in the Blippbuilder Studio. This will now be tested.

2.5. Evaluation
Due to time restriction, purposive sampling will be used. A System Usability Scale survey questionnaire will be distributed to the respondents.

3. Discussion of Findings

3.1. Analysis
The pedagogy has not been learner-centered; students are less interested in learning history subjects.The solution to the problem is to develop an Augmented Reality (AR) mobile application to enhance the pedagogical approach in teaching history and would satisfy the interest of the students in learning history.

The tool used to create the AR application was the Blippar,an augmented-reality mobile app.The instructional content was chosen on the CMO No. 20, series 2013, Readings in Philippine History specific topic on "one past, but many histories": controversies and conflicting views in Philippine history. This has four (4) sub-topics 1) Site of the First Mass, 2) Cavite Mutiny, 3) Retraction of Rizal and 4) Cry of Balintawak or Pugadlawin. The target users are secondary or tertiary students who should demonstrate the ability to formulate arguments in favor of or against a particular issue using primary courses.

3.2. Design
All the educational content was prepared; it contains different media textual, acoustic, animated, video, and imaginary, to be adapted to different learning styles.

A Use case diagram was created for the respective role of Teacher and Student in using the AR technology. The teacher is responsible for creating/editing the course materials that will be uploaded to the studio. Create the AR technology, publish, provide the code to the student, as well as save and test. The student will download and install the Blippar application into their mobile phone, open, enter the blipp code,and start blipping and interact with the AR technology. The Internet is required to use the apps.

Every registered user should be provided with a username and password for authentication purposes, and this will also serve as a security measure to the project created.

Fig. 1. is the specific activity of the teacher, using Blippbuider AR creation a simple drag-and-drop interface makes it very fast and comfortable for history teachers to create powerful augmented reality, even if the teacher has no coding skills. The teacher is responsible forsourcing, preparing, creating, editing, deleting, and uploading the course content into the AR tools. Blippar could add the power of augmented reality into the mobile app. The Augmented Reality Software Development Kit (AR SDK) allows leveraging any AR content within the existing app. The teacher can easily create these augmented reality experiences using the AR creating tools – Blippbuilder and Blippbuilder Script.History teachers have access to all functionalities without limitations.

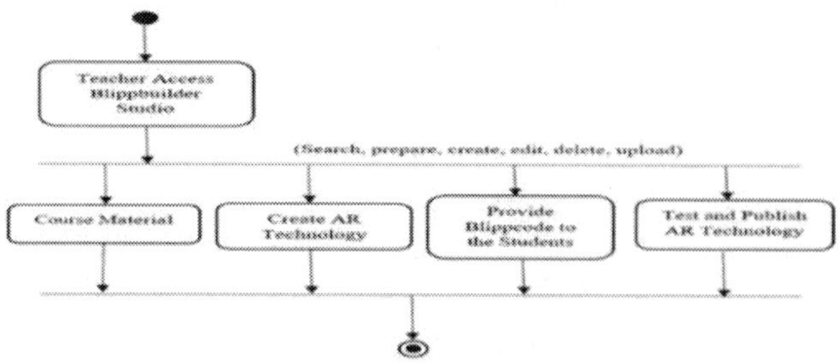

Figure 1. Activity diagram of the teacher

Another activity flow for the student using the Blippar. Student should acquire the blipp code from the teacher, go to setting menu select and enter blipp code, tap the mobile screen to start blipping and focus the mobile phone camera to the marker and start scanning, once the augmented reality overlay is already visible on the screen user can now remove camera away from the marker. One thing good about this Blippar is you do not need to keep on focusing on the tag to see the content of the AR technology. Interact with the AR content by tapping on the thumbnail on the screen.

Fig. 2. shows the architecture of the system model. The application was developed using Blippbuilder and Blippar (SDK). Blippbuilder is used to make the application, and the Blippar SDK is used to link the augmented reality camera and the image and the image tracker. The tracker is used to detect the picture which is stored in the database. Assets and scripts were working along with entree to the object database. Open the application using a mobile phone, scan the image. The scanned image is sent to the picture capturing model for identification. If the icon is detected, it is transported to the processing model. The processing model then finds the marker post and carry it to the tracker model; it then identifies the marker and sent it to the rendering model. Rendering model using picture detection will mark a virtual object to the marker, and hence, the augmented video or image will be exhibited.

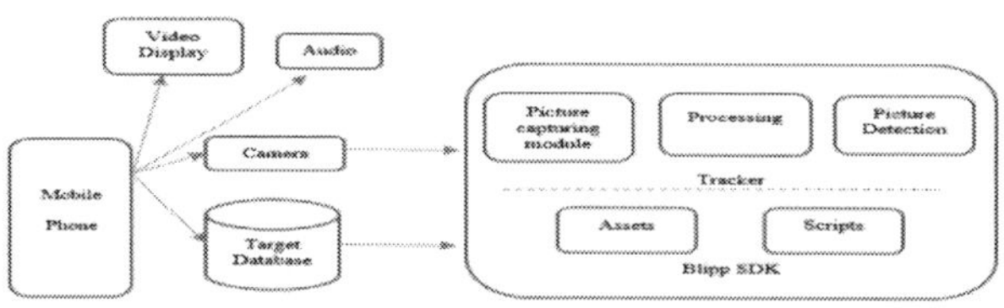

Figure 2. The architecture of the System Model

The Philippine flag will be used as a marker. By scanning the image using Blippar, the application should display corresponding thumbnails of Philippine history sub-topics. Tapping the thumbnail of a particular topic, an augmented video will be played, or history with animated text accompanied by the narrative.

3.3. Development

The process begins by selecting the Create Blipp option that is visible in Fig. 3. When this option is selected, it is necessary to upload a trigger image that will serve as the marker.

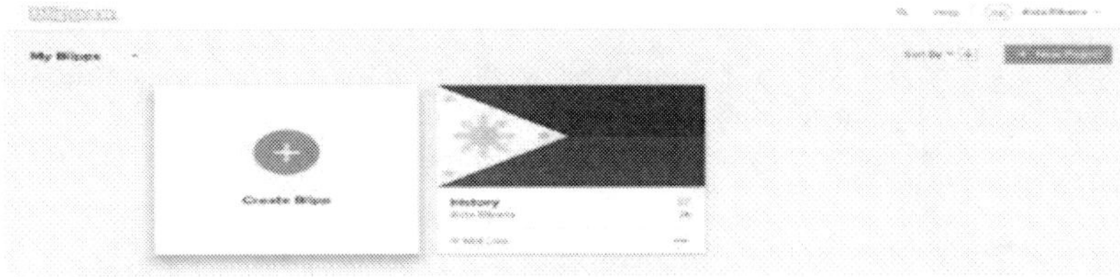

Figure 3. Create a Blipp

The next step is to add an overlay to the marker. Overlays are implemented one by one; it could be adding created multimedia elements. Fig. 4 shows how the first overlay has been implemented.

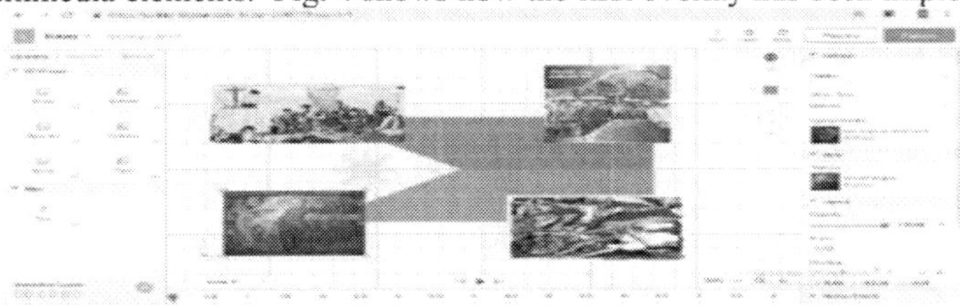

Figure 4. Overlays implemented in Blippbuilder studio

Later on, adding the overlays, the actions are put to them. All overlays are covered so that they do not appear when a mark image is scanned. Scanning the marker image using the Blippar thumbnails of the historical sub-topics should appear. These four overlays and a marker image are presented in Fig. 5.When the overlay is tapped, the augmented video is displayed.

Figure 5. Marker image with four overlays

3.4. Implementation

After finishing the blip with history sub-topics in Blippbuilder studio, created AR was tested for its functionality. The Blippar application was installed on a mobile phone, an internet connection was established. Blippar was opened; this was positioned in front of the flag image on the laptop screen. Start the blipp, tap the screen, start scanning the image. To interact with the content of the AR, the thumbnail should be touched, and the content will be displayed, as shown in Fig. 6.

Figure 6. The result of tapping the thumbnail

3.5. Evaluation

The researcher conducted a System Usability Scale survey on a purposive sampling of (30) students and 3 instructors. To rate the students' and instructors' responses, a five-scale Likert scale (Strongly Agree {5}, Strongly Disagree {1}) was used. Based on the result, it has an average of 78.27, which was converted into a percentile rank of 78%, and with a grade of B, as shown in Fig. 7. AR technology is assumed to be usable and acceptable.

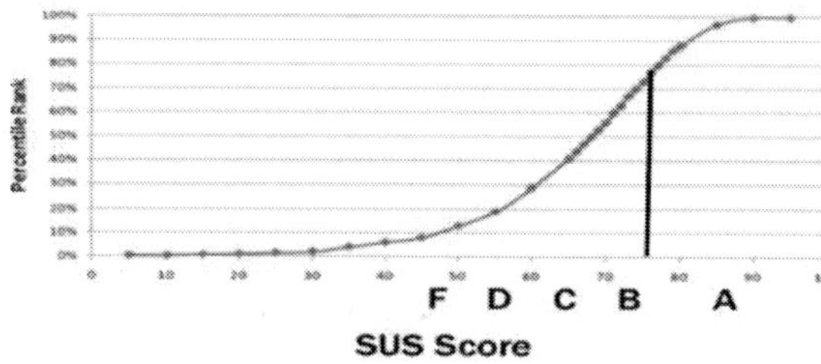

Figure 7. The result of tapping the thumbnail

4. Conclusion

A form of instructional media material for teaching history subject is the result of the research. This paper intends to arouse students' interest and encourage them to learn history through augmented images and videos as the result of a portable application of augmented reality. This is a great contribution to the educational domain through augmented reality and improving the instructional approach of teaching.

References

[1] Algawari F, Alslamah W, Alhabib A, Altehaid A and Ibrahim D 2018 Applying augmented reality technology for an e-learning system *World Acad. of Sci. Engg. and Tech. Int. J. of Comp. and Info. Engg.* Vol **12** No 3

[2] Klimova A, Bilyatdinova A and Karsakov A 2018 Existing teaching practices in augmented reality *Proc. Comp. Sci.* **136** pp 5-15

[3] Badni K 2018 Augmenting history: Case study measuring the motivation of students using augmented reality apps in history cases *World Acad of Sci., Engg. and Tech. Int J. of Educ. and Pedag'l. Sci.* Vol **12** No 6

[4] Layona R, Yulianto B and Tunardi Y 2018 Web-based augmented reality for human body anatomy learning *Proc. Comp. Sci.* **135** pp 457-64

[5] Salinas P and Pulido R 2015 Visualization of conics through augmented reality *Procedia Comp. Sci.* **75** pp 147-50

[6] Raghaw M, Paulose J and Goswami B 2018 Augmented reality for history education *Int. J. of Engg. and Tech.* Vol **7** Iss 2.6 pp 1-5

[7] Kysela J and Storkova P 2015 Using augmented reality as a medium for teaching history and tourism *Procedia – Soc. and Behav'l Sci.* **174** pp 926-31

[8] Izza N and Fuada S 2014 Augmented reality applications in learning history of Indonesia *Int. Sem. of Education Technology (ISET) 2014* Semarang June 2014

[9] Gutierrez J, Molinero M, Martin O and Medina C 2015 Augmented reality technology spreads information about historical Graffiti in the temple of Debod *Proc. Comp. Sci.* **75** pp 390-97

[10] Johnston E, Richison M and Vujnovic M 2015 Augmented Asbury park: Disrupting the present with remnants of history in augmented reality *Proc. of the 21st Int. Symp. on Elect. Art ISEA2015* ISSN: 2451-8611 ISBN: 978-1-910172-00-1

[11] Hawkinson E and Artemciukas E 2018 Simplified mobile AR platform design for Augmented tourism *World Acad. of Sci., Engg., and Tech. Int. J. of Soc. and Bus. Sci.* vol **12** No 1 2018

[12] Kurniawan M, Diana S and Witjaksono 2018 Human anatomy learning system using augmented reality on mobile application *Proc. Comp. Sci.* **135** pp 80-8
[13] Castellanos A and Perez C 2017 New challenge in education: Enhancing student's knowledge through augmented reality *Augmented Reality* ed J M Ariso p 273 (Berlin: De Gruyter)
[14] Augmented Reality & Computer Vision Company 2019 (Online) Available: https:www.blippar.com/ (Accessed: 20-March-2019)
[15] Pantelica A and Vukovac D 2017 The development of educational augmented reality application: A practical approach *Proc. of ICERI2017 Conf. 16th-18th* November 2017 pp 8745-52 DOI: 10.21125/iceri.2017.2399

Development and Usability Evaluation on Individualized Reading Enhancing Application for Dyslexia (IREAD): A Mobile Assistive Application

M A P Burac[1], J Dela Cruz[2]

[1]College of Information Technology and Engineering International School of Asia and the Pacific Penablanca
[2]College of Information Technology and Computer Science, University of the Cordilleras, Baguio City

E-mail: Cagayan,iamgelotron21@gmail.com, delacruzpen@gmail.com

Abstract. Mobile Applications Development has been the trend for the previous year to revolutionize the teaching and learning process. These technologies were designed to improve the quality of education for all types of learners, more specifically for learners who have learning difficulties like dyslexia. Learners having this learning disorder oftentimes have difficulty in reading and writing. The use of conventional instruction is challenged for this learner. With the advent of technology, specifically mobile assistive application development helps implement more dynamic, effective and efficient teaching strategies, thus provides an opportunity to improve the learning of the students. This study aimed to develop a mobile assistive application name Individualized Reading Enhancing Application for Dyslexia (IREAD) which primarily implements text-to-speech technology to enhance the reading capability of learners with dyslexia. Unity 3D is utilized in the development of the application and web support is present to manage learning resources of the application. Moreover, usability evaluation is performed to determine the overall usability of the application, with the following dimensions effectiveness, efficiency, and quality of support, ease of learning and satisfaction. The 10 participants who were teachers teaching in specialized education answered the usability questionnaire. Results show that the overall usability of the IREAD application is immensely positive.

1. Introduction

The growth in mobile application technologyseemingly provides opportunities in optimizing the Information and Communication Technology (ICT) tools in improving the teaching and learning process with appropriate pedagogy. Among its advantages of using technology in teaching and learning is more personalized learning that would cater to the diverse needs and learning styles and capabilities of learners, thus accelerate and promote learning. The advancement of technology especially in the growth of mobile technologies paved way for the innovation of numerous learning solutions that can be adapted and utilized for specialized education, which focused on the learning pattern of every student. This is also advantageous to the teachers as they can easily collect, synthesize and analyze the voluminous amount of data, and swiftly have feedback on the students' performance and achievements [1]. Mobile Learning combines personalized learning wherever and whenever it is accessed. It provides a different educational paradigm which alters the nature of knowledge and is focused on the user's experience of learning through mobile devices. It gives a wide range of educational and learning material in a uniquely engaging manner, giving them the freedom to choose what information to deal with which can enrich their knowledge and improve their skills. Mobile Educational Applications motivate children and engage their attention while focusing on problems, improving their memory, reading and writing skills [2]. It has been stressed out that m-learning had created a method of learning that showed differences compared to conventional methods commonly used in teaching and learning [3][4]. Integrating m-learning in a classroom environment would be an effortless action if teachers or educators do not grab the opportunity to use mobile technology in a classroom setting. However, most of the Mobile Applications available in the digital stores did not consider this as a priority, yet they tend to focus only on commercial value. Mobile Application with relevant learning theory is to be used for the sake of dyslexic students learning [4].

There is already a variety of purposes in utilizing technologies especially for individuals who have dyslexia, these technologies provide supports for content area learning and skills practice which enables performing functions that are usually challenging. There was researches that suggest that computer-based programs help younger students learn to read which provides a more holistic approach to reading thus, yielded a better result [5].On the other hand, it was found out that acceptance and readiness among select teachers of dyslexic students at a mild level in a dyslexia center in Klang Valley provide significant information on the impact of M-learning to that teacher perceived, the frequency of utilization of relevant Mobile Application in the classroom setting. Integrating m-learning for dyslexic students requires a big change in attitude and moral support towards dyslexic students themselves. Developing an engaging m-learning environment may be inside or outside the classroom is the actual challenge. Thus, it suggests that every stakeholder in the field should suggest a suitable approached to be implemented to uphold the objective of implementing M-learning for dyslexics [4].

With the above-mentioned statements, this study aims to develop the Individualized Reading Enhancing Application for Dyslexia (IREAD) a mobile assistive application and evaluate its usability to determine the user experience and eventually underscore the attainment of the goals in the development of the application is achieved.

2. Literature Review

Rahim et. al [6] developed a prototype of a mobile application to assist reading in dyslexic children name CinTA. The features provided by the app are modules to learn alphabets and writing. Addressing the instruction, a video is presented in the application to teach the students with several special writing techniques. Teaching alphabets in the application has been categorized through vowels and consonants to which, in each of the categories, the audio of the sound and phonics of the alphabet will be played to the children. However, the application doesn't provide assistive technology to properly help the learners, like text-to-speech technology.

Skiada et. al [2] developed a mobile application named easyLexia which focuses on teaching learners with difficulties with the use of Mobile Learning. The main features of their application composed of a word finder - where children identify the correct letters which address the gaps and choose the best answer from among alternatives, choose it- which enables the children to listen and recognize the work given and lastly, Sound finder, which consists of sentences with missing words while combining phonological recordings of the missing word. Nonetheless, the application doesn't have data logs that track the records of the learners. This allows the educators and parents to assess the learning acquired of the learner using the application.

The literature review conducted by the researchers highlighted some research gaps in the development of an assistive application for dyslexics. The inclusion of text-to-speech technology and data logs is essential for the application. This is to provide historical data analysis which underscores the enhancement in the reading proficiency of dyslexic learners. The application shall be following the specific needs of dyslexics. These are the primary motivation of the researchers to develop the IREAD Mobile Assistive Application.

3. Methodology

The consideration of the use of agile software development methodology for the development of the Individualized Reading Enhancing Application for Dyslexia (IREAD) is in congruence with Agile Patil, Vaishnavi & Panicker, Sanjana & Kv, Maitreyi [7] conclusion that there has been a significant increase in number of mobile application development projects, as the demand for mobile applications is growing and the potential number of different mobile applications is virtually unlimited. Several published pieces of research confirm that Agile methodology practices are naturally fit for the development of mobile applications.

3.1. IREAD Development

The initial stage of the methodology has been initiated through the conduct of an in-depth interview with the Special Education Teachers of Tuguegarao West Central School and the School Principal to elicit the instructional materials used in teaching dyslexic students as well as their strategies in delivering the instruction to the learners. A literature review from published articles and journals from different databases such as EBSCOhost, Elsevier, and Procedia were also being done to support the current practices in addressing the needs of dyslexic students using traditional and ICT based delivery of instruction. Thorough document analysis was also performed to develop the architectural framework of the mobile assistive application for learners with dyslexia. It composed two components, the Mobile Platform, and Web platform. These platforms are designed to address the diverse need of dyslexic students.

The researcher build-up the IREAD, a mobile assistive application using unity 3D, which is a cross-platform game engine that is easy and has a user-friendly development environment.The sample user interface design of the IREAD app is seen in Figures 1, 2, and 3. The IREAD Mobile Application composed of three modules, which includes a learning module consisting of reading lessons and writing lessons, the evaluation module, and the history/reports module. The application was installed in a smartphone operating with android OS.

Fig 1. IREAD Main Menu

Fig 2. IREAD Learn Module

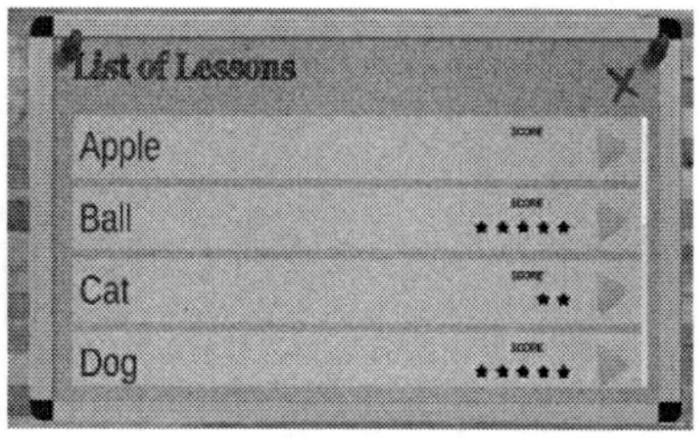

Fig 3. IREAD Evaluation Module

3.2. IREAD Usability Evaluation

Usability evaluation is a vital criterion to understand and analyze usability issues in any system or application. Usability issues are identified through an evaluation conducted with users [8]. The International Organization for Standardization (ISO) 9241-210 standard [9] and the ISO and International Electrotechnical Commission (IEC) 9126 standard [10] guided the initial user experience design of the IREAD application. Based on these standards, the researcher operationalized usability as the degree to which a user of the application can achieve the goals of the IREAD protocol with effectiveness, efficiency, quality of support, Ease of learning and satisfaction.

The usability questionnaire is composed of 33Likert type questions in which the participants were asked for their level of agreement on the usability of the IREAD app. Specifically, it has 10 questions for the effectiveness of the application, 9 for efficiency, 3 for quality of support, 4 for ease of learning and 7 for user satisfaction. Quantitative analytics was utilized to quantify the data gathered to answer the questions of the usability of the IREAD app. Cronbach's alpha was used to test of internal consistency of the usability survey tool. The results of the test show that all categories were above .70 alpha value which indicates that the items in the questionnaire are reliable. Table 1 shows the details of the test.

Table 1. Questionnaire Reliability Scale Index Using Cronbach's alpha

Category	**Number of items**	**Cronbach's Alpha**
Effectiveness	10	.925
Efficiency	9	.836
Quality of Support	3	.936
Ease of Learning	4	.833
Satisfaction	7	.709

The participants of this study were the 10 teachers of the west-central elementary school, who has direct experience in teaching dyslexic students. The teachers utilized the IREAD app to test the content and overall functionality of the application. After utilizing the application, the usability questionnaire was distributed for the usability evaluation.

4. Results and Discussion

Table 2 shows that all teachers teaching in special education classes were female and most of them graduated with their post-graduate degrees. This attests that women characterized patience and care which is needed in handling learners with special needs. Also, the teacher's educational qualification meets the standards for them to provide quality education to the young.

Table 2. Frequency and Percentage Distribution of the Respondents Profile

Demographic Characteristics	**Categories**	**Frequency**	**Percentage**
Gender	Male	0	0%
	Female	10	100%
HighestEducational Attainment	Bachelors' Degree	1	10%
	Masters' Degree	9	90%

4.1. Findings on the level of usability of the Individualized Reading Enhancing Application for Dyslexia Application

Table 3. The effectiveness level of IREAD App

Item	**Weighted Mean**	**Descriptive Interpretation**
The IREAD Application is easy to use.	4.80	Strongly Agree
The IREAD Application is simple to use.	4.80	Strongly Agree
The IREAD application lets learners do reading and writing lessons.	4.90	Strongly Agree
People using it once or many times would like it.	4.30	Strongly Agree
Mistakes can be fixed quickly and easily.	4.20	Agree
The learners can use it well every time.	4.20	Agree

The effectiveness of the mobile application is one of the key considerations for evaluation to meet the expected characteristics and functions of the application. As gleaned in table 3, most of the respondents strongly agree that the IREAD application lets learners do reading and writing lessons, it is simple and easy to use, and it is understandable.This implies that the IREAD application is effective in delivering its desired functions. Relative to Kumar and Karie [11] study which a mobile application was developed for children with speech disabilities. Its features comprise a graphical user interface that implements categorized images and a keyboard. The user clicks on the image and the system sends the image to a phrase library to convert it to a suitable text. Then, the text-to-speech engine processes the text and produces the sound as an output.

Table 4. The efficiency level of IREAD App

Item	Weighted Mean	Descriptive Interpretation
The IREAD User Interface provides appropriate content/information for children/learners	4.90	Strongly Agree
The IREAD application is easy to learn for children/learners	4.90	Strongly Agree
The IREAD application provides useful and interesting learning activities for children	4.80	Strongly Agree
The IREAD application provide appropriate progress report/evaluation result for assessment of performance in a given activity	4.30	Strongly Agree
The IREAD application provide appropriate controls	4.40	Strongly Agree
The IREAD application provides easy readability for children/learners	4.20	Agree

As gleaned in table 4, the respondents' revealed a high level of agreement on the efficiency of the IREAD application. It provides a well-designed user interface to which, its content and information were sufficient that enables learners to learn efficiently. The application can be understood by the learners as it provides an appealing and satisfaction menu for navigation. The reading and writing activities included in the application were interesting and is useful for the learnings. This implies that the application can be considered efficient.Castañeda D. and Cho M. [12] stated that the use of mobile applications in enhancing students' learning in Spanish classrooms runs high and the results of their study showed that the activities helped students improve not only their accuracy but also their confidence in conjugating Spanish verbs. Thus, mobile application utilization in the classroom enhances students learning.

Table 5. Quality of Support level of IREAD App

Item	Weighted Mean	Descriptive Interpretation
The instructions and messages are easy to understand.	4.30	Strongly Agree
The messages to fix problems are clear.	4.20	Agree
The instructions and messages are clear.	4.20	Agree

Table 5 shows the quality of the support level of the IREAD application. The participants strongly agreed that the application instructions and messages are easy to understand. On the other hand, clarity of the instructions and messages needs to be improved to fully have an exceptional quality of support.Shadiey R.,Hwang W., and Huang Y. [13] Smartphones, mobile phones, and personal digital assistants were the technologies mostly used, whereas the English language is the most common target language. They also found out that many of the studies did not apply newly acquired knowledge by the students in solving their real-life problems or recently developed intelligent technologies for language were overlooked.

Table 6. Ease of Learning level of IREAD App

Item	Weighted Mean	Descriptive Interpretation
You quickly became good at it.	4.30	Strongly Agree
You easily remember how to use it.	4.30	Strongly Agree
It is easy to learn to use it.	4.30	Strongly Agree
You learn to use it quickly.	4.40	Strongly Agree

As shown in table 6, the participants strongly agree that the IREAD application exhibits an easy learning environment. This is to support the demanding needs of learners with special needs. Kokkalia G. and Drigas A. [14] concluded that there is a need to support on the design of educational mobile applications for intervention and diagnosis of the learning difficulties in the kindergarten while enhancing the abilities of teachers through training. The effect in the development of preschoolers having such difficulties could be best supported with mobile teaching.

Table 7. Satisfaction level of IREAD App

Item	Weighted Mean	Descriptive Interpretation
The app provides clear and understandable navigation keys such as back/next buttons to move to the previous/ next screen	4.40	Strongly Agree
The educational content matches with my course curriculum	4.70	Strongly Agree
The learning approach and activities in-app were interesting and I learned from them	4.70	Strongly Agree
You are happy with the app.	4.50	Strongly Agree
You would tell a friend about the app.	4.50	Strongly Agree
The app is fun to use.	4.50	Strongly Agree
This app works the way you would want it to work.	4.40	Strongly Agree

As can be seen in Table 7, the participants establish a high satisfaction level. The primary goal of the development of the IREAD application is to support the learners with special needs where its learning content is based on the curriculum provided by the department of education. The utilization of a mobile assistive application provides an innovative way of teaching;thus, it provides engaging and interesting activities.

Table 8. Summary of Usability level of IREAD App

Category	Weighted Mean	Descriptive Interpretation
Effectiveness	4.60	Strongly Agree
Efficiency	4.55	Strongly Agree
Quality of Support	4.23	Strongly Agree
Ease of Learning	4.32	Strongly Agree
Satisfaction	4.52	Strongly Agree
Categorical Mean	**4.44**	**Strongly Agree**

The overall usability results of the IREAD application are presented in table 8 to which the participants strongly agreed to all usability dimensions. Innovation in the mobile app has raised interest to apply the m-learning concept among teachers becauseit facilitates teaching and learning, as claimed by [15]. Then, they must get readyto utilize the technology associated with m-learning in the teaching and learningprocess towards dyslexic students.

5. Conclusion

Based on the results of the study, the researcher can attest that Individualized Reading Enhancing Application for Dyslexia (IREAD), a mobile assistive application that can support dyslexic learners can be an alternative in teaching reading and writing lessons for learners with special needs. With the design considerations that specifically address the needs of the learners, it provides individualized learning which composes interesting and engaging activities. The developed application incorporated text-speech technology to aid students in learning how to read. It supports the repeated instruction method of teaching by teachers.

The usability evaluation of the application yielded a positive result, which fulfills the requirements of dyslexic students. The participants immensely agree that the application is effective and efficient in delivering its design and functionalities. Relatively, it is evident that the participants in satisfied with the IREAD app. It can be recommended that the application will be utilized by the school to enhance the reading proficiency of dyslexic learners.

References

[1] Madeiraa J *et al* 2015 Assistive Mobile Applications for Dyslexia *Procedia Computer Science* **64** pp 417–24

[2] Skiada R *et al* 2014 EasyLexia: A Mobile Application for Children with Learning Difficulties *Procedia Computer Science* **27** pp 218–28

[3] Nawi A, Hamzah M and Rahim A 2015 Teachers acceptance of mobile learning for teaching and learning in Islamic education: a preliminary study *Turkish Online J. of Distance Education (TOJDE)* ISSN 13026488 Vol **16** No 1 Article 12 DOI: 10.17718/tojde.30611

[4] Zain N Z and Mahmud M 2018 Acceptance and readiness of mobile Learning integration among teachers of Dyslexic students: a preliminary study *The Int. J. of Multimedia & Its Applications (IJMA)* Vol **10** No 6 December 2018. DOI: 10.5121/ijma.2018.10613

[5] Dawson K *et al* 2016 How mobile app developers conceive of dyslexia and what it means for mobile app users *Interaction Design and Architecture(s) J. - IxD&A* No 28 pp 69-84

[6] Rahim S K N A, Nasrudin N H, Azmi A Z, Junid R A, Mohamed Z and Abdullah I I B 2018 Designing Mobile Application for Dyslexia in Reading Disorder Problem *Int. J. of Academic Research in Business and Social Sciences* **8(1)** pp 628–46

[7] Vaishnavi P, Sanjana P and Maitreyi Kv 2016 Use of Agile Methodology for Mobile Applications *Int. J. of Latest Technology in Engineering, Management & Applied Science*

[8] Hashim N L and Isse A 2019 Usability Evaluation Metrics of Tourism Mobile Applications *J. of Software Engineering and Applications* **12** pp 267-77 DOI: 10.4236/jsea.2019.127016

[9] International Organization for Standardization 2009 *Ergonomics of human system interaction-part 210: Human-centered design for interactive systems (ISO 9241-210; formerly known as 13407)* (Geneva Switzerland)

[10] International Organization for Standardization/International Electrotechnical Commission 2001 *A ISO/IEC 9126-1 standard, software engineering, product quality, part 1: Quality model* (Geneva, Switzerland: Author)

[11] Kumar A and Karie N M 2014 Development of basic learning and communication system for school children with speech disabilities using mobile platform *Int. J. of Advanced Research in Computer Science* vol **5** no 6

[12] Castañeda A and Cho M 2016 Use of a game-like application on a mobile device to improve accuracy in conjugating Spanish verbs *Computer Assisted Language Learning* **29** no 7 pp 1195-204 DOI: 10.1080/09588221.2016.1197950

[13] Shadiev R, Hwang W and Huang Y 2017 Review of research on mobile language learning in authentic environments *Computer Assisted Language Learning* **30** 3-4 pp 284-303 DOI: 10.1080/09588221.2017.1308383

[14] Kokkalia G K and Drigas A S 2016 *Mobile Learning for Special Preschool Education* DOI: 10.3991/ijim.v10i1.5288

[15] Johnson L, Adams S and Cummins M 2012 Mobile apps *The NMC Horizon Report: 2012 Higher Education Edition* (Austin, Texas: The New Media Consortium)

IOP Conf. Series: Materials Science and Engineering **803** (2020) 012016 doi:10.1088/1757-899X/803/1/012016

Mobile Devices and Mobile Apps Use among Indonesian College Students

Ahmad R. Pratama

Department of Informatics, Universitas Islam Indonesia, Sleman, DIY 55584 Indonesia

E-mail: ahmad.rafie@uii.ac.id

Abstract. This article adds more insights into the previous study about the use of mobile apps on smartphones and tablets among Indonesian college students based on a survey of undergraduate students at one private university in Indonesia. Data analysis includes two-sample tests of proportion and multiple correspondence analysis (MCA). The findings confirm social media as not only the most frequently used app category, but also with the highest level of engagement. While most students use their mobile device mostly for social media, it is regular and heavy use of educational apps that are more closely tied to spending more than five hours per day on mobile device whereas more frequent mobile gaming is associated to more hours spent on mobile device. Students from IT major were found to be an anomaly among other STEMM disciplines. Not only they are more likely to spend more hours on their mobile device, but also, they tend to be more socially active, at least digitally and not necessarily physically in daily life. The findings in this study give a better understanding of how Indonesian students use their mobile device that was already uncovered in the previous study by answering some of the remaining questions.

1. Introduction

For the first few decades of their introduction in the 1970s, mobile devices were used only as a voice and text-based communication device. The landscape changed significantly with the release of smartphone and then followed by tablet PC, especially with two major platforms in Apple iOS and Google Android, released in 2007 and 2008 respectively. Mobile devices, especially smartphone, are now the most popular personal computing devices and the trend is clearly in favor of mobile devices to be the most preferred personal computing devices for many [1]. An archipelagic nation of Indonesia, also happens to be the fourth in terms of population size worldwide, is a promising market of mobile technology and has been leading in terms of mobile device utilization worldwide. Its mobile connection per capita is well ahead many major countries like the United States of America, France, Australia, Japan, or South Korea [2]. This is interesting given that digital divide that has been widely accepted as a worldwide phenomenon for the past two decades usually put the developing countries behind their developed counterparts [3]. Also, considering the two-way relationship between ICT infrastructures and utilization, in which mobile technology is a major part of it, and education where higher participation rate in education leads to more utilization of ICT that further leads to better education outcomes [4], it is necessary to get regular updates in this research area.

As the literature suggests, people in many countries, including the United States had already spent more than five hours per day on mobile device in average, mostly to access social media, entertainment, and games by 2016 [5]. Furthermore, the previous study shows that more than 95% of the students in a private university in Indonesia already own smartphones in the same year [1], which is significantly more compared to the overall Indonesian population at 21% as well as the American population at 72% for the same time period [6]. Also, it was found that 55.49% of the them spent more than five hours a day on their mobile devices with five significant predictors of spending more hours per day on mobile device (i.e., sex, purchasing price of mobile device, regular use of social media, regular use of mobile games, and negative attitudes toward ICT) [7]. However, some questions were left unanswered, especially regarding the roles of the other variables of interest such as educational apps use, major, place of origin, and year in college. This study is meant to follow it up by answering the remaining questions in order to give a more comprehensive understanding in that matters. As the literature suggests, the use of mobile device can help improve the quality of education [8-10], especially when the focus is on emphasizing the opportunity of transforming education with the use of mobile learning [11-12].

2. Methodology

This study involves college students attending a private university in Indonesia as its participants. An online survey was conducted within the first quarter of 2016, filled out by the students in class settings and delivered in their own native language of Bahasa Indonesia. A total of 201 students filled out the survey, but after dealing with missing values, only 179 students included in this study. They ranged from 19 to 25 years of age ($M = 21.41$, $SD = 1.54$). Participants were then categorized based on their sex (102 males, 56.98% and 77 females, 43.02%), self-perceived ICT adoption level representing their attitude toward ICT (38 early adopters, 21.23%, 122 majorities, 68.16%, and 19 laggards, 10.61%), year in school (26 1st year students, 14.53%, 30 2nd year students, 16.76%, 56 3rd year students, 31.28%, 38 4th year students, 21.23%, and 29 students in their 5th year or above, 16.20%), socioeconomic status (SES) measured by the purchase price of their devices (35 high SES, 19.55% and 144 middle-to-low SES, 80.45%).

All respondents in this study were asked 1) how many hours a day they spend on their mobile device, and 2) how often they use apps in three separate categories: social media, educational, and mobile gaming. The survey listed up to 25 most frequently downloaded apps in Indonesia, in both iOS and Android platforms for each category. Subjects were asked to indicate whether they have installed the apps on their mobile device or not, and how often they use each of those apps (monthly, weekly, daily) if installed. For each of the three apps categories included in this study (educational apps, mobile games, and social media apps). Students who use at least one app in the category at least once a week are considered regular users of that category. Those who use at least one app in the category daily are considered heavy users. All heavy users are regular users, but not all regular users are heavy users. All analysis is done for both heavy and regular use as a form of sensitivity analysis.

Data analysis in this study involves descriptive statistics, two-sample tests of proportion, and multiple correspondence analysis (MCA). A series of two-sample tests of proportion is used to compare apps use in three categories, while logistic regression method is used to estimate the frequency of apps use to identify any significant predictors of regular and heavy use in each app category. MCA is used to analyze the multivariate relationship between app use and time spent on mobile devices. This method is selected because of the following reasons. First, it is specifically designed for analyzing nominal and categorical data, like how principal component analysis (PCA) works for analyzing continuous or numerical data. Second, it can be used to detect the underlying structures in the dataset without having to satisfy a certain assumption like it is in any regression analyses. And third, it will be more helpful to have a good visualization of the data with points cloud of individuals in a two-dimensional Euclidean space produced by MCA. The MCA analysis was done with R 3.4.3 in RStudio with the help of `FactoMineR`, `factoextra`, `ggplot2`, and `ggrepel` libraries. All four variables of interest are set as active categories while the other relevant variables from the survey such as demographic data are used as supplementary variables.

After several iterations of MCA, each with different model specifications, two final models were selected. The main difference between the two models is that the first model (Model A) compares regular vs non-regular (referred to as light users) while the second model (Model B) compares heavy vs non-heavy (referred to as casual users) for all three apps categories. Regarding the daily time spent on mobile devices, a simpler 3-level category is used in both models. It is done by combining all the intermediary groups (i.e., 1-2 hours, 2-3 hours, 3-4 hours, and 4-five hours per day) into one (i.e., 1-5 hours) as they often overlap with each other in the visualization of other models. To produce the best visualizations and to be able to explain slightly more variance while removing no important information, some outliers, identified as the non-user and the light user of social media apps, were removed from the final models.

3. Results and Discussions

3.1. Social Media, Games, and Educational Apps

Among three apps categories in this study, social media is the most frequently used by far. Interestingly, these students play games significantly less than they use educational apps. The results from a series of two-sample tests of proportion between each pair of apps categories (i.e., education vs. games, education vs. social, and games vs. social) confirm the significant difference in use between these three categories and the results show consistency for both regular and heavy users. Table 1 summarizes the number of regular and heavy users, as well as the ratio of heavy to regular users for each app category. Social media apps have the highest ratio of heavy to regular users, indicating a very strong level of engagement since 95% of the regular users are also heavy users. On the other hand, while the ratio for educational apps and games are not as high as social media apps, they both still indicate a moderately strong level of engagement at 68% and 52%, respectively. In other words, most regular users for any categories use the apps daily instead of weekly. In the case of social media, almost everyone uses the apps daily rather than weekly.

Table 1. Regular and Heavy Users in Three Separate Mobile Apps Category

	SOCIAL	GAMES	EDU APPS
Heavy Users	168 *(93.85%)*	32 *(17.88%)*	105 *(58.66%)*
Regular Users	177 *(98.88%)*	61 *(34.08%)*	155 *(86.59%)*
Heavy to Regular Ratio	.95	.52	.68

3.2. Apps Use and Time Spent on Mobile Devices

In the previous study on the same data set, five significant predictors of time spent on mobile devices were found by using OLS regression method. Regular use of social media and mobile games significantly increase the amount of time spent on mobile device, whereas negative attitudes toward ICT significantly decrease it. Meanwhile, females and the owner of most expensive devices spent more hours on their mobile device than males and the owner of less expensive devices [7]. However, some gaps remained, especially regarding the roles of the other independent variables such as educational apps use, major, place of origin, and year in college.

In this study, MCA is used to try finding the answer to those questions. The first MCA visualization approach used is the biplot of active categories for each model. As shown in Figure 1 and Figure 2, the first two principal dimensions can explain 40.18% of the total variances for model A and 39.20% of the total variances for model B, respectively. In both models, social media use is the barycenter of the plot due to removal of the outliers. The first principal dimension separates game use as it progressed linearly from non-gamer on the right to light gamer and then regular gamer on the left for Model A and to casual gamer and then heavy gamer on the left for Model B. In terms of education apps use, both principal dimensions play their part as the second dimension separates the educational apps users and the non-users, whereas the first dimension separates the light users from the regular users in Model A and the casual users from the heavy users in Model B. As for the daily time spent on mobile device, the first dimension plays a much stronger part in Model B than in Model A. The time spent progresses linearly from right to left in Model B, but not in Model A.

Based on the findings from the previous study, social media use contributes the most to the amount of time spent on mobile devices, but that information might be incomplete, and it can be complemented by looking at all the active categories in this study at once. There is a strong indication that those who spend more than five hours per day on their mobile device are the regular (including heavy) users of educational apps. Also, more frequent mobile gaming leads to more hours spent on mobile device, which is explained well by the first principal dimension in both models and this is true for about 21% of the students. The MCA Factor Maps as shown in Figure 3 for Model A and Figure 4 for Model B give a better visualization on how closely associated both regular and heavy use of education apps as well as casual gaming are to more than five hours spent per day on mobile devices. The factor maps also make it clear that each group of students are well separated to each other based on all four active categories used in the MCA.

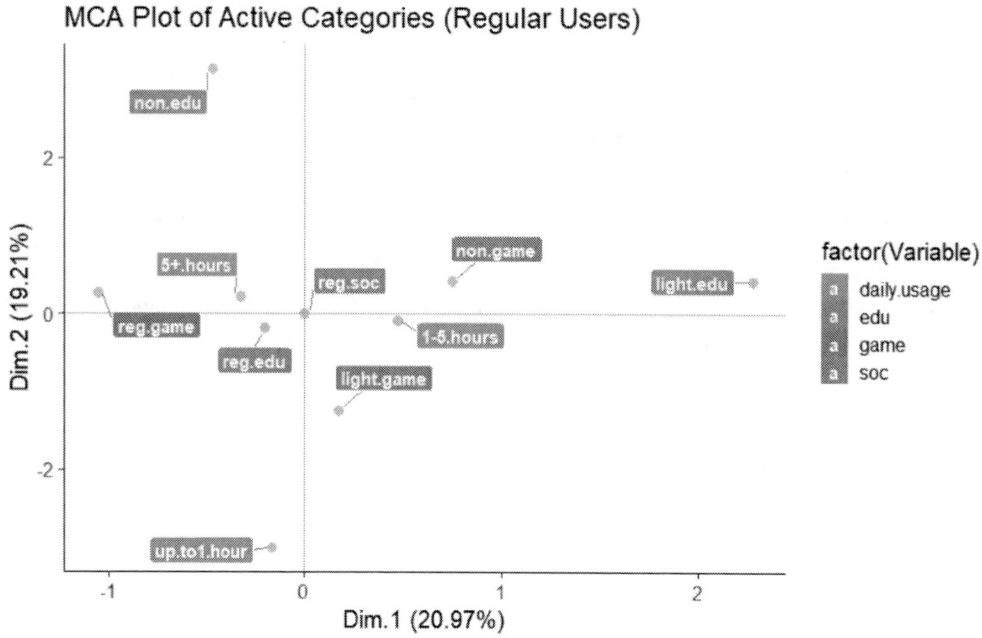

Figure 1. MCA Biplot of the first 2 principal dimensions of active categories for Model A

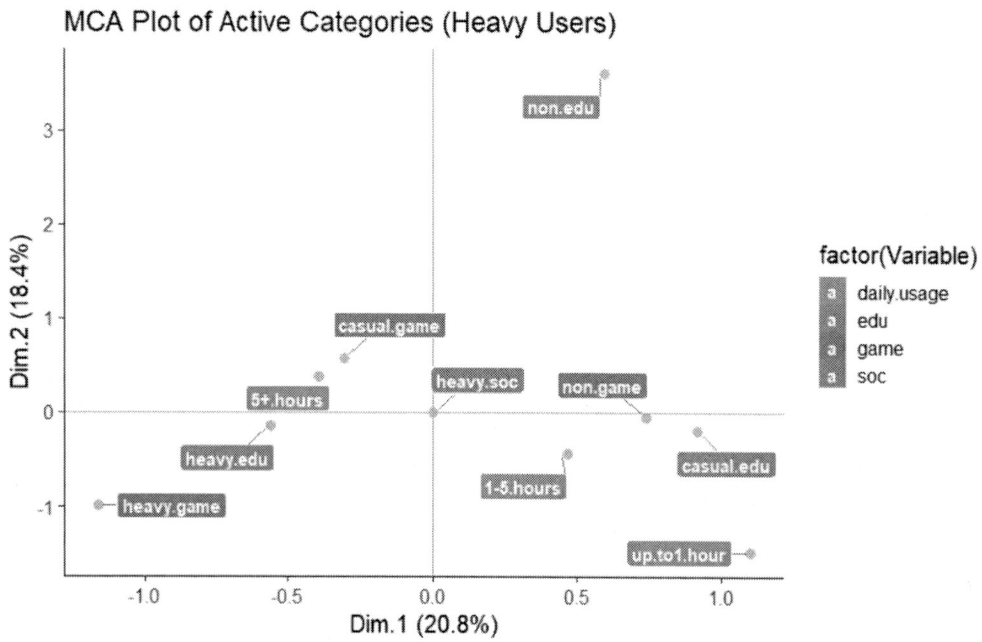

Figure 2. MCA Biplot of the first 2 principal dimensions of active categories for Model B

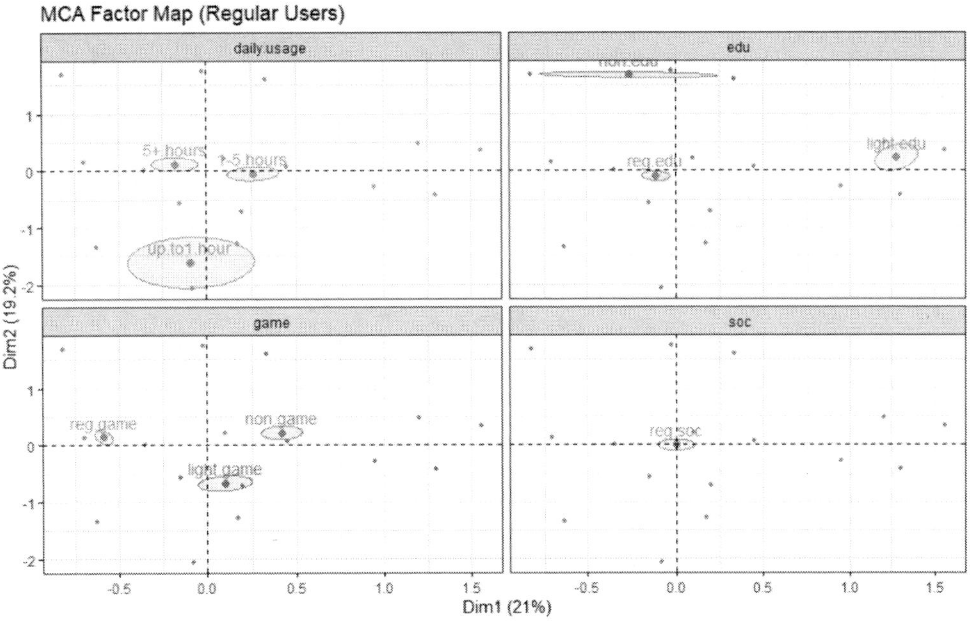

Figure 3. MCA Factor Map of active categories for Model A

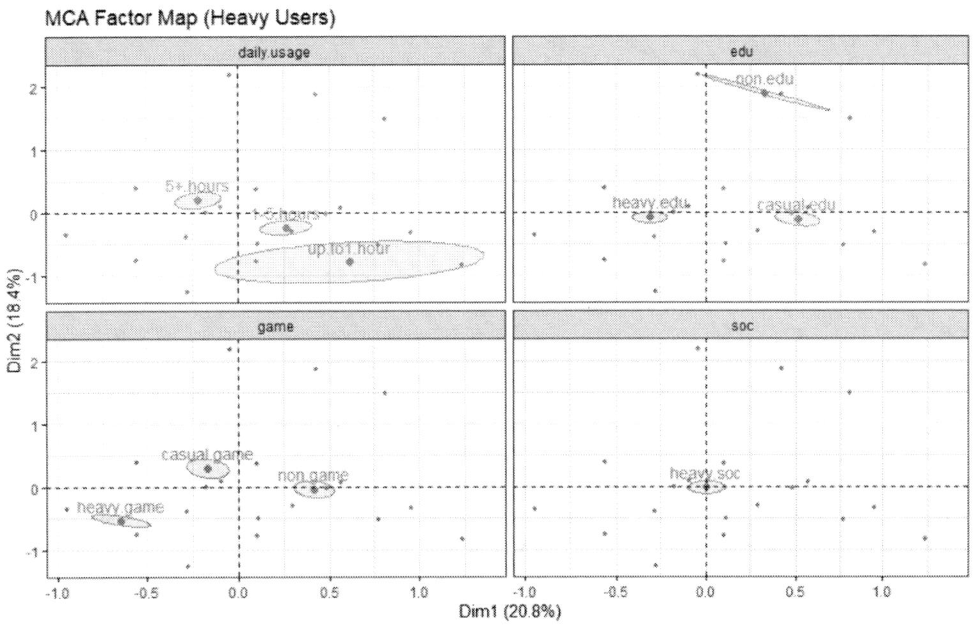

Figure 4. MCA Factor Map of active categories for Model A

The last visualization of MCA used in this study is the biplot of supplementary variables as shown in Figure 5 for Model A and Figure 6 for Model B respectively. Six variables of interests set as the supplementary variables are attitudes toward ICT (early adopter, majority, laggard), major (Computer Science or Information Technology major, other STEM major, Health Science & Medicine major, and

Social major), place of origin (Java, non-Java), socioeconomic status (SES) measured by purchasing price of mobile device (high, middle-to-low), sex (male, female), and year in college (1st year, 2nd year, 3rd year, 4th year, and 5th year or above). After several runs of MCA, many of these variables were no longer in their original scale of measurement as they were contracted into fewer and simpler categories such as place of origin and SES in order to produce a better visualization that can help with the interpretation. From the findings of the previous study, some of these variables were confirmed to have significant impact on daily time use, such as sex, SES, and (negative) attitudes toward ICT whereas the rest was found to have no significant impact whatsoever. The MCA biplots of supplementary variables for both models show that is not necessarily true. As it turned out, the first two principal dimensions clearly separate students based on their place of origin (Java vs non-Java) and major (CS/IT and Social vs other STEM and Health/Medicine) in addition to confirming the findings from the previous study for all the confirmed significant factors like sex (male vs female), SES (high vs middle-to-low), and (negative) attitudes toward ICT (laggard vs majority and early adopter). In terms of year in college, the biplot separates the older students in their 4th year or above to the younger students in their first or second year only for heavy users (Model B) while yielding no clear pattern for the regular users (Model A).

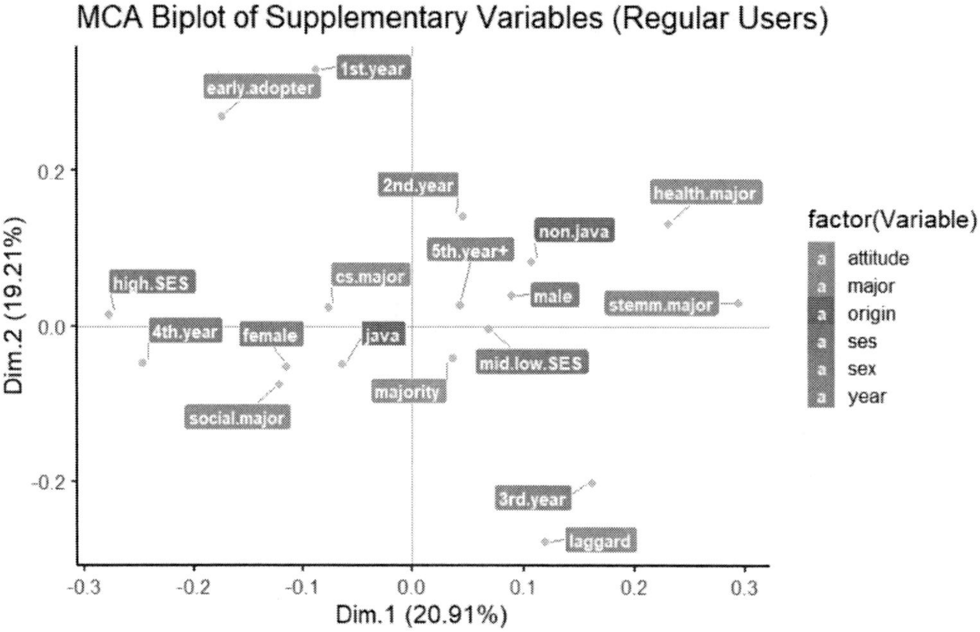

Figure 5. MCA Biplot of the first 2 principal dimensions of supplementary variables for Model A

In both models, the more negative value for the first principal dimension is associated with longer duration of daily mobile device use and so is the more positive value for the second principal dimension. The biplots indicate that students taking either social or computer science/information technology major are more likely to spend more hours on their mobile device compared to students from other STEM major and students from health sciences and medicine. Likewise, students from Java island are more likely to spend more hours on their mobile device than students from outside Java island. These two findings complement all of the confirmed findings from the previous study that females and students with high SES spend more hours on their mobile device than males and students with middle-to-low SES or that students who consider themselves as laggards spend less hours on their mobile device than students who consider themselves as part of the (early and late) majority or the early adopter.

In terms of major, it is interesting to see the difference between students from Social Science who spend more time on their mobile devices and students from science, technology, engineering and mathematics (STEM) and Medicine majors except for students from Computer Science or Information

Technology major. Apparently, IT students are the most social ones, at least digitally and not necessarily physically in real life, among students from the STEM disciplines. As for year in college, the difference is more subtle than any other variables of interest in this study. Based on the biplot of Model B, older students in their 4th year or above are more likely to spend more hours on their mobile device and to be heavy gamers than younger students in their 1st or 2nd year. However, there is no clear-cut pattern when it comes to regular use as shown in the biplot of Model A. More information is clearly needed to get a better insight on the effect of this variable to the use of mobile device and mobile apps.

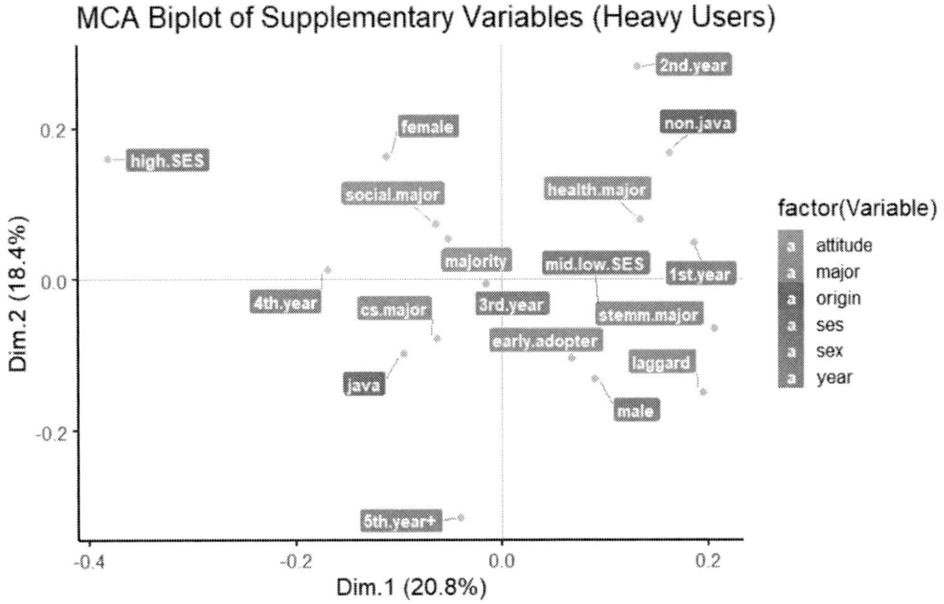

Figure 6. MCA Biplot of the first 2 principal dimensions of supplementary categories for Model B

4. Conclusions

A series of two-sample tests of proportion between each pair of three apps categories confirm the significant difference in use with social media came as the most frequently used app category where 99% use it at least once a week with 95% of them do so daily, followed by educational apps, used by 87% regularly with 68% of them do so daily, and mobile games, played regularly by 34% with a little bit than a half of them play daily. While most students use their mobile device for social media, as also confirmed in the previous study, the findings in this study show that regular and heavy use of educational apps are closely tied to the more than five hours spent on mobile device daily whereas more frequent mobile gaming may lead to more hours spent on mobile device. The MCA biplots indicate that students in social or computer science/information technology major are more likely to spend more hours on their mobile device compared to students from other STEM majors and students from health sciences and medicine major. The same is true for students from Java island who are more likely to spend more hours on their mobile device than students from outside Java island. The new findings in this study can help provide a better understanding of how Indonesian students use their mobile device that was already uncovered in the previous study by answering the remaining questions.

Just like in the previous study, despite a better insight offered by this study, it still couldn't capture a more subtle difference due to homogeneity in the sample, especially in terms of location and generation that might still be playing its part. A follow up study using better and bigger sample size from more different regions in Indonesia or a comparison with older (i.e. postgraduate) and younger (i.e. secondary school) students or better yet with overall Indonesian population might be able to overcome the limitation in this study.

References

[1] Pratama A R 2017 Exploring Personal Computing Devices Ownership Among University Students in Indonesia *Int. Conf. on Social Implications of Computers in Developing Countries* (Cham: Springer) pp 835-41
[2] Kemp S 2015 *Social, Digital & Mobile Worldwide in 2015* Retrieved from http://wearesocial.com/uk/special-reports/social-digital-mobile-worldwide-2015
[3] Pratama A R 2017 Cross-country Analysis of ICT and Education Indicators: An Exploratory Study *IOP Conf. Series: Materials Science and Engineering* vol **185** no 1 (Bristol: IOP Publishing)
[4] Pratama A R and Al-Shaikh M 2012 Relation and Growth of Internet Penetration Rate with Human Development Level from 2000 to 2010 *Communications of the IBIMA*
[5] Khalaf S and Kesiraju L 2017 US Consumers Time-Spent on Mobile Crosses 5 Hours a Day *Flurry Analytics Blog* Retrieved from http://flurrymobile.tumblr.com/post/157921590345/us-consumerstime-spent-on-mobile-crosses-5
[6] Poushter J 2016 Smartphone ownership and internet usage continues to climb in emerging economies *Pew Research Center* **22**
[7] Pratama A R 2018 Investigating Daily Mobile Device Use Among University Students in Indonesia *IOP Conf. Series: Materials Science and Engineering* vol **325** no 1 (Bristol: IOP Publishing) p 012004
[8] Fernández-López Á, Rodríguez-Fórtiz M J, Rodríguez-Almendros M L and MartínezSegura M J 2013 Mobile learning technology based on iOS devices to support students with special education needs *Computers & Education* **61** pp 77-90
[9] Hwang G J, Wu P H and Ke H R 2011 An interactive concept map approach to supporting mobile learning activities for natural science courses *Computers & Education* **57(4)**
[10] Martin F and Ertzberger J 2013 Here and now mobile learning: An experimental study on the use of mobile technology *Computers & Education* **68** pp 76-85
[11] Ally M 2009 *Mobile learning: Transforming the delivery of education and training* ed M Ally (Athabasca: Athabasca University Press)
[12] Goksu I and Atici B 2013 Need for Mobile Learning: Technologies and Opportunities *Procedia-Social and Behavioural Sciences* **103** pp 685-94

IOP Conf. Series: Materials Science and Engineering **803** (2020) 012017 doi:10.1088/1757-899X/803/1/012017

Classifying Wastes Using Random Forests, Gaussian Naïve Bayes, Support Vector Machine and Multilayer Perceptron

Ibrahim F. Hanbal[1], Jeffrey S. Ingosan[2], Neal Arden A. Oyam[3], Yafeng Hu[4]

College of Information Technology and Computer Science, University of the Cordilleras, Baguio City 2600, Philippines.

Email: [1]ibrahimhanbal@yahoo.com, [2]jsingosan@uc-bcf.edu.ph, [3]oyamneal123@gmail.com, [4] kenhuyafeng@yahoo.com.ph,

Abstract. As the population of the Philippines increases in size, so does the amount of garbage that it generates. A lot of this garbage is not properly segregated when dumped andmany citizens are not aware of the effects of unsegregated garbage which leads to them simply disposing mixed waste in dump sites. A method to help classify garbage can help address this issue. This paper presents a comparison of the efficiency of four commonly used machine learning models, Random Forests, Gaussian Naïve Bayes, Support Vector Machines and Multilayer Perceptron in classifying biodegradable and non-biodegradable waste. The data used for training and testing was collected from various sources and then augmented to increase the size of the dataset resulting in 15,000 images of biodegradable and non-biodegradable trash used for training. The four models were trained using the K-Fold cross validation technique with the dataset being split ten times. The results indicate that the four models were able to achieve a relatively high accuracy in classifying images of biodegradable and non-biodegradable trash with the Random Forests, Gaussian Naïve Bayes, Support Vector Machines and Multilayer Perceptron models achieving an accuracy of 97.49%, 81.46%, 89.51% and 96.44% respectively.

1. Introduction

Solid waste mismanagement is one of the most serious issues around the world. It can cause damage when not addressed. One way to manage solid waste is segregation. It is done by sorting different types of solid wastes. It is therefore important to identify clearly the type of wastes so that they can be disposed properly. In most developing nations, such as the Philippines, a large portion of the waste generated is composed of organic substances which means the use of incineration to dispose waste is less effective[1, 2].Properly segregating and managing garbage therefore is needed to ensure thatwaste is properly organized and handled according to the type of garbage.Many studies have been conducted on the growing amount of waste both in Asia and around the world. A study in India shows that even developing countries like China, India and Brazil, accommodating about 87% of the world human population, do not follow segregation of waste that results into large piles of mixed waste in their dump yard and landfills. The researchers conducted several experimental studies to validate the developed approach and obtained a classification success rate in the range of 85-96%[3].

Another study focused on classifying waste presented a mobile manipulator based system for automated solid waste sorting for recycling. The system is capable of automatically recognizing the recyclable materials by applying bag-of-words and K-means clustering to build a Support Vector

Machine classifier model on images obtained from an on-board thermal imaging camera. The experiment obtained a classification accuracy of 94.3% [4].

Another study presented a hybrid system where a convolutional network is trained to detect and recognize generic objects, and a Gaussian-kernel Support Vector Machine is trained from the features learned by the convolutional network. On the test set, which contains different object instances than the training set, the SVM alone yields a 43.3% error rate, the convolutional net alone yields 7.2% and an SVM on top of features produced by the convolutional net yields 5.9%[5].

One study that used machine learning to classify images of garbage used 5000 images in total in the JPG format. Each waste image is labelled as either recyclable or not for training/testing purpose, [6] and a similar study used k-means clustering and linear regression for classifying garbage. The paper focused on the issue of municipal waste prediction and evaluated a dataset containing 63,000 records [7].

Thispaper is focused on Baguio City since poor solid waste management is still a big issue within the city especially with the growing population and rising number of tourists coming to the city every year[8].The studyfocused on classifying biodegradable and non-biodegradable solid waste using multiple machine learning models and comparing the effectiveness and accuracy of these models with the use of the Scikit-learn module.

The researchers focused on four commonly used machine learning models that are available within Scikit-learn. The first model is the Random Forest Classifier. This model is created by taking a large amount of different samples of data and growing "trees" out of them. After which, if input is given to the "forest", all the trees vote on what they think the output is and the majority of the vote is considered as the possible output [9]. The second model is theGaussian Naive Bayes Classifier which is based on Bayes' theorem with an assumption of independence among predictors where this models assess that the presence of a particular feature in a class is unrelated to the presence of any other feature [10]. The third model is the Multilayer Perceptron Classifier which is a type of neural network. It has the input layer which introduces input values into the network, hidden layers which performs classification of features but hides the result and output layers which functions like the hidden layer but displays the results [11]. The fourth model is the Support Vector Machine (SVM) which is used mainly for binary classification. It finds a hyperplane that can classify new data points by seeing which side of this hyperplane it lands on even if small amount of data is used [12].

The main objectives of the study were to collect a sufficiently large dataset containing images of biodegradable waste and non-biodegradable waste. After collecting the images for the dataset, the next objective was to build and train the models with the use of Scikit-learn. After training concluded, the last objective was to evaluate the models with the use of K-Fold cross-validation.

2. Methods

The study was conducted with the use the of Scikit-learn Python module in order to preprocess, train and test multiple machine learning models onto one set of data. Scikit-learn was chosen due to its state-of-the-art implementations of many machine learning algorithms and ease of use.The module is also distributed under the simplified BSD license, encouraging its use in both academic and commercial settings [13, 14].

2.1. *Data Collection Process*

The dataset collected for the study is originally composed of 1500 images of garbage. The researchers divided the images into two classes: "biodegradable" and "non-biodegradable" with each having 750 images. These were the chosen classes as these were also used to divide the dataset in previous studies. The images were collected from several sources. First, a select number of the images were taken from the public GitHub dataset of Thung and Yang's Trashnet, which they used for garbage image classification. The dataset contained 2527 images with six classes: glass, paper, cardboard, plastic, metal, and trash [15].Around 400 of these images were collected from the six categories and used as part of the non-biodegradable class for this study. Another 300 images were created by the researchers for the biodegradable class by taking pictures of common food wastes in the local area. The remaining 800 images were taken from online sources such as Google and Bing by using a web crawler. Web crawlers are tools that navigate web pages and download content and metadata from these webpages using an algorithm as opposed to manual browsing [16].A total of 1,282 images were taken from online sources but a large number of these images were not usable for training as they contained watermarks, text, duplicates and non-garbage subjects. Because of this,they were removed from the dataset which meant only 62.40% of the images downloaded online were usable. In order to ensure that the models were not biased when trained, both classes contained an even number of 750 images.

Previous studies show that having a larger dataset to train a machine learning model generally improves the accuracy of that model. The researchers then used three parameters for the analysis of three features of the images, namely: (1) Color Histogram for colors, (2) Hu Moments for shapes and (3) Haralick Texture for textures. These parameters were used for all four machine learning models and are available within Scikit-learn.

2.2. *Data Augmentation Process*

The more data an algorithm has access to, the better the performance of that algorithm. As the original dataset was only composed of a relatively small number of images, it needed to be increased in order to properly train the machine learning models. This solution to help expand the dataset and improve the accuracy of the model is data augmentation[17].Data augmentation enhances data by deriving new data from the base data. Some simple data augmentation techniques include cropping, rotating, zooming in and flipping input images to enhance the dataset [18].

The initial 1,500 images where then augmented nine times each which resulted in 13,500 new images to be added to the dataset, with 7,500 images for biodegradable and 7,500 images for non-biodegradable. The techniques used to modify the images are: (1) increasing brightness by 100%, (2) decreasing brightness by 100%, (3) resizing the image by 50%, (4) using a grayscale filter, (5) using a sepia filter, (6) inverting the colors of the image, (7) flipping the image horizontally, (8) flipping the image vertically, and (9) rotating the image 90 degrees clockwise. The modified images were then combined with the original dataset to create the training and testing dataset used for the models.

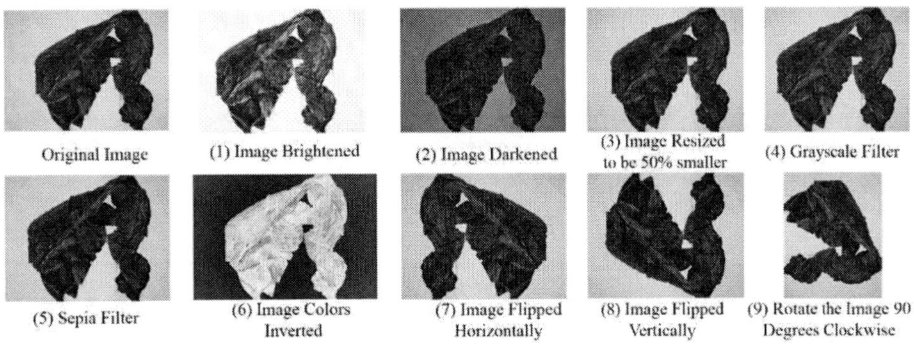

Figure 1. Sample of an original image of "biodegradable" trash and the nine modified versions of the image used for training and testing.

2.3. *Training and Testing*

The four models were trained and tested using the 15,000 images of biodegradable and non-biodegradable trash. The method for the training and testing is the K-Folds cross-validator which splits the dataset into k consecutive folds [19]. For this study, the data was split 10 times for the four models. The data was also split with 90% being part of the training set and 10% being part of the testing set. From the total images in the dataset, 13,500 images were used for training and 1,500 images were used for testing in each split. The mean results of the training and testing were then displayed for evaluation.

3. Findings

The images where labelled according to the type of garbage and uploaded for training for the four models. The images were scanned and features of the images were extracted in the process to be used for training and testing the models. After training the 15,000 images of biodegradable and non-biodegradable trash using the K-Folds cross-validator of Scikit-learn with 10 splits, the results of the training are displayed below:

Table 1. Mean Accuracy and Percentage Accuracy of the Four Models

Machine Learning Models K-Folds cross-validator accuracy		
Model	Accuracy (Mean)	Accuracy (Percentage)
Random Forest Classifier	0.974949	97.49%
Gaussian Naïve Bayes	0.814628	81.46%
Support Vector Machine	0.895151	89.51%
Multi-Layer Perceptron	0.964429	96.44%

The results of the training and testing using the K-Folds cross-validator show high accuracy values for all of the four models with the Random Forest Classifier having the highest accuracy and the Gaussian Naïve Bayes having the lowest accuracy for classification.

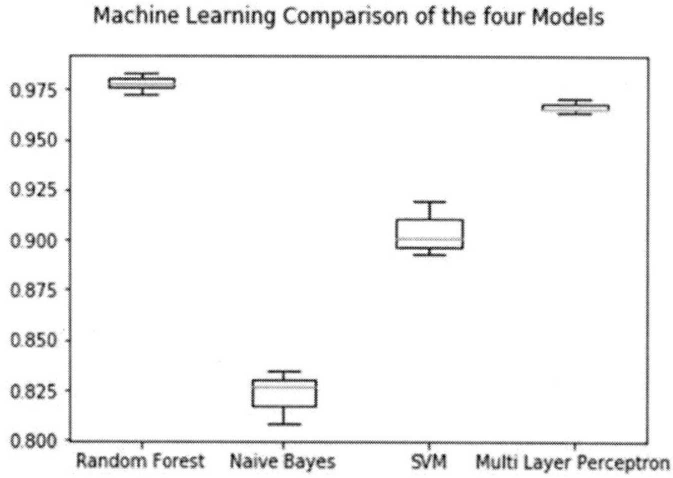

Figure 2. Boxplot showing the distribution of data for the four models.

4. Discussions

The results of the K-Folds cross-validator show that the Random Forest Classifier was able to achieve the highest accuracy at 97.49% with the Multi-Layer Perceptron (MLP) neural network achieving a similarly high accuracy with 96.44%. There are several possible reasons for the high accuracy of the Random Forest Classifier. The number of trees used for the Random Forest Classifier was 100 trees, this allowed the classifier to better take into account the features of the different types of trash for both the biodegradable and non-biodegradable classes and spread these features out among the many trees in the classifier. Another factor is the images themselves with most of them being edited or modified versions of the original 1,500 images. This creates a more diverse set of tree structures known as bootstrap aggregating [9]. The MLP neural network with a similarly high accuracy also benefitted from the augmented data set. As neural networks perform better with more data, the MLP was able to achieve a high accuracy with the large dataset of images[20].To further improve the accuracy of these models, more data can be collected and augmented to better cover a wider variety of different types of trash.

The Naïve Bayes classifier and Support Vector Machine (SVM) had lower results for their accuracy at 81.46% and 89.51% respectively. The SVMmay have had some difficulty creating an optimal hyperplane for the data as many of the images contain a lot of noise in the background. Additionally, by augmenting the datasetto increase the total number of images, this has affected the accuracy of the SVM model as the complexity of the model scales with the size of the dataset[21].One possible reason for the Naïve Bayes classifier having the lowest accuracy may be that classifying biodegradable and non-biodegradable trash relies on the features of the trash as both these classes have very distinct differences in the images of the datasets [10].

Table 2. Comparison of the accuracy of the Four Models with previous studies

Machine Learning Models accuracy comparison			
Model	Skin Detection [22]	Potential Distribution Modelling– Average over 35 datasets[23]	**Waste Classification**
Random Forest Classifier	88%	91.43%	**97.49%**
Gaussian Naïve Bayes	77%	84.14%	**81.46%**
Support Vector Machine	78%	87.23%	**89.51%**
Multi-Layer Perceptron	82%	N/A	**96.44%**

As seen in Table 2, the accuracy of the models in classifying wastes is higher, in the case of the Random Forest Classifier and MLP or quite close, in the case of the Naïve Bayes classifier and SVM classifier.

5. Conclusions

The results of the cross-validator show that the Random Forest Classifier and MLP neural network are able to classify images of trash into the biodegradable and non-biodegradable categories with a high accuracy. While the Naïve Bayes classifier and SVM classifier did not reach the same level of accuracy, the two models still showcased a relatively good accuracy. Despite the images of trash having different features, shapes and textures, along with the edited images having different sizes and colours, the augmented dataset was sufficient to train the four models to correctly classify biodegradable and non-biodegradable with a high degree of reliability.

Although the initial set of images only numbered at 750 per class, augmenting the data by editing the images manually and adding them to the original dataset proved to be fruitful as they allowed the four models to achieve a high degree of accuracy after being trained by the expanded dataset containing 15,000 objects of unedited and modified pictures of biodegradable and non-biodegradable trash.

References

[1] Alvim-Ferraz M C M and Afonso S A V 2005 Incineration of healthcare wastes: management of atmospheric emissions through waste segregation *Waste Management* vol **25** Issue 6 2005 pp 638-48

[2] Gonzales L B 2016 Urban Sprawl: Extent and Environmental Impact in Baguio City Philippines *Spatium* **1** pp 7-14 DOI:10.2298/SPAT1636007G

[3] Gundupalli S P, Hait S and Thakur A 2017 Multi-material classification of dry recyclables from municipal solid waste based on thermal imaging *Waste Management* **70** DOI:10.1016/j.wasman.2017.09.019

[4] Thakur A, Paulraj S G and Hair S 2016 Automated Municipal Solid Waste Sorting for recycling using a mobile manipulatorin *ASME 2016 Int. Design Engineering Technical Conf. (IDETC)*

At Charlotte, North Carolina, USA Volume **5A** DOI:10.1115/DETC2016-59842

[5] Huang F and LeCun Y 2006 Large-scale Learning with SVM and Convolutional Nets for Generic Object Categorization *in 2006 IEEE Computer Society Conf. on Computer Vision and Pattern Recognition (CVPR'06) New York, USA 2006* pp 284-91

[6] Chu Y, Huang C, Xie X, Tan B, Kamal S and Xiong X 2018 Multilayer Hybrid Deep-Learning Method for Waste Classification and Recycling*in Computational Intelligence and Neuroscience* vol **2018** Article ID:5060857

[7] Livani E, Nguyen R, Denzinger J, Ruhe G and Banack S 2013 A Hybrid Machine Learning Method and Its Application in Municipal Waste Prediction *Advances in Data Mining. Applications and Theoretical Aspects ICDM 2013* Lecture Notes in Computer Science vol **7987** ed P Perner (Berlin, Heidelberg: Springer)

[8] Atienza V 2011 Review of the Waste Management System in the Philippines: Initiatives to Promote Waste Segregation and Recycling through Good Governance*in Economic Integration and Recycling in Asia: An Interim Report, Chosakenkyu Hokokusho, Institute of Developing Economies 2011*

[9] Oshiro T M and Perez P S and Baranauskas J A 2012 How Many Trees in a Random Forest? *Machine Learning and Data Mining in Pattern Recognition MLDM 2012* Lecture Notes in Computer Science vol **7376** ed P Perner (Berlin, Heidelberg: Springer)

[10] Zhang H 2004 The Optimality of Naive Bayes *Proc. of the Seventeenth Int. Florida Artificial Intelligence Research Society Conf. FLAIRS 2004* **2**

[11] Ruck D, Rogers S and Kabriksy M 1989 Feature Selection Using a Multilayer Perceptron *in the J. of Neural Network Computing* **2**

[12] Smola A and Scholkopf B 2004 A tutorial on support vector regression *Statistics and Computing (2004)* **14** 199 https://doi.org/10.1023/B:STCO.0000035301.49549.88

[13] Pedregosa et al 2011 Scikit-learn: Machine Learning in Python *JMLR* **12** pp 2825-30

[14] Buitinck et al 2013 API design for machine learning software: experiences from the scikit-learn project *European Conf. on Machine Learning and Principles and Practices of Knowledge Discovery in Databases (2013)* arXiv:1309.0238

[15] Thung G and Mingxiang Y 2016 Classification of Trash for Recyclability Status *arXiv Preprint* Retrieved from: https://github.com/garythung/trashnet

[16] Shrivastava V 2018 A methodical study of web crawler in *J. of Engineering Research and Application* ISSN: 2248-9622 vol **8** Issue 11 (Part -I) Nov 2018 pp 1-8

[17] Perez L and Wang J 2017 The Effectiveness of Data Augmentation in Image Classification using Deep Learning *arXiv:1712.04621* [cs.CV]

[18] Mikołajczyk A and Grochowski M 2018 Data augmentation for improving deep learning in image classification problem*in the 2018 Int. Interdisciplinary PhD Workshop (IIPhDW), Swinoujście 2018* pp 117-22

[19] Raschaka S 2018 Model Evaluation, Model Selection, and Algorithm Selection in Machine Learning *arXiv:1811.12808* [cs.LG]

[20] Pal S K and Mitra S 1992 Multilayer Perceptron, Fuzzy Sets, and Classification *IEEE Transactions on Neural Networks* vol **3** No 5 September

[21] Burges C J C 1998 A tutorial on support vector machines for pattern recognition *Data Mining and Knowledge Discovery* **2** pp 121–67 (1998)

[22] Khan R, Hanbury A, and Stoettinger J 2010 Skin detection: A random forest approach *2010 IEEE Int. Conf. on Image Processing* pp 4613-6

[23] Lorena A C, Jacintho L F, Siqueira M F, De Giovanni R, Lohmann L G, De Carvalho A C and Yamamoto M 2011 Comparing machine learning classifiers in potential distribution modelling *Expert Systems with Applications* **38(5)** pp 5268-75

A Hybrid Approach to Representing Chessboard using Bitboard and Compact Chessboard Representation

Sigit Kariagil Bimonugroho, Nur Ulfa Maulidevi

School of Electrical Engineering and Informatics, Institut Teknologi Bandung, Indonesia

E-mail: `sigitkariagil@gmail.com`, `ulfa@informatika.org`

Abstract. A chess engine is a computer program that can perform position analysis in chess and take which are the most optimal move. The main process in a chess engine can be divided into three: move generation, searching, and evaluation. Each of these processes are interrelated and require efficient board representation methods. Some methods of representation from previous research are efficient in some processes but less efficient in others. This study proposes an efficient representation method with a hybrid approach, which combines Compact Chessboard Representation (C.C.R) and bitboard representation. The C.C.R method is used in move generation and searching processes to reach conciseness, and bitboard representation method is used in the evaluation process to make it faster. The result of the movement validity test for the hybrid representation method shows the move generation is valid, and a hybrid-based chess engine also can do searching and evaluation faster than a pure C.C.R based chess engine. **Keywords**: data representation, chessboard, hybrid, compact chessboard representation, bitboard

1. Introduction

Since the mid-1960s, researchers in computer science have called chess as a drosophila of artificial intelligence (AI) [1]. Drosophila is a genus of small fruit flies used by Thomas Hunt Morgan to study the basic mechanism of inheritance in biological science. The phrase drosophila of AI is used to show that chess, like fruit flies, is an experimental technology that is easily accessible, general, and relatively simple but can be used productively to produce valid knowledge about other more complex systems.

There are three main processes in chess engine: move generation, searching, and evaluation. Move generation is the process of generating all candidate move. Searching is the process of finding optimal move that can be taken based on evaluation score. Evaluation is the process of calculating score of a position. In each of these processes requires data representation to store position information.

An efficient chessboard representation is needed to build a good chess engine. As Reul [2] notes, the benefit of such data structures "lies in their simplicity and ease, and the possibility to implement specific and complex chess knowledge in a more compact and transparent way".

At present there are two basic representation methods used in chess engines, representation with arrays and representation with bitboards. Array representations were widely used at the beginning of computer chess research. The array representation was first mentioned by

Shannon on 1950 [3]. Then, along with the development of hardware, especially 64-bit CPUs, a bitboard representation method (a set of bit-vectors or bitmaps) was discovered. The bitboard representation method has advantages in terms of efficiency compared to the array representation method [4].

The idea of bitboard representation was developed by Atkin and Slate on 1988 [5]. Atkin and Slate's research describes an approach using 12 64-bit integers to represent 6 types of black chess pieces and 6 types of white chess pieces. Research by Adelson-Velskii et al. on 1970 [6] developed the same idea separately at the same time as Atkin and Slate's research. The advantage of this method is that the evaluation process can be carried out quickly using bit operations [4]. This approach is widely used in chess engines until now.

Some research attempts to improve the performance of bitboard representations. Research R. M. Hyatt on 1999 [7] succeeded in increasing the performance of bitboard representation at the move generation stage by the bitmap rotation method. However, this method is not good when used in the search stage because it is less concise [8]. Therefore, Vuckovic on 2012 [8] proposed a new method of representation that is more concise, namely Compact Chessboard Representation (C.C.R). The C.C.R method can provide good performance at the search stage because it is concise. However, C.C.R is not fast enough when used at the evaluation stage. Therefore, we proposed a hybrid approach that combine C.C.R and bitboard representation to solve the inefficiency problem in chessboard representation.

2. Proposed System

The development of data representation with a hybrid approach in this research was carried out in conjunction with the development of a chess engine. The design of the developed chess engine consists of move generation, searching, and evaluation. C.C.R representation is used in the move generation and searching stages. At the evaluation stage the conversion process is carried out from C.C.R to Bitboard. The general design of a chess engine is shown in Figure 1.

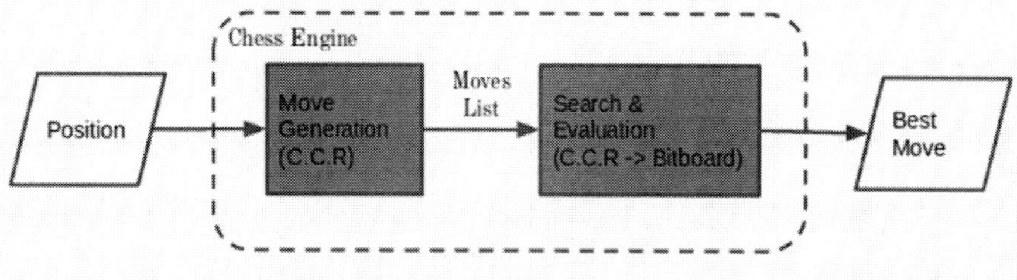

Figure 1. Chess engine general design

2.1. Move Generation

In this research, move generation process uses the C.C.R representation to store the position that the move will be raised for. C.C.R representation was chosen because it is concise. Position representation using the C.C.R representation only requires 32 bytes of memory while a representation using bitboard requires a minimum of 96 bytes of memory. Input from the move generation process is the position of the pieces represented in C.C.R. Output of the move generation process is a list of movement candidates. Illustration of input and output of the move generation process is shown in Figure 2.

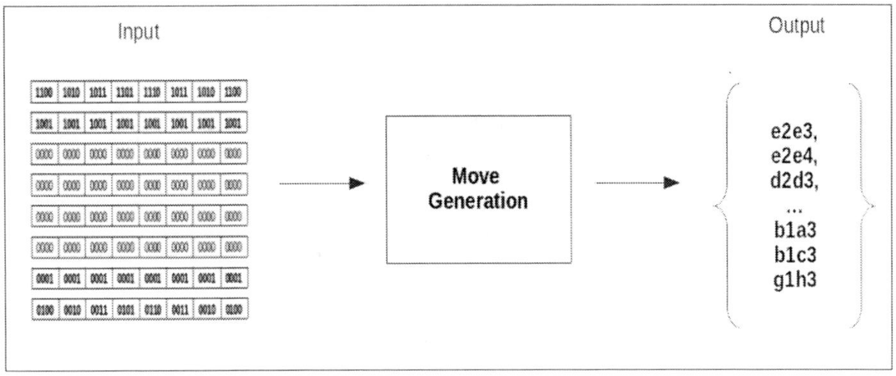

Figure 2. Input and output of move generation process

2.2. Search

Searching process is carried out by tracing the movement candidates from the move generation process. Input from the searching process is a movement candidate, and the output is a chosen move. Illustration of input and output searching process is shown in Figure 3.

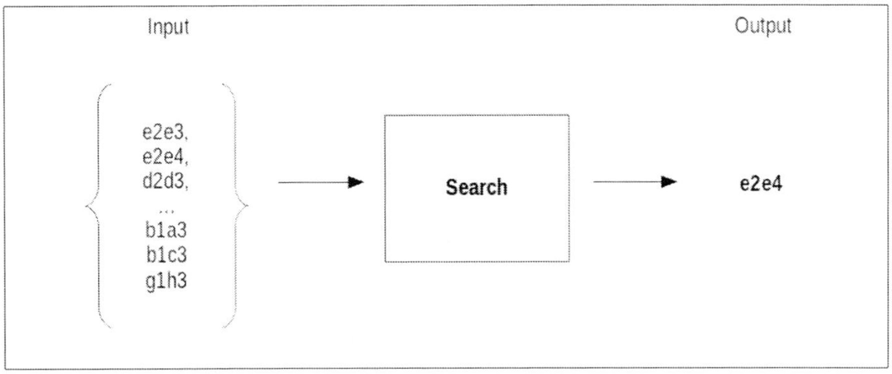

Figure 3. Input and output of searching process

Searching process in this study uses the Alpha-Beta algorithm. Alpha-Beta method is a simpler and more efficient than the Minimax algorithm. This is because the Alpha-Beta algorithm trims sub-trees that have no effect on the final grade [9, 10].

The results of a comparison of the search process for the sample position in Figure 4 using the Alpha-Beta and Minimax algorithms are shown in Table 1. Comparison results show that the Alpha-Beta and Minimax algorithms provide the same output. However, number of nodes traversed by the Alpha-Beta algorithm is less than the Minimax algorithm. From these results, it can be concluded that the Alpha-Beta algorithm is more efficient than the Minimax algorithm when used in the searching process in a chess engine.

2.3. Evaluation

Evaluation process is done to calculate the heuristic value of a chess position at the terminal node of the search process. Heuristic value used in this study is the material value and position value. Material value is calculated based on the number of pieces left and the weight of each piece. The position value is calculated based on the position held by the piece. For example, a pawn in the middle position will be given a higher value than a pawn in the back. Input from

IOP Conf. Series: Materials Science and Engineering **803** (2020) 012018 doi:10.1088/1757-899X/803/1/012018

Figure 4. Chess position

Table 1. A comparison table for searching algorithm

	Alpha-Beta	**Minimax**
Best Move	g3g6	g3g6
Node visited	39511	113843

the evaluation process is a position in the representation of C.C.R. Output of this process is the value in integer. Illustration of input and output of the evaluation process is shown in Figure 5.

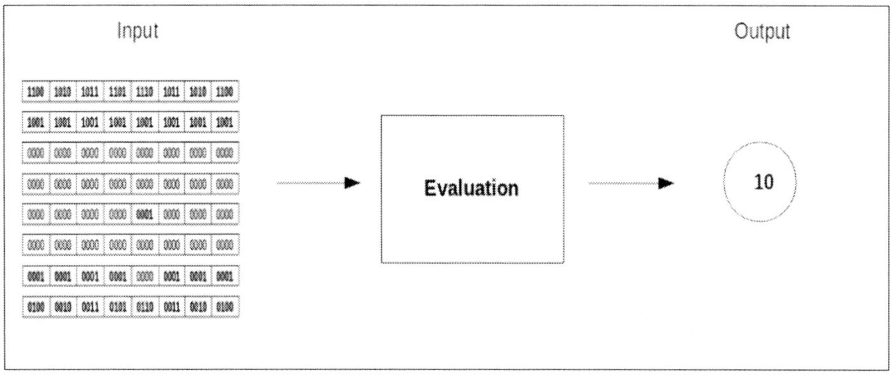

Figure 5. Input and output of evaluation process

Position evaluation process can be done by two methods. Process of evaluating the position of the first method is done by calculating the heuristic value directly from the representation of C.C.R. Second method evaluation process is done by calculating the heuristic value from the bitboard representation. Both methods are applied in this study. This is done to compare the performance of hybrid representation-based chess systems with C.C.R-based chess systems only. Position in the bitboard representation is obtained by performing the C.C.R position conversion process.

3. Result

Evaluation process of the proposed method was conducted by two methods, search speed test and chess match test. Speed test is used to measure the speed of searching process. Chess match is used to find out the effect of the representation developed on the overall engine performance.

3.1. Search Speed Test
Search speed test is carried out with three scenarios. The three scenarios represent the condition of the board at the opening, middle, and end game. These three scenarios are used to test the consistency of the search process speed from the beginning to the end of the game. The speed test results are shown in Table 2. Speed are shown in nodes per second. The speed in nodes per second is measured by counting the number of nodes visited by the machine during the search process divided by the length of the processing time.

Table 2. Search speed test in node per second

Position	**Bitboard**	**C.C.R**	**Hybrid**
Opening	521225	618475	662625
Middle	350175	601750	659525
End Game	1029150	799075	788800

3.2. Matches against other computers
The chess match test is carried out by playing a chess match against the bitboard based engine and the C.C.R based engine. The results of the match test are shown in Table 3.

Table 3. Matches result

Opponent	**Win**	**Draw**	**Lose**
vs Bitboard	3	1	6
vs C.C.R	6	2	2

4. Conclusion
The utilization of the representation method that combines the C.C.R and bitboard methods has succeeded in improving chess engine performance in terms of speed and memory usage in the opening and middle game. C.C.R representation method is best used in the process of move generation and searching, while the bitboard representation is good for the evaluation process. However, the hybrid representation method has not been able to provide optimal results in terms of speed in the end game compared to bitboard. This is because at the end of the game many empty squares are still checked by the hybrid representation method. Therefore, future research can focus on increasing the speed of the search in chess engine in the end game. This is because the engine based on hybrid representation method already has good performance in the opening and middle game. However, it is still not good in the end game. One approach that can be proposed is to use representations that dynamically adjusted. For example, the hybrid representation is used during the opening and middle game. Then, the bitboard representation is used fully when entering the end game.

Reference
[1] Ensmenger N 2012 *Social studies of science* **42** 5–30
[2] Reul F 2009 *New architectures in computer chess* Ph.D. thesis series: TiCC Ph.D. Series Volume: 6
[3] Shannon C E 1950 *The London, Edinburgh, and Dublin Philosophical Magazine and Journal of Science* **41** 256–275
[4] Hyatt R 2004 *University of Alabama at Birmingham*

[5] Atkin L and Slate D 1988 Computer chess compendium ed Levy D (Berlin, Heidelberg: Springer-Verlag) chap Chess 4.5-The Northwestern University Chess Program, pp 80–103 ISBN 0-387-91331-9 URL `http://dl.acm.org/citation.cfm?id=61701.67010`

[6] Adel'son-Vel'skii G M, Arlazarov V L, Bitman A, Zhivotovskii A and Uskov A 1970 *Russian Mathematical Surveys* **25** 221

[7] Hyatt R M 1999 *ICGA Journal* **22** 213–222

[8] Vučković V 2012 *Yugoslav Journal of Operations Research* **22** 265–284

[9] Campbell M S and Marsland T 1983 *Artificial Intelligence* **20** 347 – 367 ISSN 0004-3702 URL `http://www.sciencedirect.com/science/article/pii/0004370283900012`

[10] Pearl J 1982 *Communications of the ACM* **25** 559–564

IOP Conf. Series: Materials Science and Engineering **803** (2020) 012019 doi:10.1088/1757-899X/803/1/012019

Predicting Macronutrient of Baby Food using Near-infrared Spectroscopy and Deep Learning Approach

M N Aulia[1], M L Khodra[2], A P Koesoema[3]

Institut Teknologi Bandung, Jalan Ganesha, Bandung, Indonesia

Email: [1]masyithahaulia@students.itb.ac.id, [2]masayu@stei.itb.ac.id, [3]allya.paramita@stei.itb.ac.id

Abstract. —In Indonesia, malnutrition and overnutrition can still be found especially among children and toddlers who are supposed to get adequate nutrition. The cause of malnutrition is inadequate nutrition intake, both in terms of the quantity and quality of the food. Therefore, mothers need to know whether the food consumed by children has reached the specified nutrition intake recommendation. This research develops a system to predict macronutrient content in baby food using Near-infrared Spectroscopy (NIRS) to obtain the spectral profile of the food and Deep Learning approach such as Deep-belief network (DBN) and Convolutional Neural Network (CNN) to build the prediction model. We used instant-porridge scan data from SCiO to build the model. CNN managed to give the best performance with error for carbohydrate, protein and fat 11.70%, 26.14% and 28.72% respectively.

1. Introduction

Food intake is one of the keys to have healthy life [1]. Food intake will greatly affect human health conditions especially among children and toddlers who are still growing. But in reality, malnutrition and overnutrition can still be found especially among children and toddlers who are supposed to get adequate nutrition. The cause of malnutrition is inadequate nutrition intake, both in terms of the quantity and quality of the food. According to WHO, around 45% children under 5 in the world, suffer and die from malnutrition [2]. In Indonesia, malnutrition can still be easily found. In January 2018, dozens of children in Asmat regency die from malnutrition and hundreds of them need intensive care [3].

On the other hand, it is difficult to determine whether the food consumed by the children has reached the specified nutrition intake recommendation. The ministry of health in Indonesia has issued a nutrition adequacy rate which provides the daily minimum nutrition intake for both children and adults. However, this method is complicated and time consuming as the mothers need to remember the food intake and manually calculate the nutrition intake from the food. Moreover, another concern is the quality of the food consumed by children cannot be calculated using this approach.

Predicting nutrition intake from food using technology has become a hot issue these days in food industry [4]. These researches include camera-based techniques and sensor-based techniques [1]. One of the most popular techniques is NIRS due to its ability to analyze chemical compounds in a food within a short time and its non-destructive way [5]. NIRS analyzes the spectrum of absorption or reflectance of electromagnetic waves with wavelength ranging from 700 to 2500 nm where most of chemical and biochemical compounds have unique absorption bands so that it can be used to detect macronutrient such as carbohydrate, protein and fat [1].

Figure 1. The SCiO Scanner

One of the portable NIR scanner to be easily used in daily needs is SCiO due to its handy size. Although SCiO has narrower wavelength range, but it is still suitable for daily needs [1]. SCiO can only analyze wavelength ranging from 740 to 1040 nm. They also provide mobile application so people can use SCiO to scan their food and get information about nutrition in the food using their application.

SCiO works by scanning the food to get the reflected spectrum. People can place the sample of food in a steady surface and hold the scanner about 5-10 mm above the food. If the food is too small and dry, they can place the sample within the solid sample holder for more accurate scan. SCiO will start analyzing the food when the button on the SCiO or scan button in the mobile application is pressed. The reflected spectrum can be found in the web application. These reflected spectra are different for various foods due to different absorption levels [1].

In this paper, we use SCiO as the scanner to build a macronutrient prediction system for baby food. We aim to develop the model using CNN and DBN. We use instant-porridge as the data. The instant-porridge is cooked firstly and then we use SCiO to scan the food. We use 30 instant-porridges in the market as they already provide nutrition facts printed on the package. These nutrition facts are the labels for our predicting system. We only use carbohydrate, protein and fat from the nutrition facts to train our prediction model and correlate the spectra features to their corresponding carbohydrate, protein and fat contents. Finally, we evaluate this prediction system error using RMSE and error.

2. Related Works

There are already some researches in the past that utilize NIRS and machine learning to both classify and regress.

One of the researches use SCiO to predict energy and carbohydrate content from drinks. Thong dkk. (2017) implement SVR and PLS then compared the result [1]. SVR with RBF Kernel achieves the best result compared to PLS.Khan dkk. (2018) use NIRS to predict wind power using DBN [8]. Yang dkk. (2017) also use DBN to detect counterfeit drugs [12]. Li dkk. (2017) also use DBN to predict moisture, protein and ash content from wheat [13]. Lengkey dkk. (2013) implement regression for Jarak Pagar beans to detect the level of water and fat using PLS [14]. There is also a research that implements CNN conducted by Acquarelli dkk. (2017) to classify various foods and drinks dataset [15]. Besides Thong dkk. (2017), other researches use different NIR Scanners, usually NIR scanners developed to be used by laboratories and factories which also means have wider wavelength spectrum. From those previous works, we conclude that we can build a prediction system using DBN and CNN to predict the carbohydrate, protein and fat content from food, in this case, baby food.

3. System Design

3.1. Dataset Preparation
The dataset consists of 450 instances from 30 instant-porridges. An instance is a scan from a bowl of porridge. We also observed different levels of consistency by pouring more or few waters to the porridge. We use instant-porridges for children under 3 and mostly for toddlers above 6 months as they already need complementary foods beside consuming breastmilk.

Macronutrient value from nutrition facts on the package are transformed into the percentages of its content divided by the total weight of the porridge (the weight of water and porridge). The percentages are used as labels for training phase.Table I shows the example of three instances of data from SCiO. The dimension of raw data is 330.

Table 1. Three instances of Data from SCiO

Spektrum 740	Spektrum 741	Spektrum 742	...	Spektrum 1068	Spektrum 1069	Spektrum 1070
0.8533703	0.85370379	0.85422415		0.71406316	0.71391301	0.71282065
0.7645731	0.76521258	0.76593314		0.63705830	0.63694547	0.63598849
0.8722648	0.87259958	0.87316037		0.73208251	0.73193435	0.73081943

3.2. Regression models
We use CNN and DBN as some of deep learning methods to achieve better results compared to conventional methods such as Linear Regression and PLS. By using deep learning approach, the model can learn and extract more complex information and features from the data [6].

DBN is one of semi-supervised learning techniques as it combines supervised and unsupervised learning. DBN consists of two main processes which are pre-training using RBM and fine-tuning using Backpropagation. In pre-training phase, random weights and biases will be trained to construct the input data [8]. This phase does not need any labels so it called unsupervised technique [7]. After some iterations, the trained weights and biases will be used in backpropagation and fine-tuned using labels.

CNN is a variation of Multilayer Perceptron. There are two additional layers in CNN namely convolutional layer and pooling layer. These layers extracts feature from the data to make more representative feature that can help model to learn and build better model [9]. The feature then will be trained in fully-connected layer which consists one or more hidden layers. In this research, we use 1D CNN instead of 2D CNN. In some cases, CNN can lead to overfitting. In order to prevent overfitting, we also perform dropout. By using dropout, for each training stage, some nodes are either being dropped out or remained in the network using a probability value [10].

Moreover, we also perform PCA before DBN and CNN for some scenarios. We would like to observe the effect of PCA to the performance of the system. PCA can be used to reduce the dimension of the data while also removes noise from the data [1]. PCA can also emphasize strong patterns in the spectra to create better model [11]. We only use PCA to extract features from data and we do not perform any other pre-processing techniques in this research.

3.3. Evaluation
The output of our prediction system is the percentages of macronutrient in a bowl of cooked porridge. However, by using mathematics calculation, we can also infer the needed quantity of the food to reached nutrition intake recommendation for toddlers and childrenfrom those percentages.

$$RMSE = \sqrt{\frac{\sum_n \left(y_{pred} - y_{true}\right)^2}{n}} \quad (1)$$

Trained models by CNN and DBN are compared with each other. The evaluation metrics we used are RMSE.

IOP Conf. Series: Materials Science and Engineering **803** (2020) 012019 doi:10.1088/1757-899X/803/1/012019

4. Experiments

All instant-porridges are cooked first by using warm water. We use 5 levels of water to capture the effect of water levels in spectral data by reducing about 20 ml water from the ideal water ratio that we get from the package. In this research, we assume that the nutrition facts are valid. After scanning the porridges, we then download the raw spectral data from SCiO Lab web application. In this research, we only try to predict carbohydrate, protein and fat as some of the important nutrition for toddlers beside micronutrient.

4.1. Splitting Dataset

We use 360 data for training and 80 data for testing. In this research, we also use k-fold cross validation with k=5 to evaluate the best models. The mean percentages of carbohydrate, protein and fat in training data are 0.1685, 0.0310 and 0.0191 and for testing data 0.1692, 0.0306 and 0.0195 respectively. The carbohydrate percentages range from 0.1 to 0.33, the protein percentages range from 0.016 to 0.075 and fat percentages range from 0.007 to 0.05.

4.2. Models and Parameters

In order to achieve the best model, we observe some scenarios and evaluate the result. We observe DBN and CNN architecture, range values of principal components in PCA and some values of dropout.

4.3. Architecture Observations

In this observation, we evaluate the performance of CNN and DBN. Both CNN and DBN use Adam as optimizer and the number of epochs is 500.

Figure 2 presents the result of this observations with validation data. CNN C achieves best result compared to DBN.

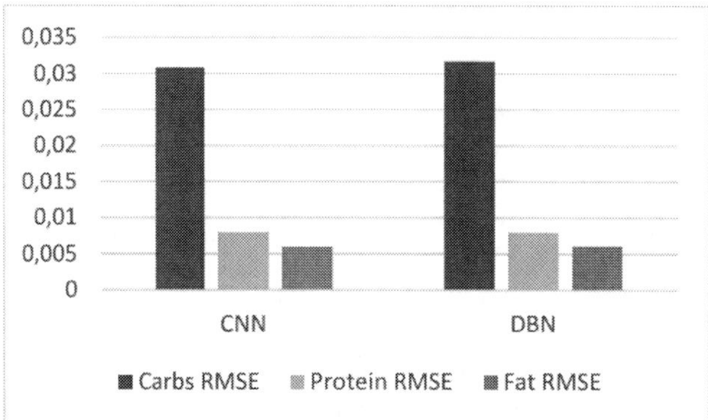

Figure 2. Macronutrient RMSE in various architectures to validation data

CNN model consists of three blocks where a block contains two convolutional layers and one max-pooling layer. The first block uses 16 3x1 filter, the second uses 32 3x1 filter and the last uses 64 3x1 filters for convolutional layer and 2x1 filter for pooling layer. All convolutional and pooling layers use 2 strides. DBN uses 4 RBM. CNN yields better result compared to DBN due to its convolutional layer that can extract feature from the input data. While processing signal data, it is important to extract feature from the input to make more representative input.

4.4. PCA Observations

We evaluate number of principal components ranging from 40 to 60 to the best model from CNN and DBN.

Figure 3 and Figure 4 reports the result of this observation with validation data. PCA successfully improve the performance for both models. The best principal components value for CNN is 43 and 44 for DBN. CNNstill achieves better result compared to DBN.

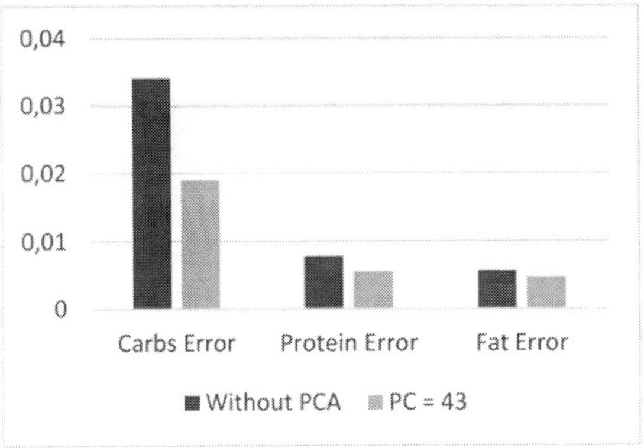

Figure 3. RMSE prediction with and without PCA of CNN to validation data

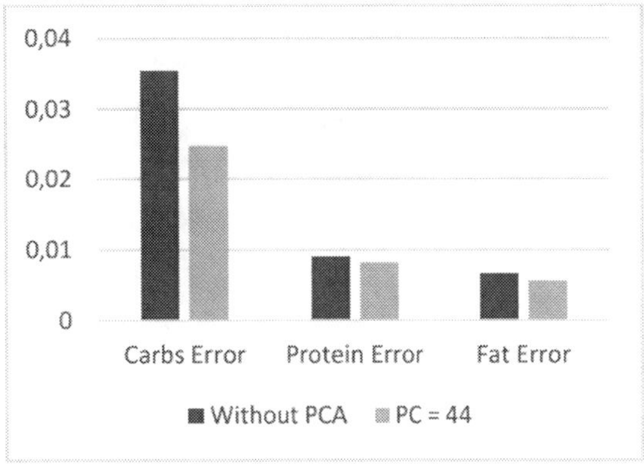

Figure 4. RMSE prediction with and without PCA of DBN to validation data

4.5. Dropout Observations

PCA successfully improves the performance of CNN but leads to overfit. In order to prevent overfitting, we try to use dropout with the observed values are 0.1 and 0.2.

Fig. 5 reports the result after using dropout after each convolutional layer and fully-connected layer to validation data. Unfortunately, dropout does not improve the result. Dropout can decrease the training performance but does not improve the validation performance. We suspect that dropout can achieve better performance in larger neural network as in this research, the neural network is quite simple and by adding dropout will make the network smaller and leads to worse performance. We conclude that another way to improve the result is by adding more training data.

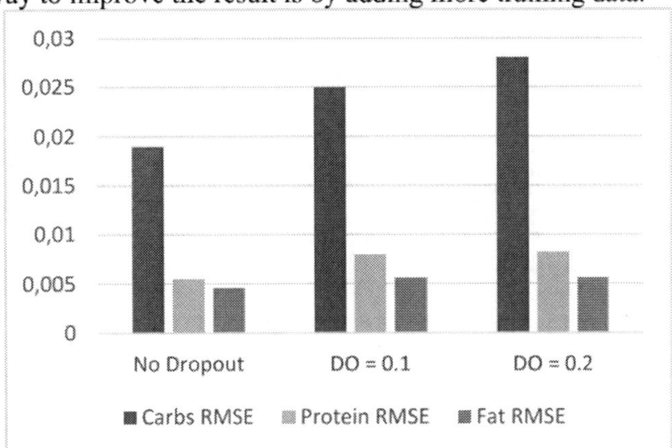

Figure 5. CNN RMSE Prediction with and without Dropout to validation data

5. Discussions and Future Work

We present the use of NIRS and deep learning specifically DBN and CNN to build a macronutrient prediction system of baby food. The experimental results show that CNN achieves best resultwith prediction errors of carbohydrate, protein and fat achieved by CNN and PCAare 11.70%, 26.14% and 28.72% respectively.This experimental result also means that the correlation between spectral input and macronutrient values is non-linear.However, in this research we can only able to obtain the percentage of macronutrients in the food. In order to achieve more specific and exact result, user needs to provide the weights of the porridge as the input. We suggest the use of camera combines with SCiO as the camera can measure the volume of the product so the system can give exact and specific result of the macronutrient content in the food.

References

[1] Thong Y J, Nguyen T, Zhang Q, Karunanithi M and Yu L 2017 Predicting food nutrition facts using pocket-size near-infrared sensor *2017 39th Annual Int. Conf. of the IEEE Engineering in Medicine and Biology Society (EMBC)* pp 742-5
[2] WHO 2017 *World Health Organization* Accessed April 4th 2018 http://www.who.int/mediacentre/factsheets/malnutrition/en/
[3] BBC 2018 *BBC Indonesia* Accessed April 4th 2018 http://www.bbc.com/indonesia/indonesia-42872190
[4] Juan X L and Ying Y B 2009 Use of near-infrared spectroscopy and least-squares support vector machine to determine quality change of tomato juice *J. of Zhejiang University SCIENCE B* **10** (6) pp 465-71
[5] Hailong W, Peng J, Xie C, Bao Y and He Y 2015 Fruit Quality Evaluation Using Spectroscopy Technology: A Review *Sensors* **15** (5) pp 11889-927
[6] Deng L and Yu D 2014 Deep learning: methods and applications *Foundations and Trends® in Signal Processing* **7** (3–4) pp 197-387
[7] Huihua Y, Hu B, Pan X, Yan S and Feng Y 2017 Deep belief network-based drug identi¯cation using near infrared spectroscopy *J. of Innovative Optical Health Sciences* **10** (2) 1630011
[8] Asifullah K, Zameer A, Jamal T and Raza A 2018 Deep Belief Networks Based Feature Generation and Regression for Predicting Wind Power *arXiv preprint arXiv:1807.11682*
[9] Acquarelli J, van Laarhoven T, Gerretzen J, Tran T N, Buydens L M and Marchiori E 2017 Convolutional neural networks for vibrational spectroscopic data analysis *Analytica chimica acta* **954** pp 22-31
[10] Srivastava N, Hinton G, Krizhevsky A, Sutskever I and Salakhutdinov R 2014 Dropout: a simple way to prevent neural networks from overfitting *The J. of machine learning research* **15** (1) pp 1929-58
[11] Jonathon S 2014 A Tutorial on Principal Component Analysis *arXiv preprint arXiv:1404.1100*
[12] Huihua Y, Hu B, Pan X, Yan S and Feng Y 2017 Deep belief network-based drug identi¯cation using near infrared spectroscopy *J. of Innovative Optical Health Sciences* **10** (2) 1630011
[13] Wenwen L, Lin M, Huang Y, Liu H and Zhou X 2017 Near infrared spectroscopy detection of the content of wheat based on improved deep belief network *J. of Physics: Conf. Series* **887** (1) 012046 (Bristol: IOP Publishing)
[14] Lengkey, Lady C E C H, Budiastra I W, Seminar K B and Purwoko B S 2013 Model pendugaan kandungan air, lemak dan asam lemak bebas pada tiga provenan biji jarak pagar (Jatropha Curcas l) menggunakan spektroskopi inframerah dekat dengan metode partial least square (PLS) *Jurnal Penelitian Tanaman Industri* **19** (4) pp 203 - 11
[15] Acquarelli J, van Laarhoven T, Gerretzen J, Tran T N, Buydens L M and Marchiori E 2017 Convolutional neural networks for vibrational spectroscopic data analysis *Analytica chimica acta* **954** pp 22-31

Orange Fruit Images Classification using Convolutional Neural Networks

Dhiya Mahdi Asriny, Septia Rani, Ahmad Fathan Hidayatullah

Department of Informatics, Universitas Islam Indonesia, Yogyakarta, Indonesia

Email: 15523033@students.uii.ac.id, septia.rani@uii.ac.id, fathan@uii.ac.id

Abstract. The quality of fruit is important to increase sales in the market. Right now, the quality selection of orange fruit is mostly still complete by humans. Several drawbacks such as inaccuracy and inconsistency in results happened due to the limitation of human perceptions. The development of computer vision, making it possible to train the computer to classify images based on specified characteristics. This paper proposes the classification model to classify orange images using Convolutional Neural Network (CNN). Five classes of orange namely good-orange-grade-1, good-orange-grade-2, immature-orange, rotten-orange, and damaged-orange are classified using deep learning CNN. Total of 1000 orange images is collected using the smartphone camera. Each class consists of 200 images which are divided into 60% as the training data, 20% as the validation data, and 20% as the testing data. K-Fold Cross-Validation method is used to validate the model. In this paper, the hidden layer of CNN consists of 256 nodes. Two activation functions, ReLU and Tanh, are employed for comparing the accuracy of classification with the Softmax classifier. The result shows that the accuracy of ReLU activation function is 96%, which is better than the Tanh activation function that gives only 93,8%.

Keywords: classification, convolutional neural network, deep learning, orange fruit

1. Introduction

The productivity of orange fruits in Indonesia is high. Badan Pusat Statistik (Central Bureau of Statistics) Indonesia [1] reported that the commodity of orange fruits ranked in the third position with a high productivity rate after banana and mango. Therefore, orange fruits have become the main commodity because of their great contribution to the national economy. In 2017, the productivity of orange fruits in Indonesia increased by 150,978 ton or about 7,50% compared with the previous year. In line with that, orange fruits have given high profit for Indonesian farmers [2].

In order to increase sales and market competitiveness of orange fruits (oranges), it is important to select the oranges with high quality. Several indicators such as size, shape, and color are used to determine the quality during the selection process. However, the selection process of the oranges is still conducted manually by humans, which is time-consuming. Some other problems such as visual limitation, fatigue, slow speed, and low precision have also happened which can make the selection's results are not in accordance with the requirements of the market. Therefore, we need to overcome those problems by proposing a model that is able to recognize the quality of the oranges automatically.

This study aims to build a classification model for orange fruit images. We propose Convolutional Neural Network (CNN) as the method to classify orange fruit images into 5 classes, namely good-orange-grade-1, good-orange-grade-2, damaged-orange, rotten-orange, and immature-orange. It has been proven that CNN shows great performance in some tasks, including object detection and image recognition [3]. Moreover, we also investigate the performance of our model using several scenarios and metrics to obtain the best model for classification.

The rest of this paper is organized as follows. Section 2 describes the related work. Section 3 describes the methodology. Section 4 describes the results and discussion. Finally, the concluding remarks are described in section 5.

2. Related Work

Research on the quality of oranges has previously been done by [4] using Learning Vector Quantization (LVQ) method and get 76% accuracy. In their research, they focused on classifying one type of orange, which is lime. Subsequent research was conducted by [5] using Naive Bayes on orange images and obtained an accuracy result of 91.6%. In 2016, research conducted by [6] on orange images obtained an accuracy result of 88%. All of those research still use traditional machine learning approach for classifying orange images. A recent study carried out by [7] also conducted orange images classification. However, they were not only utilized traditional machine learning but also utilized a deep learning approach. The study revealed that Hybrid Artificial Neural Network-Harmony Search (ANN-HS) outperformed k-Nearest Neighbors (kNN) with accuracies of 94.28% and 70.88% respectively.

Therefore, we will discuss more about the application of deep learning approach for image classification. One of the most popular deep learning methods that are suitable for image classification is CNN. CNN seeks to emulate the image recognition system in visual cortex such as humans, thus by applying CNN, machine (computer) can be able to process image information [8]. Other research stated that CNN can reduce the number of parameters and able to handle the input image deformation such as translation, rotation, and scale [9].

Research using CNN was conducted by [10] to classify advertising images. According to their experiments, the accuracy result is 86%. Other research [11] also using CNN on different images, which is chili images. Their model achieved 80% accuracy on the test datasets. Recent research by [12] conducted a classification of fruit images (apples, bananas, blueberries, kiwi, and raspberry) using CNN. The result of their study showed that the accuracy achieved 94%.

3. Methodology

In this study, we classified images of orange using Convolutional Neural Network (CNN). CNN is suitable for processing data with grid-like topologies, such as time-series data (1D grid) and image data (2D grid of pixels). CNN used a mathematical operation called a convolution, which is a type of special linear operation. The convolution network is simply a neural network that uses convolution as a substitute for common matrix multiplication in at least one layer [13]. The model proposed in this research has several stages, consisting of collecting data image, image pre-processing, training the model using CNN, tuning the parameters to produce the best model, and testing the model using new data.

3.1. Dataset

The orange images were taken using a camera phone with 16 Megapixels resolution under the room lighting environment. The orange fruit is placed on a white surface. The five classes of images, namely good-orange-grade-1, good-orange-grade-2, damaged-orange, rotten-orange, and immature-orange are used (shown in figure 1). The pre-processing steps including cropping and resizing the object into 32×32 pixel.

We have collected 1000 images consists of 200 images for each class. The k-Fold Cross-Validation is used to distribute the dataset and validate the model. We use $k = 5$, which means that for each class the data will be divided into five folds (F1, F2, F3, F4, and F5). Each fold in each class will consist of 40 data. The distribution of these datasets is as follows: data number 1-40 will be categorized as group

F1, data number 41-80 as group F2, data number 81-120 as group F3, data number 121-160 as group F4, and data number 161-200 as group F5. In addition, the illustration of k-Fold Cross-Validation with $k = 5$ is shown in figure 2. There are five data sharing scenarios: D1, D2, D3, D4, and D5. For each scenario, 60% of data will be used as the training data, 20% will be used as the validation data, and 20% will be used as the testing data.

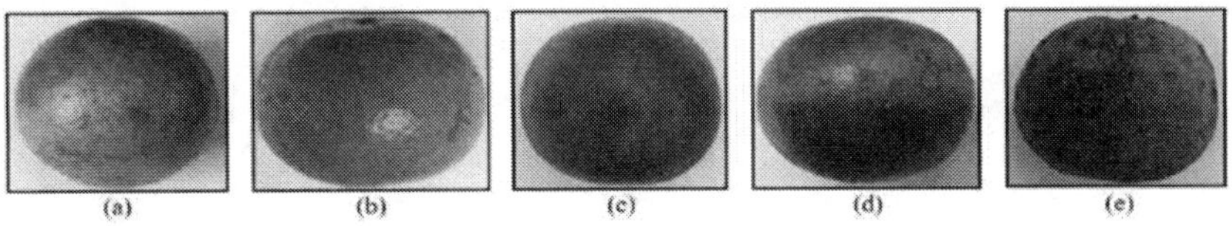

Figure 1. The sample of orange images (a) good-orange-grade-1, (b) good-orange-grade-2, (c) immature-orange, (d) rotten-orange, (e) damaged-orange.

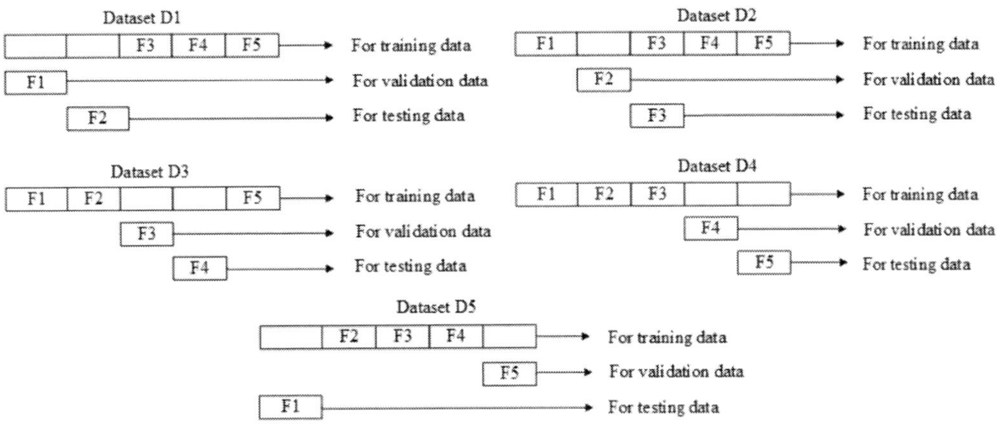

Figure 2. The illustration of k-Fold Cross-Validation with k=5 [14].

3.2. Model Building

Figure 3 shows our proposed CNN architecture with several different layers, including convolution layer, pooling layer, dropout layer, flatten layer, and layer dense.

Pooling layer aims to reduce the dimensions of the feature map (downsampling) and overcome overfitting. Dropout layer is a regularization technique in a neural network where some neurons will be selected randomly and not applied during the training process. Flatten layer is used to transform the feature map in a multidimensional array into a vector form that will be used as an input for the fully-connected layer.

The difference between traditional neural network and CNN is the convolution layers. The input of the convolution layer is an image which is used for the classification process and the output is the feature map. The convolution layer has three main processes: convolution, activation, and pooling [15]. So, in this research, the convolution process was conducted 4 times to train the model and measure the performance of the model. We used two activation functions: Rectified Linear Unit (ReLU) and Tanh to compare the performance of the model. Based on the study of Krizhevsky, ReLU is preferable, work faster than Tanh without making a significant difference to the generalization of the accuracy [16]. The size of kernel/filter used for each convolution layer is 3×3 and the size of pooling is 2×2. To keep the size of the input, the pooling is applied to every output from the convolution filter, which makes the input does not reduce drastically on every process.

The number of filter/kernel used on layer 1 and 2 are 32 filters, while the number of filter on layer 3 and 4 are 64 filters. Layer 3 and 4 need more filters because these layers need to extract the information from the image which is smaller after the process in the first and second convolution layers. The optimal neurons in the hidden layer used in this research are 256 nodes. The Softmax classifier is applied in the classification to provide more intuitive results and easy to classify probabilistic interpretations for all labels produced. Finally, the model will be implemented using the Keras package with Rstudio.

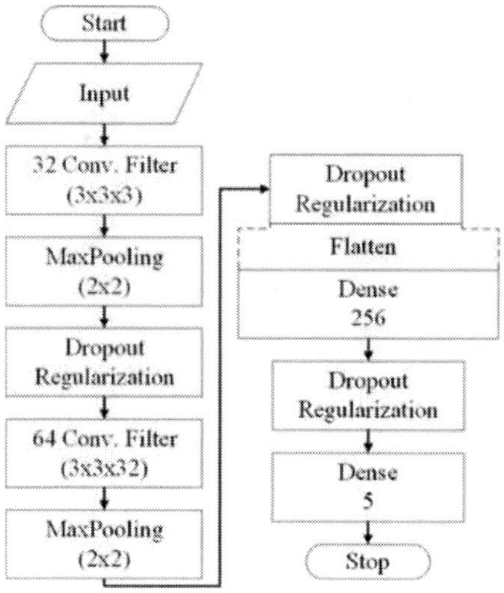

Figure 3. The convolutional neural network model.

4. Results and Discussion

In our experiments, we employed ReLU and Tanh activation function and compared their performance. We use two metrics: loss and accuracy to measure the performance. The loss function is used to penalize the output of the model. It indicates the magnitude of error of the model made on its prediction. We used categorical cross-entropy loss, which is the common setting for classification problems with multi classes. Meanwhile, the accuracy indicates how close a measurement is to the correct value. The accuracies of both activation function for each fold are shown in Table 1-3.

Table 1. The accuracy of training data for each data sharing scenario/fold.

Data Training	Number of Data	Loss Value		Accuracy Value	
		ReLU	Tanh	ReLU	Tanh
D1		0.040389	0.002499	99%	100%
D2		0.010151	0.001387	99%	100%
D3	600	0.004420	0.002139	100%	100%
D4		0.088673	0.006887	97%	100%
D5		0.020158	0.001592	98%	100%
Average		0.032758	0.002901	98.6%	100%

The loss value on the training, validation and testing data using the ReLU activation function are 0.032758, 0.239608, and 0.147128. While, Tanh activation function created loss value 0.002901 on training data, 0.198132 on validation data, and 0.204804 on testing data. Both the loss value from each activation function is relatively low. This implies that the results of the proposed models are great.

The accuracy value represents the correct classification rate. It is reasonable while getting a low level of loss and get high accuracy. Table 1 shows 98.6% and 100% accuracy on the training data for both activation functions. On the validation data, the accuracies are 92.8% and 93.2% for both activation functions (see Table 2). Meanwhile on the testing data (see Table 3), the accuracies are 96% and 93.8%

by using ReLU activation function and Tanh activation function, respectively. The more detailed performance of the model along with the increase in the number of the epoch can be seen in figure 4.

Table 2. The accuracy of validation data for each fold.

Data Validation	Number of Data	Loss Value		Accuracy Value	
		ReLU	Tanh	ReLU	Tanh
D1		0.418156	0.159932	91%	95%
D2		0.032446	0.241223	98%	90%
D3	200	0.371931	0.407679	90%	90%
D4		0.183812	0.086802	93%	95%
D5		0.191694	0.095025	92%	96%
Average		0.239608	0.198132	92.8%	93.2%

Table 3. The accuracy of testing data for each fold.

Data Testing	Number of Data	Loss Value		Accuracy Value	
		ReLU	Tanh	ReLU	Tanh
D1		0.098135	0.047535	98%	98%
D2		0.274564	0.460625	93%	86%
D3	200	0.207179	0.149645	96%	95%
D4		0.052459	0.048771	98%	98%
D5		0.103303	0.317443	95%	92%
Average		0.147128	0.204804	96%	93.8%

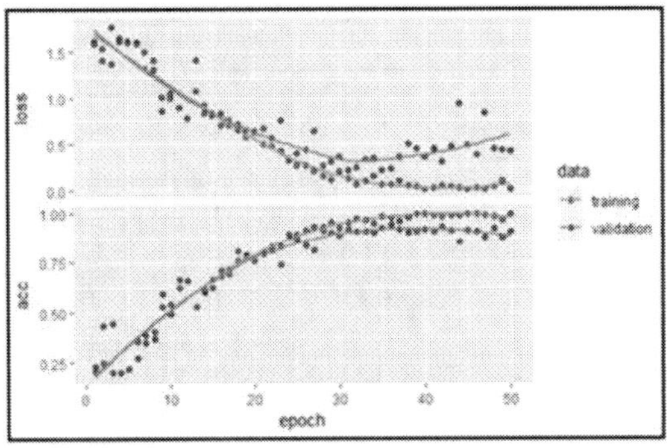

(a)

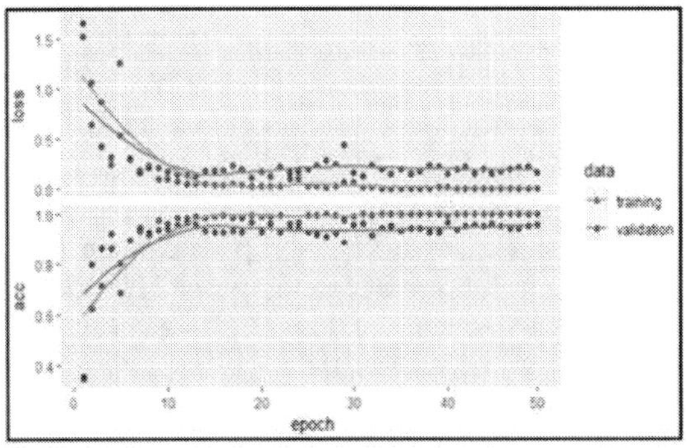

(b)

Figure 4. Loss and accuracy of data training and validation with (a) ReLU Activation Function and (b) Tanh Activation Function

In line with the increasing number of the epoch, generally, the value of the loss is approaching zero. While the value of accuracy is increasing to 1.00 (maximum value) for both data training and data validation. From this, we can conclude that the number of epoch used in the training process has a positive correlation with accuracy. Otherwise, the correlation between the number of epoch and the value of the loss is a negative correlation. Based on this, we can minimize the value of loss by expanding the number of epoch in the training process. Thus we will get a CNN model that produces high accuracy.

However, when we look more closely at figure 4(a), the performance of the training process using ReLU activation function is decreased when the number of epoch approaches 40. It is shown by the graphic trend of loss value on data validation that increases after the 40th epoch. This phenomenon implied that it is not always necessary to use a large number for the epoch. In our case, the model will give the best performance if we choose 40 as the number of the epoch.

5. Concluding Remarks

In this study, we applied Convolutional Neural Networks to classify the orange images. The CNN model consists of four layers: two pooling layers using 3×3 kernel size, a softmax layer, and a hidden layer consists of 256 nodes. In the hidden layers, we compared the use of both ReLU and Tanh activation function to obtain the best model performance. According to the experiments, the performance of ReLU is better than Tanh. Tanh produces lower accuracy when we use a new dataset. The classification results of data training by using both activation function are 98.6% and 100%. The accuracies of data validation are 92.8% for ReLU activation function and 93.2% for Tanh activation function. Furthermore, the accuracy of data testing is 96% by using ReLU activation function and 93.8% by using Tanh activation function.

References

[1] Badan Pusat Statistik 2017 Statistik tanaman buah-buahan dan sayuran tahunan Indonesia *Badan Pusat Statistik* 05120.1807 p 46

[2] Nainggolan C I, Tarigan K and Salmiah 2013 Analisis usaha tani jeruk dan faktor-faktor yang mempengaruhi penerimaan petani *J. Agriculture and Agribusiness Socioeconomics* **2**

[3] LeCun Y, Bengio Y and Hinto G 2015 Deep learning *Nature Int. J. of Science* **521** pp 436-44

[4] Romadhon A S and Widyaningrum V T 2015 Klasifikasi mutu jeruk nipis dengan metode learning vector quantization (LVQ) *Jurnal Ilmiah Rekayasa* **8** pp 121–8

[5] Agustian W, Setyaningsih S and Qur'ania A 2015 Klasifikasi buah jeruk menggunakan metode naive bayes berdasarkan analisis tekstur dan normalisasi warna *Program Studi Ilmu Komputer FMIPA - Univ. Pakuan* **2**

[6] Capizzi G, Sciuto G L, Napoli C, Tramontana E and Wozniak M 2016 A novel neural networks-based texture image processing algorithm for orange defects classification *Int. J. Computer and Science Applications* **13** pp 45–60

[7] Sabzi S, Abbaspour-gilandeh Y and Garcia-Mateos G 2018 A New approach for visual identification of orange varieties using neural networks and metaheuristic algorithms *Information Processing in Agriculture* **5** pp 162–72

[8] Wayan S E P I, Arya Y W and Soelaiman R 2016 Klasifikasi citra menggunakan convolutional neural network (CNN) pada caltech 101 *Jurnal Teknik ITS* **5** pp 65–9

[9] LeCun Y, Bottou L, Bengio Y and Haffner P 1998 *Gradient-based Learning Applied to Document Recognition* vol **86** pp 2278-324 (USA: IEEE)

[10] An T V, Tran H S and Le T H 2017 Advertisement image classification using convolutional neural network *The 9th Int. Conf. on Knowledge and Systems Engineering* **1** pp 197–202

[11] Purwaningsih T, Anjani I A and Utami P B 2018 Convolutional neural networks implementation for chili classification *2018 Int. Symp. on Advanced Intelligent Informatics (SAIN)* pp 190-4

[12] Khaing Z M, Naung Y and Htut P H 2018 Development of control system for fruit classification based on convolutional neural network 2018 *IEEE Conf. Russ. Young Res. Electr. Electron. Eng.* EIConRus.2018.8317456 pp 1805–7

[13] Goodfellow I, Bengio Y and Courville A 2016 *Deep Learning Book* (United States: MIT Press) p 326

[14] Suyanto 2018 *Machine Learning: Tingkat Dasar dan Lanjut* (Bandung: Informatika) p 350

[15] Rismiyati and SN A 2016 Convolutional neural network implementation for image-based salak sortation *2nd Int. Conf. Sci. Technol.* (ICST) ICSTC.2016.7877351 pp 77-82

[16] Krizhevsky A, Sutskever A and Hinton G E 2012 ImageNet classification with deep convolutional neural networks *Adv. Neural Inf. Process. Syst.* **8** pp 713–72

Forecasting and Pattern Analysis of Philippine Regions' Palay and Corn Production

Michelle M. Cortez

De La Salle University, Manila, Philippines

michelle_cortez@dlsu.edu.ph

Abstract. Inadequate food production has been an immediate concern that the country is trying to address. Where farmers used fertilizers in their farms in the hope of increasing their agricultural production. Though, if fertilizer application is poorly managed, instead of having an increased production it will result in the contrary. Therefore, the efficient use of fertilizer is critical and has a great impact on crop production. Thus this study aims to show the pattern of Philippine' regions crop production, specifically in rainfed and irrigated palay; white and yellow corn. This will also show which fertilizer will maximize their crop production and to seek for the most applicable association model in forecasting future crop production. Three predictive techniques were used namely, canopy clustering, Apriori association rule mining and time series forecasting models. Results reveal that all regions have a low volume of production for rainfed rice. The canopy clustering shows the pattern leading to the high production of irrigated rice for Region III. Also, Region II, Region X, and Region XII have a high volume of production of yellow corn and lastly clustering results on white corn shows Region VII has a Mid area harvested but shows Low volume of production, while Region X though have a low area harvested managed to have a Mid volume of production. The association of fertilizers to the volume of production shows that low Ammophos leads to a lower volume of production and the low Ammosul is not associated with a low volume of production hence a combination of low Ammosul and low Ammophos leads to a low volume of production. The forecasting methods' linear regression, Gaussian processes, and SMOreg are all applicable in predicting the regions' volume of production, whereas the SMOreg has the least MAE of 8.90% for Region VI.

1. Introduction

The Philippines' agricultural production is not enough to sustain the current population demand [1] in 2015 for both rice and corn [2]. Due to El Niῆo (2016), a massive dropped on crop production with 6.8% in the first half of the year [1]. To address these concerns the country's agricultural sector is introducing programs and practices to improve productivity across all commodities especially to the major crops, the rice, and corn. One of the practices that are used to improve crop production or yield is site-specific nutrient management (SSNM) which enables farmers to dynamically adjust their fertilizer usage especially during the crucial stage of crop growth [3]. The efficient use of fertilizers is critical, as if it is improperly, instead of increasing the crop's production it might result on the contrary. The application of data analytics in analyzing the production of rice and corn concerning fertilizer will help the farmers in decision making. Predicting the volume of production will allow the farmers to extract insights on what type of fertilizer and its amount will positively improve the crop's production. This paper aims to identify the current status of crops' production in the different regions by identifying the pattern for each region. This also aims to raise farmers' awareness of what fertilizer will maximize their crop production.

This paper also aims to seek what is the most applicable forecasting model to be used in forecasting future crop production.

2. Related Work

The potential of big data analytics in agriculture [4] is seeming of great importance. Where it is seen as a key factor in smart farming to address global issues such as food security and safety [4] [5]. Three levels of analytics were used in conducting analytics namely descriptive, predictive and prescriptive. This is mainly important to explain and predict patterns from the available dataset [6]. As for this study, the main focus is on predictive analytics. This level of analytics has three (3) kinds of techniques, the traditional, model-dependent and data-adaptive methods [7]. Wherein it includes techniques such as web mining, text mining, data mining, forecasting techniques [8]. Regressions such as general linear models (GLM), partial least squares (PLS), LASSO (least absolute shrinkage and selection operator) wherein the study of Mayer, et al shows both GLM and PLS shows doesn't meet the accuracy target of $\pm$ 10% error [9]. Time series has a wide range of methods such as autoregressive integrated moving average model (ARIMA), artificial neural networks (ANN) [10] and other known methods [11]. In the study of Deb, Chirag, et al, [12] different time series forecasting techniques were evaluated. The ANN can precisely map input and output relationships, perform well for non-linear time series and it is more general and flexible, but depends on weight values initialization and is difficult to generalize. The ARIMA uses historical data's log and shift, a regression model with moving average and provides confidence intervals on predictions. But tends to have difficulty in model identification, not suitable for long term prediction and doesn't fully capture the non-linear patterns of the series. While Hybrid has a complementary combination of different machine learning methods and robust for complex problems and often improves performance. But it has high model complexity, computationally intensive and often difficult to identify which methods to combine. As for this study, the main focus is on the predictive analytics utilizing the canopy clustering, association rule mining, and time series forecasting. This is also to seek the applicability of the forecasting methods' linear regression, Gaussian processes and sequential minimum optimization (SMoreg).

Canopy clustering is a very simple, fast and accurate method of grouping objects into clusters [13]. It is an algorithm that uses fast approximate distance metrics and two distance thresholds T1>T2 for processing [14]. It is mainly used to increase the efficiency of K means partitioning algorithm. And thus it is said to effectively reduce the exploration time and more effective in a given larger area. In a study, canopy clustering was employed and stated that using the algorithm the numerous expensive distance measurements can be significantly reduced by ignoring outliers [15]. sing the algorithm reduces expensive distance calculation by measuring exact distances between points in the common canopy [16]. It is mainly used in this study to see the pattern of each region's agricultural production, their area harvested, area applied and average fertilizers' quantity applied and based on the fertilizer types.

On the other hand, the process of association rule mining technique is to find all item sets that have the support greater than the minimum support, generate rules that have confidence level higher than the minimum confidence and a lift of the observed support to that expected if X and Y were independent [17]. In recent studies, the association rule mining technique was used in identifying risk factors in adverse diseases, in agricultural parameters, in the manufacturing industry's total productive maintenance strategy, and in drought monitoring to name a few. In risk factors in adverse diseases, rare association rule mining is seen capable of generating a new set of rare association rules from the updated medical databases in a single scan without repeating the same mining process [18]. In the study about the total productive maintenance (TPM) strategy, the use of association rule mining accelerates the decision to establish an appropriate TPM management strategy based on the association rules [19]. In agriculture parameters, using the mining technique states that cereal yield, agricultural land, and agricultural value-added per worker are associated to crop production index [20].

Lastly, the Time Series Forecasting uses a collection of observations that is time-dependent [21], normally done in spaced intervals (e.g. daily, monthly, quarterly, yearly). Time series forecasting plays an important role not only in agriculture but also in other fields such as economic, engineering, medicine, etc. It is a challenging yet have an important task in many domains [22]. Although there are many single models proposed for time series forecasting, these can't be applied to all situations with the desired performance [23]. In previous studies, the use of time series forecasting in a hybrid model is seen to be an effective model over the regular and randomness component of the data [24]. Similarly to energy forecasting for building, the hybrid model provides the most effective time-series technique [12]. Wang and Jiang even integrate linear and non-linear forecasting techniques to improve forecasting accuracy [25]. And, in the study of Palomares-Salas, et al, it is seen as the best model based on artificial neural networks [26].

3. Methodology

In this section, the process of data collection and preparation of the dataset were discussed.

3.1. Data Collection

Data were obtained from the countrystat.psa.gov.ph, from the two (2) databases namely the production database and fertilizer and pesticide database. The database contained the production of irrigated palay, rainfed palay, yellow corn, and white corn in all regions across the country during the years 2003 to 2018. While from the fertilizer and pesticide database only the fertilizer information for palay and corn was obtained. These attributes are area harvested, the area where fertilizer was applied, the average quantity, the different fertilizers applied (Ammophos, Ammosul, Urea, complete and others).

3.2. Data Preprocessing

This technique was imposed to get the dataset clean and transform to have an appropriate representation of crop production analysis.

3.2.1. Data Cleansing. To have consistency in the dataset, this process was performed. To properly assigned fertilizer information according to the types of corn (yellow, white) and palay (irrigated, rainfed). Fertilizer information for each crop type was obtained by splitting data for each fertilizer attributes. To complete the dataset's years for fertilizer attributes, the moving average technique was used to fill in the missing data from 2015-2019 respectively. Similarly, the same technique is used to fill in the year 2019 data for production.

3.2.2. Data Integration. Attributes from the production database and selected attributes from fertilizer and pesticides database were merged into the dataset. The different attributes obtained from both databases were used in the canopy clustering, association rules mining and time series forecasting techniques.

3.2.3. Data Transformation. To identify patterns existing in the dataset, values of the different attributes were coded as Low, Mid and High. These values were coded accordingly as shown on Table 1 and 2.

Table 1. Volume of Production & Area Harvested code

Vol. of Production (metric tons)	*Area Harvested (hectares)*	*Area Applied (hectares)*	*Code*
0 – 1141017.66	0 – 209168.69	0 – 210235.55	Low
1141017.67 – 2282035.42	209168.7 – 418337.39	210235.56 – 420471.11	Medium
2282035.43 – 3423053	418337.40 – 627506.09	420471.12 – 630706.67	High

Table 2. Fertilizer code[a]

*Ave. Quantity Applied**	*Ammosul**	*Ammophos**	*Urea**	*Complete**	*Others**	*Code*
0 – 3.60	0 – 2.44	0 – 1.15	0 – 1.25	0 – 1.15	0 – 0.28	Low
3.61 – 7.22	2.45 – 4.90	1.16 – 2.33	1.36 – 2.72	1.16 – 2.33	0.29 – 0.58	Medium
7.23 – 10.84	4.91 – 7.35	2.34 – 3.49	2.73 – 4.08	2.34 – 3.49	0.59 – 0.86	High

[a] *bags of 50 kilograms*

3.3. Canopy Clustering

Canopy clustering was used to identify the number of clusters or groups in the dataset [27]. These clusters are overlapping referred to as canopies [28]. The patterns of volume production of each commodity per region based on area harvested, area applied and average fertilizers' quantity applied and based on the fertilizer types (Ammosul, Ammophos, Urea, complete and others) as illustrated on Table 3. This clustering method is applied to see the patterns of crop production in different regions. Since the aim is to see each crop's production status, the number of seeds was adjusted and from the adjustments, 100 is chosen as the number of seed. And limit the number of clusters to 20. It is adjusted to certain values to obtain the clustering result that shows patterns for each commodity (irrigated palay, rainfed palay yellow corn, and white corn).

3.4. Association Rules Technique

In finding the frequent patterns the association rule technique was used. Wherein it also includes correlations, associations or causal structures from the data sets found in different kinds of databases [29]. The frequent pattern of each commodity attributes namely volume of production were identified using the area harvested, area applied, the average quantity applied, the different fertilizers (Ammophos, Ammosul, Urea, complete and others). These attributes are of numerical types, which were converted to binomial values and is needed to be used in association rules mining in Weka. In validating these patterns, the minimum support, confidence, and lift measures were used. Support is how frequent the attributes or items in the rule appear together. Confidence, on the other hand, indicates the antecedent of both the antecedent and consequent appeared on the same occurrence. Both were used to determine the validity of an association rule. Lastly, the lift measure evaluates the quality of the rule. It indicates the strength of a rule over the random co-occurrence of antecedent and consequent, given the individual support [30]. A rule should have an improvement that is greater than 1 to give a positive correlation [21]. The lower minimum bound support is 0.4 and the min metric set is 0.7 for the volume of production with area harvested, area applied and average quantity applied. While for the volume of production with fertilizer's Ammosul, Ammophos, Urea, Complete and Others, the lower minimum bound support is 0.5 and min metric of 0.7 respectively.

3.5. Forecasting Methods

The methods used to forecast the crops' volume of production were linear regression, Gaussian processes and sequential minimum optimization (SMoreg). Linear regression is used to summarize and study relationships between continuous variables, where one variable (x) as the independent variable and another variable (y) as the dependent variable [31]. The Gaussian process is a method for modeling probability distributions over functions [32]. It is specified by a mean and covariance functions [33]. And lastly, the sequential minimum optimization is used in solving quadratic programming (QP) problems [34]. These methods were forecasted using Weka version 3.8.2 wherein default evaluation metrics are used to evaluate the forecasting results and should have a score that is less than 100. Because a score of >=100 indicates that the prediction is not doing good and there is something wrong with the prediction [35]. In comparing the results of each forecasting method, the mean absolute error (MAE) is used. This is one of the default evaluation metrics, which is used to measure the closeness of the prediction to its outcomes [36] or to check the forecast accuracy [37] [38].

4. Results and Discussion

The following are the discussion of results of the analytics experiment.

4.1. Canopy Clustering

To identify the clusters of groups in the dataset, the canopy was employed. As illustrated in Table 3, Region III showed a High crop production in terms of irrigated palay. Such output manifests since the region has a category High area harvested, High area applied and Mid average quantity of fertilizers applied while Low in the application of Ammosul, Ammophos, and other fertilizer applied. While having Mid or average application of Urea and Complete. Another interesting pattern for Region II under the irrigated rice, it has most of the categories similar to Region III but has Mid complete fertilizers and a Mid volume of production. It is also evident that two cluster patterns emerged for Region IV-A under the irrigated rice. Wherein there were shifting of average usage of complete fertilizers from 2003-2018 but result in the same Low volume of production.

Under the rainfed rice, all regions have a low volume of production. The clusters under this crop show low area harvested, area applied, the average quantity of fertilizers applied and also for each fertilizer type and have a low volume of production for Region VII and Region IX. Three regions are clustered as a High volume of production for yellow corn namely, Region II, Region X, and Region XII. All of these three regions have Mid area harvested, and have a low application of Ammosul and other fertilizers (Others). It is also evident that despite Low area harvested for Region I and Region V their volume of production is of average (Mid) volume of production. Other scenarios are evident for the groups under the white corn, Region VII has a Mid area harvested but shows Low volume of production, while Region X though have a low area harvested managed to have a Mid volume of production.

Table 3. Clustering Method of Volume of Production

Crop	Region_ID	Area_Harvested	Area_Applied	AveQuantity Applied	Ammosul	Ammophos	Urea	Complete	Others	Volume of Prod
Rainfed Rice	Region VII	Low	Low	Low	Low	Low	Low	Low	Low	Low
	Region XI	Low	Low	Low	Low	Low	Low	Low	Low	Low
Irrigated Rice	Region I	Mid	Mid	Mid	Low	Low	Mid	Mid	Low	Mid
	Region II	High	High	Mid	Low	Low	Mid	Mid	Low	Mid
	Region III	High	High	Mid	Low	Low	Mid	High	Low	High
	Region IV-A	Low	Low	Low	Low	Low	Low	Low	Low	Low
	Region IV-A	Low	Low	Low	Low	Low	Low	Mid	Low	Low
	Region V	Mid	Low	Low	Low	Low	Low	Low	Low	Low
	Region VI	Mid	Mid	Low	Low	Low	Low	Low	Low	Low
	Region X	Low	Low	Mid	Low	Low	Low	Low	Low	Low
	Region XIII	Low	Low	Low	Low	Low	Low	Mid	Low	Low
Yellow Corn	CAR	Low	Low	Mid	Low	Low	Mid	Mid	Low	Low
	Region I	Low	Low	Mid	Low	Low	High	High	Low	Mid
	Region II	Mid	Mid	Mid	Low	Mid	High	Mid	Low	High
	Region V	Low	Low	Low	Low	Low	Low	Mid	Low	Mid
	Region X	Mid	Low	Mid	Low	Low	Low	Low	Low	High
	Region XII	Mid	Mid	Low	Low	Low	Mid	Low	Low	High
White Corn	Region VII	Mid	Low	Low	Low	Low	Low	Low	Mid	Low
	Region IX	Low	Low	Mid	Low	Low	Low	Low	Low	Low
	Region X	Low	Low	Low	Low	Low	Low	Low	Low	Mid

This method clearly shows the patterns of the volume of production for rainfed rice, irrigated rice, yellow corn and white corn commodities based on area harvested, area applied and average fertilizers' quantity applied and based on the fertilizer types (Ammosul, Ammophos, Urea, complete and others). But based on the result of this clustering, no clusters were found for the other regions (CAR, ARMM, Region VIII, and Region XIII). Looking at the dataset, CAR has a low volume of production for all commodities under this study. It was in 2013, 2014 and 2017 where they have improved production in yellow corn. Similarly, ARMM has a low volume of production for irrigated and rainfed palay. While a diverse result for white and yellow corn's volume of production. Region VIII similarly shows a low volume of production, low area harvested, low area applied, and low in terms of fertilizer usage.

These results can be used to understand the data patterns of crop production for palay and corn. As illustrated in the table it is not always the case that the area should be large to produce a high volume of production. Like in the case of Region X and Region XII which produce a high volume of production for yellow corn. Region II on the other hand though has a large harvested area for irrigated rice still has an average volume of production, which might be because of other factors such as applied fertilizers, pest infestations and other natural disasters that can affect their crops' health. High utilization of inorganic fertilizers doesn't always correlate with a high volume of production, proper usage of fertilizers should be done to have a higher volume of production.

Though this clustering method can efficiently show data patterns or clusters from huge data sets, the resulting cluster/s is a rough pre-partitioned data that should be further examined by other clustering methods. Also in this method, the number of clusters and seeds is user-defined, which means it is up to the user, how many clusters and seeds he/she wanted the output of this method to be. Criteria for grouping/cluster and the number of groups/clusters should be known to produce a meaningful classification of clusters.

Table 4. Food Production Apriori Resuts

Frequent Patterns	*Min sup*	*conf*	*lift*
MinAreaApplied ==> LowProd	0.40	0.99	1.08
LowAreaHarvested ==> LowProd	0.40	1	1.10
LowAreaHarvested, MinAreaApplied ==> LowProd	0.40	1	1.10
MinAreaApplied, Low AveQuantApplied ==> LowProd	0.40	0.99	1.09
LowAveQuantApplied ==> LowProd	0.40	0.79	1.07
MinAreaApplied, LowAveQuantApplied ==>LowProd, LowAreaHarvested	0.40	0.98	1.16
LowAreaHarvested==>LowProd, MinAreaApplied	0.40	1	1.17
LowAreaHarvested,MinAreaApplie, LowAveQuantApplied==> LowProd	0.40	1	1.10
LowAreaHarvested, LowAveQuantApplied==>LowProd, MinAreaApplied	0.40	1	1.17
LowAmmophos ==> LowProd	0.50	0.92	1.01
LowAmmosul, LowAmmophos ==> LowProd	0.50	0.92	1.01
LowAmmosul ==> LowProd	0.50	0.91	1
LowAmmosul==>LowProd, LowAmmophos	0.50	0.91	1

Table 5. Evaluation of Forecasting Method (year)

	Region	*Linear Regression*	*Gaussian Processes*	*SMOreg*
1	CAR	28.21%	29.88%	23.08%
2	Region I	21.31%	24.36%	13.34%
3	Region II	19.47%	58.59%	17.18%
4	Region III	16.90%	43.10%	14.96%
5	Region IV	12.20%	41.78%	13.00%
6	Region V	24.42%	39.56%	24.58%
7	Region VI	18.25%	18.72%	8.90%
8	Region VII	34.14%	64.39%	31.61%
9	Region VIII	20.47%	42.83%	16.62%
10	Region IX	20.06%	52.36%	17.67%
11	Region X	22.32%	28.16%	24.07%
12	Region XI	10.91%	45.21%	11.31%
13	Region XII	23.76%	33.23%	24.93%
14	Region XIII	15.98%	32.31%	14.54%
15	ARMM	25.12%	48.25%	24.08%

4.2. Association Rule Technique

To identify the best patterns to increase crops' volume of production, the Apriori algorithm was used. The class association rules generated that are not leading to a low volume of production were eliminated and only those frequent patterns that are associated with a low volume of production were retained. As illustrated in Table 4, frequent patterns were generated for a low volume of production. Wherein results indicate that minimum area applied, lower area harvested and a lower area where fertilizers are applied is linked to lower production. Similarly, the association of fertilizers to the volume of production indicates that the low Ammophos is linked to the low volume of production. Low Ammosul, on the other hand, is not associated with a low volume of production, hence a combination of low Ammosul and low Ammophos leads to a low volume of production. Though, there were no association patterns generated for the average volume of production and a high volume of production because these have a minimal number of occurrences on the dataset. This information on the association of different variables to a low volume of production can be used to understand such occurrence and could use a basis on how to improve crops' production.

4.3. Forecasting Methods

This study employed the linear regression, Gaussian processes, and SMOreg to predict the volume of production of the rice and corn commodities in the different regions with the fertilizer applied. The evaluation results showed that all forecasting methods applied were applicable in predicting the regions' volume of production as shown in Table 5. Where the least MAE of 8.90% on Region VI under SMOreg. MAE is one of the evaluation metrics that can be used in evaluating forecasting results. Though MAE shows all forecasting result which are all less than 100%, validating by also looking into other evaluation metrics such as Mean Squared Error (MSE), Root Mean Square Error (RMSE), Mean Percentage Error (MPE) and Mean Absolute Percentage Error (MAPE) might be used in forecasting time series data to further understand the forecasting model's performance.

5. Conclusion

Different data mining techniques employed in this paper may help in understanding the pattern of the volume of production for irrigated rice, rainfed rice, yellow corn and white corn in the country's different regions. The results in canopy clustering show that most of the clusters are producing a low volume of production as affected by the area harvested, applied, the average quantity applied and fertilizers. These results will also enlighten the current situation on the food production of the country. As it clearly shows the resulting patterns per commodity under this study. The results from the clustering method show that it is not only the farm size that leads to a high volume of production, but there are also other factors that the local government unit, organizations, agricultural officers and farmers should look into to improve crop production. Future studies should have the resulting clusters from this type of method be examined further by using other clustering methods such as K-Means or the Hierarchical clustering method. The Apriori association model shows similar results with the canopy clustering. It shows that low Ammophos leads to a low volume of production and the combination of low Ammophos of Ammosul similarly leads to a low volume of production. The results of the Apriori only show the association of other variables to a low volume of production, it is because most of the regions have a low volume of crop production for palay and corn. Also, the high volume of fertilizers doesn't always contribute to crop production, hence proper management of these fertilizers should be imposed. Having these results, crop production is still an area of concern that we should look into, on how it could be improved despite the effect of different factors. All forecasting models, on the other hand, were applicable in forecasting the regions' volume of production and were SMOreg has the least mean absolute error. Though to further check the result of forecasting methods future works could also validate by utilizing other evaluation metrics.

References

[1] Briones R, Antonio E, Habito C, Porio E and Songco D 2017 "*Food Security and Nutrition in the Philippines*" (Philippines: Brain Trust, Inc)

[2] Philippine Statistics Authority 2016 "*Food Sufficiency and Security*" (Quezon City: PSA)

[3] IRRI nd "How to manage soil fertility" *Rice Knowledge Bank* [Online] Available: http://rkb.irri.org/step-by-step-production/growth/soil-fertility [Accessed 23 November 2018]

[4] Pham X and Stack M 2018 "How data analytics is transforming agriculture" *Business Horizons* no **61** pp pp 125-33

[5] Wolfert S, Ge L, Verdouw C and Bogaart M-J 2017 "Big Data in Smart Farming - A review" *Agricultural Systems* pp 69-80

[6] Sharma R, Kamble S S and Gunasekaran A 2018 "Big GIS analytics framework for agriculture supply chains: A literature review identifying the current trends and future perspectives" *Computers and Electronics in Agriculture* **155** pp 103-20

[7] Balaji P B and Dakshayini M 2018 "Performance Analysis of the Regression and Time Series Predictive Models using Parallel Implementation for Agricultural Data" *Procedia Computer Science* **132** pp 198-207

[8] Sharma R, Kamble S S and Gunasekaran A 2018 "Big GIS analytics framework for agriculture supply chains: A literature review identying the current trends and future perspectives" *Computer and Electronics in Agriculture* **155** pp 103-20

[9] Mayer D, Chandra K A and Burnett J R 2019 "Improved crop forecasts for the Australian macadamia industry from ensemble models" *Agricultural Systems* **173** pp 519-23

[10] Park K-J and Son S-Y 2019 "Forecasting Analysis Based on Multi-scale and Multi-time with Uncertainty" *IFAC Papers Online: Conf. Paper Archive* pp 318-23

[11] Yasasin E, Prester J, Wagner G and Schryen G 2020 "Forecasting IT security vulnerabilities –An empirical analysis" *Computers & Security* pp 1-23

[12] Deb C, Zhang F, Yang J, Lee S E and Shah K W 2017 "A review on time series forecasting techniques for building energy consumption" *Renewable and Sustainable Energy Reviews* **74** pp 902-24

[13] Sharma S, Tiwari R, Shukla A and Yadav J 2014 "Canopy Clustering Based Multi Robot Area Exploration" *Third Int. Conf. on Advances in Control and Optimization of Dynamical Systems* Kanpur, India

[14] Mahout 2014 "Canopy Clustering" *The Apache Software Foundation* [Online] Available: https://mahout.apache.org/users/clustering/canopy-clustering.html [Accessed 6 December 2018]

[15] Kumar A, Ingle Y S, Pande A and Dhule P 2014 "Canopy Clustering: A Review on Pre-Clustering Approach to K-Means Clustering" *Int. J. of Innovation & Advancement in Computer Science* vol **3** no 5 pp 22-9

[16] Sabena S, Yogesh P and Saimaresh L 2011 "Image Retrieval using Canopy and Improved K mean Clustering" in *Int. Conf. on Emerging Technol. Trends (ICETT)*

[17] Sayad S 2010 "Association Rules" *An introduction to Data Science* [Online] Available: https://www.saedsayad.com/association_rules.htm [Accessed 6 December 2018]

[18] Borah A and Nath B 2018 "Identifying risk factors for adverse diseases using dynamic rare association rule mining" *Expert Systems with Application* pp 233-63

[19] Djatna T and Alitu I M 2015 "An application of association rule mining in total productive maintenance strategy: an analysis and modelling in wooden door manufactuting industry" *Procedia Manufacturing* **4** pp 336-43

[20] Hira S and Deshpande P 2015 "Data Analysis using Multidimensional Modeling, Statistical Analysis and Data Mining on Agriculture Parameters" *Procedia Computer Science* **54** pp 431-9

[21] Marzan C S, Baculo M J C, Bulos R D and Ruiz Jr C 2017 "Time Series Analysis and Crime Pattern Forecasting of City Crime Data" *Proc. of the Int. Conf. on Algorithms, Computing and Systems* pp 113–8

[22] Büyüksahin Ü Ç and Ertekin S 2019 "Improving forecasting accuracy of time series data using a new ARIMA-ANN hybrid method and empirical mode decomposition" *Neurocomputing* **361** pp 151-63

[23] Hajirahimi Z and Khashei M 2019 "Hybrid structures in time series modeling and forecasting: A review" *Engineering Applications of Artificial Intelligence* **86** pp 83-106

[24] Bendre M and Manthalkar R 2019 "Time series decomposition and predictive analytics using MapReduce framework" *Expert Systems with Applications* **116** pp 108-20

[25] Wang Q and Jiang F 2019 "Integrating linear and nonlinear forecasting techniques based on grey theory and artificial intelligence to forecast shale gas monthly production in Pennsylvania and Texas of United States" *Energy* **178** pp 781-803

[26] Palomares-Salas J C, Aguera-Perez A and De La Rosa J J G M-M A 2014 "A novel neural network method for wind speed forecasting using exogenous measurement from agricultural stations" *Measurement* **55** pp 295-304

[27] O'Reilly 2018 "Chapter 3: Understanding Canopy Clustering" *O' Reilly Safari* [Online] Available: https://www.oreilly.com/library/view/apache-mahout-clustering/9781783284436/ch03.html [Accessed 4 December 2018]

[28] Mapt Beta 2018 "Canopy clustering" *Mapt Beta* [Online] Available: https://www.packtpub.com/mapt/book/big_data_and_business_intelligence/9781783555215/7/ch07lv11sec33/canopy-clustering [Accessed 4 December 2018]

[29] TechnoPedia 2018 "*Association Rule Mining*" [Online] Available: https://wwwtechopedia.com/definition/30306/association-rule-mining [Accessed 4 December 2018]

[30] Oracle 2018 "Data Mining Concepts: Apriori" *Oracle Help Center* [Online] Available: https://docs.oracle.com/cd/B28359_01/datamine.111/b28129/algo_apriori.htm#CHDCDJEF [Accessed 4 December 2018]

[31] Pennsylvania State University nd "*What is Simple Linear Regression?*" [Online] Available: https://online.stat.psu.edu/stat501/lesson/1/1.1 [Accessed 25 November 2018]

[32] Lee H 2008 "*Gaussian Process*" [Online] Available: http://cs229.stanford.edu/section/cs229-gaussian_processes.pdf [Accessed 4 December 2018]

[33] Mackay D J 1998 "Introduction to Gaussian processes" *NATO ASI Series F Computer and Systems Sciences* **168** pp 133-166

[34] Big Data Mining & Machine Learning 2016 "*Sequential Minimal Optimization*" [Online] Available: http://www.big-data.tips/sequential-minimal-optimization [Accessed 4 December 2018]

[35] Pentaho Platform 2014 "*Time Series Analysis and Forecasting with Weka*" [Online] Available: https://wiki.pentaho.com/display/DATAMINING/Time+Series+Analysis+and+Forecasting+with+Weka#TimeSeriesAnalysisandForecastingwithWeka-Confidenceintervals [Accessed 3 November 2019]

[36] Adetiloye A A T 2017 "Chapter 8 - Predicting Short-Term Congested Traffic Flow on Urban Motorway Networks" *Handbook of Neural Computation* (Cambridge: Academic Press) pp 145-65

[37] Watson N 2012 "*Using Mean Absolute Error for Forecast Accuracy*" [Online] Available: https://canworksmart.com/using-mean-absolute-error-forecast-accuracy/ [Accessed 3 November 2019]

[38] Vandeput N 2019 "*Forecast KPI: RMSE, MAE, MAPE & Bias*" [Online] Available: https://medium.com/analytics-vidhya/forecast-kpi-rmse-mae-mape-bias-cdc5703d242d [Accessed 3 November 2019]

[39] Briones R, Antonio E, Habito C, Porio E and Songco D 2017 "*Food Security and Nutrition in the Philippines*" (Manila: Brain Trust, Inc)

[40] Reid J 2012 "*What are Gaussian processes?*" [Online] Available: https://pythonhosted.org/infpy/gps.html [Accessed 4 December 2018]

[41] Weka 2017 *weka.classifiers.functions.SMOreg* (Hamilton, New Zealand: The University of Waikato)

In-store Customer Shopping Behavior Analysis by Utilizing RFID-enabled Shelf and Multilayer Perceptron Model

Ganjar Alfian[1], Muhammad Syafrudin[2], Jongtae Rhee[2], Pavel Stasa[3], Agus Mulyanto[4], Agung Fatwanto[4]

[1] Nano Information Technology Academy, Dongguk University, Seoul, Korea
[2] Department of Industrial and Systems Engineering, Dongguk University, Seoul, Korea
[3] Department of Economics and Control Systems, Faculty of Mining and Geology, VSB–Technical University of Ostrava, Czech Republic
[4] Teknik Informatika, Universitas Islam Negeri Sunan Kalijaga, Indonesia

Email: ganjar@dongguk.edu, udin@dongguk.edu, jtrhee@dongguk.edu, pavel.stasa@vsb.cz, agus.mulyanto@uin-suka.ac.id, agung.fatwanto@uin-suka.ac.id

Abstract. Understanding customer shopping behavior in retail store is important to improve the customers' relationship with the retailer, which can help to lift the revenue of the business. However, compared to online store, the customer browsing activities in the retail store is difficult to be analysed. Therefore, in this study the customer shopping behavior analysis (i.e., browsing activity) in retail store by utilizing radio frequency identification (RFID)-enabled shelf and machine learning model is proposed. First, the RFID technology is installed in the store shelf to monitor the movement tagged products. The dataset was gathered from receive signal strength (RSS) of the tags for different customer behavior scenario. The statistical features were extracted from RSS of tags. Finally, machine learning models were utilized to classify different customer shopping activities. The experiment result showed that the proposed model based on Multilayer Perceptron (MLP) outperformed other models by as much as 97.00%, 96.67%, 97.50%, and 96.57% for accuracy, precision, recall, and f-score, respectively. The proposed model can help the managers better understand what products customer interested in, so that can be utilized for product placement, promotion as well as relevant product recommendations to the customers.

1. Introduction

Understanding customer shopping behavior will provide useful input for the managers to improve the effectiveness of marketing and service quality. For online store, customer shopping behavior can be analyzed based on click streams (pages visited) and customer shopping carts (purchased products) data [1]. However, retail stores still lack effective method to comprehensively identify shopping behaviors. The retail stores only provided sales history data, which difficult to understand the customer behavior before they leave the store, such as what products the customer browse or interested in. Therefore, it is necessary to provide a solution to understand customer behavior in retail store.

Radio frequency identification (RFID) is well-known auto identification technology that has been applied to many areas especially for item-level in the retail store. Athauda et al [2] proposed RFID based smart shopping trolley. The UHF antenna is mounted to trolley so that tagged products can be traced in real-time. Benes et al [3] utilized RFID to detect customer movement. By utilizing tagged cart, it is possible to track movement and the time spent of customers. Syaekhoni et al [4] utilized RFID for analyzing shopping path of customers in retail store. By combining shopping path with purchasing data, the clustering analysis can be applied to group the customers and uncover the interesting characteristics of each group. Regarding customer browsing detection, Choi et al [5] and Zhou et al [6] utilized RFID to understand several customer activities such as picking up, browsing, returning, interest in and match up. They utilized count or total time of being read for tag [5] and phase readings [6] as input to distinguish different type of customer behavior.

Furthermore, received signal strength (RSS) information and machine learning models have been utilized to detect tagged product movement type in retail store. Hauser et al [7] utilized RSS of RFID tag to obtain different pattern of tag movement so that it can help retail store to prevent shop theft. In addition, the RSS of RFID tag could be utilized as input for automated checkout systems by correctly detecting the items that leaving a store [8].

Nevertheless, there is no study about customer shopping behavior analysis (i.e., customer browsing activity) by utilizing RSS of RFID tag as an input for machine learning models. Therefore, in this study we proposed multilayer perceptron model to identify customer shopping behavior with the RSS value as an input. In addition, the statistical features were extracted from RSS of tags to be utilized as attribute for classifiers.

2. Methodology

In our study, the customer shopping behavior in retail store, i.e., customer activity when browsing products were investigated. Firstly, the RFID-enabled shelf was installed in the retail store. The RFID-enable shelf consists of single RFID reader and an antenna facing directly to the products where passive tags were attached on them. Secondly, receive signal strength (RSS) from the tags were collected for different customer behavior scenario. The statistical features were extracted from received signal strength (RSS) of tags. Finally, several machine learning algorithms were applied to distinguish different customer behavior on the tagged product. Experiments were carried out in a laboratory environment as a typical retail store scenario. Figure 1a shows the possible customer behavior that feasibly occur on products, such as the product is being browsed by customer and no behavior (customer does not have attention to the product). Figure 1b shows an example of tagged product is being browsed by customer. In our scenario, the time needed by the customers to browse each product is approximately less than 15 seconds and during each session, RSS of tag products were gathered for further analysis.

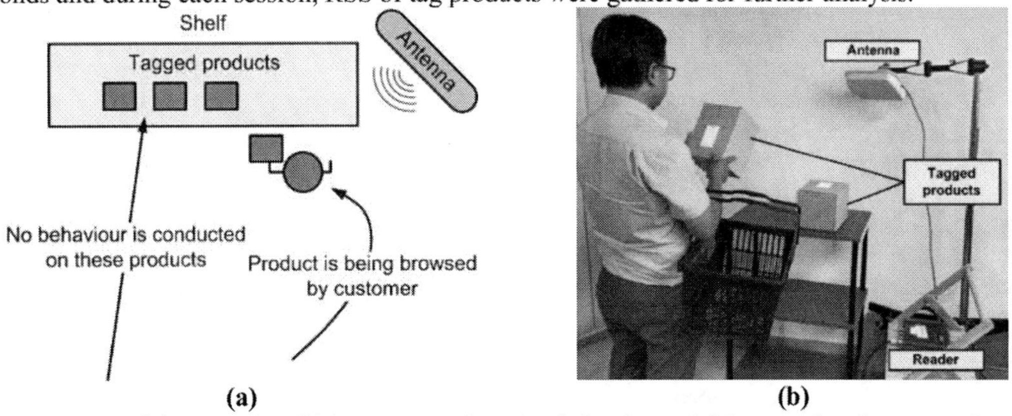

(a) **(b)**

Figure 1. Possible scenario of **(a)** customer shopping behavior and **(b)** example of customer is browsing the product.

We considered two type of tag reads, they are "no behavior" and "browsing" tag. "No behavior" represents situation where customer does not have any attention to the product, while "browsing" reveals the product is being browsed by customer. Single RFID reader ALR-9900+ and linear antennas ALR-9610-AL with 5.90 dbi Gain are utilized in this experiment [9]. In addition, UHF passive tags (model 9662, Alien H3) were attached to all products. We developed a gathering program based on Java programming language and installed in host computer connected to the reader. During each gathering session, the reading data were stored as CSV file for further analysis. In total, 109 unique data readings were collected. The total number of data readings for "no behavior" and "browsing" are 62 and 47 respectively. For a gathering session, the received data or collection of tag event is expressed as follows:

$$X = \{\{ID, T_1, RSS_1, Ant_1, Class\}, \ldots, \{ID, T_i, RSS_i, Ant_i, Class\}\} \quad (1)$$

where X denotes the data received by reader for tag ID, and $i = 1, 2, \ldots, n$. The parameter n represents the total number of tags occurs during a gathering session. Ant_i refers to the antenna which receives the tag data in the i_{th} time, while RSS_i denotes signal strength of the received data in the i_{th} time. The *Class* represent the class label of readings, where its value is either 0 (for "no behavior") or 1 (for "browsing").

Figure 2 shows example of RSS readings for a typical customer shopping behavior in retail store. For "no behavior", tag has relatively constant RSS, since the distance between the antenna and tagged product is fixed (Figure 2a). Figure 2b showed the product is being browsed by customer. The RSS decreases when tag stay away from antenna, achieving minimum when the tag is farthest from antenna or closest to the customer. When the product is returned to the original place, the RSS value increases and become constant again. As compared to the "no behavior" tag, the "browsing" tag tends to exhibit larger variance of RSS. However, different type of situation might arise in real case. Figure 2c showed the signal from "no behavior" or un-moved tag to antenna was blocked by the movement of other product. The RSS values decreases as other product move in between line of sight (LOS) of tag and antenna. Finally, figure 2d showed the tagged product is being browsed by customer but returned to the different place which is farther from antenna. The RSS value decreases as tag move away from antenna and generate lower RSS when it is placed farther from antenna. These conditions generate complex dataset of customer shopping behavior and appropriate attributes need to be extracted from time series dataset as input for classifiers.

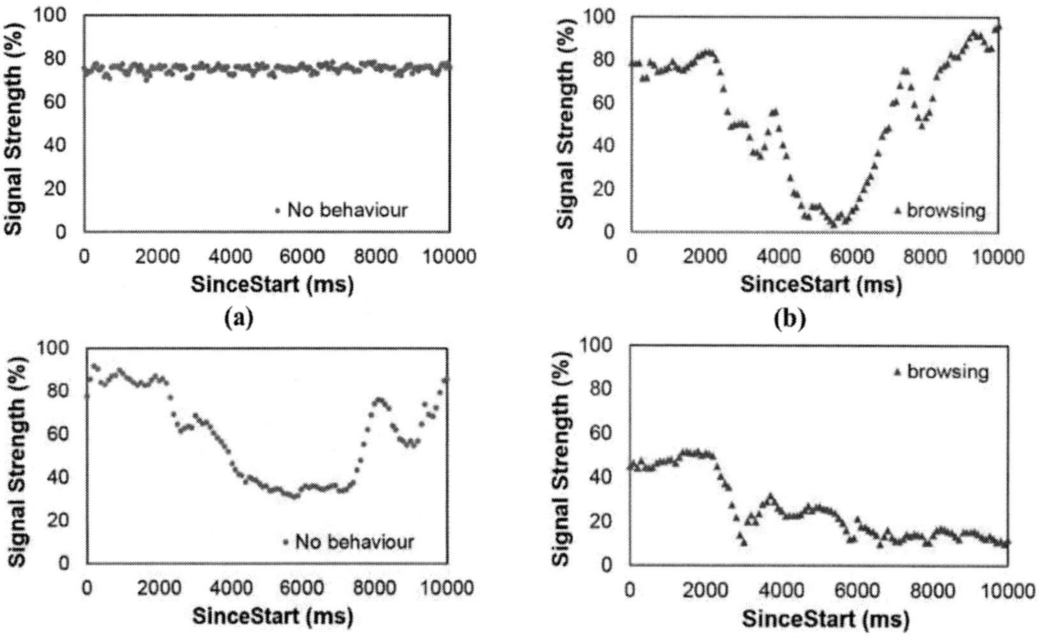

Figure 2. Example of RSS readings: **(a)** no behavior is conducted by customer to the product, **(b)** product is being browsed by customer, **(c)** signal of un-moved product is blocked by the movement of other product, and **(d)** product is being browsed and returned to different place by customer.

Since RSS information depends on the distance between the antenna and tag, closer tags generate larger RSS. Therefore, statistical information from RSS provided important information to differentiate between moves and un-moved tags and have been utilized in previous studies [10,11]. Table 1 shows the proposed statistical features extracted from the RSS of single antenna.

Table 1. Statistical attributes extracted from a gathering session.

Attribute no	Attribute name	Description
1	min	Minimum signal strength during a gathering session.
2	max	Maximum signal strength during a gathering session.
3	mean	Average signal strength during a gathering session.
4	std	RSS standard deviation during a gathering session.
5	diff	Difference between highest and lowest signal strength during a gathering session.
6	median	Middle value signal strength during a gathering session.
7	kurtosis	Indicates if the signal strength distribution is heavy- or light-tailed, relative to normal.
8	skew	Distribution asymmetry of signal strength during a gathering session.
9	count	Total number of reads for the tag during a gathering session.

The preprocessing step needs to be conducted to convert tag readings into input matrix X and output vector Y, so that conventional machine learning model can learn and predict the outcome. Finally, given n different total tag occurrence in each gathering session, 9 (nine) total number of statistical features and m total unique tag readings data, the input X can be derived by creating the $[m \times 9]$ matrix

$$X = \begin{bmatrix} \min(RSS_{1,1}, \dots, RSS_{1,n}) & \max(RSS_{1,1}, \dots, RSS_{1,n}) & \dots & \text{count}(RSS_{1,1}, \dots, RSS_{1,n}) \\ \min(RSS_{2,1}, \dots, RSS_{2,n}) & \max(RSS_{2,1}, \dots, RSS_{2,n}) & \dots & \text{count}(RSS_{2,1}, \dots, RSS_{2,n}) \\ \vdots & \vdots & \vdots & \vdots \\ \min(RSS_{m-1,1}, \dots, RSS_{m-1,n}) & \max(RSS_{m-1,1}, \dots, RSS_{m-1,n}) & \dots & \text{count}(RSS_{m-1,1}, \dots, RSS_{m-1,n}) \\ \min(RSS_{m,1}, \dots, RSS_{m,n}) & \max(RSS_{m,1}, \dots, RSS_{m,n}) & \dots & \text{count}(RSS_{m,1}, \dots, RSS_{m,n}) \end{bmatrix} \quad (2)$$

and the $[m \times 1]$ output vector Y.

$$Y = \begin{bmatrix} Class_1 \\ Class_2 \\ \vdots \\ Class_{m-1} \\ Class_m \end{bmatrix} \quad (3)$$

In this study, we employed Multilayer Perceptron (MLP) model to distinguish the browsing activity of customer on the product. The MLP is a class of feedforward artificial neural network (ANN) with one input layer, one or more hidden layers, and one output layer. The backpropagation algorithm is utilized to train the MLP [12,13]. Net input was calculated by multiplying each input and its corresponding weight, and then summed. Each unit in the hidden layer took net input and then applied an activation function. Backpropagation compared the prediction result with the target class value and modified the weights for each training tuple to minimize mean squared error between prediction and target values. This process was iterated multiple times to produce optimal weights, providing optimal predictions for the test data. Furthermore, we employed feature selection method based on extremely randomized trees (Extra-Trees) algorithm [14] to remove irrelevant features and applied the final relevant attribute to the MLP, so that we expect to improve accuracy of proposed model.

The data preprocessing, feature selection and machine learning models were implemented in Python V3.6.6 and Scikit-learn V0.19.1 [15]. We used default parameters provided by Scikit-learn to simplify experiment process. 10-fold cross-validation were employed for all classification models.

3. Result and Discussion

Table 2 compares various model performances in terms of the percentage of accuracy, precision, recall, and f-score. The machine-learning models such as Logistic Regression (LR), Decision tree (DT), Naïve Bayes (NB), Random Forest (RF), AdaBoost, Support Vector Machine (SVM) are compared with the proposed model based on Multilayer Perceptron (MLP) to distinguish the customer shopping behavior. In this scenario, statistical features are extracted from RSS of tagged products and used as input attributes for classification models. In addition, the tree-based feature selection was applied only to proposed MLP model. The findings revealed that the proposed model outperformed other models by as much as 97.00%, 96.67%, 97.50%, and 96.57% for accuracy, precision, recall, and f-score, respectively.

Table 2. Performance of classification models on customer shopping behavior.

Method	Performance Evaluation (%)			
	Accuracy	Precision	Recall	F-score
LR	93.44	98.00	87.00	91.11
DT	93.18	95.00	93.50	92.74
NB	92.18	95.00	91.00	90.83
RF	94.18	95.71	93.50	93.34
AdaBoost	93.18	95.00	93.50	92.74
SVM	91.44	98.00	82.00	87.50
Proposed MLP	97.00	96.67	97.50	96.57

Furthermore, the impact of feature selection on the accuracy of classification models are presented in Figure 3. The result showed that by utilizing feature selection method for classifiers, provided higher accuracy as compared to utilizing all attributes as input, except for LR. Our result showed that by removing irrelevant features, it could improve accuracy of classifiers. Finally, by employing tree-based feature selection, the average of accuracy is improved as much as 1.019% compared to classifiers without feature selection method.

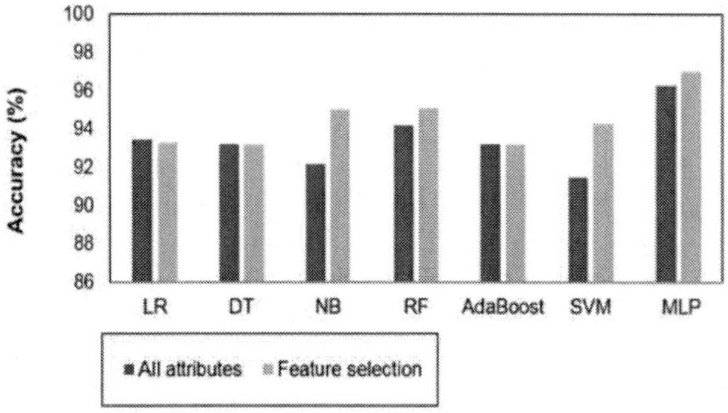

Figure 3. Impact of feature selection on classification accuracy.

The experimental results indicate that customer browsing activity on specific product can be detected by MLP model with high accuracy. By utilizing proposed model, the managers can better understand customer browsing pattern and what products customer interested in. Retail managers can use browsing and purchasing patterns for promotion as well as relevant product recommendations to the customers.

In addition, the managers might evaluate the product placement in retail store by removing the unpopular products from layout and promote other products that might boost their sales. Finally, as the interaction and quality of shopping increased, the customers' relationship with the retailer became stronger, which helped to lift the revenue of the business.

4. Conclusion and Future Works

This study proposed machine learning model to distinguish customer shopping behavior in retail store. The study utilized RFID-enabled shelf to track the movement of tagged products. The machine learning models were utilized to detect whether the tagged product was being browsed by customer or not. The RSS of tags were gathered, and statistical features were utilized as input for classification models. The proposed model based on Multilayer Perceptron (MLP) was compared with other classification models to detect the customer browsing activity on the products. The result revealed that the proposed model outperformed other models by as much as 97.00%, 96.67%, 97.50%, and 96.57% for accuracy, precision, recall, and f-score, respectively. Furthermore, the result showed that by employing tree-based feature selection, the average of accuracy was improved as much as 1.019% compared to classifiers without feature selection method.

Future study should consider more complex real situations, such as by considering other type of customer shopping behavior. Furthermore, employing different type of time series feature extraction and extending the comparison with other classification models, could be presented in the near future.

Acknowledgments

This research was financially supported by the Ministry of Trade, Industry and Energy (MOTIE) and the Korea Institute for Advancement of Technology (KIAT) through the International Cooperative R&D program (Grant number N0002301).

References

[1] Lee Y S, Yen S J, Tu G H and Hsieh M C 2004 Mining traveling and purchasing behaviors of customers in electronic commerce environment *Proc. of the IEEE Int. Conf. on e-Technology, e-Commerce and e-Service 2004 EEE '04 2004* pp 227-30

[2] Athauda T, Marin J C L, Lee J and Karmakar N C 2018 Robust Low-Cost Passive UHF RFID Based Smart Shopping Trolley *IEEE J. of Radio Frequency Identification* **2** no 3 pp 134-43

[3] Benes F, Stasa P, Svub J, Unucka J, Rhee J and Vojtech L 2017 Object Localization for Determining Customer's Behavior: Auto-ID Based Approach *Proc. of the 14th EAI Int. Conf. on Mobile and Ubiquitous Systems: Computing, Networking and Services* pp 468-77

[4] Syaekhoni M A, Alfian G and Kwon Y S 2017 Customer Purchasing Behavior Analysis as Alternatives for Supporting In-Store Green Marketing Decision-Making *Sustainability* **9** 2008

[5] Choi S H, Yang Y X, Yang B and Cheung H H 2015 Item-level RFID for enhancement of customer shopping experience in apparel retail *Computers in Industry* **71** pp 10-23

[6] Zhou Z, Shangguan L, Zheng X, Yang L and Liu Y 2017 Design and Implementation of an RFID-Based Customer Shopping Behavior Mining System *IEEE/ACM Transactions on Networking* **25** no 4 pp 2405-18

[7] Hauser M, Zügner D, Flath C M and Thiesse F 2015 Pushing the limits of RFID: Empowering RFID-based Electronic Article Surveillance with Data Analytics Techniques *Proc. of the 2015 Thirty Sixth Int. Conf. on Information Systems* Fort Worth TX USA 13–16 December 2015

[8] Hauser M, Günther S A, Flath C M and Thiesse F 2018 Towards Digital Transformation in Fashion Retailing: A Design-Oriented IS Research Study of Automated Checkout Systems *Bus. Inf. Syst. Eng.*

[9] Alien Technology 2019 *ALR-9900 Enterprise RFID Reader* http://www.alientechnology.com/wp-content/uploads/Alien-Technology-ALR-9900-Enterprise-RFID-Reader.pdf Retrieved May 30th 2019

[10] Ma H, Wang Y and Wang K 2018 Automatic detection of false positive RFID readings using

machine learning algorithms *Expert Syst. Appl.* **91** pp 442–51

[11] Alfian G, Syafrudin M, Yoon B and Rhee J 2019 False Positive RFID Detection Using Classification Models *Appl. Sci.* **9** 1154

[12] Han J, Kamber M and Pei J 2011 *Data Mining: Concepts and Techniques* 3rd ed M Kaufmann (MA USA: Publishers Burlington)

[13] Rumelhart D E, Hinton G E and Williams R J 1986 Learning representations by back-propagating errors *Nature* **323** pp 533–6

[14] Geurts P, Ernst D and Wehenkel L 2006 Extremely randomized trees *Mach. Learn.* **63** 3

[15] Pedregosa F, Varoquaux G, Gramfort A, Michel V, Thirion B, Grisel O, Blondel M, Prettenhofer P, Weiss R, Dubourg V, Vanderplas J, Passos A, Cournapeau D, Brucher M, Perrot M and Duchesnay É 2011 Scikit-learn: Machine Learning in Python *The J. of Machine Learning Research* **12** pp 2825-30

Establishing the Syntactic Rules of the Kankanaey Dialect using RNN

Laurie Lynne F. Aspiras[1], Ibrahim F. Hanbal[2], Jeffrey S. Ingosan[3], Kim Benedict P. Panlilio[4]

College of Information Technology and Computer Science, University of the Cordilleras, Baguio City 2600, Philippines.

Email: [1] laurielaspiras@gmail.com, [2] ibrahimhanbal@yahoo.com, [3] jsingosan@uc-bcf.edu.ph, [4] kimbpanlilio@gmail.com

Abstract. The diverse culture and ethnic groups in the Philippines creates a beautiful mixture of ideas, traditions, and practices but also makes it hard for researchers to keep track of them all. One integral part of any culture is language, with one of the most spoken languages in the Cordillera Administrative Region (CAR) being Kankanaey. Unfortunately, it has very little resources and documentation for it. This paper presents a corpus created for Kankanaey that contains 3412 words and was trained with a dataset containing 400 Kankanaey sentences in order to establish its syntactic rules. Data for the collected texts for Kankanaey were taken from public sources online and were organized into various categories based on the type of content. Training and testing was done to establish the syntactic rules using the Keras API. The rules were derived by having each word in the training sentences tagged with the corresponding POS tag. After tagging, the number of POS tags were then expanded to all possible combinations of the POS which resulted in the documenting of 1,722 syntactic rules for Kankanaey with the model having an accuracy of 64% when it was tested to identify the syntactic rules in 50 test sentences.

1. Introduction

Language is the core of every culture which an individual is able to create the meaning of human experience, thought, feeling, appearance, and behavior. In context, language is integrally intertwined with the concept of both the personal and the broader, societal levels as it is used to develop, strengthen, or even create a cultural, national, or other form of group identity [1]. One study found that as many as 6,000 languages worldwide are in danger of becoming extinct as the number of speakers for these languages are slowly dwindling with no new speakers being taught the language [2].

One way to address this threat is through the use of Natural Language Processing (NLP). NLP is a field concerned with the study of the syntax and semantics of a language [3, 4, 5, 6]. NLP is very important in today's globalized world as people from different backgrounds, professions, cultures and countries communicate with one another. A better understanding of all these languages will allow people to better understand one another and master these languages and dialects for various reasons [7, 8]. The first step in understanding a language is learning its grammar, or in more technical terms, its syntactic rules. Having knowledge of syntactic rules is very important especially for children as it enables them to better understand and use more complex versions of these rules as they grow up [9]. Once one is familiar with the syntax of a language, the next step will be understanding the semantics of the language.

Studies have also been conducted in using NLP tools to learn the Filipino language. As it is considered the national language of the Philippines [1] and is spoken by more than 55% of the population and a large portion of written, visual and audio content is created every day using Filipino [10]. Due to the large number of languages and dialects in the Philippines, researchers have turned to using NLP tools to help them study and synthesize the various languages and dialects in the Philippines. One example of this is a study conducted in 2003 for English-Filipino machine translation system. The system took input for a statement in the source language and produced the corresponding target language translation of the statement using a combination of rule based and corpus-based approaches [11].

A similar study was conducted in 2015 where a group of researchers created a Filipino-Maranao translator using machine translation technology. The translator was able to convert text-to-text and text-to-speech for Filipino-Maranao and vice versa with a high degree of accuracy [12]. Documenting the syntax of dialects in the Philippines is important now more than ever especially since there are many languages in the Philippines that endangered due to the shrinking number of people speaking the language [13].

This paper focused on the first step of understanding a language. The first objective was to collect and categorize texts for the Kankanaey dialect, one of the common dialects in the Cordilleras. The next objective is to tag several sentences in Kankanaey to establish the rules of the dialect. The last objective was the review the recorded rules and to determine which rules are correct and which are outliers in the data.

2. Methods

The Keras API was used to collect, tag and process the text data for training. The Keras API was utilized for the study due to its implementations of many machine-learning algorithms, its ease of use and its fast prototyping tools [14, 15]. The Keras API also has many tools from other deep learning libraries that make it very useful for building NLP applications with one of these tools being Recurrent Neural Networks (RNN).

2.1. Building of the Dictionary

Dictionary refers to a large database of lexical and grammatical resources used for study and analysis in NLP [16]. The dictionary was built in using Microsoft Excel, a spreadsheet application, and texts were collected from various sources online such as blogs, personal websites, other research papers, and translated works. The collected text was then organized into nine different categories for easier sorting and analysis. The categories are: (1) books, (2) dictionaries, (3) news articles, (4) poems, (5) religious texts, (6) songs, (7) stories, (8) web sources and (9) others. The sentences and text that were used for training and testing the model were taken from these nine categories. A large number of the sentences that were used for training and testing came from the religious text category as this contained the largest amount of collected text. A database of words with their associated tags in Kankanaey was created using the collected texts to be used for training and establishing the syntactic rules of the Kankanaey language. The words were tagged using the Parts of Speech (POS) tags of the Penn TreeBank study [17, 18, 19, 20, 21] which was a study that has analysed over 13 million words in the English language and has been used as a foundation for creating language corpora for other languages such as Chinese [22], French [23] and Turkish [24]. The Penn TreeBank POS tags that were used consisted of 36 POS tags and 12 others for punctuations and other symbols. An additional tag labelled as "NF" was also added to the tagset for words that were in the training sentences but were not found in the database of tagged words.

The dictionary containing words with their associated tags in Kankanaey was cleaned with duplicate, incorrect or misspelled words being removed [25, 26]. This resulted in a database containing 3412-tagged Kankanaey words used for NLP analysis and training. The model was trained using 400 Kankanaey sentences, which was used to establish the POS rules. After the POS patterns were compiled from these sentences, the patterns were then filtered so that only unique Kankanaey POS patterns remained.

2.2. Training the Model

Text mining is the process of extracting information and patterns from text documents [27, 28]. Text mining involves making use of information retrieval, text analysis, information extraction, clustering, categorization, visualization, database technology, machine learning, and data mining in order to gain knowledge from a collection of text [29, 30]. Text mining was done to establish the syntactic rules of the Kankanaey dialect. The dictionary containing 3412-tagged words were used as basis to scan and analyse the 400 training sentences. The model only covered 400 sentences as the hardware used to run the model would run into GPU (Graphics Processing Unit) errors when a larger number of sentences were entered. Each word in each of the training sentences were tagged with the corresponding POS tag based on the database. The words were tagged like so:

PRP|MD PRP IN|DT JJ. -> A | denotes that the word is associated with multiple POS tags.

The number of POS tags were then expanded to all possible combinations of the POS. Using the above example; we would derive these possible sentence structures.

PRP PRP IN JJ.
PRP PRP DT JJ.
MD PRP IN JJ.
MD PRP DT JJ.

These POS patterns were then recorded, resulting in 6299 possible POS patterns and 3624 possible unique POS patterns. The Keras library's LSTM Sequential Model was used to turn the sentences into POS patterns. The LSTM Sequential Model is a RNN model that has LSTM cellblocks in place of standard neural network layers. These cells have various components such as the input gate, the forget gate and the output gate. The model was trained for 48 hours with 187 epochs, as further passes resulted in no change in the model. According to the Keras documentation, an epoch is an arbitrary cutoff, generally defined as "one pass over the entire dataset", used to separate training into distinct phases, which is useful for logging and periodic evaluation." The current rules in this state are not readable and usable for analysis as the POS tags was arranged in one continuous line of text. There was a need to further process the compiled patterns to turn them into readable and usable rules, a generator was created.

2.3. Establishing the Rules

A Keras generator was then created to write POS patterns based on the trained model. It was made to run for 40 hours and generated 10,182 lines of POS patterns. Of these 10,192 POS patterns, 1,722 were unique patterns. The model was then tested using 50 new sentences using the 1,722 generated POS patterns. The model attempted to identify if the POS pattern of the sentences are in the generated POS patterns. The model was able to identify the POS patterns of 32 sentences out of the 50 test sentences which resulted in an accuracy of 64%.

3. Findings

The results of training have provided 1,722 syntactic rules in the Kankanaey language. The recorded rules indicate that the syntactic rule "PRP SBAR RB UH VB MD" has the highest occurrence with a total frequency of 220. Twenty of the recorded rules during training can be found on Table 1.

The model was then tested using 50 new sentences, which are different from the 400 training sentences, which resulted in 32 of the sentences having rules confirmed by the model and 18 having rules that were not recognized by the model. This resulted in the model having an accuracy of 64%.

4. Discussions

The results of the training indicate that 3412-tagged Kankanaey words was sufficient for training allowed for the generation of 1,722 unique POS patterns. The large number of POS patterns that came from the training was due to recording every possible combination of POS tags for the sentences. With a higher number of sentences, the model can be enhanced with a larger number of syntactic rules or higher frequencies for the recorded rules. The model was only able to read at most around 400 sentences due to limitations in the GPU used for training. A dedicated high end GPU can be utilized for future studies to allow for more sentences to be used for training and tagging. The training process can be further improved by adding a filter that will remove impossible or highly unlikely POS combinations when the POS patterns are being generated to better reflect real life sentences in Kankanaey.

Table 1. Twenty of the recorded syntactic rules and their frequency of the Kankanaey dialect

#	POS Tag Structure (Rules)	Frequency
1	PRP SBAR RB UH VB MD	220
2	DT PRP DT	211
3	PRP SBAR PRP MD VB MD	174
4	DT PRP MD PRP UH DT CC DT DT NN PRP MD	149
5	PRP MD DT PRP NN PRP UH DT CC DT SBAR NN VBD PRP UH DT PRP	111
6	IN DT NN RB UH VB MD PRP MD PRP UH DT DT IN DT NN IN PRP	108
7	DT NN PRP UH DT PRP MD CC PRP MD VBD DT NN IN PRP	107
8	PRP MD WRB DT NN IN PRP	106
9	PRP MD PRP CC DT SBAR PRP UH DT PRP MD RB UH VB MD PRP UH DT PRP UH DT NN	94
10	RB DT PRP PRP MD VB PRP IN DT PRP IN DT DT ADV	92
11	PRP MD VB CC DT DT SBAR IN DT NN	90
12	CC DT SBAR	86
13	JJ NN CC DT DT NN PRP MD PRP UH DT	80
14	JJ PRP MD CC PRP DT IN DT IN PRP IN IN PRP DT IN DT PRP MD IN DT	66
15	IN PRP DT IN PRP PRP UH DT PRP MD IN DT	62
16	CC DT SBAR DT PRP PRP CC PRP DT N DT	61
17	PRP DT PRP PRP UH DT PRP MD IN DT CC VB IN PRP NN PRP UH DT CC DT NN PRP UH DT	60
18	PRP DT NN PRP MD IN DT	59
19	RB DT PRP PRP MD VB PRP IN DT PRP IN DT DT	59
20	PRP MD DT PRP DT IN DT	58

The model was moderately successful in identifying the POS patterns of the sentences when it was tested, as its accuracy was 64%. As compared with the results of Gulordava et al., who obtained a prediction accuracy of 81% on English sentences from their test corpus. Considering that the English language is well documented while the Kankanaey dialect has a scarce of formally written documents - due to the speakers preferring oral tradition - the results were considered as moderately successful [31]. This suggests that the model is lacking in data and required more sentences for training in order to be effective. One possible reason for this is that the number of sentences used for training and testing were not sufficient to establish a large enough collection of syntactic rules for the Kankanaey language. Another limitation is that many of the sentences used for training also came from religious texts since they were the most prevalent, which have a tendency to have long and unique sentences structures which led to the model not being able to identify the more common sentences structures. Having a larger and more diverse set of sentences for training along with more generated POS patterns can increase the overall accuracy of the model when presented with new sentences to identify.

5. Conclusions

There are not a lot of studies or formal documents regarding the syntactic rules of the Kankanaey dialect, the researchers therefore conclude that this study would be one of the initial attempts to fill the distinct lack of formal documents. With the result of the study presenting that Keras API can be used to generate commonly used syntactic rules, future researchers may perform an in-depth comparison between the results of this study with those that make use of widely-known Natural Language Processing Python libraries such as NLTK, CoreNLP, Gensim, spaCy, and polyGlot.

The model can be further improved by expanding the number of sentences used for training and testing. This will increase the number of recorded rules, as this will allow for a larger amount of combinations for the POS and provide a higher frequency for the already recorded rules. A more powerful system can be utilized to train the model with a larger number of words and sentences. The recorded POS patterns can be further expanded by increasing the number of sentences that were used for training the model. Other sources of words and sentences such as village elders, local songs, hyms and stories and written works (such as old books) can be considered as another source to further expand the data of the corpus and improve the diversity of the rules in the corpus.

References

[1] Catama B et al 2018 Challenges in maintaining a heritage language *Int. J. of Research Studies in Language Learning 2018* vol **7** no 1 pp 39-61

[2] Krauss M 1992 The world's languages in crisis *Language* **68(1)** pp 4-10 https://doi.org/10.1353/lan.1992.0075

[3] Chowdhury G 2003 Natural language processing *Annual Review of Information Science and Technology* **37** pp 51-89 ISSN 0066-4200

[4] Zitouni I 2011 Introduction to Arabic Natural Language Processing *Computational Linguistics* **37** pp 623-5

[5] Lewis D and Sparck Jones K 1996 Natural Language Processing for Information Retrieval *Communications of the ACM* **39** DOI:10.1145/234173.234210

[6] Loper E and Bird S 2002 NLTK: the Natural Language Toolkit *CoRR* cs.CL/0205028 DOI: 10.3115/1118108.1118117

[7] David M K, Cavallaro F and Coluzzi P 2009 Language Policies - Impact on Language Maintenance and Teaching: Focus on Malaysia, Singapore, Brunei and the Philippines *The Linguistics J.* September

[8] Domingo D R 2016 Content Area Effectiveness: English Vs Filipino Medium of Instruction *PEOPLE: Int. J. of Social Sciences* **2(1)** pp 1514-29

[9] Menyuk P 1964 Syntactic Rules Used by Children from Preschool Through First Grade *Child Development* **35(2)** pp 533-46 doi:10.2307/1126717Fds

[10] Rau D V 1992 Language Contact in the Philippines *Philippine J. Of Linguistics* vol **23** no 1 and 2 June and December

[11] Roxas R E, Borra A, Cheng C, Lim N R, Ong E and Tan M W 2008 Building language resources for a multi-engine machine translation system *in Language Resources and Evaluation* vol **42** pp 183-95 (Berlin: Springer)

[12] Fallan J Q, Rasuman M A T, Sacil A V and Domingo I V R 2015 Filipino- maranao bi-directional language translator with text-to speech synthesizer *Int. J. of Information Technology and Business Management* 28th Feb 2015 vol **34** no 1

[13] Headland T N 2003 Thirty Endangered Languages in the Philippines *Work Papers of the Summer Institute of Linguistics, University of North Dakota Session* vol **47**

[14] Chauhan K and Ram S 2018 Image Classification with Deep Learning and Comparison between Different Convolutional Neural Network Structures using Tensorflow and Keras *Int. J. of Advance Engineering and Research Development* vol **5** Issue 02 February 2018

[15] Vidnerova P and Neruda R 2017 Evolving Keras Architectures for Sensor Data Analysis *Proc. of the Federated Conf. on Computer Science and Information Systems* pp 109–12

[16] Van Der Eijk P, Bloksma L and Van Der Kraan M 2003 Towards Developing Reusable NLP Dictionaries *COLING 1992 Proc. of the 14th Conf. on Computational linguistics* vol **1** pp 53-9

[17] Taylor A, Marcus M and Santorini B 2003 The Penn Treebank: An Overview *Treebanks: Text, Speech and Language Technology* ed A Abeillé vol **20** (Dordrecht: Springer)

[18] Chen J and Shanker V K 2004 Automated Extraction of Tags from the Penn Treebank *New Developments in Parsing Technology: Text, Speech and Language Technology* vol **23** ed H Bunt *et al* (Dordrecht: Springer)

[19] Marcus M *et al* 1994 The Penn Treebank: Annotating predicate argument structure *Proc. of the 1994 Human Language Technology Workshop* pp 110–5

[20] Marcus M, Ann Marcinkiewicz M and Santorini B 2002 Building a Large Annotated Corpus of English: The Penn Treebank *Computational Linguistics - Special issue on using large corpora: II* vol **19** Issue 2 June pp 313-30

[21] Fleming A J, Kingsbury P T, Palmer M and Marcus M 1998 Adding Semantic Annotation to the Penn TreeBank *Proc. of the second Int. Conf. on Human Language Technology Research* 2002 pp 252-6

[22] Xue N, Xia F, Chiou F-D and Palmer M 2005 The Penn Chinese TreeBank: Phrase structure annotation of a large corpus *The J. of Natural Language Engineering* vol **11** Issue 2 June 2005 pp 207-38

[23] Abeillé A, Clément L and Toussenel F 2003 Building a Treebank for French *Treebanks: Text, Speech and Language Technology* vol **20** ed Abeillé A (Dordrecht: Springer)

[24] Oflazer K, Say B, Hakkani-Tür D Z and Tür G 2003 Building a Turkish Treebank *Treebanks: Text, Speech and Language Technology* vol **20** ed Abeillé A (Dordrecht: Springer)

[25] Batista G, Prati R and Monard M-C 2004 A Study of the Behavior of Several Methods for Balancing machine Learning Training Data *SIGKDD Explorations* **6** pp 20-9 DOI: 10.1145/1007730.1007735

[26] Guyon I, Matic N and Vapnik V 1994 Discovering Informative Patterns and Data Cleaning *Advances In Knowledge Discovery And Data Mining* pp 145-56

[27] Fayyad U, Piatetsky-Shapiro G and Smyth P 1996 From data mining to knowledge discovery: An Overview *Advances in Knowledge Discovery and Data Mining* ed U Fayyad *et al* (Cambridge, Mass.: MIT Press) pp 1-36

[28] Simoudis E 1996 Reality check for data mining *IEEE Expert* **11(5)**

[29] Tan A H *et al* 1999 Text mining: The state of the art and the challenges *Proc. of the PAKDD 1999 Workshop on Knowledge Disocovery from Advanced Databases* vol **8** pp 65–70

[30] Jacobs P S 2002 Joining Statistics with NLP for Text Categorization *ANLC '92 Proc. of the third Conf. on Applied natural language processing* pp 178-85 DOI: 10.3115/974499.974532

[31] Gulordava K, Bojanowski P, Grave E, Linzen T and Baroni M 2018 Colorless green recurrent networks dream hierarchically *Proc. of the 2018 Conf. of the North American Chapter of the Association for Computational Linguistics: Human Language Technologies* vol **1** pp 1195–205

Hybridization of Genetic Algorithm and Artificial Immune System for Assignment Problem

La Ode Muhammad Farhan[1], Zainudin Zukhri[2]

Department of Informatics, Islamic University of Indonesia

Email: [1]ldmuhfarhanf@gmail.com, [2]zainudin@uii.ac.id

Abstract. Genetic Algorithms (GA) have proven its reliability to obtain near optimum for complex problems. However, to optimize solutions and prevent premature convergence, operators on the GA were highly dependent on large population sizes, causing the computational speed to be slower than newer algorithms. Therefore, in this study, we propose a way to cause a significant improvement in GA for a higher quality of solutions and lower computing costs yet easy to implement. Hence, we analyzed the clonal selection algorithm from AIS to inspire genetic processes in GA. We then compared the performance of the proposed algorithm called HGA as with the GA to solve the assignment problem. Simulation results showed that HGA performed better in case of preventing on being stuck with the local optima as it accurately obtains an average of the optimum solution up to 100% for every given dataset while also reducing computational costs as it has less generation.

1. Introduction

Genetic Algorithm (GA) is one of the evolutionary computation methods that mostly known and used in a wide range of problem areas. The application of GA has been tested in various fields, one of which is the assignment problem using the Traveling Salesman Problem (TSP) approach as in [1]. In that research, the use of GA operators has proven to solve assignment problems and provide optimal solutions. Even though it can find effective global optimum values, GA cannot guarantee that the solutions provided are not stuck at the local optimum. This can occur depending on the scale of the problem at hand. In this case, GA and other similar algorithms are very dependent on population size so that they are vulnerable to the problem of premature convergence that triggers stagnation, causes the search process to stop and leads to wasteful computing. Although these problems can be overcome with large populations, computing time will increase dramatically. Coupled with the structural complexity and complex genetic operators of GA, this can cause computational speed becomes slower compared to newer algorithms [2].

Artificial Immune System (AIS) is a bio-inspired computing area, but it is relatively new compared to GA. AIS adopts the principle of immunity or the body's defense system from attacks by harmful microorganisms such as viruses and bacteria (pathogens). To overcome the pathogen attack, AIS carries out a learning process or an adaptation of the attacking pathogen so that an optimal immune response is achieved. In its application, AIS shows that the algorithm used can be applied to solve complex problems that can be done by GA in general as in [3] and [7]. AIS and GA have a similar algorithm. This is inseparable from the principle of evolution that exists in each of these algorithms. However, the most striking difference between AIS and GA is the absence of crossover operators on the AIS because normally cells only need to duplicate themselves to increase diversity in the population. Even so, AIS is still able to produce optimal solutions and reduce computing costs. With the assumption that AIS can solve problems that can also be achieved by GA, this algorithm can be the key to solving the complexity of algorithms that GA has in general.

Several studies have been carried out to improve GA performance, including GA hybridization as in [5] and [6], and simplification of GA as in [2] and [7] but few have tried to hybridize GA and AIS. In a recent study, simplified GA was inspired by AIS to solve MST problems by removing the crossover operator [7]. The results of the study stated that the proposed method can solve the problem as GA did even without a crossover and discussed that the research needs to be done by solving other complex problems. Therefore, in this study, we conduct a hybridization of GA and AIS to solve the assignment problem. We then carried out the analysis with a comparative study between three algorithms. The first algorithm is the real coded genetic algorithm (RGA) which was used as a benchmark for comparison against the proposed algorithms. The second algorithm that we proposed is the simplification of genetic algorithms by ignoring the crossover operator (SGA). Meanwhile, the third algorithm that we proposed is the hybridization of GA and AIS (HGA) with the ideas to extract some prominent feats from the clonal selection algorithm to the GA. This comparative study is carried out by comparing the computational time and the quality of the solutions obtained from each algorithm. The data that we tested is in the form of an assignment matrix based on 10 data sets that have tested before with the assumption that many resources and activities are the same for each data set. The results of this trial are expected to contribute to the application of bio-inspired algorithms, especially GA and AIS so that the development areas in this field which cover the theoretical scope and application of theories to overcome various optimization problems can be more advanced.

2. Literature Review

In this section, we presented several works related to this study that has tried to improve GA performance by hybridizing GA with AIS and also discussed the clonal selection algorithm as one of the AIS methods that inspire the proposed method which is HGA. We then explored the nature of AIS particularly the clonal selection and analyze the prominent feats of its algorithm.

2.1. Related Works

The research on the hybridization of GA-AIS was carried out by Ali *et al.* as in [8]. The performances of the Hybrid method is tested based on eight popular mathematical benchmark functions. The Hybrid includes two processes; firstly running the AIS to earn the desired solution. After that, GA is run with the initial population taken from the solution from the last generation of AIS with the purpose that Hybrid makes GA enters the stage of standard solutions rapidly and accurately higher compared with GA initialized population at random. Moreover, the results after 20 times trials based on the minimum value of objective function and time for each method indicate that Hybrid GA-AIS is performs better compared with GA and AIS separately as it obtained the optimum solution with acceptable time. Although the proposed method has better performance, it is not simple to apply given the complexity of the algorithm and the number of parameters that need to be determined.

In another study, Zukhri [7] proposed a simplified genetic algorithm (GA_0) to solve the MST problem by adopting AIS operator. In this case, the complexity of real coded GA (GA_1) is simplified by setting aside crossover operators based on the similarity between the operators of GA and AIS. Moreover, in AIS, the cloning operator can be said to belong to the selection operator so that there are only two operators, namely selection and mutation. Hence, the difference in GA leaves only the crossover operator. Assuming that AIS can solve complex problems even without crossover operators, this certainly can be applied to GA. By consider this analogy, a trial to compare the performance of GA_0 and GA_1 were performed on 3 data sets with consecutive nodes of 6, 8, and 13. The trial took random parameter values that normally used with the condition that both algorithms were tested using the same parameter values. From the results of the two-sample t-test, it can be said that there is no difference in the quality of the solution produced between the two algorithms so it can be concluded that GA_0 considered being well implemented in other complex problems and also brings convenience when determining parameters. Thus, in this study, we tested this approach and also implemented it on the HGA that we are proposing to solve complex assignment problems.

2.2. *Clonal selection algorithm*

There are three main theories on the immune system that underlie the birth of the field of Artificial Immune Systems, namely the Immune Network Theory, the Negative Selection theory, and the Clonal Selection theory [9]. With the similar structure of the algorithm between GA and clonal selection algorithm, the clonal selection is easy to implement and uses for hybridization with GA.

2.2.1. Biological explanation. Clone selection is a term used to describe the basic response of the adaptive immune system to antigenic stimuli. The idea of this theory states that only cells that can recognize antigens with high affinity can reproduce by multiplying themselves (proliferation) while other cells that cannot recognize antigens will not be preserved. In this case, they will destroy (apoptosis). Furthermore, after proliferation, the formed cells will be subject to mutations with a fairly high probability (hypermutation). Cells that have an increased affinity will be differentiated to form an immune response while cells that have decreased affinity will be destroyed [10]. Thus, unlike the simple genetic algorithm which generally avoids high mutation rates, the mutation in clone selection is the main operators that forming population. These characteristics will be further explored in the proposed HGA. Meanwhile, the clonal selection process, in general, can be seen in Figure 1 below.

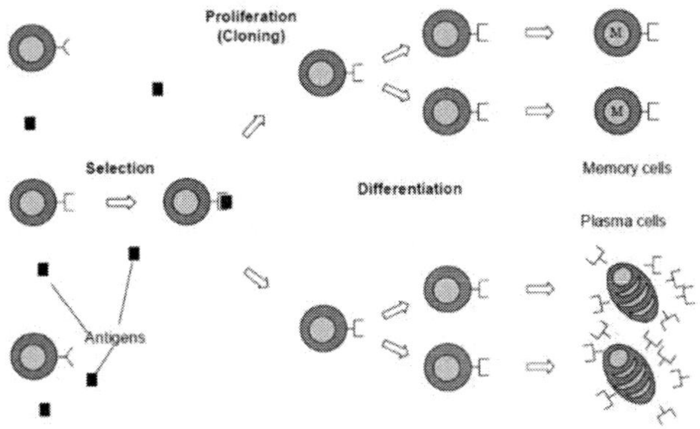

Figure 1. The Mechanism of clonal selection for the immune system
(Liang Hongtao and Kang Fengju, 2016:3461)

2.2.2. Analysis of algorithm. In order to solve global optimization problems, the hybrid algorithm should strike a balance in exploration (global search) and exploitation (local search) [11]. Several studies support this approach and apply the GA hybridization method with local search [6, 8, 12]. Although GA can find optimum global, in many occasions, it has a proclivity to being stuck on local optima. Several reasons might explain this behavior such as weak initial population, highly dependent on population size, and lack of local search. Meanwhile, the clonal selection in AIS has local search attributes that can be used in GA to balance the global search. Therefore, we are using this capability of AIS to inspire GA in certain parts of genetic operators.

In the clonal selection algorithm, there are three main operators used, namely selection, cloning, and mutation. The immune response obtained in the form of a group of antibodies represents a solution. Meanwhile, the antigen acts as a fitness function so that the strength of the receptor binding (affinity) to the antigen represents the quality of the solution. Furthermore, clonal selection has the capability of local search, it can be seen in the cloning process (clone pool). In this process, the algorithm exploits its antibodies whereas the best clone from a particular antibody selected from the mutated clone pool and had a chance to replace the current antibody if the affinity is higher. The clone pool selection has a similar function in the mutation of GA. The difference is GA uses the low probability of mutation and no selection thus allowing it to move further into other search areas while keep maintains diversity at an acceptable rate. However, mutations in GA often lead to unfavorable search points, causing slow convergence. Therefore, in this study, we modified the mutation in GA to mimic the local search as in clone pool selection. This approach is used on HGA as shown in Figure 4.

3. Methods

To complete this research various stages are carried out as follows:

- Conduct a comparative study between GA and AIS.
- Model RGA, SGA, and HGA algorithms to solve assignment problems.
- Collect datasets in the form of assignment matrices of varying sizes.
- Implement model design in the form of simulation software.
- Conduct experimental testing of simulation software that has been developed to given datasets and performs results analysis.
- Conduct a comparative study of the effect of the proposed model on the performance of GA.

3.1. RGA for solving assignment problem

Assignment problems are complex problems that can be used in experiments to compare performance between optimization algorithms. The type of assignment problem taken in this study is a one-to-one assignment problem. In this type, there are n resources and n activities. To solve this problem, a real coded genetic algorithm (RGA) is modeled from chromosome representation to genetic operators. The RGA model is used as a basis for modeling the proposed algorithm such as SGA and HGA.

3.1.1. Chromosome Representation. To solve the one-to-one assignment problem, this study proposes a combinatorial chromosome representation in the form of positive integers whose values range from 1 to n (the total number of genes). Each gene (g_i) on a chromosome represents a resource while the gene index represents an activity. The chromosome representation can be seen in Figure 2 below.

activities-j:	1	2	3	4	5	....	n
resources-i:	g_1	g_2	g_3	g_4	g_5	....	g_n

Figure 2. Proposed chromosome representation

3.1.2. Fitness function. The objective function is used to calculate the total cost obtained from all assignments of resources to activities. If the costs required by the resources in i-th gene (g_i) to carry out the i-th activity are represented as C_{ig_i}, then the objective function to determine the total cost can be formulated as in equation (1) below.

$$f(x) = \sum_{i=1}^{n} C_{ig_i}(x) \qquad (1)$$

The fitness function can be determined based on the optimization problem. If the assignment problem uses a function for maximization, then the fitness function used can be defined directly from the objective function as in equation (1). Whereas if the assignment problem uses a function for minimization, the fitness function described as eval(x) used is the opposite of the objective function as shown in equation (2) below.

$$\text{eval}(x)=\frac{1}{f(x)}=\frac{1}{\sum_{i=1}^{n} C_{ig_i}(x)} \tag{2}$$

3.1.3. RGA operators. In this study, any method will be chosen to determine the effect of simplification and modification made on SGA and HGA operators. However, the chosen method is the same for all three algorithms tested. The methods to be used are roulette wheel selection combined with elitism, order based crossover using absolute replacement, and order based mutation.

3.2. SGA

In this study, a simplified genetic algorithm (SGA) is proposed and tested for its performances. SGA previously proposed and tested by Zukhri to solve the minimum spanning tree problem (MST) [6]. This algorithm is tested again for a different problem, namely the assignment problem. The algorithm is modeled by eliminating the crossover operators, so SGA only uses the selection and mutation operators to manipulate the population.

3.3. HGA

HGA is modeled by removing the crossover operators and mimicking local search as in clone pool selection. Moreover, a fairly high probability of mutation is applied in HGA. Since local search is applied to mutation, more selection is used causing vast convergence. To balance the global search with local search, reinitiate population is activated when the average fitness of the population is below the certain threshold. Therefore, the entire operation on the HGA can be seen in Figure 3 below.

```
Pseudocode of HGA
1. Encode solution space
2. Set pop_size, chrom_size, max_gen, reinitiate threshold ,
   mutation_rate, Gen = 0
3. Initialize population P randomly
4. For each individual i ϵ P: calculate fitness (i)
5. While Gen < max_gen
   * do selection
   * do mutation
   * Test: Test whether the average fitness of population P below the
     reinitiate threshold. If so, reinitiate P. If not, continue using
     current P.
   * do replacement
   * save current best solution
6. Return the best solution
```

Figure 3. Pseudocode of HGA

Henceforth, local searches based on clone pool selection applied in genetic operator. The algorithm is added to mutation after the clone is mutated. The pseudocode of an additional algorithm in HGA mutation can be seen in Figure 4 below.

Pseudocode of additional algorithm in mutation
1. Calculate the fitness of chromosomes
2. Calculate the fitness of clone
3. If (fitness_clone > fitness_chromosome)
   Replace the chromosome with the clone if better

Figure 4. Pseudocode of an additional algorithm in HGA mutation

4. Test Cases

The analysis to compare the three algorithms starts with determining the data sets and parameters to be used. Testing is done to get results in the form of solution quality and computational costs. The testing is done using java programming language. Meanwhile, all the executions were progressed using a computer with an Intel (R) HD Core ™ i5-5200U CPU that works to a frequency of clock of 2.20 GHz, 4 GB of RAM Memory with Windows 8 operating system.

4.1. Data sets

The data sets that will be used in this study are one-to-one assignment problem data sets with a matrix size from 6x6 to 15x15. The 6x6 up to 10x10 matrix are primary data sets that have previously been tested in various studies. While 11x11 up to 15x15 matrix are engineered data sets (dummy) in which the values are randomly generated from 40 to 60.

Table 1. Special parameters of algorithms

Method	Parameters		
	mr	cr	$r\theta$
RGA	0.05	0.5	-
SGA	0.05	-	-
HGA	0.75	-	0.01

Table 2. General parameters of algorithms

Matrix size	Opt. solution (99.5 %)	Population size	Maximum of generation	Number of elites (RGA & SGA)	Number of elites (HGA)	Number of trial
6x6	105,21	20	1000	4	1	349
7x7	95,11	20	2000	4	1	349
8x8	33,06	25	3000	4	1	349
9x9	625,32	40	4000	6	1	349
10x10	387,45	60	5000	8	1	349
11x11	471,84	75	6000	10	1	349
12x12	519,87	80	7000	12	1	349
13x13	543,08	100	8000	12	1	349
14x14	589,17	150	9000	15	1	349
15x15	616,32	200	10000	15	1	349

4.2. Parameters of each algorithm

In this study, the value of each parameter is taken arbitrarily. There are two types of parameters that will be used, namely special parameters and general parameters. In the special parameters as shown in Table 1, each algorithm has its value. In this case, the operator used determines what parameters are used. Whereas in the general parameters as shown in Table 2, all algorithms assign the same value but will change as the size of the assignment matrix increases.

4.3. Experiment and results

In this study, each assignment matrix will be subjected to 349 trials for performances testing. The number of trials is based on the Isaac and Michael Tables. First, the average computational time will be measured and then normalized as in equation (3) whilst given 1000 ms as the estimated time to get the optimum value.

$$\text{Normalized exec. time} = \frac{\text{Average execution time}}{1000}.100\% \tag{3}$$

Second, the average quality of the solution will be calculated based on the maximum generation that has been set. Furthermore, the average quality of the solution for each assignment matrix will be measured for its accuracy to the global optimum value as in equation (4) below.

$$\text{Accuracy of the solution} = \frac{\text{Max. solution - Solution}}{\text{Max. solution - Min. solution}}.100\% \tag{4}$$

The results of experiments to measure computational time and the quality of solutions can be seen as in Table 3 below.

Table 3. Comparison of results from experiments

Matrix size	Average accuracy of the solution (%)			Average normalized execution time per 1000 ms (%)		
	RGA	SGA	HGA	RGA	SGA	HGA
6x6	100,00	100,00	100,00	0,77	0,60	0,42
7x7	100,00	100,00	100,00	0,71	0,61	0,49
8x8	100,00	100,00	100,00	0,98	0,82	0,55
9x9	100,00	100,00	100,00	0,86	0,72	0,58
10x10	100,00	100,00	100,00	3,78	3,07	1,41
11x11	100,00	100,00	100,00	7,54	5,82	1,90
12x12	99,80	99,78	100,00	49,95	28,15	3,15
13x13	100,00	100,00	100,00	13,72	10,29	1,44
14x14	99,96	99,86	100,00	33,72	25,92	2,43
15x15	99,92	99,85	100,00	38,34	36,35	1,97

4.4. Comparative analysis

After the experimental results are obtained, a statistical test of two-sample t-tests is then performed to determine the effect of the removal of the crossing and hybridization operators on the quality of the RGA solution. In this case, given a hypothesis (H0) based on statistical data, there is an assumption that the quality of the solution produced by SGA and HGA is the same as the quality of the solution produced by RGA. Meanwhile, the samples to be compared with two-sample t-tests are the 12x12, 14x14, and 15x15 matrices. This is occurred because the other data samples have obtained an average solution quality of up to 100%. The results based on the H0 can be seen in Table 4 below.

Table 4. Two-sample t-test results between

Method Compared	Dataset	Two sample t test ($\alpha = 0.01$)		Two sample t test results
		Statistical t value	2-tailed critical t value	
SGA and RGA	12x12	0,099845	2,582912	H0: $\mu1 = \mu2$ accepted
	14x14	3,055085	2,582912	H0: $\mu1 \neq \mu2$ rejected
	15x15	5,112955	2,582912	H0: $\mu1 \neq \mu2$ rejected
HGA and RGA	12x12	-8,49995	2,582912	H0: $\mu1 \neq \mu2$ rejected
	14x14	-2,66886	2,582912	H0: $\mu1 \neq \mu2$ rejected
	15x15	-9,67696	2,582912	H0: $\mu1 \neq \mu2$ rejected

5. Discussions

The results of experiments and analysis of the three algorithms namely RGA, SGA, and HGA can picture the performance of each algorithm in solving a variety of complex problems, especially the assignment problem that was tested in this study. From the analysis result of the quality of the solution, there is no significant difference between RGA, SGA, and HGA over 1000 generations although HGA provides the best average solution quality. Furthermore, based on the results of the two-sample t-test conducted to prove the H0 hypothesis, it can be assumed that for relatively small-sized datasets there are no significant differences whereas for relatively large-sized datasets its estimated that there will be differences in the quality of the resulting solutions. Henceforth, conducting test at larger data sizes with greater complexity of problems is also necessary.

In another experiment, HGA provides a significant reduction in terms of time complexity. In this test, an estimated maximum of 1 second is given to find the global optimum solution. As a result, HGA obtained the lowest percentage in approaching 1 second compared to the other two algorithms, followed by SGA and then RGA. In this case, HGA can reduce the complexity of RGA time significantly because it has the lowest percentage of time usage and also the lowest average optimum generation, while SGA can also reduce the time complexity of RGA even though the average optimum generation owned by SGA is greater than RGA.

The excellent performance of HGA depends on the mutation process where the clone pool selection is added to the mutation operator to mimic local search. Also, mutation on HGA used fairly high probability on mutation rate. Therefore, it can be assumed that the higher the mutation probability, the faster the convergence will occur which will affect the decrease in time complexity. For further research, a simplification can also be made on the HGA by eliminating the selection operator so that only mutation remains.

6. Conclusions

The Hybridization of Genetic Algorithm and Artificial Immune System can be effectively carried out and successfully tested to solve complex problems such as one-to-one assignment problems. The solution obtained for all datasets with matrix sizes ranges from 6x6 to 15x15 can reach global optimum values. Meanwhile, GA-AIS hybridization (HGA) can also simplify the process of RGA while providing a fairly high increase in performance compared to the RGA or SGA methods. HGA can significantly reduce the time complexity, has the highest quality of the solution and has the lowest optimum generation. Also, HGA is easy to implement because it doesn't need the crossing operator and the whole process of AIS to hybridize.

References

[1] Zukhri Z 2004 Penyelesaian Masalah Penugasan dengan Algoritme Genetika *Seminar Nasional Aplikasi Teknologi Informasi* pp 51-8

[2] Ngamtawee R and Wardkein P 2014 Simplify and Improve RGA for Parameter Optimizations *Adv. in Electrical and Computer Eng.* vol **14** no 4 pp 55-64

[3] Coello C C, Rivera D C and Cortes N C 2003 Use of an Artficial Immune System for Job Scheduling *Int. Conf. on Artificial Immune Systems*

[4] Idris I 2012 Model and Algorithm in Artificial Immune System for Spam Detection *Int. J. of Artificial Intelligence & Applications* vol **3** no **1** pp 83-94

[5] Ahmed Z and Habeebvulla M 2014 A Hybrid Genetic Algorithm for the Quadratic Assignment Problem *Proc. of the 2nd Int. Conf. on Emerging Research on Computing, Information, Communication and Applications* (Amsterdam: Elsevier) pp 916-22

[6] Ariani A K and Wayan F M 2017 Hibridisasi Algoritme Genetika dan Simulated Annealing untuk Optimasi Multi-Trip Vehicle Routing Problem with Time Windows *Jurnal Pengembangan Teknologi Informasi dan Ilmu Komputer* vol **1** no **12** pp 1608-15

[7] Zukhri Z 2017 Penyelesaian Masalah Pohon Rentang Minimum Dengan Algoritme Genetika yang Disederhanakan Suatu Adopsi Operator Dari Sistem Kekebalan Buatan *Prosiding Seminar Nasional Inovasi dan Aplikasi Teknologi di Industri*

[8] Ali M O, Koh S P, Chong K H and Yap D F 2010 Hybrid AIS-GA based on mathematical test function *Proc. of 2010 IEEE Student Conf. on Research and Development* pp 256-61

[9] Purbasari A and Supriana I 2011 Clonal Selection Algorithm: Bio Inspired Algorithms Sebagai Solusi Persoalan Kompleks *Konferensi Nasional Sistem dan Informatika* pp 329-34

[10] Timmis J, Hone A, Stibor T and Clark E 2008 Theoretical Advances in Artificial Immune Systems *Theoretical Computer Sci.* (Amsterdam: Elsevier) pp 11-32

[11] Anu *et al* 2013 Hybridizing GA with Hill Climbing in Replacement Operator *Int. J. of Adv. Research in Computer Sci. and Software Eng.* vol **3** issue **12** pp 270-8

[12] Rahmi A, Mahmudy W F and Anam S 2017 Hibridisasi Algen dan Variabel Neighborhood Search Pada Optimasi Biaya Distribusi *Jurnal Teknologi Informasi dan Ilmu Komputer* vol **4** no **2** pp 87-96

[13] Hongtao L and Fengju K 2016 Adaptive Chaos Parallel Clonal Selection Algorithm for Objective Opimization in WTA Application *Optik* (Amsterdam: Elsevier) pp 3459-65

Establishing the Syntactic Rules of the Kankana-ey Dialect using TensorFlow

Angeli A. Cabilitazan[1], Jeffrey S. Ingosan[2], Jason Irabagon[3], Philip Irving G. Jacinto[4], Jan Jefferson O. Lam[5]

College of Information Technology and Computing Science, University of the Cordilleras, Baguio City, Philippines.

Email: [1] g.anncabilitazan96@gmail.com, [2] jsingosan@uc-bcf.edu.ph, [3]trize8081@gmail.com, [4] philsicom@gmail.com, [5] janjeffersonlam@gmail.com,

Abstract. Kankana-ey is a widely used dialect in the northern region of the Philippines. Unfortunately, there are no documented studies on the syntactic rules of this dialect. This study explored the development of a corpus for the Kankana-ey dialect. Further, the corpus was then used to establish the syntactic rules of Kankana-ey. A Kankana-ey version of the bible, dictionaries, news articles, songs and various online resources were used to collect words for the corpus of the Kankana-ey dialect. These identified words were also tagged using the parts of speech tags of the Penn TreeBank. Using the corpus and TensorFlow, 320 Kankana-ey sentences were analysed to determine the syntactic rules. In addition, 80 sentences were used to test the accuracy of the identified rules. At the end of the study, the created corpus has 3,412 tagged Kankana-ey words, while the analysis of the syntactic rules resulted to 1,722 rules. Testing also showed a 60% accuracy of the syntactic rules. In conclusion, the high number of identified rules from the 320 sentences was due to multiple Kankana-ey words having different possible tags. This also resulted to the low accuracy of the syntactic rules

1. Introduction

One particular dialect known in the Cordillera Administrative Region is the Kankana-ey. Kankana-ey is a South-Central Cordilleran dialect spoken on the northern part of Luzon in the Philippines. On the 1990 to 2003 census, there are around an approximate of 240,000 native speakers cited and is also closely related with the Ilocano dialect. Alongside Ilocano, it is widely used in the Mountain Province and northern part of Benguet Province. Visitors of these provinces, such as linguists, tourists, researchers, teachers, advertisers, journalists and reporters, find it hard to communicate with the locals especially when these locals only speak Kankana-ey.

The only means to understand each other is through translators. However, human translators may not be readily available and there are still no available portable and electronic translators. The study on syntactic rules of the Kankana-ey language is also needed to be able to identify the rules that can be used for electronic translators [3,7,10].

Natural Language Processing (NLP) during the last two decades has made tremendous leaps in terms of language processing such as syntax analysis. However, current technologies would still require further studies of languages and dialects used today [3].

Researches stated the need to focus on studies in NLP. Knowledge acquisition, interaction with multiple underlying systems and partial understanding has been cited as each addresses a critical challenge in using the techniques developed to address problems encountered in NLP [3,7,11,12].

It is important to note that in the study of language, such as English, grammatical or syntactically valid does not imply sensible, or semantically valid statements.

This study aimed to do a basic syntactic analysis of the Kankana-ey dialect. Specifically, this study sought the creation a corpus for the Kankana-ey dialect; use the corpus to identify the syntactic rules; and determine the accuracy of the identified syntactic rules.

2. Review of Related Literature

In learning a language, especially during the early years, data that are not structured in any way are studied. After amassing this massive information, the learner processes and gives meaning to these based on the implications as to where these can be applied. Machine in turn also learn the same as people do, they follow a similar path where machine processes that unstructured data into a somewhat meaningful information and determines relevant connections to comprehend the context [12, 14, 15]. In the most recent decade, a sensational move in the NLP research has prompted the pervasiveness of extremely enormous scale uses of factual techniques, for example, AI and information mining. Normally, this additionally opened the path to learning and streamlining techniques that comprise the center of current AI, most eminently hereditary calculations and neural systems [2, 7].

As mentioned by D.B. Jones and H. Somers, classifiers could be applied on the training and testing of data, as there are several classifiers available. In tagging parts of speech, several techniques could be used and applied. Existing programs pick the word from the corpus and based on the most probable grammatical category a grammatical pattern is then created giving sense to the sentence and allowing a deeper understanding on how the sentence is created and used [16, 17].

In the study of corpus-based translation, the first thing to note is what is the meaning of corpora and encoding [16]. Part of the concern of the study is the availability of the data whether printed or in electronic form. With the available data, annotations are made in order to create a stable, flexible and accessible corpus resource to be used in language/translation studies [18, 19].

Content generated today on the web has substantially reached a point where a need for sensible computation and information aggregation has reached a critical point as the trend today by big data experts and the growth of new fields in data science. As NLP research shifts from lexical semantics to compositional semantics, it offers various new insights in the development of new NLP technologies [13, 14, 17].

Using RNN in the study, tags where given to the words using the Penn Treebank Parts-of-Speech tags. With that the creation of grammar rules was done, and the model was trained predicting the next set of P.O.S. based on the grammar rules created from the number of occurrences it made during the training/learning phase [18, 19].

In training a model, recently in the areas of machine learning and NLP a new field of study called learning to rank has emerged. In some sense it refers on how to rank the techniques used in machine learning and NLP based on practicality of the techniques used and applied in practice. This applies to the study, where a need to create a ranking system is thought of. Using supervised learning training data were used as input and allows the system to generate a ranking model based on that training data. It would then help improve the performance of the algorithm used [3, 15, 16, 19, 20].

This will help non-expert clients effectively recover data without the need to become familiar with any extra coding languages or projects. The examination information incorporates regular language messages gathered from different sources, both on the web and disconnected, to cover on both formal and semi-formal language level. By utilizing normal language handling and procedures, for example, semantic examples, philosophy, and intuitive discussion framework, this model can investigate the culmination and significance of common language explanations just as it enables clients to alter the fragmented or flawed proclamations, and improve the model by including new words, sentence punctuation and semantic examples for increasingly exact outcomes [10].

C. Tapsai [9] mentioned in their research the proposed usage of several algorithms in determining the relativity of the word used in the sentence to the meaning it would convey when used in a sentence using Word Usage Frequency (WUF). This is to determine how words are used and how frequent they appear in a sentence allowing a more reliable dictionary which significantly raises the efficiency and precision all the way reducing the dictionary size by a significant amount of margin. Similar to this study is the usage of WUF in determining the frequency of words used allowing one to determine how often this kind of sentence structure is used on the Kankana-ey dialect [5, 17, 20].

3. Methodology

The study was conducted with the use of TensorFlow. The researchers used Recurrent Neural Networks (RNN) to accomplish the needed task for identifying the syntactic rules or pattern in the Kankana-ey dialect.

3.1. Creating the Corpus

Various texts from the internet such as dictionaries, poems, and articles were collected. The texts that were collected were categorized as (1) books, (2) dictionaries, (3) news article, (4) poems, (5) religious text, (6) songs, (7) stories, (8) web sources, and (9) others. Unique words from these collected texts were listed using Microsoft Excel. Cleaning of data was also done by removing redundant data and checking for the spelling of the words. The Parts of Speech tags from the Penn TreeBank were used to tag the listed words. It is also important to note that several words were tagged under several parts of speech. To further validate the corpus, Kankana-ey speakers were asked to review the corpus.

Sentence:	She sells seashells on the seashore.
POS Tag Structure:	PRP -> VBZ -> NNS -> IN -> DT ->

Figure 1. Sample sentence with its corresponding POS Tag Structure

3.2. Model Training

A total of 400 sentences were taken from the same texts used to create the corpus of Kankana-ey words. 80% of these sentences were used to determine the syntactic patterns of the Kankana-ey language. RNN was used through TensorFlow to identify these syntactic patterns. The remaining 20% of the sentences were used to check the accuracy of the identified syntactic patterns. Figure 1 shows a sample syntactic rule.

Table 1. A slightly more complex table with a narrow caption.

Syntactic Rule	Frequency
VB – NN – JJ	300
NN – VB – JJ	280
PRP – VBZ – NNZ – IN – DT – NN	243
PRP – SBAR – PRP – MD – VB – MD	184
RB – VB – JJ	179
NN – JJ – VB	188
NN – VB – RB	187
RB – DT – PRP – PRP – MD – VB – PRP – IN – DT – PRP – IN – DT – DT	155
VB – JJ – RB	149
PRP – SBAR – PRP – MD – VB – MD	131
IN – DT – NN – RB – UH – VB – MD – PRP – MD – PRP – UH – DT – DT – IN – DT – NN – IN – PRP	107
PRP – MD – DT – PRP – DT – IN – DT	90
CC – DT – SBAR – DT – PRP – PRP – CC – PRP – DT – N – DT	84
CC – DT – SBAR	50

4. Findings

A total of 3,412 words were identified and tagged during the creation of the Kankana-ey Corpus. These tagged words came from the texts that were collected. During the extraction of the syntactic rules, 1,722 rules were identified from the 320 evaluated sentences. Table 1 shows examples of the identified syntactic rules. The sentence pattern of VB-NN-JJ has the highest frequency of 300.

The researchers used 80 sentences for testing. 320 sentences were also used to train the model. The accuracy of the model is only 60%.

5. Discussions

The 3,412 words of the created Kankana-ey corpus has words with more than one tag. This is due to the possibility of a word to be used as different parts of speech. For example, a word can be a verb and another part of speech depending on how this is used. This is also common in other dialects or languages such as English. In English for example, fly can be used as a verb or a noun.

From the training process to extract the syntactic rules, a total of 1,722 rules were generated from just 320 sentences. This was encountered due to the fact that words from the corpus could have different parts of speech. This resulted to a sentence having different syntactic rules. In reality, the 1,722 rules are deemed valid on the basis of the arrangement of words but not necessarily on the semantics. The accuracy of the testing at 60% and can still be improved by increasing the training and testing data.

6. Conclusions

The corpus of the Kankana-ey dialect is composed of words that can be used as different parts of speech. This can thus result to a corpus with a greater number of words than the actual number of uniquely spelled words. As a consequence, the use of this corpus on the extraction of the syntactic rule of a sentence will lead to at least one rule. Although, some of these rules may not be acceptable if semantics is to be considered. As a general conclusion a certain level of semantic analysis may be useful in the study of syntactic rules.

References

[1] Zaccone G 2016 *Getting Started with TensorFlow* (Birmingham: Packt Publishing)

[2] Gelbukh A 2005 Natural Language Processing *Fifth Int. Conf. on Hybrid Intelligent Systems* p 1 DOI:10.1109/ICHIS.2005.79

[3] Zong Z and Hong C 2018 On Application of Natural Language Processing in Machine Translation *2018 3rd Int. Conf. on Mechanical, Control and Computer Engineering (ICMCCE)* pp 506-10 DOI: 10.1109/ICMCCE.2018.00112

[4] Xihong Wu, Meng Zhang and Xiaojun Lin 2011 Parsing-based Chinese word segmentation integrating morphological and syntactic information *2011 7th Int. Conf. on Natural Language Processing and Knowledge Engineering* pp 114-21 DOI:10.1109/NLPKE.2011.6138178

[5] Allen J L 2003 Operators and Clause Structure in Kankana-ey *Philippine J. of Linguistics*

[6] Zhao L and Li F 2008 Statistical Machine Learning in Natural Language Understanding: Object Constraint Language Translator for Business Process *2008 IEEE Int. Symp. on Knowledge Acquisition and Modeling Workshop* pp 1056-9 DOI: 10.1109/KAMW.2008.4810674

[7] Kumar P, Srivastava S and Joshi M 2015 Syntax directed translator for English to Hindi language *2015 IEEE Int. Conf. on Research in Computational Intelligence and Communication Networks (ICRCICN)* pp 455-9 DOI: 10.1109/ICRCICN.2015.7434282

[8] De Silva D *et al* 2008 Sinhala to English Language Translator *2008 4th Int. Conf. on Information and Automation for Sustainability* pp 419-24 DOI: 10.1109/ICIAFS.2008.4783983

[9] Lane R and Bansal A 2017 An Adaptive Machine Translator for Multilingual Communication *2017 IEEE 26th Int. Conf. on Enabling Technologies: Infrastructure for Collaborative Enterprises (WETICE)* pp 21-3 DOI: 10.1109/WETICE.2017.53

[10] Tapsai C 2018 Information Processing and Retrieval from CSV File by Natural Language *2018 IEEE 3rd Int. Conf. on Communication and Information Systems (ICCIS)* pp 212-6 DOI: 10.1109/ICOMIS.2018.8644947

[11] Bates M and Weischedel R M 1993 *Challenges in Natural Language Processing* (Cambridge: Cambridge Press)

[12] Ishaq M 2019 *What are some of the challenges we face in NLP today* [online]

https://medium.com/datadriveninvestor/what-are-some-of-the-challenges-we-face-in-nlp-today-2e9d94da1f63

[13] Cambria E and White B 2014 Jumping NLP Curves: A Review of Natural Language Processing Research *IEEE Computational intelligence magazine* vol **9** no 2 pp 48-57 DOI: 10.1109/MCI.2014.2307227

[14] Sag I A, Baldwin T, Bond F, Copestake A and Flickinger D 2002 Multiword Expressions: A Pain in the Neck for NLP *Proc. of Int. Conf. on Intelligent Text Processing and Computational Linguistics* ed A Gelbukh (Berlin, Heidelberg: Springer) pp 1-15

[15] Li H 2011 *Learning to Rank for Information Retrieval and Natural Language Processing* (California: Morgan Claypool)

[16] Jones D B and Somers H L 1997 *New Methods in Language Processing* (London: UCL Press)

[17] Zanentin F 2011 *Hardwiring Corpus-Based Translation Studies: Corpus Encoding Corpus Based Translation Studies* (London: Bloomsbury Academics)

[18] Yao K, Zweig G, Hwang M, Shi Y and Yu D 2013 Recurrent Neural Networks for Language Understanding *Int. Symp. on Computer Architecture* DOI: 10.13140/2.1.2755.3285

[19] Arenas M, Maturana F, Riveros C and Vrgoc D 2016 A framework for annotating CSV-like data *Proc. of the VLDB* **9** 11 pp 876-87

[20] Tapsai C, Meesad P and Haruechaiyasak C 2016 TLS-ART: Thai Language Segmentation by Automatic Ranking Trie *The 9th Int. Conf. Autonomous Systems*

[21] Marcus M P, Santorini B and Marcinkiewicz M A 1993 *Building a large annotated corpus of English: the Penn Treebank* Retrieved from https://catalog.ldc.upenn.edu/docs/LDC95T7/cl93.html

[22] Granger S 2004 Computer learner corpus research: Current status and future prospects *Applied Corpus Linguistics: A Multidimensional Perspective* vol 52 pp 123–45

[23] Chun S 2011 Learners as corpus researchers *Foreign Languages Education* vol **18** no 3 pp 73-103

[24] Mason O 2008 Developing Software for Corpus Research *Int. J. of English Studies* vol **8** no 1 pp 141-56 DOI: 10.6018/ijes.8.1.49141

[25] Granger S, Gilquin G and Meunier F 2015 Introduction: Learner corpus research – past, present and future *Cambridge Handbook of Learner Corpus Research* ed S Granger *et al* (Cambridge: Cambridge University Press) pp 1-5

Determining the usage feasibility of battery for disaster mitigation devices using Fuzzy Inference with Tsukamoto

Agus Darmawan[1], La Ode M. Farhan[2], Sabri Kurniadi[3], Elyza G. Wahyuni[4]

Department of Informatics, Islamic University of Indonesia

Email: [1]agusdar97@gmail.com, [2]ldmuhfarhanf@gmail.com, [3]sabrikurniadi16@gmail.com, [4]elyza@uii.ac.id

Abstract. Batteries have become an electrochemical device that plays an essential role in human life. Lately, batteries have also been used in disaster research, especially volcanic disaster mitigation in Indonesia. In addition, some electronic devices require a high amount of batteries thus these devices can remain alive for the continuation of mitigation research. Therefore, we research the fuzzy inference system to determine the usage feasibility of battery in Mount Merapi. The system is used for the simulation of the application of battery monitoring which is done periodically. As a result, users can monitor and evaluate the use of a high amount of batteries at several installation points easily and cost-effectively. Moreover, from the results of the test cases using the fuzzy inference system and fuzzy manual calculations, it can be concluded that both calculations produce the same output of fuzzy value for two test cases with a different dataset.

1. Introduction

Batteries have become an electrochemical device that plays an essential role in human life, almost all electronic and transportation devices use the battery as a device that can continue to conduct electricity. Lately, the battery has also been used in disaster research, especially volcanic disaster mitigation in Indonesia. Battery plays a role in supplying electricity to electronic equipment used by BPPTKG (Geological Disaster Technology Research and Development Agency) in monitoring the condition of Mount Merapi in Indonesia. Besides, some electronic devices require a high amount of batteries thus these devices can remain alive for the continuation of mitigation research.

Battery power will decrease until it runs out if used continuously. Therefore, the battery needs to recharge to maintain power in the normal ranges. This process can damage the battery if cannot be used properly. Cell components in the battery cannot last long in temperatures that are too high or temperatures that are too low. Also, some factors may affect the downgrade of battery usage feasibility such as voltage or current that is incompatible with the initial length or the charging time that is above or below the normal time required.

Correspondingly to prevent the use of batteries in vain, we conduct a study to determine the feasibility status of battery usage using the fuzzy inference method. Also, to make it easier to monitor, an expert system will be developed which has the functions to monitor several battery points installed around disaster mitigation areas. Therefore, the data of battery condition in the form of temperature, voltage, current and charging time will be read over when the charging process occurs and transmitted after the charging process is complete. However, in this study the problem is constrained to the implementation of methods to determine the feasibility status of battery use and the development of a prototype system for monitoring thus we use the simulated data in the process. Hence, this study is used to provide a concept of maintaining the continuity of batteries condition thus it is not easily damaged and it is suitable for reuse.

2. Literature Review

The research on monitoring and determining the battery conditions has been carried out by several studies. Agustian [1] tried to design a tool based on the internet of things to monitor the condition of dry batteries in motor vehicles in the form of temperature, voltage, and current. The lowest recommended voltage level is a minimum of 10.5 volts. While for batteries that are in good condition, the battery voltage will range from 12.3 - 12.6 volts. For battery temperature, it takes 30 degrees Celsius as a maximum value when the battery is in charging or the battery condition is being used. Furthermore, the warning will occur when battery conditions are not within normal limits. Although the tool can read and maintain the conditions of battery usage, it has not been embedded with a fuzzy inference system to determine the usage feasibility of battery so that its use is still limited to current data reading and more depends on the subjectivity of the user.

In another study, Novia [2] developed a tool that can monitor battery conditions and can prevent the battery from overheating due to overcharging by using a fan that has been controlled by a fuzzy logic system. This tool is designed based on fuzzy logic, based on 3 sensors that will detect the conditions of temperature, current, and battery voltage. Although the tool is already embedded with fuzzy logic and can prevent overheat due to the fan feature, it has not yet determined the usage feasibility for the future use of the battery. Moreover, using the fuzzy logic to control the state of charge (SOC) is also studied by Dwiono and Taufiq [3] which research the fuzzy performance controller for battery charging. This study also explains the results of the design and testing of acid battery charging using a fixed current with the fuzzy controller which is then compared to the PID controller. The state of charge ratio is calculated using a fuzzy logic system with input variables are temperature and battery voltage, then the fuzzy controller is used using two input variables namely the error and integral error variables and one output variable in the form of a duty cycle weight. The results stated that the fizzy controller gain less error than the PID. One in common to previous studies is that the fuzzy system to determine the usage feasibility of the battery has not yet been developed.

Lately, the study to determine the feasibility of an electronic device, especially the usage feasibility of battery is quite in demand. For example, in 2019, Wang et al conducted a study that estimates the State of Function (SOF) of power lithium-ion batteries [4]. In that study, 3 fuzzy variables are used such as battery status when charging (SOC), battery health status (SOH) and Maximum C-rate (charge and discharge Rate) using the Fuzzy c-Means Clustering algorithm method (FCM). The results obtained from that study aim to estimate the magnitude of the percentage or the feasibility of using lithium-ion batteries before being marketed or leaving the factory.

As the development of the Internet of Things (IoT) architecture is increasing, several studies including the previous which have been explained have suggested the real-time monitoring system on battery conditions. While others enhanced the system or tools by adding embedded fuzzy logic to control and maintain the state of charge to normal conditions. However, the expert system which uses fuzzy inference to determine the usage feasibility of battery and also provides monitoring of many batteries at the same time has not yet been developed in regards to mitigation researches. Therefore, we developed a system that was used to facilitate the monitoring of several batteries and determine its condition at several points in the Merapi volcano region. The final result of this research is to provide a simulation or prototype of a monitoring system that embedding the fuzzy inference to determine the usage feasibility of the battery periodically.

3. Methods

3.1. Fuzzy Inference with Tsukamoto

Tsukamoto is a vague conclusion drawing method where each consequent to IF-THEN rules needs to be represented by a vague set with a monotonous membership function. As a result, the inference results from each rule are given explicitly (crisp) based on α-predicate (fire strength). The final result is obtained by using a weighted average [5]. Tsukamoto is easy to implement yet still powerful thus makes it usually suitable for simple problems and calculations that are not too complicated. While Mamdani is suitable for intuitive problems and Sugeno is suitable for problems that handle control. As for the flow of inference process with Tsukamoto can be seen in Figure 1 below.

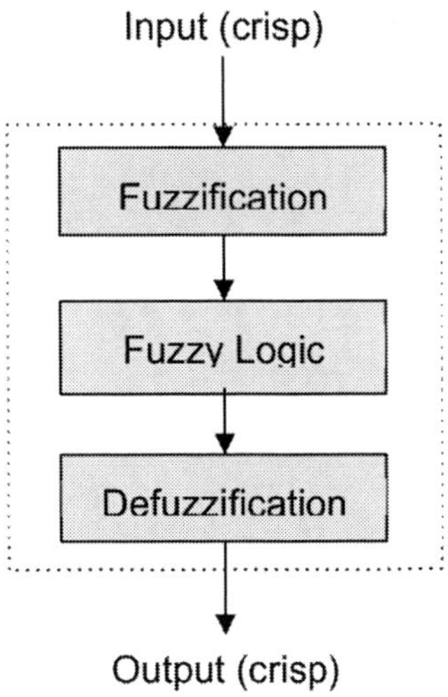

Figure 1. The flow of inference process with Tsukamoto

The design process in the Tsukamoto fuzzy inference method has ease compared to other methods such as Mamdani and Sugeno. This is based on the implications and aggregation process that does not exist. In addition to its simplicity, several studies can prove the superiority of the Tsukamoto method compared to other methods as in [6] and [7].

3.2. Fuzzy Variables and Fuzzy Sets

In this study, we use 4 fuzzy variables as input which taken from an expert in volcanic disaster mitigation at BPPTKG. Those variables are temperature, voltage, current, and charging time which used to determine the usage feasibility of a battery. Each of the variables consists of several fuzzy sets in the specified domain. Meanwhile, detailed information about these fuzzy variables with respectively corresponding fuzzy sets can be seen in Table 1 below.

Table 1. Detailed information of fuzzy variables and fuzzy sets

Fuzzy Variable	**Fuzzy Sets**	**Explanation**
Temperature	COLD, NORMAL, WARM	The battery temperature on charging
Voltage	LOW, NORMAL, HIGH	The battery voltage on charging
Current	POOR, STRONG	The current power on charging
Charging Time	QUICK, SLOW	The time needed until the charging process is complete

3.3. Membership Functions

The membership function is used to map inputs to the degree of membership which symbolized with $\mu(x)$. In this study, the membership function used is a combination of a triangle shape curve and a shoulder shape curve. There are 5 membership functions which consist of 4 membership functions to map the input in the form of crisp to the degree of the membership and a membership function on the feasibility variable to map the degree of membership into the output in the form of crisp.

3.3.1. The Input of temperature. The first curve of the membership function is a function of the temperature fuzzy variable in °C consisting of three sets with their respective domains namely COLD [0,22], NORMAL [20,30], and WARM [28,35]. The membership functions for the temperature fuzzy variable can be seen in Figure 2 below.

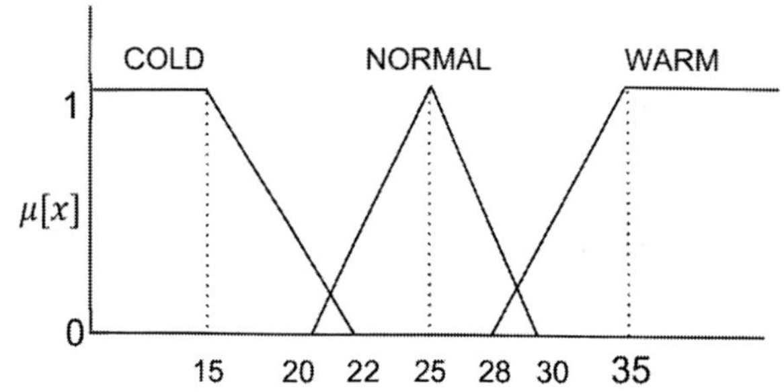

Figure 2. Membership functions for temperature fuzzy variable

3.3.2. The Input of voltage. The second curve of the membership function is a function of the voltage fuzzy variable consisting of three sets with their respective domains namely LOW [0, 12.2], NORMAL [12, 13], and HIGH [12.8, 13.8]. The membership functions for the temperature fuzzy variable can be seen in Figure 3 below.

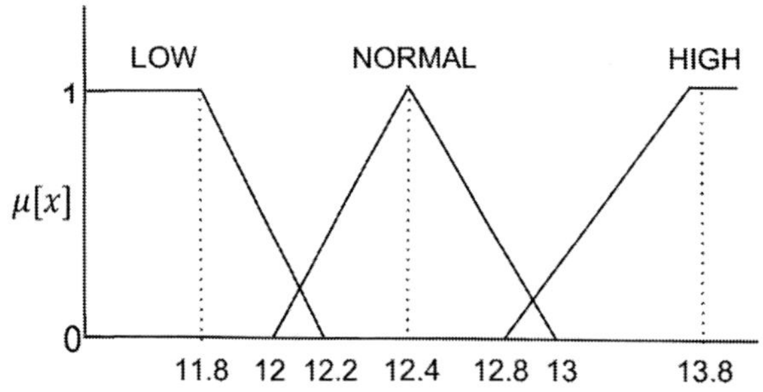

Figure 3. Membership functions for voltage fuzzy variable

3.3.3. The Input of current. The third curve of the membership function is a function of the current fuzzy variable consisting only two sets with their respective domains namely POOR [0, 12] and STRONG [8, 24]. The membership functions for the current fuzzy variable can be seen in Figure 4 below.

3.3.4. The Input of charge time. The fourth curve of the membership function is a function of the charge time fuzzy variable in hour consisting only two sets with their respective domains namely QUICK [0, 2.5] and SLOW [1.6, 10]. The membership functions for the charge time fuzzy variable can be seen in Figure 5 below.

3.3.5. The Output of usage feasibility. The last curve of the membership function is a function of the usage feasibility fuzzy variable in percent consisting only two sets with their respective domains namely NOT FEASIBLE [0, 50] and FEASIBLE [25, 100]. The membership functions for the usage feasibility fuzzy variable can be seen in Figure 6 below.

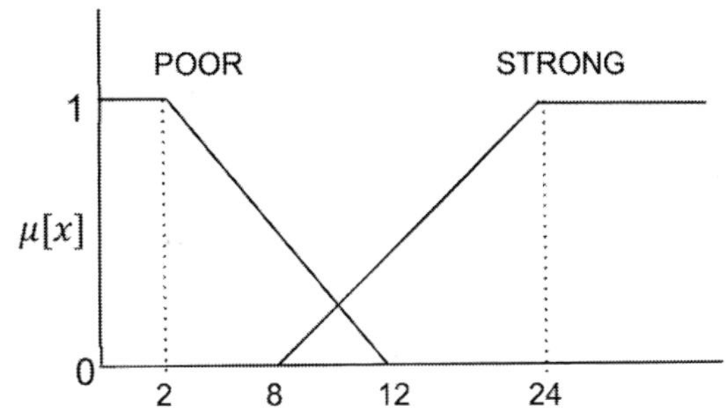

Figure 4. Membership functions for current fuzzy variable

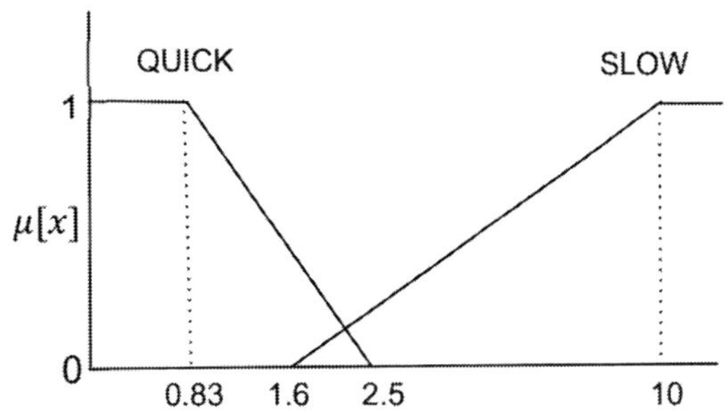

Figure 5. Membership functions for charge time fuzzy variable

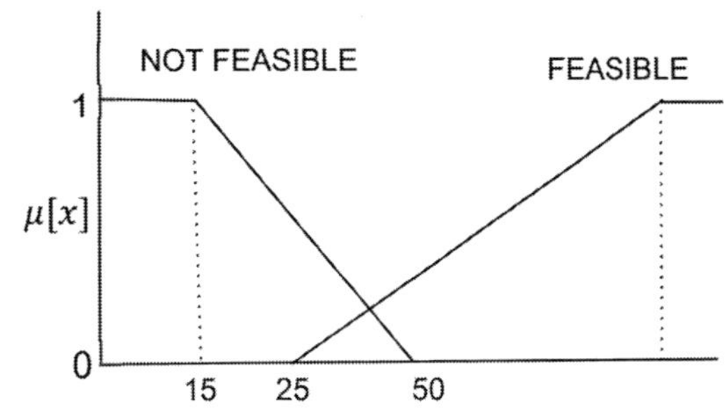

Figure 6. Membership functions for usage feasibility fuzzy variable

3.4. Fuzzy Rules.
There are several fuzzy rules compiled based on the result of an interview with the expert on volcanic disaster mitigation. In this study, there are 4 fuzzy variables which consist of 3 temperature sets, 3 voltage sets, 2 current sets, and 2 charge time sets respectively. Each variable gets one allotment of occurrences in one rule with the fuzzy operator used is an intersection in which order does not matter. From this definition, there are 36 combinations of rules that allowed to use in the inference process. However, in this study only used 6 rules that can be seen in Table 2 below.

3.5. Defuzzification
Defuzzification is the calculation process to get the crisp output value of the fuzzy value symbolized by z. Deffuzification on Tsukamoto is usually obtained using a weighted average function where there is z_i which is a result of defuzzification of fuzzy values obtained from the inference process for each rule (α-predicate). Therefore the value of z is a weighted average obtained from one or more rules that can be formulated in equation (1) below.

$$z = \frac{\sum_{i=1}^{n} \alpha_i z_i}{\sum_{i=1}^{n} \alpha_i} = \frac{\alpha_1 z_1 + \alpha_2 z_2 + \ldots + \alpha_n z_n}{\alpha_1 + \alpha_2 + \ldots + \alpha_n} \tag{1}$$

Meanwhile, to get the value α_i, the intersection operator is used. If the result of the intersection is symbolized as $\mu_{A \cap B}$, α_i can be obtained by equation (2) below.

$$\alpha_i = \min(\mu_A[x], \mu_B[y]) \tag{2}$$

Table 2. Fuzzy rules of inference

Symbol	Rule
R1	IF Temperature is NORMAL AND Voltage is NORMAL AND Charge Time is SLOW AND Current is STRONG THEN Usage Feasibility is FEASIBLE
R2	IF Temperature is COLD AND Voltage is NORMAL AND Charge Time is SLOW AND Current is STRONG THEN Usage Feasibility is NOT FEASIBLE
R3	IF Temperature is NORMAL AND Voltage is NORMAL AND Charge Time is QUICK AND Current is POOR THEN Usage Feasibility is NOT FEASIBLE
R4	IF Temperature is WARM AND Voltage is NORMAL AND Charge Time is SLOW AND Current is POOR THEN Usage Feasibility is FEASIBLE
R5	IF Temperature is WARM AND Voltage is NORMAL AND Charge Time is SLOW AND Current is STRONG THEN Usage Feasibility is NOT FEASIBLE
R6	IF Temperature is COLD AND Voltage is NORMAL AND Charge Time is QUICK AND Current is STRONG THEN Usage Feasibility is FEASIBLE

Afterward, the value of z_i which is defuzzification will be determined based on the membership function of usage feasibility. Mapping α_i to z_i is done based on consequent in certain rules. If consequent is FEASIBLE, the function used can be seen as in equation (3), but if it is NOT FEASIBLE then the function used can be seen as in equation (4). In the equations below, a is prefer to the left side of the domain and b is prefer to the right side of the domain.

$$z_i = (b-a) \times \alpha_i + a \tag{3}$$

$$z_i = b - \alpha_i \times (b-a) \tag{4}$$

3.6. System Design

A web-based system will be created to provide an interface for users, especially the BPPTKG, who want to monitor the condition of battery usage easily. This system is designed with the PHP 7 programming language and MySQL DBMS. The features contained in this system include rule management, membership function management, installed battery condition monitoring, and battery data input form. In addition, the data manipulated in the system is not real-time so it is still a simulation. In this case, the data is still inputted and updated manually by the user to test the reliability of the fuzzy inference system. Meanwhile, database implementation can be seen in Figure 7 below.

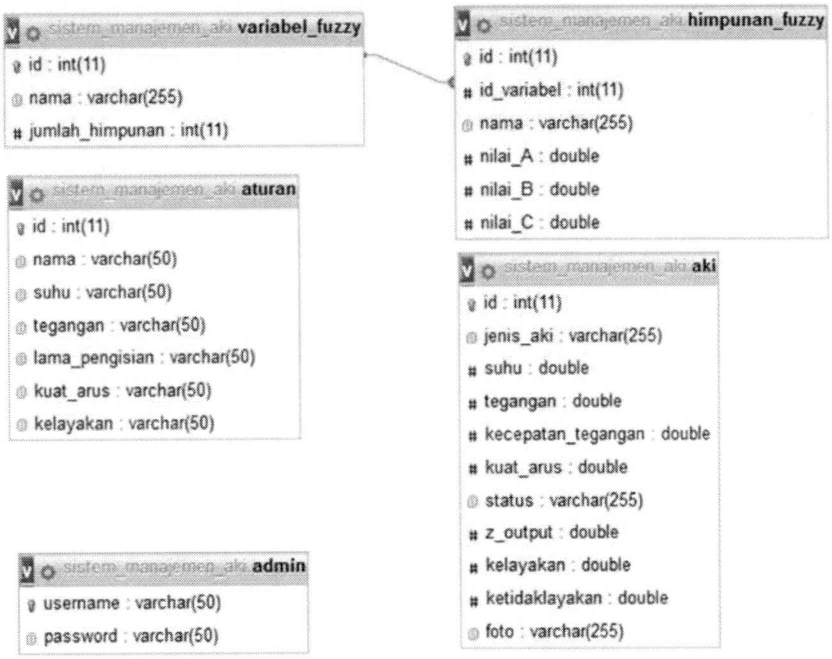

Figure 7. Database implementation of web-based fuzzy inference system

4. Result and Discussion

4.1. Testing stage

At this testing stage, calculation using fuzzy inference is done manually and used validated data from an expert who knows the characteristics of the battery. After that, the fuzzy inference system is coded and executed to compare the results with manual calculation. Moreover, the data used is also data that is relevant to the level of influence of the battery condition on the indicators that are the variables of the system calculation process. Two test cases are used as shown in Table 3.

Table 3. Test case

Name	Temperature (°C)	Voltage	Current	Charge time
Test case 1 (TC1)	35	12.4	20	10
Test case 2 (TC2)	26	12.4	12	2

To get the output value feasibility in the form of crisp (z_i), it is necessary to map the degree of inference result members (α_i) into the usage feasibility membership function. After that, a weighted average is calculated to get the final value of usage feasibility (Z). The Z value is used to determine the degree of membership of FEASIBLE and NOT FEASIBLE as shown in Table 4.

4.2. Result Stage

From the results of testing the data above, it can be seen the status of the battery's usage feasibility. Meanwhile, the results of calculations using the fuzzy inference system can be seen in Figure 8 below.

Table 4. Inferences Results

Test Cases	Usage Feasibility (Crisp)						Z (Weighted Average)	Degree of Membership	
	Z1	Z2	Z3	Z4	Z5	Z6		FEASIBLE	NOT FEASIBLE
TC1	25.00	50.00	50.00	25.00	23.75	25.00	23.75	0.00	0.75
TC2	43.75	50.00	50.00	25.00	50.00	25.00	43.75	0.25	0.18

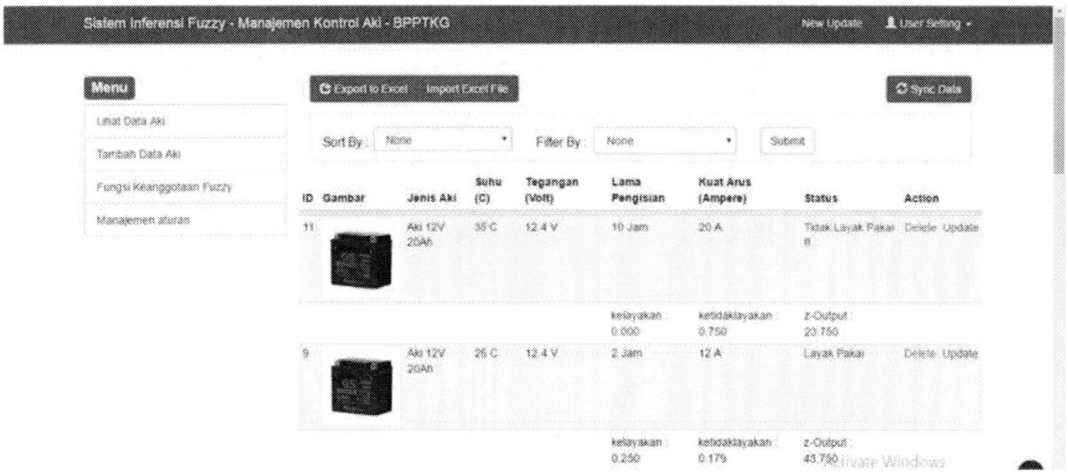

Figure 8. Results by system and test case validation

In this study, the test cases are done twice using data on different battery conditions which in this case is TC1 and TC2. From Figure 8 above, it can be seen that the tests that done manually by tabulating and the tests that done by generating the values on the system have the same output of fuzzy value. As shown by the results, the first battery (TC1) is not suitable for the use of mitigation research while the second battery (TC2) is suitable for the use of mitigation research. Moreover, the weighted average value obtained is 23,750 for TC1 and 43,750 for TC2. After the conversion to the degree of usage feasibility is done, it showed that TC1 gets a membership degree of 0.00 for the FEASIBLE set and 0.75 for the NOT FEASIBLE set while TC2 obtains a membership degree of 0.25 for the FEASIBLE set and 0.18 for the NOT FEASIBLE set. Although this research can provide a good simulation to determine the usage feasibility of batteries with a fuzzy inference system, this research is far from appropriate to be applied in the real environment because real-time monitoring has not yet been developed.

This research can be further developed by adding a control system with fuzzy inference and automatic charging that implanted in the battery so that it can make decisions directly. In this way, the control system can automatically check the battery voltage when charging which then processes the reading results in the control system where the fuzzy inference program is coded. In addition, these control systems can also be used to record the length of time of charging and the time of battery usage which will be used as a variable to determine the feasibility of using a battery. The control system calculation results will then be reported to the system interface owned by BPPTKG, and BPPTKG as the decision-maker will follow up on the use of the battery.

5. Conclusion

The system created using Tsukamoto's fuzzy method is useful to assist the Yogyakarta Geological Disaster Technology Research and Development Agency (BPPTKG) in determining the usage feasibility status of batteries as energy suppliers for electronic disaster mitigation equipment at Mount

Merapi. This study uses 4 variables including temperature, voltage, charging time, and current strength with 2 expected outputs that are feasible to use and not feasible to use. In addition, there are 6 rules used in this fuzzy system based on interviews with experts. From the results of the test cases using the fuzzy inference system and fuzzy manual calculations, it can be concluded that both calculations produce the same output of fuzzy value for two test cases with a different dataset.

References

[1] Agustian L 2015 *Rancang Bangun Sistem Monitoring Kondisi Aki Pada Kendaraan Bermotor* Unpublished Bachelor's Thesis (Pontianak, Indonesia: Universitas Tanjungpura Pontianak)

[2] Novia 2018 *Rancang Bangun Alat Monitoring Proses Charging Battery Berbasis Fuzzy Logic* Unpublished Bachelor's Thesis (Balikpapan, Indonesia: Politeknik Negeri Balikpapan)

[3] Dwiono W and Taufiq A 2017 Performances of Fuzzy Controller for Battery Charging in Constant Current Mode *SENATEK* pp 19-24

[4] Wang D, Yang F, Gan L and Li Y 2019 Fuzzy Prediction of Power Lithium Ion Battery State of Function Based on the Fuzzy c-Means Clustering Algorithm *World Electric Vehicle J.* vol **10** no **1** pp 1-13

[5] Kusumadewi S and Purnomo H 2010 *Aplikasi Logika Fuzzy untuk Pendukung Keputusan* (Yogyakarta: Graha Ilmu)

[6] Saepulla A and Wahono R S 2015 Comparative Analysis of Mamdani, Sugeno And Tsukamoto Method of Fuzzy Inference System for Air Conditioner Energy Saving *J. of Intelligent System* vol **1** no **2** pp 143-7

[7] Wantoro A and Priandika A T 2017 Komparasi Perhitungan Pemilihan Mahasiswa Terbaik Menggunakan Metode Statistik Klasik dengan Logika Fuzzy Tsukamoto dan Mamdani) Studi Kasus : Stmik Teknokrat *Seminar Nasional Teknologi Informasi* pp 25-32

IOP Conf. Series: Materials Science and Engineering **803** (2020) 012027 doi:10.1088/1757-899X/803/1/012027

Convolutional neural network acoustic model for robust Indonesian speech recognition in noisy environment

M J Budiman and D P Lestari

School of Electrical Engineering and Informatics, Bandung Institute of Technology Bandung, Indonesia

E-mail: marjer17dim@gmail.com

Abstract. Noise causes the decreasing accuracy of automatic speech recognition (ASR). Several techniques have been developed and proposed to overcome this problem. Using artificial neural network (ANN) as acoustic model is one of the techniques. Convolutional neural network (CNN) is a variant of ANN that has been used for acoustic modeling. Another approach is to do pre-processing to the speech signal or to the extracted acoustic feature from speech signal, such as cepstral mean and variance normalization (CMVN). On this work, CNN acoustic models were trained by using CMVN pre-processed acoustic feature to make a noise-robust speech recognition system. Two group of models were made, each to handle 2 kinds of noise (babble noise and street noise). Those acoustic models were tested with noisy speech at different SNR (signal-to-noise ratio) value. Testing results from CNN acoustic models were compared with the ones from Gaussian Mixture Model-Hidden Markov Model (GMM-HMM) acoustic models. Testing results showed the increasing accuracy scores of acoustic models when models were trained using more variation of training data. CNN acoustic models that were trained using FBANK feature have higher accuracy scores than GMM-HMM models that were built using the same feature.

1. Introduction

Noise is some unwanted component in a signal that can cause degradation of automatic speech recognition (ASR) accuracy [1]. Noise causes mismatch between training data and recognition environment. Models which were trained using clean speech have good accuracy score in clean environment, but give poor performance in noisy one. With signal-to-noise ratio (SNR) gets higher, the performance gets worse. Type of noise present in the environment also affect the performance of ASR. Multi-style training is a method that tries to minimize the mismatch, by adding various training data to the training corpus [2]. Another way to improve recognition accuracy is to pre-process acoustic feature, such as applying cepstral mean and variance normalization (CMVN) to acoustic feature.

Deep neural network (DNN) based acoustic models match and even outperform the state-of-the-art performance of Gaussian mixture model (GMM) ASR system. It is known that DNN based models are more robust to the presence of noise [3]. Convolutional neural network (CNN) as one type of DNN has been developed as acoustic models [4] and give better performance in noisy environments [5]. In [5], CNN based acoustic models have 3.9% relative error reduction on average when compared with DNN (feed-forward neural network) based acoustic models.

In this study, we build CNN based acoustic models to improve recognition accuracy in noisy environments. We begin with building speech corpora by augmenting clean speech corpus with 2 types of noise (babble noise and street noise), with each corpus has a certain SNR value. Those corpora are used to train triphone GMM-HMM acoustic models which act as baseline in this study. Also, by using the corpora and GMM-HMM models, we build CNN based acoustic models. The acoustic features extracted from the corpora is pre-processed by applying CMVN to them. By the end of this study, we obtained a significant decrease in the value of WER from CNN acoustic models, compared to our baseline.

The following sections are organized as follows. Section 2 contains brief explanation about acoustic feature, CMVN, and CNN in general, and CNN as acoustic model. The next section explains more about our CNN acoustic model experiments in this research. The experiments are described step by step, from speech corpus acquisition, corpus augmentation, until the training of CNN acoustic models. The final section discussed about result and analysis based on the experiment results.

2. Feature Extraction and CMVN

2.1. Feature Extraction

Feature extraction is one of few components that form ASR. In feature extraction, speech signal is processed by some operations to obtain acoustic features. The acoustic features then used in model training and decoding process. There are many kinds of acoustic feature used in acoustic modeling, such as MFCC, PLP, FBANK, etc.

In GMM-HMM based acoustic modeling, mel-frequency cepstral coefficient (MFCC) is the most commonly used acoustic feature in GMM-HMM. MFCC processing [6] is shown in Figure 1. Each frame in speech signal produce 12 MFCCs, followed by 1 energy coefficient, resulting with 13 coefficients. Those 13 coefficients often accompanied with 13 delta coefficients and 13 delta-delta coefficients. MFCC is chosen because of uncorrelated characteristics in each of its coefficients. Uncorrelated nature of those coefficients reduces the number of GMM parameters that need to be learned when training acoustic model [6].

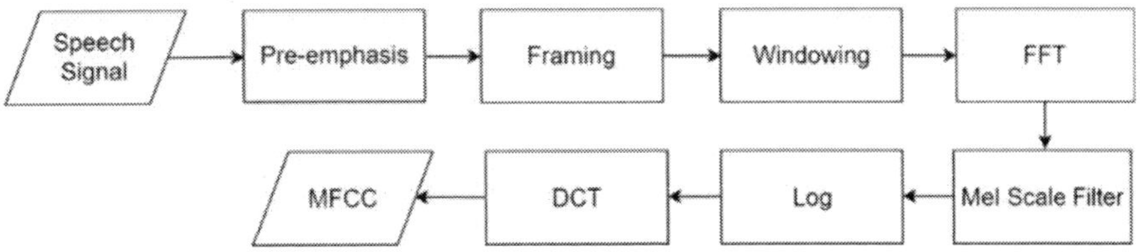

Figure 1. Steps to obtain MFCC feature

Another type of acoustic feature is filterbank (FBANK) feature. Producing filterbank feature from speech signal is done similar to producing MFCC, without performing discrete cosine transform (DCT) [4]. FBANK feature is often called mel-frequency spectral coefficient (MFSC). Another difference is the number of mel-scale filters used in producing FBANK; it is common to use 40, unlike MFCC which usually uses 23 filters. Without DCT, the resulting FBANK coefficients are highly correlated, and thus are used in DNN based acoustic modeling, because DNN can take advantage of highly correlated inputs. The scheme to produce FBANK is shown in Figure 2.

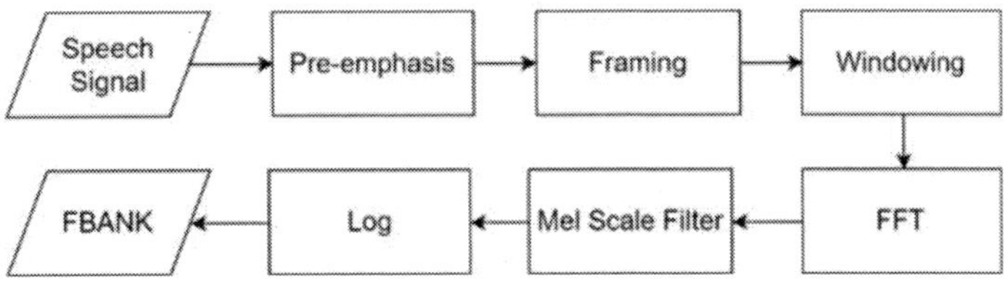

Figure 2. Steps to obtain FBANK feature

2.2. CMVN

Both types of feature can be pre-processed, by applying CMVN. Steps of CMVN begin with subtracting mean of all coefficients from all coefficients. Then, the each of the resulting coefficients is divided with the standard deviation (square root of variance) of all subtracted coefficients. Formula (1) shows the coefficients subtraction and formula (2) shows the division of the coefficients,

$$X[k] = C[k] - \mu_C \qquad (1)$$

$$Y[k] = \frac{X[k]}{\sigma_X} \qquad (2)$$

with C is the initial coefficients, μ_C is the mean of coefficients, and σ_X is the standard deviation of the subtracted coefficients.

3. Convolutional Neural Network

Convolutional neural network (CNN) is a variant of neural network, that is inspired from how human vision system works [7]. It is firstly used in image classification task. CNN can take inputs in the form of vectors (1 dimensional), matrices (2 dimensional), and higher dimensional data structures. The difference between CNN and plain neural network is the emergence of convolution layer and pooling layer. Usually a CNN contains some convolution layers, some pooling layers, and some fully connected layers.

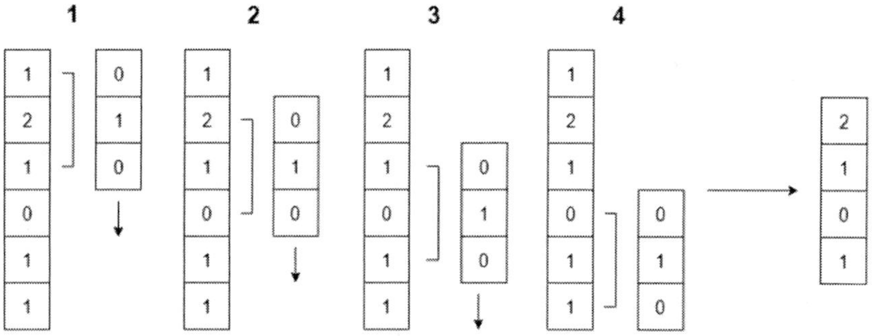

Figure 3. Illustration of convolution layer

In convolution layer, there are data structures that are called filters/kernels. Each kernel can take the form of a vector, matrix, or data structure with higher dimension. These filters and the inputs of convolution layer are used as operands to convolution operation. Each element in the inputs is multiplied with the respecting element in the kernels, then the results is summed. Many kernels can be used for one single input. Convolution layer makes the dimension of output is smaller. The illustration of how convolution layer works is shown in Figure 3. On Figure 3, the kernel is a vector that has 3 elements, those are <0, 1, 0>, so the kernel size is 3. After doing convolution operation of the first region of input, the kernel is moved along a certain axis for 1 unit. The number of units is called stride.

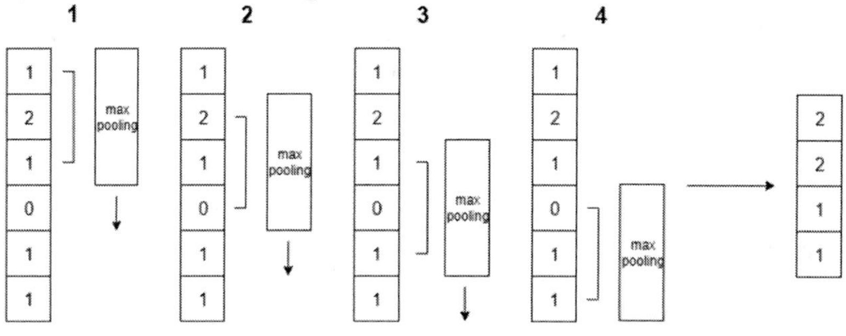

Figure 4. Illustration of pooling layer

The next layer of CNN is the pooling layer. In this layer, reduction of input dimension is done again, using some aggregate function, over some region of the input. Common functions used in this layer is max (max-pooling) and average (average-pooling). Similar to convolution layer, there is kernel size (instead in this layer it's called pool size) and a certain stride value. Figure 4 shows the illustration of a max-pooling layer, with pool size 3 and stride 1.

The final layers of CNN are fully connected layers. Inputs to the fully connected layers must be in the form of a vector. The all outputs from the previous layer is flattened to form a single vector. A particular activation function can be applied to the output of convolution, pooling, or fully-connected layer. The simple topology of CNN is shown in Figure 5.

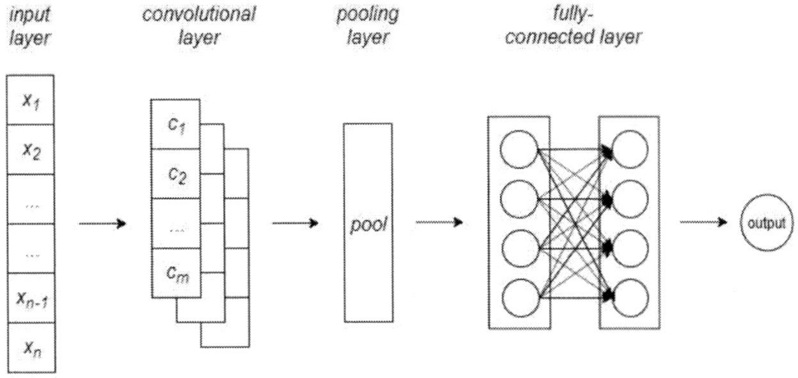

Figure 5. A simple topology of a CNN

Acoustic model in [4] used CNN with one convolution layer, one pooling layer, and 2 fully connected layers. The details of each layer are described in Table 1. The complete topology is shown in Figure 6.

Table 1. Details of each CNN layer

Layer	Description
Convolution	• Number of filters: 150 • Filter size: 8 • Stride: 2
Pooling	• Pool size: 6 • Stride: 2
Hidden	• Number of neurons: 1024 • Activation function: ReLU
Output	• Number of neurons: number of states in HMM • Activation function: softmax

Training of a DNN or CNN based acoustic models, requires that each feature vector (input) has a certain label, that is the HMM state id associated with that frame. To assign labels on all frames of input, alignment must be done using a previously trained GMM-HMM acoustic model. So, to build a CNN based acoustic model, a GMM-HMM acoustic model must be made first.

4. Experiment

4.1. Data Acquisition and Augmentation

Feature extraction is one of few components that form ASR. In feature extraction, speech signal is processed by some operations to obtain acoustic features. The acoustic features then used in model training and decoding process. There are many kinds of acoustic feature used in acoustic modeling, such as MFCC, PLP, FBANK, etc.

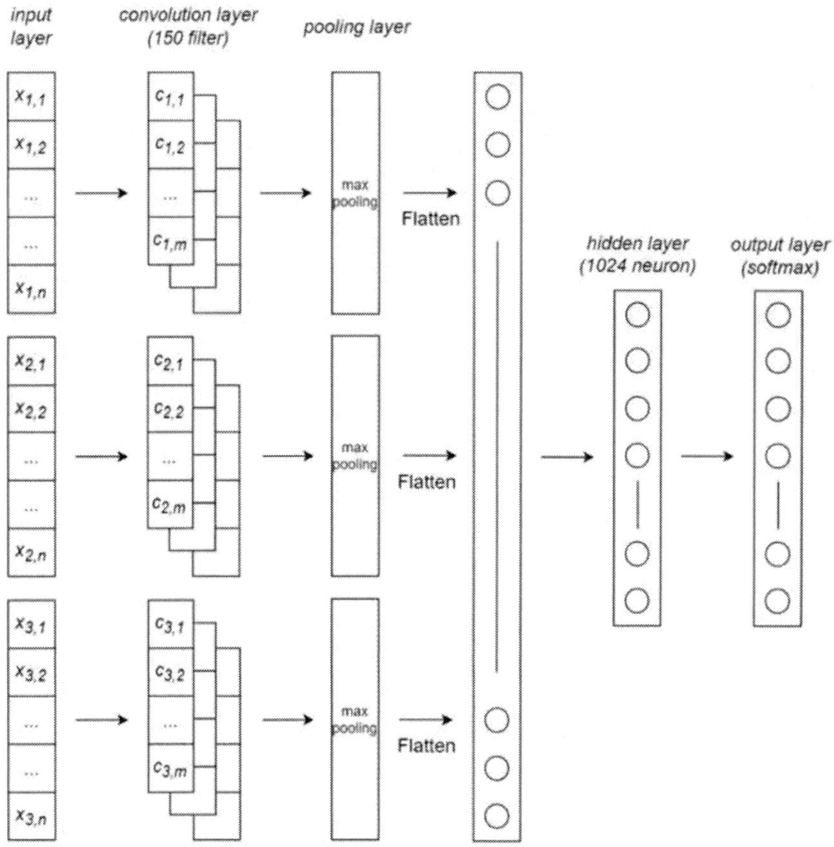

Figure 6. Topology of CNN for speech

Testing corpus, obtained from PT. Prosa Solusi Cerdas is also a clean speech corpus, recordings of dictated speech from 30 people, each of them uttered 20 Indonesian sentences. Total duration of the corpus is about 57 minutes. The difference is each sentence they read is only comprised of digits. The digits contained in each sentence are limited to 0 until 9. The reason why testing corpus only contains reading of digits, is to maximize the accuracy of ASR. Lesser the number of vocabularies in the lexicon makes accuracy of ASR gets higher. Focus of this research is to measure accuracy of ASR in noisy environment, so any other factor that degrades accuracy should be minimized. Transcripts and lexicon are also included in this corpus.

Aside from clean speech, recordings of noise were also gathered. These recordings were used to augment the clean speech corpus, to yield new corpus with a certain noise. There are 2 noise recordings gathered, babble noise and street noise. Each of them downloaded from www.orangefreesounds.com and has duration of 90 seconds.

Data augmentation is done with a tool called Sox. Sox can do some manipulation to audio files, such as mixing them, trimming, adjusting volumes, etc. For each utterance in clean speech corpus, its duration is calculated. Then a segment with having such duration is sliced from noise recording. The segment volume is adjusted to obtain a certain SNR value, then it is mixed with clean speech utterance.

Two groups of corpora are yielded from the augmentation process, and are shown in Table 2 and Table 3. First group is corpora that contain babble noise (identified with "BB" infix) and second group is corpora that contain street noise (identified with "SR" infix). SNR value of each corpus is shown in the suffix of the corpus name. Suffix "20_10_00" means a corpus made from combination of SNR 20 dB, 10 dB, and 0 dB corpora.

Table 2. Corpora for training

No.	Corpus Name	No.	Corpus Name
1	TR_CLEAN	5	TR_SR_20
2	TR_BB_20	6	TR_SR_20_10
3	TR_BB_20_10	7	TR_SR_20_10_00
4	TR_BB_20_10_00		

Table 3. Corpora for testing

No.	Corpus Name	No.	Corpus Name
1	TS_CLEAN	7	TS_SR_20
2	TS_BB_20	8	TS_SR_15
3	TS_BB_15	9	TS_SR_10
4	TS_BB_10	10	TS_SR_05
5	TS_BB_05	11	TS_SR_00
6	TS_BB_00		

4.2. Baseline

Baseline for this experiment is obtained from a collection of triphone GMM-HMM acoustic models. Those models are built using the training corpuses, each corpus becomes one model. FBANK acoustic feature is used to build these models. CMVN is not applied for the acoustic feature. Kaldi speech recognition toolkit is used to do feature extraction, CMVN, and GMM-HMM modeling. The models are used to recognize (decode) audio files from testing corpuses, then the WER is computed by comparing recognition result and transcripts of testing corpuses. Decoding is also done by Kaldi. Table 4 shows average WER for each SNR value of testing corpus, and Table 5 shows average WER for each training corpus.

Table 4. Average WER on each SNR value (baseline)

	SNR Values					
Models	**Clean**	**20 dB**	**15 dB**	**10 dB**	**5 dB**	**0 dB**
Babble Noise	18.91	24.60	29.16	41.70	63.98	85.05
Street Noise	25.88	27.70	38.88	63.43	86.45	92.67

Table 5. Average WER on each training corpus (baseline)

	Training Corpora			
Models	**Clean**	**20 dB**	**20, 10 dB**	**20, 10, 0 dB**
Babble Noise	72.68	45.36	29.98	27.58
Street Noise	79.33	55.11	45.35	43.54

4.3. CNN Modeling

CNN acoustic models in this research are built using FBANK acoustic feature. CMVN was applied on the extracted feature vectors. GMM-HMM models from baseline is used for the alignment process. Alignment is done using Kaldi. The training scheme of CNN acoustic models is shown in Figure 7.

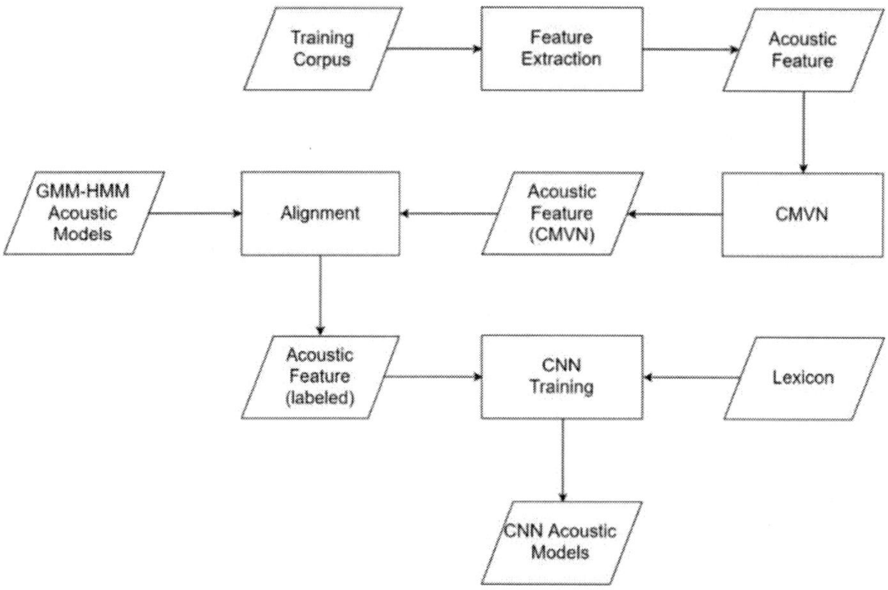

Figure 7. Training scheme of CNN acoustic models

CNN is built using topology from [4]. Keras library is used to compose the CNN, layer by layer. To connect CNN built with Keras and alignment process by Kaldi, an interface from https://github.com/dspavankumar/keras-kaldi is used. In the beginning the interface was made to build DNN based acoustic models, so it was modified to accommodate CNN acoustic models.

CNN training is using holdout validation schema, with 20% data is used as validation set, and 80% data as training set. The training is using hyperparameter as described in Table 6.

CNN models are used to decode the same testing corpus, then WER scores are also measured. The result is shown in Table 7 (comparison for each SNR value) and Table 8 (comparison for each training corpus).

Results from Table 4 and Table 7 show that CNN acoustic models, when tested with 0 dB SNR corpus give 20.89% (babble noise model) and 3.9% (street noise model) error reduction compared to respective GMM-HMM models. In other SNR value, the error reduction ranges between 48.2 – 88.4% (babble noise model) and 7.8 – 88.1% (street noise model). These results thus prove that CNN acoustic models built with CMVN pre-processed features have better accuracy in noisy environment. The results from Table 5 and Table 8 show that by using more training data and more variation of SNR value in training data, the WER gets lower. CNN models trained with the largest corpus give

17.8 % (babble noise model) and 14.4% (street noise model) error reduction. From results in all tables, it is obvious that babble noise models have lower error rate than street noise models.

Table 6. Details of CNN hyperparameters

Hyperparameter	**Value**
Learning rate	0.1
Batch size	512
Momentum	0.5
Loss function	Sparse categorical cross-entropy

Table 7. Average WER on each SNR value (CNN)

	SNR Values					
Models	**Clean**	**20 dB**	**15 dB**	**10 dB**	**5 dB**	**0 dB**
Babble Noise	2.20	2.82	5.46	13.05	33.12	67.28
Street Noise	3.07	13.34	28.51	53.30	79.68	89.02

Table 8. Average WER on each training corpus (CNN)

	Training Corpora			
Models	**Clean**	**20 dB**	**20, 10 dB**	**20, 10, 0 dB**
Babble Noise	23.24	20.46	19.82	19.10
Street Noise	48.47	46.78	41.22	41.48

5. Conclusion

In this research we build CNN acoustic models to build robust Indonesian acoustic models in noisy environment. By using CMVN and multi-style training, CNN models we build give lower WER on all SNR value of testing data, compared to GMM-HMM acoustic models. On the noisiest test data (0 dB), we get 20.89% error reduction from babble noise models and 3.9% error reduction from street noise models.

Based on our experiment results, we encourage the future research on noise-robust Indonesian ASR to develop large vocabulary ASR with robust-to-noise characteristics. We suggest to add more data to the training set and use testing set that contains large vocabulary. We also suggest to test acoustic models in real environment or using recordings from real environment, not the ones from augmentation process.

Acknowledgements
We thank PT. Prosa Solusi Cerdas and other parties who contributed in this experiment or during the process of working on this paper. We thank everyone who gives their time to record their voice.

References

[1] Benesty J, Chen J, Cohen I and Huang Y 2009 *Noise Reduction in Speech Processing* (New York: Springer)

[2] Li J, Deng L, Gong Y and Haeb-Umbach R 2014 Overview of noise-robust automatic speech recognition *IEEE/ACM Transactions on Audio, Speech, and Language Processing* **22** pp 745–77

[3] Li J, Deng L, Gong Y and Haeb-Umbach R 2016 *Robust Automatic Speech Recognition A Bridge to Practical Application* (Cambridge: Academic Press)

[4] Abdel-Hamid O, Mohamed A, Jiang H, Deng L, Penn G and Yu D 2014 Convolutional neural networks for speech recognition *IEEE/ACM Transactions on Audio, Speech, and Language Processing* **22** pp 1533–45

[5] Huang J, Li J and Gong Y 2015 An analysis of convolutional neural networks for speech recognition *2015 IEEE Int. Conf. on Acoustics, Speech, and Signal Processing* pp 4989–93

[6] Jurafsky D and Martin J 2008 *Speech and Language Processing: An Introduction to Natural Language Processing, Computational Linguistics, and Speech Recognition* (New Jersey: Prentice Hall)

[7] Goodfellow I, Bengio Y and Courville A 2017 *Deep Learning* (Cambridge: MIT Press)

[8] Lestari D P, Iwano K and Furui S 2006 A large vocabulary continuous speech recognition system for Indonesian language *15th Indonesian Scientific Conf. in Japan* pp 17–22

The Identification System of Acne Type on Indonesian People's Face Image

Agus Darmawan[1], Annisa Rositasari[2], Izzati Muhimmah[3]

Department of Informatics, Islamic University of Indonesia

Email: [1]agusdar97@gmail.com, [2]16523011@students.uii.ac.id, [3]izzati@uii.ac.id

Abstract. We report the results of the detection of types of acne on the human face that are identified directly by a specialist skin and genitals. The proposed algorithm has been tested on a standard image and gives very reasonable results, there are 5 types of acne identified, namely Blackhead Blackheads, Blackhead Blackheads, Papules, Pustules and Nodules. Work completed using PYTHON.

1. Introduction

Digital Image Processing (Digital Image Processing) is a technique for processing or taking an image from the real world to be seen, changed and stored into a computer, the stored digital image is usually in the form of a combination of RGB colors (Red, Green and Blue)) which can be referred to as pixels, the denser the number of pixels, the higher the resolution of the image. Digital image processing is very widely used from various fields, ranging from security to the medical field, one of its uses in the medical field is to detect diseases that exist in parts of the human body.

Acne is one type of skin disease that can affect the shape of a person's surface, the occurrence of acne is caused by the blockage of hair follicles by dead skin cells and oil under the outer layer of skin, causing inflammation of the skin called acne [1] . Factors causing acne are genetic, hormonal, stress, Propionibacterium acne microorganisms, environmental factors and the use of cosmetics that are not appropriate to their skin type [2]. With these various factors can cause various types of acne such as Blackheads, Whiteheads, Papules, Pustules and Nodules. Blackheads are small, black bumps that often appear on the nose area. Blackheads occur due to open hair follicles clogged with oil. Blackheads look like black spots but do not cause pain, this type of acne is commonly called mild acne because it does not cause inflammation that results in redness of the facial skin.

Whiteheads are a type of acne that occurs because the pores are blocked by oil and dead skin cells. Blockage of oil and dead skin cells is able to cover the entire top surface of the pores because that is the type of acne that is difficult to treat. Whiteheads look like white bumps but are small. Whiteheads often occur in women of any age at puberty, menstruation, pregnancy and menopause. Papules are acne that appear on the underside of the skin surface, if touched like a bulge is dense and painful. The surrounding skin area is red and swollen. This type of acne papules is often called inflammation because of irritation that can damage the surrounding skin. Pustules are acnes that have bumps, the tops of inflamed reddish-colored pus-filled skin. This type of acne occurs because the clogged pores are infected by bacteria. Nodules are acnes that cause pain, this type of acne starts from clogged pores and is infected with bacteria. The infected bacteria enter the surface of the skin and then damage the tissue and cells underneath, resulting in the pores become red and swollen. Types of acne Nodules will appear bumps, if the lump has deflated, acne scars appear usually black or dark.

Acne growth is very susceptible to oily facial skin types, on average Indonesian skin types are oily skin types with tropical country conditions, so that there are many cases of acne easily appearing on one's facial skin. The average Indonesian does not pay too much attention to the condition of his facial skin, as a result many people do not know the condition of facial skin and what types of acne are on his facial skin. From this problem, an analysis of the types of pimples suffered by a person can be done by using digital image processing techniques to detect pimples on facial skin and the classification of these types of acne.

Detection of the type of acne that will be done, will only be used on images with Indonesian facial skin objects, the limitation of this study is because the average facial skin color of Indonesian people has a brown tan range. The image that will be processed to detect pimples uses an image that has a face facing forward or toward the camera with even lighting. The type of camera to take a minimum camera using a smartphone camera. The author ensures that it is acne (not blackheads) by using feature extraction. Feature extraction is a stage which is done to obtain certain characteristics or information from an object. Information obtained from an object can be used to recognize similar objects or to distinguish them from other objects. Detection of features based on color and reading results from experts. But it must still be considered a problem in the lighting because of the different light colors, different color values of the object.

2. Literature Review

Acne treatment on the face is something that is important to do. But there are still many ordinary people who cannot identify acne or acne-like diseases because of their limited knowledge about it. In 2019, Kusbianto et al conducted a study using the Forward Chaining method to assist the identification process by matching the symptoms of a disease according to existing regulations [3]. In this research, a desktop-based system, designed using the Visual Basic programming language and XAMPP Database. The system contains seven disease data, 32 disease symptom data and 18 disease treatment data. The results obtained from this study aim to help ordinary people take the first steps in treating acne to prevent errors and provide information about the disease. zits. At the time of this study, the system was tested by experts and users. The final result of this study is 83.3% of the system is quite effective in identifying acne on the face.

Face Detection is very popular in today's society. Many applications have been developed to be able to detect faces. Face detection itself is useful for knowing the color of a person's facial skin because facial skin is one of the important things and is often noticed, especially by women. In 2010 Aldasouqi et al conducted a study that detected human faces in an image [4]. In that study, there are five algorithms used for face detection such as Principle Component Analysis, Linear Discriminator Analysis, skin color, wavelet and artificial neural networks using the Hue, Saturation, and Value (HSV) color method. The results obtained from these studies aim to classify the face criteria through a model of facial skin color that matches the HSV value. At the time of this study, the system was still being worked out using MATLAB. The final result of this research is to use the HSV color model to increase face detection around 12-28s detection speed and performance by 93-100%. These results are very reasonable to be used for further system development.

In 2012 Han et al conducted a study of lighting for facial recognition [5]. In that research, illumination preprocessing is divided into three categories, namely: Gray level transformation, Gradient or edge extraction and Reflectance plane estimation. The results obtained from these studies aim for facial recognition, a better visualization effect after preprocessing lighting does not imply higher recognition accuracy. From this research, the final result is a holistic localization with illumination preprocessing approaches such as HE, GIC, LTV and TT can improve the existing performance.

3. Methods

3.1. Segmentation

Segmentation is an image processing technique that can be used to separate objects that will be detected with the background of the object, this technique utilizes colors and shapes of objects to be able to distinguish objects from their background, so that it will produce images that focus on the object [6].

3.2. Morphology
Morphology is an operation that can normalize the shape and expansion of object readings, in this operation can process gray and binary images to be able to expand or narrow the readings of the object, using two basic operations, namely erosion operations that reduce the number of pixels against pixels that become a barrier between each object and also a dilation operation that will increase the number of pixels to the pixels that are boundaries between each object [7].

3.3. Edge Detection
Edge detection is one of the image processing techniques to get or find out the farthest points located on an object, but it can also be used to analyze the shape of the detected object because this technique will only display shapes in the form of dots that follow the shape of objects and display edges from the object that was detected [7]. One method that can be used for edge detectors is laplacian.

3.3.1. Laplacian. Laplacian is an image processing technique that can be used to detect edges of objects, this technique uses mathematical convolution operations to produce the images needed in the next process [8], an example of the results of laplacian processing as follows.

4. Result and Discussion

4.1. Training Stage
Before the system design process is carried out, the author collects data for facial images with a acne face, the data will be used as reference material for the types of acne and the characteristics of acne, then from the objects identified as acne will be used as a benchmark of the system to detect acnes from

other images, namely the color intensity of each type of acne. In this study the authors used 36 acne facial images as training data.

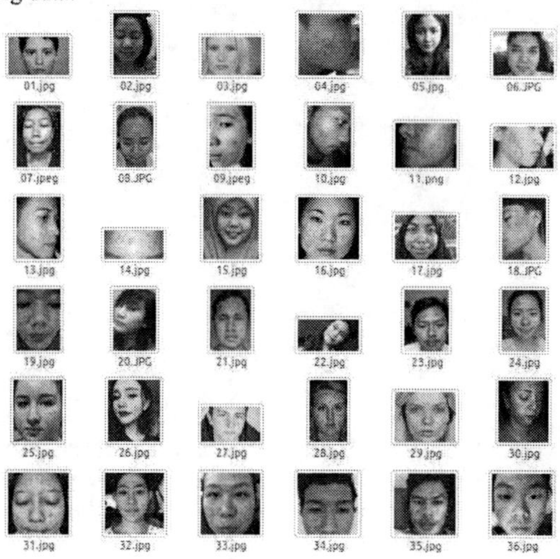

4.2. Testing Stage

At this testing stage, using image data from Indonesian faces, judging from the characteristics of Indonesian skins, which on average have a range of tan color and oily skin types, to ensure that the results of acne detection using the system are of true value, observational data will be used from the dermatologist as a benchmark for the analysis of the types of acne, so the development of this system also involves experts in terms of the observation of the type of acne, namely Jodi, Rosmelia, and Nafiah.

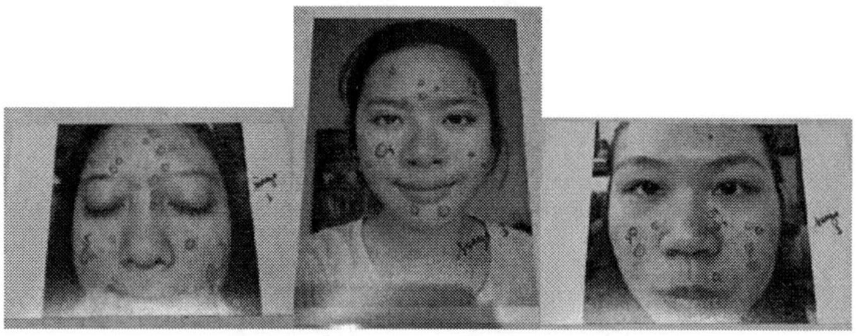

IOP Conf. Series: Materials Science and Engineering **803** (2020) 012028 doi:10.1088/1757-899X/803/1/012028

Doctor identification results are based acne type code	
Acne Type Code	Details
1	Blackhead Komedo
2	Whitehead Komedo
3	Papul
4	Pustul
5	Nodul

From these data will then be detected by using a system of acne, the results of the detection of acne will then be taken the color value of the zits that are detected, so that the type of acne can be determined based on the colors that arise, the system will work using image processing methods with the following flow.

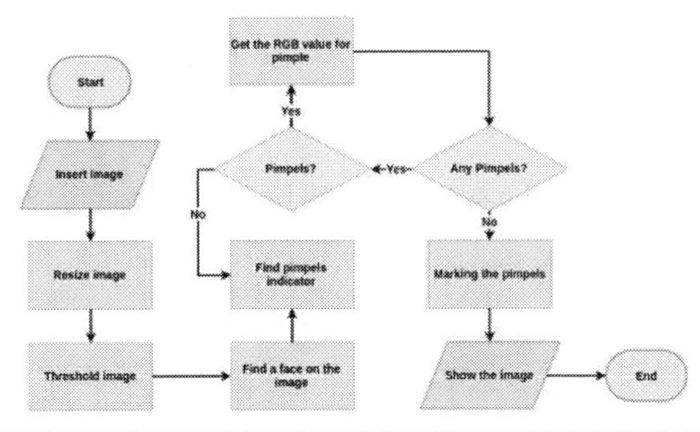

4.3. Result Stage
From the results of detection using the system, for testing conducted on the 4th image obtained the number of points indicated as zits as much as 216 points. And for these points have the following color values.

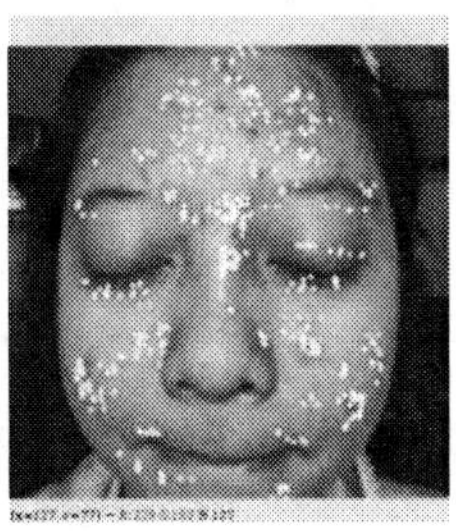

Spots On The Face	Spots RGB On The Face			Type Identification
	R	G	B	
3	145,48	156,01	209,74	3
4	149,74	162,16	213,12	3
14	139,51	146,97	186,94	4
37	155,01	163,13	233,05	5
53	188	194,09	247,33	3
89	131,11	145,55	207,54	3
100	159,16	175,33	225,17	3
120	146,92	160,12	220,14	3
148	187,01	198,99	246,57	3
150	165,67	179,87	236,86	3
204	174,5	188,99	238,33	3

5. Conclusion

The developed system tries to detect zits based on the dots that are indicated as zits on the face, then the system will distinguish the types of zits that are detected based on the value of the RGB color of each point. From the results of the analysis of acne detection, some points that are indicated as pimples on facial images are dominated by Papul acne types, but at some points there are also Pustules and Nodules, and from each type of acne has a combination of different RGB color values.

References

[1] James W D 2005 Acne *The New England J. Of Medicine* vol **352** no 14 pp 1463-72

[2] Manarisip C K, Kepel B J and Rompas S S 2015 Hubungan stres dengan kejadian acne vulgaris pada mahasiswa semester V (lima) Program Studi Ilmu Keperawatan Fakultas Kedokteran Universitas Sam Ratulangi Manado *Jurnal Keperawatan* vol **3** no 1

[3] Kusbianto D, Ardiansyah R and Hamadi D A 2017 Implementasi sistem pakar forward chaining untuk identifikasi dan tindakan perawatan jerawat wajah *Jurnal Informatika Polinema* vol **4** no 1 pp 71-80

[4] Aldasouqi I and Hassan M 2010 Human Face Detection System Using HSV *Proc. Of 9th WSEAS Int. Conf. on Circuits, Systems, Electronics, Control & Signal Processing (CSECS'10)* pp 13-6

[5] Han H, Shan S, Chen X and Gao W 2013 A comparative study on illumination preprocessing in face recognition *Pattern Recognition* vol **46** no 6 pp 1691-9

[6] Setiawan M T and Firdausy K 2016 Aplikasi Penghitung Telur Ikan Gurami menggunakan Deteksi Blob berbasis Android *Seminar Nasional Aplikasi Teknologi Informasi(SNATi) Yogyakarta* pp 28-31

[7] OpenCV 2019 *Open Source Computer Vision Library* Retrieved from https://opencv.org/about/

[8] HPR2 2003 *Laplacian/Laplacian of Gaussian* Retrieved from http://homepages.inf.ed.ac.uk/rbf/HIPR2/log.htm

Dual Security System for Room Access Control Using RFID at Islamic University of Indonesia (UII)

Bayu Aprilananda Sujatmoko[1] and Ari Sujarwo[2]

[1,2] KH. Mas Mansyur Building 2nd floor. Universitas Islam Indonesia. Jl. Kaliurang KM. 14,5 Sleman Yogyakarta.

Email: [1]bayusujatmoko@gmail.com

Abstract. RFID (Radio Frequency Identification) technology is chosen because of the unique ID on each tag that can be used to identify and validate users by matching it to the account that has been registered in the database. On the other hand, RFID is a relatively inexpensive technology and is not a tool that can stand alone but RFID will be very useful if it is developed into a whole system by implementing various components. The problem of this research is how to make a room security system using RFID to limit someone's access rights when he wants to enter a room such as a laboratory and other important rooms by involving various actors in the authorization. Authorization is given to someone who really has access to a particular room through a centralized room access control system. The user's identity card is used as a tool to authenticate when accessing a room using RFID technology. The research method was conducted using one of the approaches of the System Development Life Cycle method, which is the waterfall method. This research consisted of several stages: system requirements analysis, design, and implementation. The results of this study showed that the room security system is expected to be used as a room access control by involving various actors in granting room access and monitoring the number of users who access a room centrally.

1. Introduction

At present most of human activities use computer technology in its application. The need for the continuous availability of real-time data is a fundamental requirement in life in the information age as it is today to analyse the data into useful information [1]. According to Few, a dashboard is used to display information related to activities carried out by a system where the dashboard is typically manifested in a website that has a database that allows for continuous updating [2]. We can monitor all activities through the dashboard and, if needed, make a decision. Each agency must have several rooms with certain functions such as the financial division room, manager's room, IT division room, and server room. Certainly, some of these rooms have security; therefore, not everyone can enter the room but only people with authorization can do. In this case, information technology is needed to secure space and monitor anyone who gets access to enter the room through a dashboard equipped with RFID (Radio Frequency Identification) technology.

At UII there are more than 30 thousand active users consisting of various groups such as lecturers, students, employees and so on. Each user has a card as the identity of the academic community at UII. On each card there is an RFID tag containing a unique ID that can be used for authentication using an RFID reader. Passive RFID tags are chosen because of their relatively thin size, light weight and relatively low price compared to other tags. UII uses a Microsoft Active Directory database to manage user accounts. Microsoft Active directory uses the LDAP network protocol as a standard for communication between devices. This research used RFID technology and LDAP protocol.

In general, RFID consists of two parts: RFID reader and RFID tag card. RFID reader is used to receive data emitted from RFID tag cards. It has a terminal ID that is used as a unique RFID Reader identity. The purpose of this RFID system is to send data from tags, which are then read by an RFID reader and processed by a computer application. Data transmitted earlier can contain various information, such as unique IDs or other information [3]. Data from each user's RFID tag can be stored in an Active Directory database system as an implementation of the LDAP (Lightweight Directory Access Protocol) protocol. LDAP is a computer network protocol for lightly accessing a user's account directory.

The combination of various components, as explained above, can create a room security system that can certainly involve several stakeholders at UII. A backend system is needed to manage room access control and a dashboard is needed to record system activity when performing room authentication by the user. Therefore, this room security system is called a dual security system.

2. Literature Review

2.1. Authentication
Authentication is a process of checking the identity of a communication system user when logging into a system. Users who pass identity checking are users who are officially registered on the system or who have authority on a system. The use of authentication systems is expected to form a special system that can be used by people having the right to use [4].

2.2. Backend technology uses the LDAP protocol
LDAP, Lightweight Directory Access Protocol, is an application protocol that is used to control the directory services that run on the TCP/IP protocol. As its name implies, LDAP is a protocol for accessing a user account directory lightly. As per the X.500 standard, LDAP manages directory entries in a hierarchical form [5].

The LDAP structure is not composed of columns and rows like normal databases, making it easy to enter data details in an organized form but rather as a directory. Directory is a structured service that is arranged logically and hierarchically. LDAP is often used as a tool to authenticate various systems such as computers or printers according to the structure used. This application protocol allows us to search for an organization, individuals and other resources such as files or devices on a network.

2.3. Room Security Strategies Using RFID (Related Research)
Research conducted by Gyanendra Kumar Verma entitled A Digital Security with Door Lock System Using RFID Technology [6] explained that RFID technology can be implemented in various security applications such as asset tracking, people tracking, inventory detection, and access control applications. This study more focused on the RFID implementation as a digital security tool that will provide authentication to someone who has the right to access. RFID technology was chosen because this technology does not require high cost in its application.

Research conducted by Kumar is on how to secure a room by giving access rights to someone authenticated using real-time RFID-tags here. RFID-tags used were passive RFID-tags considering that they did not require batteries, were lighter and certainly cheaper than active RFID-tags. Passive tags have a fast response with RFID-readers when they touch or are within a few millimetres of the reader.

The system is managed centrally related to operations and transactions that occur on RFID. RFID is a comprehensive system because it cannot stand alone but requires several other components in its implementation, namely RFID-tag, RFID-reader and backend system. This door security system is controlled and implemented by an RFID-reader, which will authenticate and validate users who will enter through a door.

2.4. Room Access Control Business Process (Related Research)
A study entitled Development of an RFID Based Access Control System in the Context of Bangladesh [7] described how to create a digital access control system that can be used to protect a room or area

IOP Conf. Series: Materials Science and Engineering **803** (2020) 012029 doi:10.1088/1757-899X/803/1/012029

where only registered and authenticated users can enter to. The aim of the study was to build an affordable security system and fits the perspective of developing countries like Bangladesh.

This system was built associated with a centralized system based on client-server and sub-systems to ensure system integrity. The sub-system association will log activity and manage the check-in and check-out status of users visiting a room. The RFID system ideally consists of RFID-reader, RFID-tag, and application-backend, which are carried out for data management.

3. Research Methods

The research method was carried out using one of the approaches of the System Development Life Cycle method, which is the waterfall method [8]. This research consisted of several stages: system requirements analysis, design and design, and implementation.

The hardware to build a space security backend application system required 1 server with the Linux/ Ubuntu operating system, RFID-tag (user card / smartcard), RFID-reader, master card editor, master card and USB Reader.

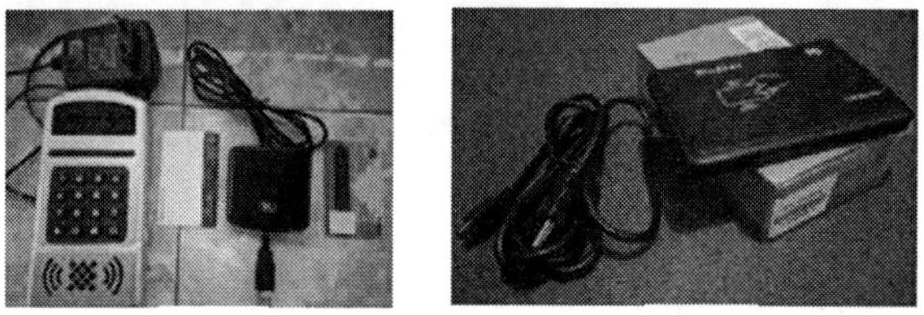

(a) (b)

Figure 1. (a) Hardware requirements, from left to right: RFID-reader, RFID-tag (smartcard), master editor card, and master card. (b) USB RFID Reader

Four main functionality requirements were developed in this study, including room user authentication, card association application with LDAP account, room access control application, and visualization of recording system or log activity. Room user authentication application has a function to provide an access to users when they are going to use or enter a room. The card association application with an LDAP account has a function to register the user's card UID to the LDAP account, which will be used for authentication purposes when registering a new user. The room access control application, meanwhile, has a function to register room numbers in an LDAP account when adding a new room access. The last functionality requirement is to record and visualize the system activities (logs). The system is expected to have an ability to record transactions occurred during the system in use.

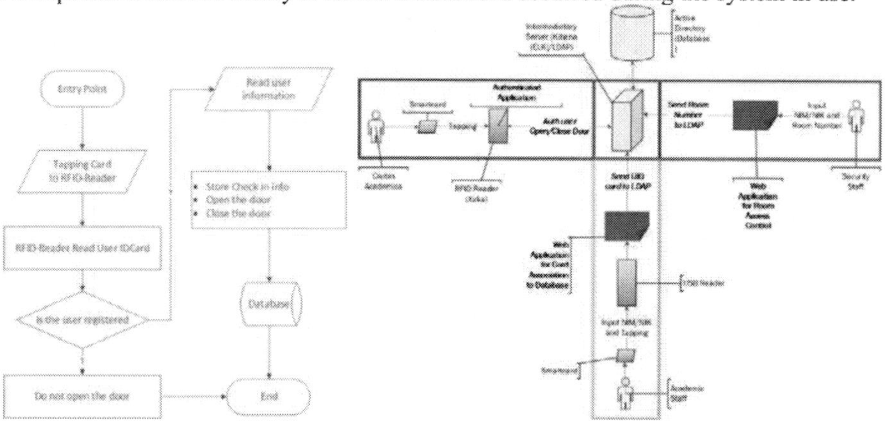

Figure 2. User Check-in Process (left) Room Security System Architecture Using RFID (right)

The design was carried out to determine the hardware and software specifications in accordance with the needs. This activity determined the overall system architecture and the mechanism of the system. The design of space security using RFID technology at least was focused on four very fundamental issues: audit, management, access control, and authentication [9]. Therefore, to build an integrated space security system, RFID technology cannot stand alone to meet the resolution of these four very fundamental problems; it requires other elements such as databases, application back ends, RFID-Reader, and RFID-tags that are integrated with one another. The other becomes a complete system [6].

Based on the explanation above, the researchers developed a room security system using RFID technology. This RFID technology was integrated with the LDAP protocol in Active Directory, which contained all user account information at the institution. This aimed to simplify the authentication process considering the central user authentication when accessing a room.

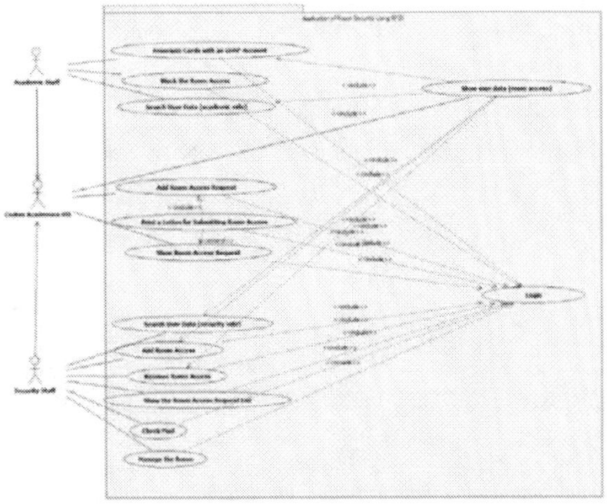

Figure 3. Use case Diagram of Room Security Using RFID

4. Discussion and Results

The room access control application was made based on the web enabling it to be accessible for every user. The application consisted of three systems: the application for room access submission, the application of the user card association into the database and the application for room access. The application for room access submission can be accessed by all UII academics who have already conducted card associations. They can make requests for room access, view room access provided and print a letter for room access application. The card association application can only be accessed by academic employees. Academic staff can associate cards to register new users and block an access to user rooms. The application for granting room access can only be accessed by security employees. Security personnel can add or delete the users' room access, check the room access filing letters and manage the room numbers.

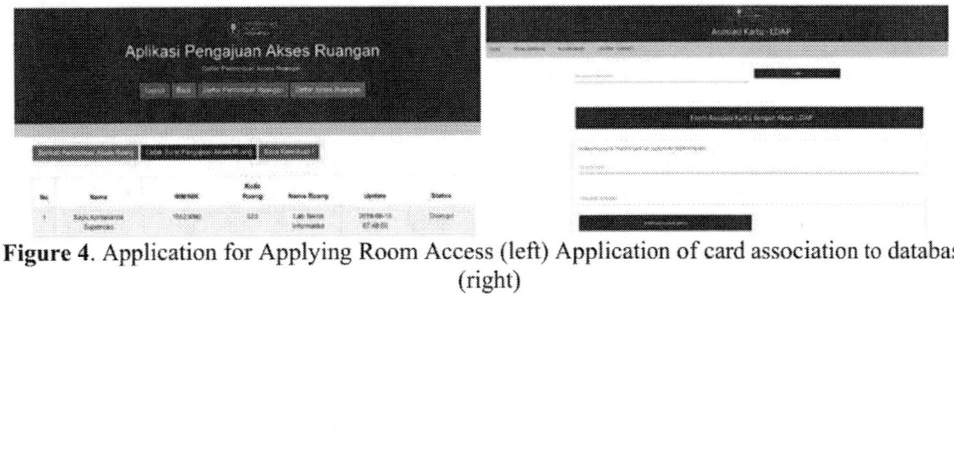

Figure 4. Application for Applying Room Access (left) Application of card association to database (right)

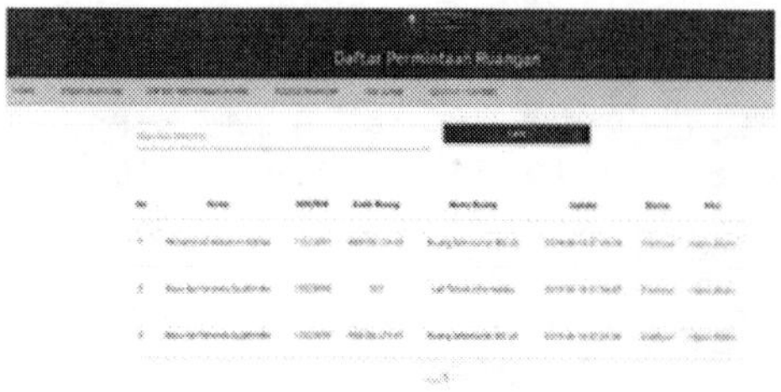

Figure 5. Room Access Control Application

The visualization log of the RFID reader is displayed in graphical form to display the number of authenticated users, valid users or invalid users when authenticating to a particular room. The graphs are displayed in the form of time series, which in this study have been displayed by hour and day to observe the trends when using a particular room.

Figure 6. System Log Visualization (left) Auth User from RFID Reader (right)

Based on the explanation explained above, the concept of dual security for securing the room, which is the result of this research is the authorization of a room involving several actors in the academic community of UII, namely academic staff and security officers. This is done to ensure the data integrity that the room access truly belongs to the UII academic community so that it is safer. The submission of room access is carried out from bringing an ID card to an academic employee to associate a card into the database to request a room access from a security officer. In addition, this study also produced a system that can be used as a room access control and to monitor the number of users accessing a room centrally in which it would make it easier for the officers to manage access to a room.

Figure 7. Dual Security System Concept for Securing Room at UII

5. Conclusion

This dual security system for providing a room access utilizes RFID technology that has been designed with a multi-stakeholder approach, taking into account the functional and hardware requirements to meet the massive needs of higher education institutions. By involving various actors in the UII academic community, certainly the granting of access to this room will be safer because of the data integrity from users because they have to go through several business processes to get an access to a room.

This research also delves deeper into the potential of RFID technology if it is used to secure a university scale room. The features possessed by RFID technology with little development in the backend application have been implemented to secure a room in this research institution.

References

[1] Martin N, Bergs J, Eerdekens D, Depaire B and Verelst S 2018 Developing an emergency department crowding dashboard: A design science approach *Int. Emerg. Nurs.* vol **39** pp 68–76

[2] Few S 2013 Information dashboard design: displaying data for at-a-glance monitoring *Inf. dashboard Des. Displaying data at-a-glance Monit.* p 246

[3] Budiharjo S and Milah S 2014 Keamanan Pintu Ruangan Dengan Rfid Dan Password Menggunakan Arduino Uno *J. ICT Penelitian dan Penerapan Teknol.* pp 28–34

[4] Sujarwo A 2010 Implementasi Network Storage dan Internet gateway Menggunakan Autentikasi OpenLDAP *Seminar Nasional Aplikasi Teknologi Informasi 2010* p 25

[5] Setiawan M A 2016 *Directory Services* (Yogyakarta: Informatika UII)

[6] Verma G K and Tripathi P 2010 A Digital Security System with Door Lock System Using RFID Technology *Int. J. Comput. Appl.* vol **5** no 11 pp 6–8

[7] Shafin M K *et al* 2015 Development of an RFID based access control system in the context of Bangladesh *ICIIECS 2015 - 2015 IEEE Int. Conf. Innov. Information Embed. Commun. Syst.* March 2015

[8] Dora S K and Dubey P 2013 Software Development Life Cycle (SDLC) Analytical Comparison and Survey on Traditional and Agile Methodology *National Monthly Refereed J. of Research in Science & Technology* **2(8)** pp 22–30

[9] Rieback M R, Gaydadjiev G N, Crispo B, Hofman R F H and Tanenbaum A S 2006 A Platform for RFID Security and Privacy Administration *Auditing* pp 89–102

Data augmentation on spontaneous Indonesian automatic speech recognition using statistical machine translation

P N Hadiwinoto and D P Lestari

School of Electrical Engineering and Informatics, Bandung Institute of Technology Bandung, Indonesia

E-mail: patrick.nugroho64@gmail.com

Abstract. Language model plays an important role in decoding process of the automatic speech recognition. The accuracy of spontaneous speech recognition is still very low compared to dictated speech of the Indonesian automatic speech recognition. It is due to the lack of the number of spontaneous data. Collecting spontaneous data is also difficult to do, so one of the candidate solutions is to augment data from existing spontaneous data. In this research, experiments are conducted on language models to improve the accuracy of spontaneous Indonesian speech recognition by conducting data augmentation. Data augmentation in this research is done by using statistical machine translation named 'Moses'. Language modeling technique used here is n-gram. GMM-HMM is used for acoustic modeling. First, spontaneous text corpus is added to the text corpus, then the data augmentation is conducted. When the language model is formed from the addition of a spontaneous text corpus, there is an increase in accuracy of 3.59% relative to the baseline. When data augmentation is done on language model there is an increase in accuracy of 2.74% relative to the baseline. However, this decrease is considered not significant compared to the effort required in collecting spontaneous data manually.

1. Introduction

Language model plays an important role in increasing the accuracy of automatic speech recognition. Language model maps the probability of the occurrence of a word. Language model is focusing on the grammatical aspect of a language rather than the condition of the speech signal. Spontaneous speech has caught the attention of many automatic speech recognition (ASR) researchers recently. It is because of the recognition accuracy of spontaneous speech is extremely poor compared to dictated speech. Word Error Rate (WER) shows how much error occurred in recognizing speech. It is counted based on how many words are wrongly recognized by the system compared to the number of words in the testing data. Indonesian automatic speech recognition (ASR) researches have been focusing on improving the accuracy of spontaneous speech recognition [1, 2]. Nowadays, Indonesian ASR used n-gram to build language model [1, 2]. The accuracy of recognizing dictated speech is already high in Indonesian ASR. In other hand, when recognizing spontaneous speech, the accuracy drops significantly. Hypothetically, the language model of spontaneous speech is not built well in the current Indonesian automatic speech recognitions. It is due to the lack of number of spontaneous speech data. Current Indonesian ASR's training data mostly consist of dictated speech. It causes problems when recognizing spontaneous speech because of the differences in the characteristic between dictated and spontaneous speech [3].

Spontaneous speech has its own characteristics, some of them are the filled pause, repetition, and revision [4]. In this research, only filled pause will be the focus of the experiments. There are 8 kinds of filled pause that are handled in this research. Those filled pauses are "eu", "euh", "eum", "gitu", "gitu yah", "kan", "tuh", and "yah". We do not handle other kinds of filled pause, such as 'napas' ('breath'), 'decak ('click'), or 'tawa' ('laughter'). We also do not handle repetition and revision.

The acoustic modeling technique used in this research is the Gaussian Mixture Model (GMM)-Hidden Markov Model (HMM). We prefer GMM-HMM over Deep Neural Network (DNN)-HMM technique due to the number of data. The number of data required for DNN-HMM technique is quite lot. Unfortunately, the number of data available for this research is limited. So, the GMM-HMM is the best choice.

N-gram technique is chosen over deep learning-based technique for language modeling due to the acoustic modeling technique chosen. For GMM-HMM acoustic modeling technique, the state-of-the-art technique for language modeling is n-gram. N-gram is effective on many speech recognition domains. The experiments conducted in this research revolve in language models. There are three experiment scenarios conducted in this research. The acoustic model of these three are all the same, which is triphone model of GMM-HMM. The variables here are the language models used. The first scenario, known as the 'baseline' (considered as the standard scenario of the research) consists of the basic text corpus, which are 'Tala' text corpus plus the transcript of speech training data. We add spontaneous text corpus in the second scenario. Text corpus for language model used in the third scenario, known as the 'proposed method' (which is the core of the research) consist of Tala + transcript of speech training data + spontaneous text corpus used in second scenario + the result of data augmentation of all.

The following sections are organized as follows. Section 2 contains brief explanations about n-gram and also n-gram as language modeling technique. The next section explains about statistical machine translation used for data augmentation in this research. The experiments are described in the next section. Later, the final section discusses about result and analysis based on the experiment results

2. N-gram and GMM-HMM Technique

2.1. Description of n-gram

N-gram is a statistical language modeling technique that assigns probability to word sequences given some prior probabilities. N-gram is considered the state-of-the-art of language modeling technique, especially in Indonesian automatic speech recognitions [1, 2]. Equation (1) shows the formula of the Maximum Likelihood Estimation of n-gram.

$$P(w_n|w_{n-N+1}^{n-1}) = \frac{C(w_{n-N+1}^{n-1}w_n)}{C(w_{n-N+1}^{n-1})} \tag{1}$$

In this equation, $P(w_n|w_{n-1})$ is the probability of word w_n following word w_{n-1}; N is the value of the gram: 2 for bigram, 3 for trigram, and so on; $C(w_{n-1}w_n)$ is the number of occurrence of word w_{n-1} followed by word w_n; $C(w_{n-1})$ is the number of occurrence of word $C(w_n)$ in the training corpus.

2.2. GMM-HMM technique

Gaussian Mixture Model (GMM)-Hidden Markov Model (HMM) is one of the most well-known acoustic modeling technique used in any automatic speech recognition. HMM is actually probabilistic model that consists of transition probability and observation probability. Transition probability is the value of probability of moving from one state to another state. Observation model is the value of the probability in each state. Fig 1 shows the HMM with 5 states.

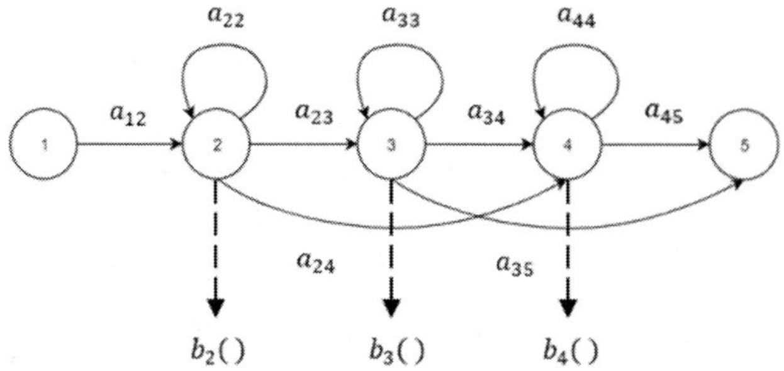

Figure 1. HMM of a word with 5 states

Every emitting state (state 2, 3, and 4 in the Figure 1) has its own GMM. It means every word can consist of one or more phoneme. Every phoneme has its own phone HMM just like Fig. 1. Every phone HMM has emitting states and non-emitting states (state 1 and 5 in Figure 1). Every emitting state has its own GMM.

3. Data augmentation

We use statistical machine translation to augment data. The purpose of data augmentation is to increase the number of spontaneous data to enrich the characteristic of spontaneous speech in language model used for training. We use 'Moses' statistical machine translation to do the data augmentation. Figure 2 from [6] shows the general architecture of statistical machine translation.

There are two main process in machine translation: training and decoding (testing). Training process needs parallel corpus, which is a pair of source and target text with its respective language. For example, if you make a machine translation of French-English, then the source language is French and the target language is English. But in this research, the source language is dictated Indonesian language (Bahasa) and the target language is spontaneous Bahasa. Before used, every text corpus must be pre-processed first which includes tokenization, truecasing, and cleaning process.

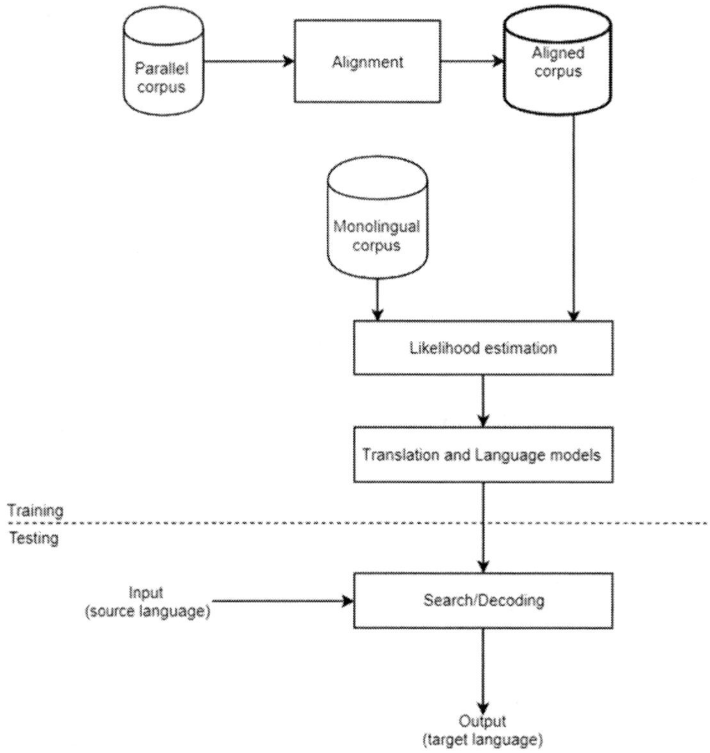

Figure 2. General architecture of statistical machine translation [6]

Training process uses count-based probability calculations on the dataset to train two statistical models: translation model and language model. Translation model is obtained by training parallel corpus of the source and target datasets. Language model is extracted from training on monolingual corpus, which is the target dataset.

Decoding process is actually the process of searching for possible translations and score them, then maximize the obtained score to get the most probable one given both the translation and language models. Algorithms used for decoding are varied, such as depth-first search (DFS), breadth-first search (BFS), or greedy search. The state-of-the art of decoding algorithm in statistical machine translation is beam search. It can prune some candidates with higher cost and instead following path with lower cost. It can speed up the decoding process.

4. Experiment

4.1. Data acquisition

First, data from prior researches are collected. Training data for this research made use of data from [5]. The testing data for this research comes from [1]. The testing data consists of two kinds of data: dictated and spontaneous. The complete text training corpus summary for this research is on the Table 1. The TR_DIKT is the transcript of the speech data from [5]. Table 2 shows the speech training corpus used in this research.

Table 1. Text training corpus

Name	Size (words)	Sentence Style
TR_DIKT	78,107	Dictated
TEXT_TALA	10,299,778	Dictated
TEXT_TAMB_SP	44,213	Spontaneous

Table 2. Speech training corpus

Name	Duration
AUDIO_TR_DIKT	14 hours 25 minutes (20 speaker, 6,679 utterances)

The complete text testing corpus summary for this research is on the Table 3. Table 4 shows the speech testing corpus used in this research. Both dictated and spontaneous data are from [1]. The text data is the transcript of the speech data.

Table 3. Text testing corpus

Kind of testing data	Size (words)
Dictated	5,446
Spontaneous	8,159

Table 4. Speech testing corpus

Name	Duration
AUDIO_TS_DIKT	53 minutes 45 seconds (20 speaker, 600 utterances)
AUDIO_TS_SPON	1 hour 14 minutes (20 speaker, 595 utterances)

4.2. Language model training

In this research, we conduct three experiment scenarios with its own language model each. Table 5 shows the scenario of language model experiment.

Table 5. Language model experiment scenario

No	Language model	Text corpus used
1.	NGRAM_LM_BASE	TEXT_TALA + TR_DIKT
2.	NGRAM_LM_TAMB_SP	TEXT_TALA + TR_DIKT + TEXT_TAMB_SP
3.	NGRAM_LM_PROPOSED	TEXT_TALA + TR_DIKT + TEXT_TAMB_SP + result of data augmentation

These three language models are used for the scenario of developing automatic speech recognition. Table 6 shows the scenario of ASR development.

Table 6. Automatic Speech Recognition (ASR) scenario

No.	Acoustic model	Language model	Lexicon (words)
1.	Triphone GMM-HMM	NGRAM_LM_BASE	63,976
2.		NGRAM_LM_TAMB_SP	74,531
3.		NGRAM_LM_PROPOSED	75,931

To build the ASR, we use Kaldi [7]. Kaldi is a well-known automatic speech recognition toolkit. It can accommodate whether acoustic modeling used is n-gram or deep learning-based, such as Long-Short Term Memory (LSTM). In triphone model training phase, we have to input the parameter for the training. There are two of them, 'num_leaves' and 'tot_gauss'. Based on the experiments, the configuration which gives the least WER is '1,000' and 'tot_gauss' = 16,000. The result of the first experiment scenario (the baseline) is given in the following Table 7.

Table 7. Testing result of first scenario

Testing Data	Perplexity	WER (%)
Dictated	127,412	9.93
Spontaneous	759,883	42.25

The result of the second experiment scenario is given in the following Table 8.

Table 8. Testing result of second scenario

Testing Data	Perplexity	WER (%)
Dictated	135,901	8.80
Spontaneous	840,462	38.66

The result of the third experiment scenario is given in the following Table 9.

Table 9. Testing result of third scenario

Testing Data	Perplexity	WER (%)
Dictated	128,281	7.34
Spontaneous	852,025	39.51

The result of comparing the WER score of these three scenario of dictated testing data is given in Figure 3.

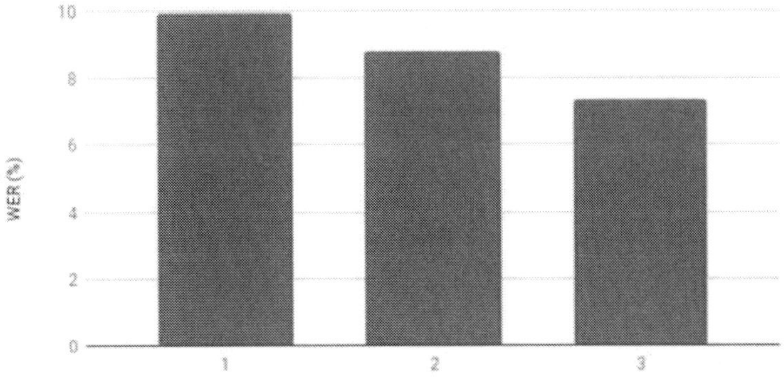

Figure 3. WER of three experiment scenarios of dictated testing data

As seen in Figure 3, the WER score of every scenario keeps decreasing in every step. It means the addition of spontaneous data also improves the recognition of dictated speech. The result of comparing the WER score of these three scenario of spontaneous testing data is given in Figure 4.

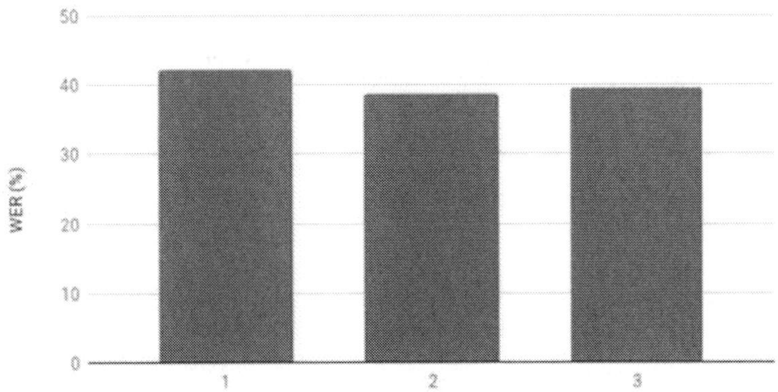

Figure 4. WER of three experiment scenarios of spontaneous testing data

The following Table 10 shows two sentences when recognized with the first, second, and third scenario ASR. 'Ref' is the real sentence, while 'Hyp 1' is hypothetical text recognized by first scenario, 'Hyp 2' is hypothetical text recognized by second scenario, and 'Hyp 3' is hypothetical text recognized by third scenario. The '***' in 'Hyp' sign means that that word is not recognized by that system (there is a deletion). If it happens in 'Ref' it means that there is insertion. That is why the value of WER can be more than 100%, because of the formula is adding all substitution, deletion, and insertion and then dividing all of them with the number of word in test data.

Table 10. Decoding result from three scenarios

No.		Decoding Result
1.	Ref	jadi <euh> kuliahnya yang penting asal datang
	Hyp 1	jadi telah kuliahnya yang penting ada tamu
	Hyp 2	jadi <eu> kuliahnya yang akan datang
	Hyp 3	jadi <eu> kuliahnya yang penting astaga
2.	Ref	<napas> nah di situ kita <napas> menggelar <napas> <euh> kain ceritanya nih jadi dukun tapi <euh> enggak enggak sampai yang harus ruis ruis gitu ya <napas> misalnya kepala dipakein apa gitu enggak <napas>
	Hyp 1	*** *** dari situ kita *** menggelar *** ar kain ceritanya *** jadi dukun *** *** amda asal ayam *** harus berselisih *** *** *** telah bisanya kepala dipakai apalagi *** ternak ***
	Hyp 2	*** *** dari situ kita *** menggelar *** h kain berceritanya ini jadi dukun tapi *** *** karena sampeyan mau harus ruis ruis itu ya bisa saja kepala dipakein kapal itu *** ***
	Hyp 3	*** *** dari situ kita *** menggelar *** h kain berceritanya ini jadi dukun tapi kita ga ga sampai yang harus bis bis itu ya bisa saja kepala dipakein kapal itu *** ***

When comparing the old and new corpus (the one before and after data augmentation), it is seen that there is an addition of 36,929 new words in the new corpus. The total number of the addition of eight filled pauses/fillers "eu", "euh", "eum", "gitu", "gitu yah", "kan", "tuh", and "yah" is 35,109 words. It means 95% of the addition revolves around these eight words. From these 35,109 new words, the filler "eum" is the most dominant with 9,702 addition or 27.63% of the total. After "eum", "gitu" comes second with 23.60% and "yah" with 13.41%. The filler "eu" is 12.63% and "euh" is 10.93%. The bottom three are "tuh" with 4.97%, "gitu yah" with 4.01% and the last one is "kan" with 2.82%.

5. Conclusion

In this research, we use n-gram to build an Indonesian language model for Indonesian spontaneous automatic speech recognition. Language model is the subject of the experiment in this research. There are three experiment scenarios in this research. The first one is baseline, the second one is with the addition of spontaneous text corpus which was from [8]. The third one is the proposed method of this research which conducted data augmentation of the language model of the second scenario.

There is an increase of accuracy of recognizing spontaneous testing data of 2.74% relative to the baseline for the proposed method. When tested to dictated testing data, there is an increase of 2.59%. Actually the accuracy in recognizing spontaneous data is higher in second scenario compared to the third one. The second scenario gives an increase of 3.59% relative to the baseline in recognizing spontaneous testing data.

If we look back to the original motivation of doing data augmentation, which is to minimize the effort required in collecting spontaneous data manually, the decrease is still considered reasonable. The fact that there is an increase of accuracy by doing data augmentation gives hope in improving spontaneous Indonesian automatic speech recognitions' performance. For future works, the researchers need to do pre-processing to the spontaneous data used for data augmentation, do more development on lexicon, and also do post-processing of the corpus after augmented. Last but not least, the size of parallel corpus needs to be increased in order to get a better translation model.

References

[1] Hoesen D, Satriawan C H, Lestari D P and Khodra M L 2016 Towards Robust Indonesian Speech Recognition with Spontaneous-Speech Adapted Acoustic Models *Procedia Compute Science* vol **81** pp 167-73

[2] Yuwan R and Lestari D P 2018 *Deep Neural Network Based Acoustic Modeling for Indonesia Spontaneous Speech Recognition with Active Learning* unpublished

[3] Nakamura M, Iwano K and Furui S 2008 Differences between acoustic characteristics of spontaneous and read speech and their effects on speech recognition performance *Computer Speech & Language* **22(2)** pp 171-84

[4] Jurafsky D and Martin H 2008 *Natural and Language Processing: An Introduction to Natural Language Processing, Computational Linguistics, and Speech Recognition* (New Jersey: Prentice-Hall)

[5] Lestari D P and Furui S 2010 Adaptation to Pronunciation Variations in Indonesian Spoken Query-Based Information Retrieval *IEICE Transactions on Information and System* **E93.D(9)** pp 2388-96

[6] Maazouzi Z E, Mohajir B E E and Achhab M A 2017 A systematic reading in statistical translation: From the statistical machine translation to the neural translation models *J. of Information and Communication Technology* **16(2)** pp 408-41

[7] Povey D *et al* 2011 The Kaldi Speech Recognition Toolkit *IEEE 2011 Workshop on Automatic Speech Recognition and Understanding* (Hawaii: IEEE Signal Processing Society) pp 1-4

[8] Zakiah I and Lestari D P 2019 *Transfer Learning from News Domain to Lecture Domain on Automatic Speech Recognition* unpublished

Acknowledgments

Thanks are due to PT. Prosa Solusi Cerdas and other parties who contributed in this research or during the process of working for this paper.

IOP Conf. Series: Materials Science and Engineering **803** (2020) 012031 doi:10.1088/1757-899X/803/1/012031

A Model of Computation-based Naming System for Musical Elements of Java Traditional Song

Arry Maulana Syarif[1,2], Azhari[1], Suprapto[1], Khafiizh Hastuti[2]

[1]Department of Computer Science and Electronics, Universitas Gadjah Mada
[2]Department of Computer Science, Universitas Dian Nuswantoro

E-mail: arry.maulana.s@mail.ugm.ac.id

Abstract. In a research which involves mathematics and music, information of pitch, such as note, bars, melody line, beat, tempo, and others, was used as dataset to be processed by computer in generating or predicting musical objects. Helmholtz notation and scientific pitch notation used in western music are formal notation writing systems that have been applied in the computer science research as in the topics of algorithmic composition or music information retrieval. A Model of computation-based naming system for musical elements of Java traditional song called *gendhing scientific pitch notation* or GSPN was proposed to provide a method of information data recording of pitch in a term that can be read and write by human and computer. The method was conducted in two phases, which were Identification of musical elements and formulation of computation-based naming system. The model was successfully implemented in a simple computer program that has a task to read a data in GSPN format and to write by generating a music sheet based on the input data.

1. Introduction

Pitch is related to wavelengths and frequencies, and different pitches in western music are named with letters A, B, C, D, E, F, and G as natural notes [1]. An octave is a set of eight musical notes, for examples from A to next A, or from C to next C. The note names are repeated across the octaves, from the lowest to the highest note. There are two formal systems used for naming notes in a particular octave, which are Helmholtz notation and scientific pitch notation. Helmholtz notation uses CC-C-c-c^{i}-c^{ii}-c^{iii} for lowest to the highest C, and lower octave is named with CCC, while higher octaves are named with higher numbers. Scientific pitch notation names octaves with numbers, for instance, starting from C_1 for the lowest C, so the higher C are C_2, C_3, and so on.

Helmholtz notation and scientific pitch notation have been applied in the topics of algorithmic composition or music information retrieval which are parts of a computer science research. The development of these research topics in various types of music genre, both in western music or traditional music, became a motivation to develop a model of computation-based naming system to be applied in *karawitan*, a traditional music from Java, Indonesia. *Karawitan* uses *gamelan* for the music instrument and *gendhing* as the song.

Both Helmholtz notation and scientific pitch notation are used in western music, and through the globalization, the notation technology of western music has been adopted in *karawitan*, including the use of song formulation for *gendhing* [2]. The notation naming system in *karawitan* is still needed to be improved in using as computation-based. For instance, the note names in notes in *gendhing* uses number 1, 2, 3, 5, 6 for musical scale of *slendro*, and 1, 2, 3, 4, 5, 6, 7 for musical scale of *pelog*. The higher note is marked with a dot mark above the note number, and for the lower note, the mark is below the note number. This naming system cannot be read by computer as different with scientific pitch notation which uses number after letter to mark higher or lower notes.

A computation-based notation naming system for musical elements of *gendhing* called *ghending scientific pitch notation* (GSPN) was proposed in order to support a computer music research with *karawitan* as the research object. The purpose of GSPN is to provide a method of information data recording of pitch in a term that can be read and write by human and computer.

2. Related Works

In a research which involves mathematics and music, information of pitch, such as note, bars, melody line, beat, tempo, and others, was used as dataset to be processed by computer in generating or predicting musical objects. Twelve attributes of harmonic function, such as tonic, dominant, supertonic, submedian, and others, were used by [3] as dataset to automatically harmonize chorales in style of J.S. Bach, while information of chord and bars was used by [4] as dataset to automatically generate a melody.

The use of machine learning approach to generate musical objects involving information of musical elements from a collection of songs used as dataset. The information is then represented to achieve knowledge to generate a musical objects or to make a prediction of musical objects. Note pattern and tempo from classic music was used by [5] to generate an original music using Long-Short Term Memory method, while Eigenfeldt dan Pasquier [6] using corpus from different composer as dataset.

Research which involves mathematics and *karawitan* has been conducted in several decades by [7-10]. Statistic approach by [7], grammar method by [8-9], rule-based method by [10] were used to formulate rules of composition generation, while used rule-based method. An automatic composition generation was developed by [11] using rule-bases method to define constraints which was then used by Genetic Algorithm to generate a note sequence of *gendhing*. All these researchers used note sequences of *gendhing* as dataset, and only note numbers used as the information. On the other hand, other musical elements which are also important to support automatic music generation or to make a prediction of musical objects are still not formulated in the research with *karawitan* as an object.

3. Proposed Method

A computation-based notation naming system for musical elements of *gendhing* was developed and named with *gendhing scientific pitch notation* (GSPN). The proposed method was divided into two phases, which were identification of musical elements and formulation of GSPN. The first phase was conducted by examining a *gendhing* music sheet to identify the musical elements. The second phase was conducted by analyzing the existing note formulation of *gendhing* and scientific pitch notation formulation to define GSPN.

3.1. Musical Elements Identification

The identification of musical elements of *gendhing* was conducted using theory of *karawitan* and analysing *gendhing* music sheets collected from www.gamelanbvg.com. In *karawitan,* musical scale is called *laras*, and there are two types of musical scales in *karawitan*, which are *laras slendro* and *laras pelog*. *Laras slendro* contains five notes of 1, 2, 3, 5, and 6, while *laras pelog* contains seven notes of 1, 2, 3, 4, 5, 6, and 7 with different wavelengths and frequencies. Both of *laras slendro* and *laras pelog* have three types of musical modes or called as *pathet*. Musical mode is defined based on the dominant notes in a composition. *Laras slendro* contains of musical modes of *pathet nem, pathet sanga,* and *pathet manyura*; *laras pelog* contains of musical modes of *pathet lima, pathet nem,* and *pathet barang*. Figure 1 shows an illustration of musical scale and mode in *karawitan*.

IOP Conf. Series: Materials Science and Engineering **803** (2020) 012031 doi:10.1088/1757-899X/803/1/012031

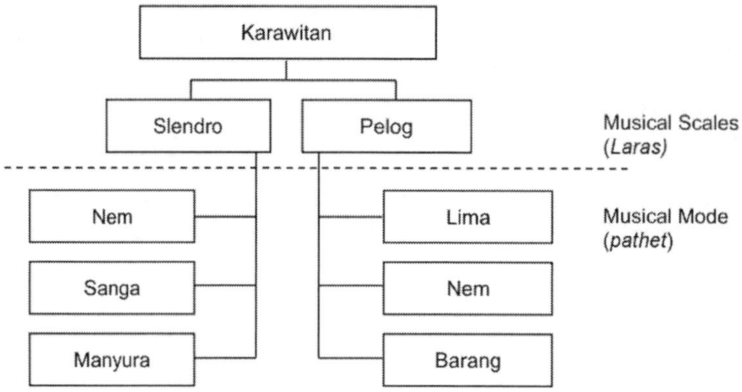

Figure 1. Musical scale and mode in *karawitan*

A *gendhing* has a song skeleton that has a role as a chord in western music, in which notes of song skeleton were played to accompany the main melody. Figure 1 shows an example of a music sheet of *gendhing* entitled *Ladrang Gayung* played in *laras slendro* and *pathet nem*, with symbol *a* indicates the song skeleton parts, and symbol *b* indicates the main melody parts.

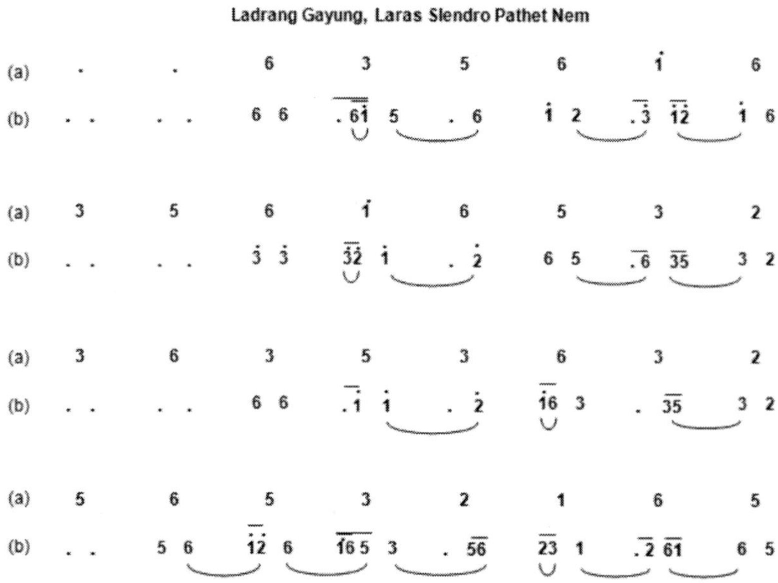

Figure 2. An example of a music sheet of *gendhing*

Theory of *karawitan* was used to analyze the structure of musical elements of *gendhing*. A *gendhing* consists of a note sequence arranged in bars called *gatra*, where bars are arranged in melody lines. In the song skeleton, a bar contains four beats, and a beat can filled with a note or more which define the rhythm. A beat filled with a note defines rhythm at 1/1, a beat filled with two notes defines rhythm at 1/2, a beat filled with four notes defines rhythm at 1/4, and so on. In parts of main melody, a horizontal line above a note indicates the value of 1/2 for the note, and double horizontal lines above a note indicate the value of 1/4 for the note, while a note without a horizontal line mark has a value of 1. Legato mark is put below the note.

3.2. Formulation of Gendhing Scientific Pitch Notation

The formulation of *gendhing scientific pitch notation* was started from the musical scale and musical mode. For the musical scales, *laras slendro* is abbreviated with S, and *laras pelog* is abbreviated with *P*, while musical modes in each musical scale is labelled with number. Alphabet sorting was applied to define the number for each musical mode. In the *laras slendro,* the number to label *pathet manyura* is 1, *pathet nem* is 2, and *pathet sanga* is 3, while in the *laras pelog*, the number to label *pathet barang* is 1, *pathet lima* is 2, and *pathet nem* is 3. The label of musical scales and musical notes is shown below:

S1: laras slendro-pathet manyura.
S2:laras slendro-pathet nem
S3: laras slendro-pathet sanga
P1: laras pelog-pathet barang.
P2: laras pelog-pathet lima.
P3: laras pelog-pathet nem.

Rhytm which has values of 1/1, 1/2, 1/4, 1/8, and 1/16 is labelled with number 1, 2, 3, 4, 5. Hence, given a name of *Ladrang Gonjang: S1-2*, it means a *gendhing* entitled *Ladrang Gonjang* played in musical scale of *laras slendro,* and in musical mode of *pathet manyura*, and in the rhythm of 1/2.

Laras slendro consists of five notes of 1, 2, 3, 5, 6, while *laras pelog* consists of seven notes of 1, 2, 3, 4, 5, 6, 7. There is a dot notation called *pin* in both of musical scales. The dot notation was converted into number of 0 for it can be treated as other notes do. Union operation was implemented to label the notes, with *T* stands for notes, *S* stands for *laras slendro*, and *P* stands for *laras pelog*, the label for notes is as follow:

$P = \{0, 1, 2, 3, 4, 5, 6, 7\}$
$S = \{0, 1, 2, 3, 5, 6\}$
$T = P \cup S = \{0, 1, 2, 3, 4, 5, 6, 7\}$

There is pitch region that has a role as octave in western music. The pitch region consists of low pitch marked with a dot below a note, middle pitch, and high pitch marked with dot above a note. The low pitch was labelled with *a*, and high pitch was labelled with *b*. The middle pitch and note of 0 was set without label addition:

Low pitch (marked with a dot below the note): 1a, 2a, 3a, 4a, 5a, 6a, 7a
Middle pitch: 1, 2, 3, 4, 5, 6, 7
High pitch (marked with a dot above the note): 1b, 2b, 3b, 4b, 5b, 6b, 7b
Dot notation: 0

A note marked with single horizontal line or double horizontal lines indicate the value of the rhythm. Label *A* and *B* was used to indicate a value of a note, as follows:

Value of 1/1: 1, 2, 3, 4, 5, 6, 7
Value of 1/2: 1A, 2A, 3A, 4A, 5A, 6A, 7A
Value of 1/4: 1B, 2B, 3B, 4B, 5B, 6B, 7B
...

Legato mark was labelled with *x* for a note where the legato mark is started, n label *y* for a note where the legato mark is ended.

Value of 1/2: 0x, 1x, 2x, 3x, 4x, 5x, 6x, 7x
Value of 1/4: 0y, 1y, 2y, 3y, 4y, 5y, 6y, 7y

Based on the description above, the formulation of GSPN are:

Musical scale = {P, S} //{*pelog, slendro*}
Musical mode (S) = {1, 2, 3} // { *manyura, nem, sanga*}
Musical mode (P) = {1, 2, 3} // {*barang, lima, nem*}
Rhythm = {1, 2, 3, 4, 5} // *{1, 1/2, 1/4, 1/8, 1/16}*
Notes = {0, 1, 2, 3, 4, 5, 6. 7} // *{., 1, 2, 3, 4, 5, 6, 7}*
Pitch region = {a,' ',b} // *{low pitch, middle pitch, high pitch}*
Note value = {' ', A, B, C, ...} // *{1, 1/2, 1/4, 1/8, ...}*
Legato = {x, '', y} // *{start legato, no legato, end legato}*

Note was written in a formula: *note + pitch region + note value + legato*. Below is an example of the implementation of GSPN for a melody line of a *gendhing* entitled *Ladrang Gayung* played in *laras slendro* and *pathet nem*.

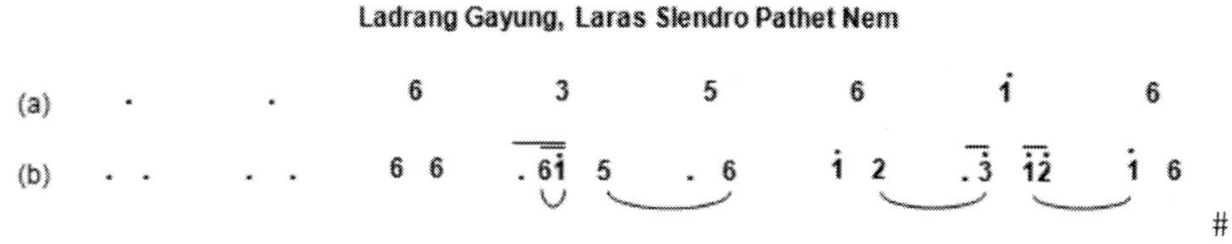

Ladrang Gayung: S1-2
(a) 0063561b6
(b) 0000660A6Bx1bBy5x0A6Ay1b2x0A3bAy1bAx2bA1by6

Figure 3. An example of the implementation of GSPN

4. Result and Analysis

The implementation was conducted by writing 30 note sequences of a *gendhing* in the GSPN format. A computer program to read and to display a note sequence of a *gendhing* was developed to evaluate the GSPN formulation. The task of the program was to read a note sequence written in GSPN format and to display the note sequence in original sheet music.

The procedure of read and display a note sequence of the computer program was defined by creating variables containing values of musical elements of *gendhing* to convert all labels in GSPN format into number. This to make all information of musical scale can be calculated in a computation process.

GSPN: pitch region = (0, a, "", b)
Program: pitch region = (0, 1, 2, 3)

GSPN: note value = ("", A, B, C, ...)
Program: note value = (1, 0.5, 0.25, 0.13, ...)

...

Below is an example of the implementation of GSPN for a *gendhing* entitled *Ketawang Barikan* played in *laras pelog* and *pathet lima*.

Ketawang Barikan: P2-1
//input note sequence of main melody in GSPN format
00000023x0053y02x0A3Ay10000555Ax1bAy6005302x0A3Ay10000555A6A4x05y61b01b
Ax2bA1bA6Ay50000762Ax5Ay4x052Ax3Ay101x2B1B6aAy5ax0A6A1y00111Ax2Ay1002
302x0A3Ay10000555Ax1bAy6005302x0A3Ay1
//extracted note sequence

0, 0, 0, 0, 0, 0, 2, 3, 0, 0, 5, 3, 0, 2, 0, 3, 1, 0, 0, 0, 0, 5, 5, 5, 1, 6, 0, 0, 5, 3, 0, 2, 0, 3, 1, 0, 0, 0, 0, 5, 5, 5, 6, 4, 0, 5, 6, 1, 0, 1, 2, 1, 6, 5, 0, 0, 0, 0, 7, 6, 2, 5, 4, 0, 5, 2, 3, 1, 0, 1, 2, 1, 6, 5, 0, 6, 1, 0, 0, 1, 1, 1, 2, 1, 0, 0, 2, 3, 0, 2, 0, 3, 1, 0, 0, 0, 0, 5, 5, 5, 1, 6, 0, 0, 5, 3, 0, 2, 0, 3, 1

//data of region pitch
0, 0, 0, 0, 0, 0, 2, 2, 0, 0, 2, 2, 0, 2, 0, 2, 2, 0, 0, 0, 0, 2, 2, 2, 3, 2, 0, 0, 2, 2, 0, 2, 0, 2, 2, 0, 0, 0, 0, 2, 2, 2, 2, 2, 0, 2, 2, 3, 0, 3, 3, 3, 2, 2, 0, 0, 0, 0, 2, 2, 2, 2, 2, 0, 2, 2, 2, 2, 0, 2, 2, 2, 1, 1, 0, 2, 2, 0, 0, 2, 2, 2, 2, 2, 0, 0, 2, 2, 0, 2, 0, 2, 2, 0, 0, 0, 0, 2, 2, 2, 3, 2, 0, 0, 2, 2, 0, 2, 0, 2, 2

//data of note value
1, 1, 1, 1, 1, 1, 1, 1, 1, 1, 1, 1, 1, 1, 0.5, 0.5, 1, 1, 1, 1, 1, 1, 1, 0.5, 0.5, 1, 1, 1, 1, 1, 1, 1, 0.5, 0.5, 1, 1, 1, 1, 1, 1, 1, 0.5, 0.5, 1, 1, 1, 1, 1, 1, 0.5, 0.5, 0.5, 0.5, 1, 1, 1, 1, 1, 1, 0.5, 0.5, 1, 1, 1, 0.5, 0.5, 1, 1, 1, 0.25, 0.25, 0.5, 1, 0.5, 0.5, 1, 1, 1, 1, 1, 0.5, 0.5, 1, 1, 1, 1, 1, 1, 0.5, 0.5, 1, 1, 1, 1, 1, 1, 1, 0.5, 0.5, 1, 1, 1, 1, 1, 1, 1, 0.5, 0.5, 1

//data of legato
1, 1, 1, 1, 1, 1, 1, 2, 1, 1, 1, 3, 1, 2, 1, 3, 1, 1, 1, 1, 1, 1, 1, 2, 3, 1, 1, 1, 1, 1, 1, 2, 1, 3, 1, 1, 1, 1, 1, 1, 1, 1, 1, 2, 1, 3, 1, 1, 1, 2, 1, 1, 3, 1, 1, 1, 1, 1, 1, 1, 2, 3, 2, 1, 1, 2, 3, 1, 1, 2, 1, 1, 3, 2, 1, 1, 3, 1, 1, 1, 1, 2, 3, 1, 1, 1, 1, 1, 1, 2, 1, 3, 1, 1, 1, 1, 1, 1, 1, 2, 3, 1, 1, 1, 1, 1, 1, 2, 1, 3, 1

Based on the input data, the computer program calculates the information of every note, and generates display as illustrated in Figure 4 using procedures below:

```
// Array extracted note sequence
For (i=0; i < extracted note sequence.length; i++)
    if (extracted note sequence[i])
    {
        data of region pitch [i]
        data of note value [i]
        data of legato [i]
        set display note sequence
    }
}
```

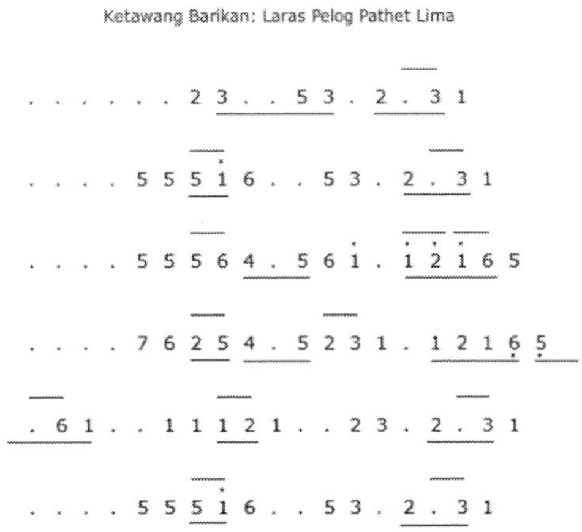

Figure 4. A music sheet generated by computer program

5. Conclusion

Knowledge representation of a song is not restricted to the notes number only, but other musical elements which construct a song are also important for research that involves mathematics and music. A Model of computation-based naming system for musical elements of Java traditional song called *gendhing scientific pitch notation* or GSPN was proposed in this study in order to provide a method of information data recording of pitch in a term that can be read and write by human and computer.

The model was successfully implemented in a simple computer program that has a task to read the data in GSPN format and display a music sheet based on the input data. It still takes an effort to manually convert an original music sheet into GSPN data format. An image processing approach can be used to overcome this problem, in which a pattern recognition method can be implemented to capture an original music sheet and then automatically convert to the GSPN data format.

For further research, the proposed model can be implemented to manage data in GSPN format to represent a mapping code for machine learning approach in generating a composition or providing a prediction of musical objects.

References

[1] Jones C S 2005 *Understanding Basic Music Theory* The Open University of Hongkong

[2] Hastuti K, Fanani A Z, Syarif A M and Mulyana A R 2019 Globalization, Information and Communication Technology, and The Existence of Culture and Its Inheritors: A Case Study on Javanese Karawitan *Proc. of the 1st Int. Conf. on Interdisciplinary Arts and Humanities (ICONARTIES) Yogyakarta, Indonesia*

[3] Hild H, Feulner J and Menzel W 1991 HARMONET: A Neural Net for Harmonizing Chorales in the Style of J.S.Bach *Proc. of Advances in Neural Information Processing Systems* **4**

[4] Pachet F and Roy-Markov P 2011 Markov constraints: steerable generation of Markov sequences *Constraints* Vol **16**

[5] Svegliato J and Witty S 2016 Deep Jammer: A Music Generation Model *Small* vol **6** p 67

[6] Eigenfeldt A and Pasquier P 2010 Realtime Generation of Harmonic Progressions Using Controlled Markov Selection *Proc. of the First Int. Conf. on Computational Creativity*

[7] Surjodiningrat W, Khandelwal V K and Soesianto F 1977 *Gamelan dan Komputer: Analisa Patet dan Komposisi Gending Jawa Laras Slendro* (Yogyakarta: Universitas Gadjah Mada)

[8] Becker J and Becker A 1982 *A Grammar of the Musical Genre Srepegan* (Texas: University of Texas Press_

[9] Hughes D W 1988 *Deep Structure and Surface Structure in Javanese Music: A Grammar of Gendhing Lampah* (Illinois: University of Illinoi Press)

[10] Hastuti K, Azhari, Musdholifah A and Supanggah R 2016 Building Melodic Feature Knowledge of Gamelan Music Using Apriori Based on Functions in Sequence (AFiS) Algorithm *Int. Review on Computers and Software* Vol **11** No 12

[11] Hastuti K, Azhari, Musdholifah A and Supanggah R 2017 Rule-Based and Genetic Algorithm for Automatic Gamelan Music Composition *Int. Review on Modelling and Simulations* Vol **10** No 3

Recommending Tourism Destination using Collaborative Filtering

Rahmatina Ari Apriliana, Ahmad Fathan Hidayatullah, Almed Hamzah

Department of Informatics, Universitas Islam Indonesia, Jalan Kaliurang KM 14,5, Sleman, Yogyakarta Indonesia, 55584

E-mail: almed.hamzah@uii.ac.id

Abstract. Tourism is among the interesting area of research as more people traveling to various destinations worldwide. With the vast amount of places to visit, there is a problem for tourist to determine which place is most suitable for them in terms of their preferences. However, it can be solved by the means of technology. This paper propose the use of recommender techniques to help tourist in selecting the most preferable destinations. A web-based application was built as an interface with the tourist. The result shows that the system is functioning as it expected. Future directions are discussed.

1. Introduction

Recommender system has been used in many different areas. In e-commerce industry, it helps customer to dive into a plenty of products choice and find the most suitable one [1]. In education sector, such system can be adopted to helps students navigate better and find the most suitable works as it needed to increase their performances [2].

In tourism, this kind of system could be implemented to help tourist selecting the destination. This is important because the huge number of various places can make it difficult for people to find the most suitable places that fit their needs and preferences. Moreover, they will be overwhelmingly flooded with the information about the place they can visit [3].

There are various recommender system was built in order to fulfill specific needs. Based on previous publications, the recommender system was used to list the most prefered activities [4], helps group of tourist to decide the destinations [5], provide recommendation of restaurant based on tourist location [6], manage loyalty of tourists based on their first visit data [7]. The present study focus on cultural and nature tourism located in Yogyakarta, one of the primary destination in Indonesia. This study aims to adopt collaborative filtering in recommender system that provide recommendation for tourists regarding to the destination that suits their needs and preferences. Based on preliminary observation, item-based collaborative filtering was chosen as a main method to select the most appropriate destination based on visitor's ratings.

The case study was conducted in Yogyakarta. It has various attractive places including beaches, mountain, culture, and culinary spot. Based on the data retrieved in 2018, there are more than a hundred destinations found in this special province. The number is burgeoning about 10 % comparing to a year before [8]. Moreover, the number of tourists visit this special province is increasing too. The number of the visitors has been increasing every year as well since 2013. It can be seen in Figure 1. The recommender system will be beneficial for the tourist that has intention to visit this area.

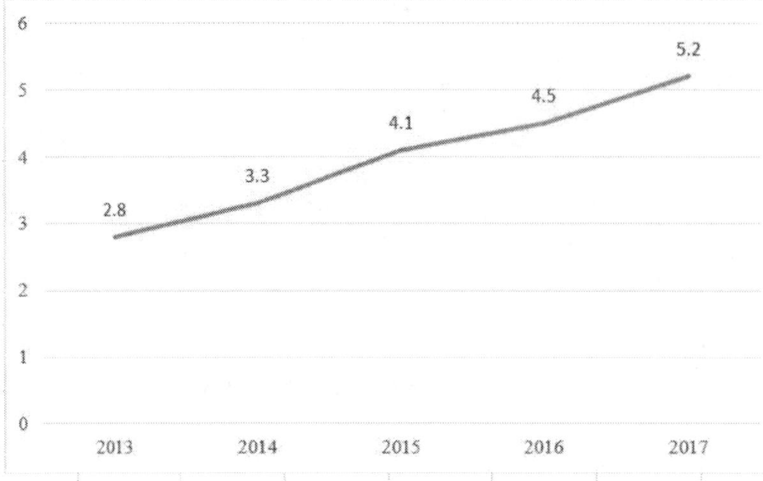

Figure 1. Tourist Arrivals in Yogyakarta

This paper consists of five sections. After introduction, section two describes the literature review. Section three explains the methods for the current research. Section four explains the prototype of system and user interface design. Section five ends this paper with concluding remarks.

2. Related Works

The use of technology in tourism sector is growing rapidly as there were many publications on this topic. Technology, in this case integrated platform, has been used to manage tourism resources with which it can be more sustainable and increase the profit eventually [9]. In addition, it can be used in a development planning of tourism area [10]. With the technology, it is easier to identify and mapped the area and its vicinity. Furthermore, an electronic tour guide system can assist tourist during their visit a destination [16]. Collaborative filtering is the most common method in the development of recommender system. In this technique, system will generate a prediction about particular user preferences based on other user rating [11]. Based on previous works, it has been used in following area: entertainment [11], education [14], and also commerce [13].

In entertainment, with the combination of social relationship and tags provided by the user, collaborative filtering can then generate the recommendation of item (movie) that most relevant to the particular user [11]. The idea is finding the other user preferences which similar to targeted user then generate the recommendation of products that might become interest of the targeted user [12]. In educational sector, collaborative filtering has been used in providing recommendation of relevant learning object for students. This applies especially in online learning setting [14]. The goal is to increase the students engagement to the course with which their academic performance will significantly improving. In commerce, collaborative filtering can be used to predict user's preferences based on their purchase history [13]. Considering the vast use of collaborative filtering in many areas mentioned above, this study propose the application of the method in tourism context.

3. Methods

This study's main objective is to develop a recommender system for tourist in selecting their destination based on other's rating. A case study was conducted at Yogyakarta, one the most popular tourist destination in Indonesia. The data was collected in 2018. We have collected the data in 2018 and obtained 36 tourism destinations as our dataset. To achieve this objective, there are three phases of study are conducted, i.e. (1) initial observation, (2) system development, (3) evaluation. Initial observation focus on gaining the requirements of system, collect documentation about the destinations, and retrieve training data. In this study, the documents are collected from the government's website that related to tourism sector. Moreover, to obtain training data set, there were eight respondents participate on the preliminary observation. In system development phase, the classic waterfall approach was used to built the recommender system. With this approach, there are four steps of development, namely analysis of requirements, design system architecture, implementation, and testing. Regarding to the evaluation phase, there were two steps of evaluation. Firstly, blackbox testing to identify any defect in the application. Secondly, user testing to identify whether the application have met the user's requirements and achieve its main objective.

Figure 2 describes the steps that must be conducted in order to produce the recommendation for the user. Overall, there are four stages in the process. To begin, a user provide rating for particular destination. Following this, the system will calculate the similarity of this user with the existing data from other user. This process will produce several destinations that might become the interest of user. Next, based on the result from previous step, system will predict the most relevant destinations for the targeted user. Finally, a list of recommended destinations displayed on the screen.

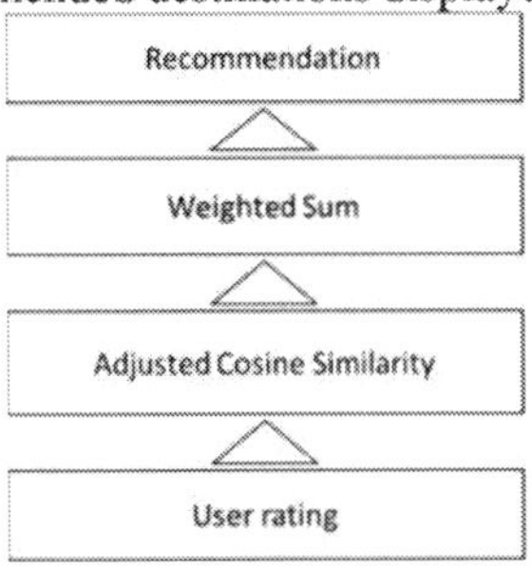

Figure 2. Research Steps

3.1. *Item-based Collaborative Filtering*

Collaborative filtering is a mechanism to generate the recommendation of particular product based on preferences of user that stored in a database [15]. The recommending mechanism is based on other's user rating. Before receive the recommendation, users should give a rating to a particular destination. After that, system will generate the recommendation based on his/her rating and other user's rating. To generate the recommendation, the system follow two steps of calculation. Initially, adjusted cosine similarity were used to examine the similarity of one place to another. Secondly, weighted sum were used to select the destinations that match the user's characteristics based on rating he/she gave to the system. Below is the equation that has been used to generate the recommendation.

$$sim(i,j) = \frac{\sum_{u \epsilon U}(r_{u,i} - \bar{r}_u)(r_{u,j} - \bar{r}_u)}{\sqrt{\sum_{u \epsilon U}(r_{u,i} - \bar{r}_u)^2}\sqrt{\sum_{u \epsilon U}(r_{u,j} - \bar{r}_u)^2}} \quad (1)$$

Below is the formula of weighted sum to predict the preferences of particular user.

$$P(u,j) = \frac{\sum_{i \epsilon I}(R_{u,i} x S_{i,j})}{\sum_{i \epsilon I}|S_{i,j}|} \quad (2)$$

3.2. *Evaluation Techniques*

The evaluation phase of system consists of two activities with different objectives. First, blackbox testing to identify any error in the system's main functionality. A table comprises testing scenarios was prepared. Second, user acceptance testing to identify whether the system is fit with the user's needs. A survey was conducted to achieve this goal. The respondents will be asked to use the application and then fill in the questionnaire. The questionnaire comprises nine questions with five-points likert scale.

4. Results

The overall functionality of system is depicted in Figure 3 below. There are four main features that available for the tourist, namely finding destinations, view destinations' profile, rate the destinations, and receiving recommendation from the system. With the search feature, a user can find the destinations based on certain keyterms. In addition, they can also view the profile of destinations including name, location, and ticket price. After that, they can provide rating for particular destination. They can rate the destinations on the range between 1 to 5. The rating will be used to calculate and generate the recommendation.

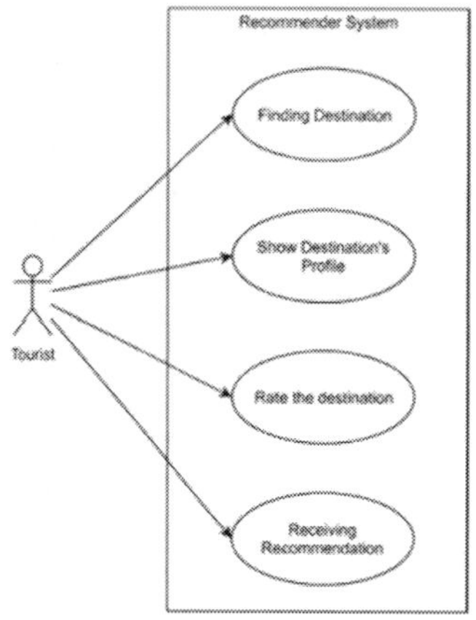

Figure 3. Use Case Diagram

Based on the use case diagram, a recommender system for tourist destination was built. Below are interface from the main features of the system. Figure 4 shows the homepage of the system which display the destinations along with its popularity based on ratings provided by all users.

From the homepage, individual user can provide rating for particular destination by clicking on the card and fill choose the rating between 1 to 5. The most preferable place is represented by number 5. After receiving rating from user as depicted in Figure 5, the system will be able to generate recommendation for this user based on the similarity with other user as depicted in Figure. 6.

The process from receiving rating to generating recommendation consists of two steps. Initially, the calculation of similarity between the place rated by user with other places that already rated by another user. This is done by using adjusted cosine similarity.

There are two steps of evaluation of the system, blackbox testing and user acceptance testing. For blackbox testing, there are four scenarios to test whether the four features designed on the previous steps yield the output as expected. All the scenarios are described in Table 1.

User acceptance test was held by means of survey to several respondents. In this test, seven questions were designed to gather the user's view about the system after they use it. Each question is measured with the Likert's Scale from 1 to 5 expressing to which extent the respondent agree with the statement. The respondents are people aged between 16 to 25 years old. There are 15 people involved as a respondent. They are the person who has an experience visiting some tourist destinations. The result of survey is described in Table 2. Generally speaking, the respondents have a good experience with the system as they are answering between 4 to 5 for the majority of the questions. However, there are some factors from the system that need to be improved, i.e user interface, information about destinations, and accuracy.

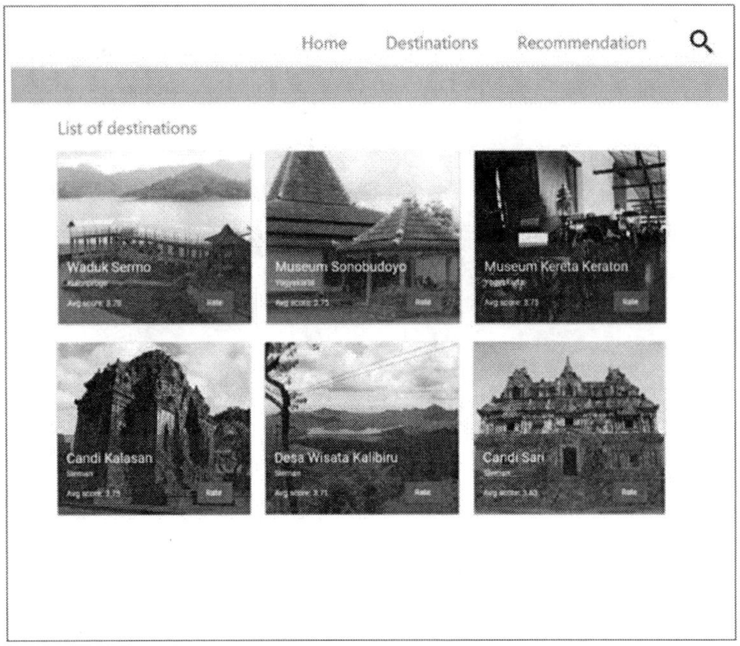

Figure 4. Homepageofrecommendersystem

Figure 5. User provide rating

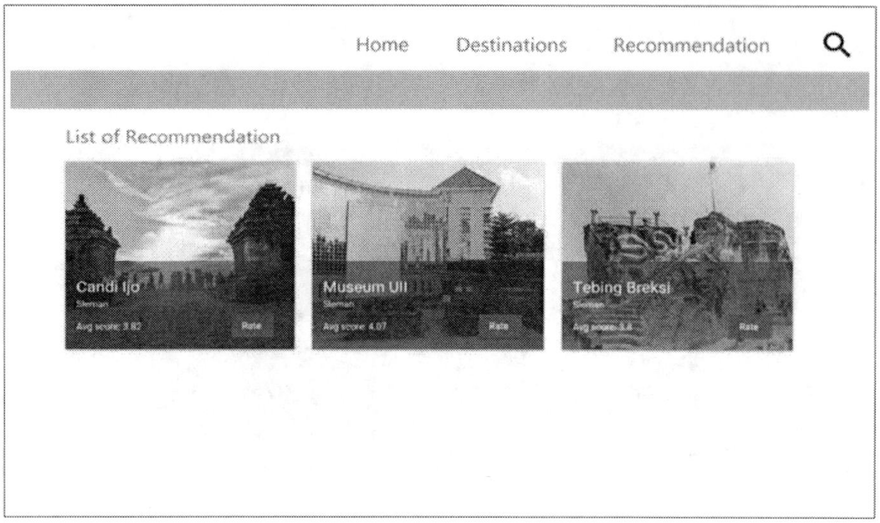

Figure 6. System generate recommendation

Table 1. Blackbox Testing.

Feature	Test case	Result
1	When user click "Rate" button, the button will be activated and rate option for particular destination will appear	Valid
2	When user enter a search term and click "Search" button, system shows the appropriate destination based on the term	Valid
3	When user click the card of particular destination, system shows the detailed profile of the destination	Valid
4	When user navigate to recommendation page, system shows three recommended destinations	Valid

Table 2. User Acceptance Test.

No	Questions	1	2	3	4	5
1	The recommender system helps me to find destinations easily				8	7
2	The user can provide rating to particular destination easily				6	9
3	The recommender system has good user interface			1	8	6
4	The recommender system could provide useful information about destinations			3	3	9
5	The recommender system helps user to find suitable destinations				9	6
6	The recommendation is fit to the user preferences			2	8	5
7	The recommendation feature is helpful				9	6

5. Conclusion

Recommender system in tourism is an interesting research topic as there are many advantages that can be obtained. In this study, such system was designed and developed to provide the tourist an aiding tools to decide the most interesting tourism destination that relevance to their preferences. To built the recommendation mechanism, a collaborative filtering was used to find the similarity of one tourist's preferences to another tourists.

Based on the testing phases, the developed recommender system still has several drawbacks. Therefore, a further development is required to improve the recommender system. There are three potential work could be done. Initially, increase the usability of the system by repairing the user interface design. A usability testing is suitable on this context. In addition, the information of tourist destination is quite minimum. Hence, a more detailed information about the destination need to be gathered. Finally, there is a need for improvement on the recommendation mechanism in order to increase its accuracy.

References

[1] Lu J, Wu D, Mao M, Wang W and Zhang G 2015 Recommender System Application Developments: A Survey, *Decision Support Systems* **74** (Amsterdam: Elsevier) pp 12-32

[2] Brusilovsky P, Sosnovsky S and Shcherbinina O 2004 QuizGuide: Increasing the Educational Value of Individualized Self-Assessment Quizzes with Adaptive Navigation Support *World Conference on E-Learning in Corporate, Government, Healthcare, and Higher Education* **1** pp 1806-13 DOI:10.1.1.91.326

[3] Nilashi M, Bin Ibrahim O, Ithnin N and Sarmin N H 2015 A Multi-Criteria Collaborative Filtering Recommender System for the Tourism Domain Using Expectation Maximization (EM) and PCA-ANFIS *Electronic Commerce Research and Applications* **14** 6 (Amsterdam: Elsevier) pp 542-62 DOI:10.1016/j.elerap.2015.08.004

[4] Garcia I, Sebastia L and Onaindia E 2011 On the design of individual and group recommender systems for tourism *Expert systems with applications* **38(6)** pp 7683-92

[5] Nguyen T N and Ricci F 2017 A chat-based group recommender system for tourism *Information and Communication Technologies in Tourism 2017* (Cham: Springer) pp 17-30

[6] Noguera J M, Barranco M J, Segura R J and MartNez L 2012 A mobile 3D-GIS hybrid recommender system for tourism *Information Sciences* **215** pp 37-52

[7] Colomo-Palacios R, Garca-Pealvo F J, Stantchev V and Misra S 2017 Towards a social and contextaware mobile recommendation system for tourism *Pervasive and Mobile Computing* **38** pp 505-15

[8] DIY D P 2017 *Panduan Destinasi Wisata DIY* Retrieved from http://visitingjogja.com

[9] Avdimiotis S and Christou E 2008 G.I.S Applications In Tourism Planning; A Tool For Sustainable Development Involving Local Communities *Yugoslav J. of Operation Research* **18** 2 pp 261-72 DOI:10.2298/YUJOR0802261J

[10] Jovanovi V 2016 The Application of GIS and Its Components in Tourism *Yugoslav J. of Operations Research* **18** 2

[11] Ma T, Zhou J, Tang M, Tian Y and Al-dhelaan A 2015 Social Network and Tag Sources Based Augmenting Collaborative Recommender System *IEICE Transactions on Information and Systems* **E 98 D** Issue 4 pp 902-10 DOI:10.1587/transinf.2014EDP7283

[12] Subramaniyaswamy V, Logesh R, Chandrashekhar M, Challa A and Vijayakumar V 2017 A Personalised Movie Recommendation System Based on Collaborative Filtering *Int. J. of High Performance Computing and Networking* **10** 12 (New York: Inderscience Publishers) pp 54-63

[13] He R and McAuley J 2016 Ups and Downs: Modeling the Visual Evolution of Fashion Trends with OneClass Collaborative Filtering *Proc. of the 25th Int. Conf. on World Wide Web, Int. World Wide Web Conferences Steering Committee* pp 507-17

[14] Khribi M K, Jemni M, Nasraoui O and Hussein T 2008 Automatic Recommendations for E-Learning Personalization Based on Web Usage Mining Techniques and Information Retrieval *IEEE Int. Conf. on Andvanced Learning Techniques* **8** 4 pp 241-5 DOI:10.1109/ICALT.2008.198
[15] Sarwar B, Karypis G, Konstan J and Riedl J 2001 Item-Based Collaborative Filtering Recommendation Algorithms *Www* pp 285-95 DOI:10.1109/ICMLC.2009.5212320
[16] Chu T-H, Lin M-L, Chang C-H and Chen C-W 2011 Developing a Tour Guiding Information System for Tourism Service Using Mobile GIS and GPS Techniques *Advances Information Sciences and Service Sciences* **3** 6 (The International Association for Information, Culture, Human and Industry) pp 49-58

IOP Conf. Series: Materials Science and Engineering **803** (2020) 012033 doi:10.1088/1757-899X/803/1/012033

Populating Databases Automatically with Realistic Data Using Metadata-Driven Methods

K S Rebolledo, J S Arroyo, and JV Roig

Asia Pacific College, 3 Humabon Place, Magallanes, Makati City, Philippines 1232

E-mail: krisr@apc.edu.ph, jeniffera@apc.edu.ph, and jvroig@apc.edu.ph

Abstract. This research investigated the feasibility and utility of a metadata-driven "fake data" generator. The researchers created a prototype for a data generator whose job is to be pointed to a database and then generate fake but appropriate and realistic data. Using metadata analysis, the generator was able to make autonomous and smart decisions as to how to populate the various tables and fields it encounters, populating them with realistic data. This promises great utility to programmers, QA testers, and database administrators for various use cases such as testing, debugging, database design validation, load testing, and performance scaling.

1. Introduction

If you need hundreds or more of sample data per table, it is too time-consuming to populate the database manually. "Faker" libraries are available as the intermediate solution to this problem [1]. "Faker" is the almost-uniform name of such libraries in a lot of platforms, from Perl [2] to PHP [3] to Ruby [4] to Python [5]. They are all very similar in function as well – they are a library that provides an API for programmers to easily generate fake data.

"Easy", of course, is relative. If you have nothing, then having access to a faker library is cool. However, once you try to use faker libraries for multiple tables and multiple projects, it breaks down bad. You must write code for each table, meaning you will have to decide what kind of data it will write per table. No programmer in his right mind would want to write code using any faker API for dozens of tables for each of his projects. No database admin would likewise do so just to verify and validate his designs, since his designs may iterate very fast.

There have been publications in the past that studied different methodologies for dynamically generating test data [6][7][8][9]. However, none of these have specifically examined the feasibility of a metadata-driven approach, which is what the researchers propose in this paper.

This research proposes a metadata-driven methodology to remove the need for tedious programming for the generation of fake data and matching them to the respective fields. Specifically, this means that determining what sorts of fake data is appropriate to create for the tables and fields in a database is determined automatically by analyzing the metadata of the database. Metadata refers to information stored in the database that describes the characteristics of the various tables and fields in the database itself – simply put, "data about data"[10].

The researchers created a proof of concept of a metadata-driven data generator that can autonomously produce realistic fake data to populate a database. The programming language used was PHP, and the database was SQLite3.

The researchers named this metadata-driven data generator proof of concept "Automatic Data Populator" (referred to as ADP from here on). The applications of ADP are varied: (1) Programmers can use it to prepare their application for a demo to customers. (2) Testers can more effectively test systems with the ability to create arbitrary amounts of data. (3) Database administrators and various sorts of development teams can use it to immediately verify and validate that their database design achieves or retains the desired performance characteristics even when loaded with tons of data well before this database design actually makes it to production and holds actual production data. (4). It would be a very useful simulation tool for database-driven systems that can help system designers identify potential bottlenecks by playing around with parameters. For example, is our system design capable of handling our customers if tables X, Y, and Z balloon to 10M records? What if it is only table X that balloons in size, but goes up to 20M records? (5) It can also be used to test hardware limitations and determine the capability of existing hardware deployments for expected volume (such as expected customer growth), and from there, make an informed decision whether more powerful hardware will be needed and when.

2. Proposed Design

2.1. Components

2.1.1. Library and API, and GUI. The research team picked up the common features of the faker libraries and identified a list of at least 100 common "data" that goes to databases – for example, human names, human first names, human last names, titles and honorifics, countries, cities, mobile numbers, email addresses, color, size, etc.

2.1.2. Metadata-Driven Analysis. With the GUI, library, and API in place, the final component is a feature for metadata-driven analysis. This metadata-driven analysis is the central component of the design, and is responsible for automating what would otherwise be tedious work – instead of users selecting what type of data is appropriate for every field in every table in every project they work on, the data generator itself can determine with reasonable accuracy what the appropriate data is for each field. The user can then just fine-tune the automated selection as desired, significantly cutting down the required effort.

2.2. Why the Metadata-Driven Method Works

Analyzing metadata works because database design is not random – database designers do not just arbitrarily or randomly choose names, data types, and other attributes for their tables. Therefore, those design decisions – table and field names, data types, and other attributes – can serve as hints as to the intention of the designer and inform us with reasonable accuracy what type of data should go into them. For example, a field with a data type of "date" tells us unambiguously that we should supply it with random dates, whereas a field with a data type of "int" tells us it only expects whole numbers.

Another field may be defined to accept decimal points, and if its name is something related to currency (e.g., "price", "cost", "amount", etc.), then that means we should be generating monetary amounts for that field (e.g., "123.45").

With this design, combining an internal repository and common algorithms of faker libraries with a component that does metadata-driven analysis to automatically determine appropriate data types, the resulting system is expected to be far less tedious to work with and require minimal-to-zero coding, allowing non-programmers like QA testers and database administrators to use for various use cases such as creating huge amounts of data for testing, server sizing, measuring performance scaling, and validating the performance of different possible database designs.

3. Implementation

Figure 1 illustrates a high-level view of how ADP autonomously produces realistic fake data.

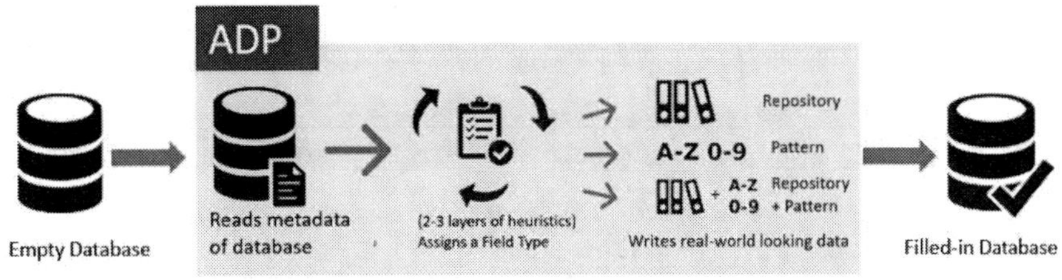

Figure 1. The Automatic Data Populator (ADP) Process.

3.1. Reading the Metadata of the Database

After ADP is pointed to the database (meaning ADP was given access), ADP will look for the information schema where the information about all tables and columns is stored. Then it will query the following; (1) table names in the database; (2) column names per table; (3) data type of each column; (4) character maximum length of each column; (5) column key (indicates if the column is a primary key); (6) extra (indicates if the column is auto-incremented); (7) Based on the values gathered, ADP will assign a field type to each column.

3.2. Assigning a Field Type

Field types are labels that the research team created and are based on types of data that are commonly stored in databases. ADP will apply 2-3 levels of heuristics on each column to determine the appropriate field type. The ADP heuristics are as follows; (1) Look at the data type. If it is a date, then assign Date. Else, go to the next level; (2) Look at the column name and length. Usually, the information given on this level is enough for ADP to match it to one of the 100+ common field types it recognizes; (3) If ADP encountered an integer data type and an id in the column name, it would find out if the column is a primary key and auto-incremented. Table 1 demonstrates the field types that ADP assigned to each column.

After all field types are complete, ADP will store all information gathered into its internal database. Then the user can view all assigned field types in the GUI. If any of ADP's default choices are incorrect, the user can edit them.

3.2.1. Categories of Field Types. Each field type is under 1 of these 3 categories: repository, pattern, or both."Repository" means ADP will choose a random value from the appropriate repository file for that field type. Repository files are just CSV files that contain thousands of sample information – for example, to generate a human given name, ADP will randomly choose one from thousands of names from the human names CSV file. "Pattern" means ADP will generate data based on a specified pattern – for example, to generate a realistic-looking TIN, ADP will follow the pattern for TINs: xxx-xxx-xxx Some field types are a combination of both, simply called "Repository and Pattern". This means ADP will apply a pattern using information from a repository file. For example, to generate a postal address, ADP will follow this pattern: street number, street name, village, city. Then ADP will choose an item from the repository file of each segment.

3.2.2. Writing Realistic Data. ADP will rely on the field type assigned to each column to produce the appropriate fake data. Since ADP will already have access to the target database (a requirement for ADP to be able to read metadata), ADP will write the fake data directly into the target database. By default, ADP will generate 10 – 100 rows (random) for each table. Users can toggle this in the GUI to specify a greater range or an exact number of rows (for example, to generate exactly 1 million rows of fake data, a user can simply indicate 1,000,000 for both the minimum and maximum rows textboxes).

Table 1. ADP recognizes over 100 field types. Below are examples of the field types assigned by ADP after reading the metadata.

Column Name	Data Type	Length	PK	AI	Field Type
first_name	varchar	255	No	No	Human Given Name
Gender	varchar	255	No	No	Gender
Birthday	Date	255	No	No	Date
Tin	varchar	255	No	No	Tax ID Number
Address	varchar	255	No	No	Postal Address
Price	double	-	No	No	Price
customer_id	Int	11	Yes	Yes	Empty Value

4. Results and Discussions

The research team created three tables to test the proof of concept.

ADP scanned *customer*'s metadata, shown in Table 2, and assigned a field type to each column, then inserted 23 rows of realistic fake data. ADP detected that *customer_id* is auto-incremented so ADP did not give it any value. The *company_name* values consist of an English noun or a human name, a generic industry type, and a company type. The *tin* values are all numeric values with the xxx-xxx-xxx pattern. The *address* values have a random street number, a street name and a barangay name taken from one of several repositories (names of heroes, festivals, virtues, etc.), and a city name from a list of actual cities in Metro Manila. The *contact_details* values have 11 random digits that start with "09". Finally, the *contact_person* values are combinations of given names and family names. The actual results are
shown in Figure 2.

Table 2. The first table called *customer* has the structure below.

Column	Type	Extra
customer_id (PK)	int (11)	AUTO_INCEREMENT
company_name	varchar (255)	
tin	varchar (255)	
address	varchar (255)	
contact_details	varchar (255)	
contact_person	varchar (255)	

Table 3. The second table called *employee* has the structure below.

Column	Type	Extra
emp_id (PK)	int (11)	
first_name	varchar (255)	
middle_name	varchar (255)	
last_name	varchar (255)	
gender	varchar (255)	
civil_status	varchar (255)	
birthday	date	
hiring_date	date	
sss	varchar (255)	
tin	varchar (255)	

ADP scanned *employee*'s metadata, shown in Table 3, and assigned a field type to each column, then inserted 21 rows of realistic fake data. ADP detected that *emp_id* is a primary key but is not auto-incremented so ADP generated unique ID numbers with a yyyy-xxxxxx format. The *first_name* values have 1-3 given names. The *middle_name* and *last_name* values have one family name. The *gender* values are either Male or Female. The *civil_status* values are Single, Married, Separated, Divorced, Lifetime Partner, or Unkown. The *birthday* and *hiring_date* values are random dates in ISO format. The *sss* values are all numeric values with the xx-xxxxxxx-x format. Finally, the *tin* values look similar to the *tin* values of *customer*. The actual results are shown in Figure 3.

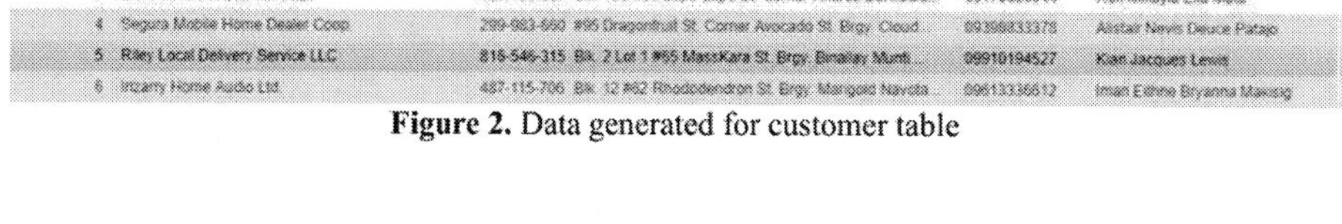

customer_id	company_name	tin	address	contact_details	contact_person
1	Remain Bed And Bath Ltd.	240-497-078	#29 Bittemut Hickory St. Brgy. Gray Birch Pasay C...	09170646475	Martha Pearson
2	Leonard Transportation Services - Other Ltd.	017-906-127	Bk. 1 Lot 6 #96 Antuwanga St. Corner Gumamela St. ...	09573774951	Isabel Hadrien De la Cruz
3	Emotion Grocery Store Corp.	189-436-007	Bk. 135 #90 Lapu-Lapu St. Corner Andres Bonifacio...	09173685344	Ren Mikayla Ella Mata
4	Segura Mobile Home Dealer Coop.	299-983-660	#95 Dragonfruit St. Corner Avocado St. Brgy. Cloud...	09398833378	Alistair Nevis Deuce Patajo
5	Riley Local Delivery Service LLC	815-546-315	Bk. 2 Lot 1 #65 MassKara St. Brgy. Binalay Munti...	09910194527	Kian Jacques Lewis
6	Irizarry Home Audio Ltd.	487-115-706	Bk. 12 #62 Rhododendron St. Brgy. Marigold Navota...	09613336612	Iman Ethne Bryanna Maksig

Figure 2. Data generated for customer table

emp_id	first_name	middle_name	last_name	gender	civil_status	birthday	hiring_date	sss	tin
2017-000001	Charlize Angelina	Abaya	Mccaffrey	Male	Life Time Partner	1995-01-18	1993-06-26	20-4862354-4	523-309-006
2017-000002	Jelina Qadira	Love	Atkins	Female	Divorced	1977-06-04	1956-02-08	71-5967503-3	576-549-793
2017-000003	**Mora**	**Barcelona**	**Yuna**	**Female**	**Divorced**	**2003-11-01**	**1968-06-22**	**10-5334565-3**	**594-519-475**
2017-000004	Erik Daenerys Jaheim	Catacutan	Franco	Female	Divorced	2016-08-21	2013-10-27	38-6076647-2	522-901-841
2017-000005	Lorelle Natalija Swanhild	Yates	Momoko	Male	Separated	1950-08-22	1977-06-18	78-8304689-7	982-371-879
2017-000006	Chace Rada Burgess	Navarro	Verduzco	Male	Life Time Partner	1951-02-10	2003-06-13	13-5710281-7	836-128-386

Figure 3. Data generated for employee table

What would happen if the column names are not human-readable? In Table 4, the researchers duplicated *employee*'s structure but changed the names of *first_name*, *middle_name*, and *last_name* to *fn*, *mn*, and *ln*, respectively.

Table 4. The structure of the third table with the modified columns marked with *.

Column	Type	Extra
emp_id (Primary)	int (11)	AUTO_INCEREMENT
fn *	varchar (255)	
mn *	varchar (255)	
ln *	varchar (255)	
contact_details	varchar (255)	
contact_person	varchar (255)	

Since ADP could not recognize *fn*, *mn*, and *ln* (they are not human-readable names, not descriptive, and could mean anything), it produced a text extracted from a corpus. This is a natural limitation of metadata analysis. Since ADP could not make sense of the metadata, it could not assign the correct field type. User intervention is needed so ADP can still write realistic data to these columns. The user can change the field type through the GUI. As for the other columns, the data produced are still appropriate and realistic. The actual results are shown in Figure 4.

emp_id	fn	mn	ln	gender	civil_status	birthday	hiring_date	sss	tin
2018-034580	dolor sit. Amet powder powder cake dessert sweet ro...	fies, narwhal Williamsburg hashtag Blue Bottle fac...	s hungry meow sleep nap for hunt by meowing loudly...	Male	Divorced	1998-07-15	1991-06-26	72-9796550-0	476-623-536
2018-191890	d Future organic seitan readymade single-origin co...	ort ribs biltong sausage filet mignon cupim. Ham h...	clothes. Chase mice i like big cats and i can not	Female	Single	1968-07-24	2008-10-05	44-7245767-8	111-058-084
2018-278864	gerbread lollipop chocolate bar jelly chups chups...	bear claw apple pie fruitcake chocolate cake bear...	in face eat and than sleep on your face. Find em...	Male	Life Time Partner	1993-10-10	1999-06-06	96-6230768-7	673-036-912
2018-465361	pie pie cotton candy souffAC Fruitcake sweet rol	private parts with tongue then lick owner's face...	y's. Jean shorts gluten-free swag bicycle rights b...	Female	Not known	2010-08-27	1970-03-26	57-3793830-6	510-105-961
2018-537035	nd in front of the computer screen. Sleep on keybo...	anish bear claw danish Cupcake ipsum dolor sit a...	ly forage typewriter. Plaid pickled chia, heirloom...	Male	Widowed	2009-10-28	2010-03-20	98-5867515-3	762-363-040
2018-623480	e ribs flank filet mignon shoulder jowl. Leberkas...	cotton candy lemon drops Liquorice sesame snaps...	YOLO four loko banh mi pug readymade taxidermy XD...	Male	Single	1987-02-23	1958-09-20	40-1882423-3	161-668-632

Figure 4. Data generated for employee table with some column names that are not human-readable

5. Conclusions and Recommendations

The ADP proof of concept system presented here shows the advantages of a metadata-driven data generator for database systems: (1) Removes the need for programming work (as opposed to using one of the many faker libraries available for various platforms); (2) Allows non-developers like testers and database admins to also use such a system for their different use cases; (3) Eliminates much of the tediousness involved in having to choose appropriate field types for every table and field.

5.1. Limitations

There are some observed limitations in the proof of concept. These limitations, however, are not due to the nature of metadata-driven analysis, but simply limitations of the current state of the proof of concept. To further explore the discover the limits of using metadata-driven analysis for automatic creation of fake but appropriate and realistic data, these limitations in the proof of concept should be addressed in future studies.

5.1.1. No knowledge of relationships. The paragraph text follows on from the subsubsection heading but should not be in italic. The research team will also have to investigate making the created library a bit smarter than generic faker libs – it has to be able to accept and understand relationships and use existing database information as a repository source. For example, it should be capable of receiving instructions such as "Fill up the transaction tables with random items from the recently populated `item` table, but only use items that are marked active (`item`.`active`='Y')". That statement, of course, is not actual syntax, but rather intent. We want the system to be able to accommodate such intent, in order to make the fake data as real-world as possible, which means it must have knowledge of relationships and can query existing tables.

5.1.2. No knowledge of field interdependence. For example, given name should match the gender. Currently, ADP treats each field independently, so while ADP may randomly choose a name that is generally male for a "first_name" field (e.g., "Michael"), ADP may end up choosing "Female" for its random value in the "gender" field.

No ability to learn or remember the common conventions of a user / team / company. ADP currently functions very well when all table names and field names are mostly human-readable but will not fare very well when the table and field names are not very human-readable and instead follow company-specific or team-specific conventions. For example, "first_name", "fname", "f_name" will be recognized as needing a human name; however, if the field is simply called "fn" as per a team's convention, ADP would not be able to recognize that as the convention for "first_name". An improvement would be for ADP to learn these conventions each time a user "corrects" one of ADP's guesses, such as when the user assigns "human given name" to the field with the name "fn". The next time ADP encounters an "fn" field, ADP should remember that it should assign "human given name" to it.

To evaluate the accuracy of the methodology of this research, the authors plan a follow-up study using participating software teams for various use cases mentioned in Section 1.1 of this paper. This paper also has not elaborated on the heuristic rules due to length considerations; the resulting paper of the follow-up study can include these rules.

References

[1] Nimesh R 2013 *Simplifying Test Data Generation with Faker* Retrieved from https://www.sitepoint.com/simplifying-test-data-generation-with-faker/
[2] Curtis B nd *Faker* Retrieved from https://github.com/stympy/faker
[3] Zaninotto F nd *Faker* Retrieved from https://github.com/fzaninotto/Faker
[4] Gutmann L 2008 *#126 Populating a Database* Retrieved from http://railscasts.com/episodes/126-populating-a-database
[5] Faraglia D nd *Django-faker* https://github.com/joke2k/django-faker
[6] Pargas R P, Harrold M J and Peck R R 1999 Test-data generation using genetic algorithms *Software Testing Verification and Reliability* **9(4)** pp 263-82
[7] Emmi M, Majumdar R and Sen K 2007 Dynamic test input generation for database applications *Proc. of the 2007 Int. Symp. on Software testing and analysis* pp 151-62
[8] Allyn G and Jung T 2006 *U.S. Patent Application* No. 11/102,486
[9] Chow R and Golle P 2009 Faking contextual data for fun, profit, and privacy *Proc. of the 8th ACM workshop on Privacy in the electronic society* pp 105-8
[10] Hoffer J A, Ramesh V and Topi H 2013 *Modern database management* 11th ed (Boston: Pearson Education)

IOP Conf. Series: Materials Science and Engineering **803** (2020) 012034 doi:10.1088/1757-899X/803/1/012034

Knowledge management system (KMS) framework for outcomes-based education sustainability

D H Galeon[1], **T D Palaoag**[2]

[1] College of Computing Sciences, Ifugao State University, Lamut Ifugao, 3605, Philippines
[2] College of Information Technology and Computer Science, University of the Cordilleras, Baguio City, 2600, Philippines

E-mail: galeon.dh@gmail.com, tpalaoag@gmail.com

Abstract. Etched in official memoranda, the Philippine higher education institutions (HEIs) are in their transformational outcomes-based education (OBE) as a government-prescribed educational reform towards quality and sustainability. The educators in these HEIs have a positive attitude and knowledge towards OBE principles and implementation however they lack resources. The paper aims to understand the knowledge management (KM) practices of the Ifugao State University (IFSU) and design a KMS framework considering the most critical factors affecting its implementation, which shall serve as a reference to its development. Using a descriptive-analytical method, it shows that the institution has a strong KM culture but needs to advance their KM strategy and implement modern KM technologies. It also shows that the stakeholders are excellent in socializing and control activities but need improvement and support in networking, encountering, collaborating and coordinating activities. Thus, the proposed KMS framework was designed to build up their KM practices. It primarily informs the stakeholders on whose knowledge, what type of knowledge, how, and why they need to manage knowledge. Future research includes the development of a KMS prototype based on the framework and tests its applicability for the community of practice.

1. Introduction

1.1. Definition, background, and application
It is now a policy of the Philippine Commission on Higher Education (CHED) to require, support and monitor the establishment of an OBE in all its HEIs offering computing programs as a way of re-orienting them to quality, excellence, and sustainable education. In effect, this necessitates major stakeholders of education to work together in following a continuous cycle of articulating desired outcomes; aligning teaching, learning and assessment; systemizing data collection on learning; and systemizing continuing quality improvement [1] which are collaborative, flexible, trans-disciplinary, outcomes-based, empowerment-oriented approach to learning [2] as they quest for quality and excellence. Currently, information technology (IT) educators in the Philippine HEIs have a positive attitude towards OBE [3] however they have limited instructional facilities and learning materials and lack well-trained teachers who could help realize the objectives of the new curriculum [2]. On the contrary, Malaysian universities had already developed meaningful and interactive learning activities, environments, materials, systems in support of their OBE implementation [4]. OBE centers curriculum, evaluation, and reporting on higher-order exit outcomes. Curriculum development ought

to be profoundly comprehensive, participative and democratic [5]. Likewise, curriculum mapping requires a ton of contemplations including but not limited to collaboration [6], institutional support [7], leadership, communication in exchanging information and coordination to come up with more centralized and less departmentally based curricula [8]. In fact, faculty and curriculum and instruction domains have the highest weight in the outcomes-based quality assurance (OBQA) program accreditation instrument for Philippine state universities and colleges which emphasizes their importance [9]. While outcomes assessment provides feedback regarding the adequacy of the instructing and learning rehearses [10], it is considerably more complex and tedious as it demands more time [11], it ought to be a continuous, iterative and coordinated process [12] that should keep on being a part of worldwide monitoring of quality education [13]. Further, outcomes assessment will finally result in decisions and activities with respect to CQI [14] in the course and program level which is believed to be the driving force to keep institutions headed straight toward curriculum excellence [15]. Philippine HEIs have to develop their system of assessing and evaluating outcomes since it has not been included in the CHED's policy. Considering that the OBE processes involve collaboration, communication, coordination and control activities among the stakeholders in HEIs KMS implementation in the institution is a dire need. Moreover, as knowledge and skillsets are rapidly changing in the field of IT due to ever-expanding technology and product innovation base, it is considered necessary to develop a knowledge management framework for computing programs curriculum development which is both industries relevant as well as supportive of the growth of new knowledge and new technologies [16]. The paper focuses on understanding the KM practices of the IFSU which serve as a basis for developing a KMS framework for the institution considering the most critical factors affecting its implementation. The framework shall provide the reference for decisions about the implementation and application of KM in the institution.

1.2. Related studies

KM is composed of people, knowledge, and sharing [17]. It is a socio-technical procedure [18] where social association, correspondence, and cooperation are in its center [19]. It is believed that the success of KM depends on knowledge sharing [20]. Today the pattern is toward KM by social technologies (ST) [21] such as social networks, social media, and KMS since they are instrumental and have provided an effective channel for the KM processes [22]. In fact, KM and STs equally require each other in order to succeed [23] where one reinforces the other and vice versa [24]. It is believed that the group feature of STs is a total KM package [22]. KM tools incline to develop towards a professional direction, and KM application development is an issue-oriented field [25]. Without a strategy and an organizational culture based on continuous learning, sharing and collaboration, the use of these tools brings few benefits to HEIs [26]. Such strategy includes having a clear objectives and goals [27] conformed with the organization's vision and management [28], having a consistent and thorough engineering for determining the different segments of an organization's knowledge domains and making these knowledge targets [29], and the cultural parts of the organization must be recognized and the KM practices must be well-matched with that culture [30]. The fact that in pursuit of excellence, the Baldrige excellence framework has KM as one of its core components [31], there was no universally accepted framework or methodology to guide KM implementation exists [30]. KMS framework for HEIs should support the core features of infrastructure, content and portal, collaboration and learning, social capital, expertise and communities, business intelligence, and integration and measurement [32]. KMS framework shall consist of knowledge retrieval, storage, sharing, and publishing modules [33] and built with web-based technologies [34]. It should compose of intelligent systems [34], group decision-making approach, multiple criteria decision-making techniques, advanced communications [35], faculty contact tracking system, knowledge base [36], data mining [37], and collaboration and workflow systems [26].

2. Methodology

A descriptive-analytical and survey method was utilized in this paper to understand the KM practices of the IFSU. It focuses and investigated the level of the organizational awareness and commitment, culture, people, technology, and organizational structure factors in facilitating KM and knowledge acquisition, development, distribution, measurement and review processes[38]; the level of stakeholders' personal knowledge sharing along the areas of socializing, networking, practicing, storytelling, encountering [39], communicating, collaborating, coordinating and controlling [40]. The survey questions used in this paper were based on various studies. The questions are modified to fit with its application in OBE. The questions used to investigate the critical KM factors was formed based from the study of Obaide (2004); the personal knowledge sharing practices of the stakeholders was based from the study of Panahi (2014); and the stakeholder's perception on the outcome of their knowledge sharing activities was based from the study of Galeon et al. (2019). In the analysis of the data, a baseline value of 51% is taken to portray the majority opinion against the KM practices. Any practices reported strongly agree and agree for its implementation and critical and important for its importance by less than 51% of the respondents are not accounted for. The product of the investigation aids as an input for the design of a KMS architectural framework for the institution. Thirty-four participants composed of the directors from various service departments, the information technology education (ITE) faculty members and the senior students from the College of Computing Sciences are selected using purposive judgment non-probability expert sampling and are able to provide data for analysis. The participants composing of administrative officials (35%), faculty members (33%), students (32%); below 25 years old (33%), 26-40 years old (35%), 41-55 years old (26%), above 55 years old (6%); high school graduate (29%), baccalaureate (3%), masters (47%), doctorate (21%) are selected since they are observant, of interest, reflective member of the institution, who know much about the institution's KM practices and are both able and willing to share their knowledge. Research analysis was performed by using statistics, tables, timelines, and charts.

3. Findings and Discussion

3.1. Level of institutional KM practices

From questionnaire findings, it remarkably shows that the majority of the participants are fully aware of the importance of investing resources in the field of KM. This holds true to all of the KM practices' categories (see Figure 1). However, the institution is still exploring the opportunities for a variety of KM practices and KM tools. From this study, it became evident that the stakeholders' positive behaviors towards KM were fully exploited. Nevertheless, there is an opportunity to build up their commitment to a program of KM improvement and invest in the right kind of technology and use it effectively enough to support KM. To improve their strategy and to invest in their KM tools with modern technologies, details shows that key performance indicators for KM shall be in place; there shall be defined responsibilities and a budget set for the KM initiative; there shall be a senior level ongoing review of the effectiveness of KM in the institution; OBE knowledge as resources shall be inventoried or recognized and some measure of value shall be attached to each; modeling systems, decision support systems, and artificial intelligence shall be implemented; directories of experts shall be available; resources shall be documented and maintained; and search for resources shall be easier. Such improvements shall increase their performances with their KM practices along with the acquisition and development, and measurement and review.

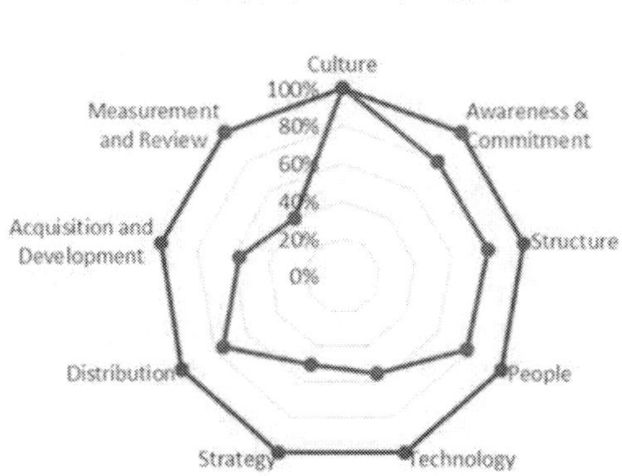

Figure 1. Overall view of questionnaire responses for the KM practices

3.2. Level of the stakeholders' personal knowledge sharing practices

Research findings evidently show that the majority of the participants communicate and exchange experiences, beliefs, and ideas via informal dialogues and conversations and they define who and what resources to follow to a great extent (see Figure 2). Whereas it is striking that locating eminent experts to track and follow their experiences or opinions while developing collaborative relationships, coordinating their activities, and collaborating among each other to create, modify and disseminate resources are practiced to the less extent. The details show that the stakeholders needed further support in their KM practices of: interacting, participating and networking to other individuals to create and share resources and discuss best practices; locating eminent experts in the field who are willing to help; documenting and storing professional resources; broadcasting or publishing resources or sharing particular skills, tips, ideas, and thoughts; searching resources; and reporting unusual practices and developing discussions about them.

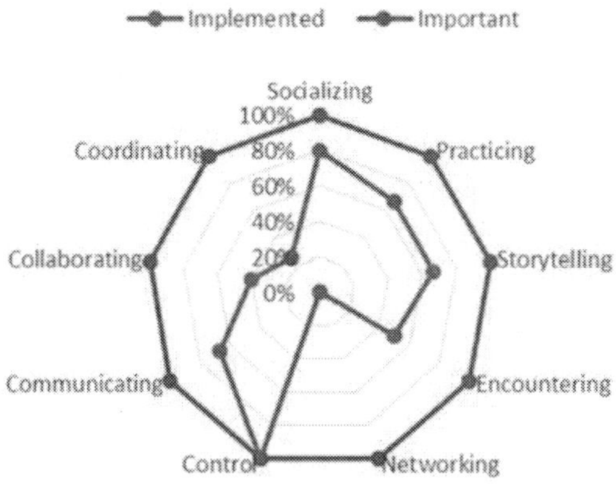

Figure 2. Overall view of questionnaire responses for the KS practices

It was also revealed that the stakeholders will more likely to continuously share knowledge since the majority of the respondents confidently perceived that they have gained positive experience in it. Specifically, there is strong support for KS among the stakeholders, they understand the importance of sharing knowledge effectively, they know how they benefit from sharing knowledge, and they share knowledge easily. Likewise, the majority of the respondents have a positive response to the precursors of KS including love, care and trust is fostered among stakeholders in the institution; employees feel secure about their work contract - the institution makes it attractive to stay and long term employment is encouraged; everyone is willing to give advice or help on request to anyone else in the institution; and individuals are visibly rewarded for sharing resources.

3.3. A KMS framework for OBE sustainability

The questionnaire findings show that the institution is still exploring the opportunities for a variety of KM practices and KM tools. As an immediate need, the findings indicate the necessity to establish a knowledge network, facilitate collaboration among the stakeholders and build an institutional memory to support their quality OBE processes. Considering the current level of their KM practices, the KMS architectural framework presented in Figure 3 is organized in four main layers: sustainable OBE; knowledge system; system interface, and people.

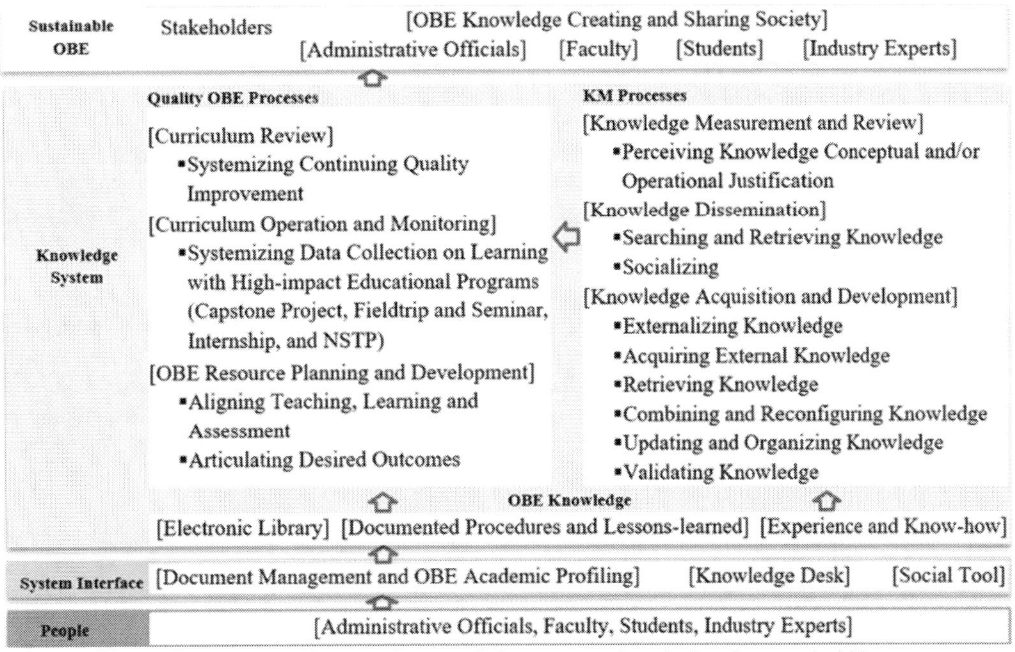

Figure 3. A KMS architectural framework for outcomes-based education sustainability

The "sustainable OBE" layer informs the stakeholders to the main intent of their KM initiative which is to adopt and implement the CHED OBE policy through a knowledge-creating and sharing society. It positions the university towards quality and excellence in the curriculum.

The "knowledge system" layer informs the stakeholders of what type of knowledge is managed in the quality OBE processes and how will it be managed through the KM processes. While research findings show that the institution's resources shall be documented, stored in a database, inventoried or recognized, the framework includes a system of collecting data on learning for its easy analysis and assessment. The framework also includes the system of planning and development of resources since studies show that the Philippine HEIs still lack resources such as training to the educators in curriculum development. To uplift the KM practices of the institution, the proposed framework supports the following vital KM processes of socializing, searching and retrieving knowledge, and knowledge measurement and review. These processes allow the stakeholders to continue their knowledge sharing activities and improve their networking, collaborating and coordinating activities. The resources being organized shall be identified into an electronic library, documented procedures and lessons-learned, and experience and know-how. The electronic library contains codified resources including yet not constrained to policies, manuals, registries, guidelines, reports, and expert profiles. Documented procedures contain information and bits of knowledge about techniques and conditions that are expected to achieve a certain OBE task, alterations to existing methods and solutions for issues experienced while performing a specific errand just as best practices. It gives direction, proposals, and reference material to facilitate human execution of the undertakings. Lessons-learned documentation covers the full and detailed depictions of the identification and the solutions of obviously clarified issues. Experience and know-how are codified learning as emotional bits of knowledge, instincts, and hunches that may start revisions to existing strategies and solutions.

The "system interface" layer serves as the window for the stakeholders to apply the KM processes in performing the quality OBE processes. This layer has components of document management (DM) and academic profiling (AP), knowledge desk (KD), and social tool (ST). The DM provides tools for storing and retrieving resources such as articles, manuals, policies and the likes. The AP provides a tool for the stakeholders to view summaries of their resources and practices for decision making. To facilitate collaboration among the stakeholders, the KD serves as an interactive learning and co-production environment that provides the stakeholders with authoring, assessment and expert profiler tools. The implementation of an ST, specifically a chat and discussion forum system, intends to facilitate communication among the stakeholders allowing them to interact, participate and network to other individuals to share and discuss their personal experiences and lessons learned. It also allows them to synchronize and integrate their activities, responsibilities, and command and control. Further, it facilitates the preservation of discourses and communication among the stakeholders while acquiring knowledge, externalizing experiential knowledge, coordinating, and attending discussions and knowledge dissemination activities via online meetings and online chat rooms.

The "people" layer informs the stakeholders of whose knowledge is managed. It shows four sets of knowledge givers. The administrative officials function as decision-makers in charge of dispatching services and/or planning KM strategies and policies. The faculty works as proficient knowledge makers who produce OBE resources. The students also work as knowledge producers who produce HIEP contents and performances. The industry experts serve as coworkers to the administrative officials, faculty, and students. These group of stakeholders forms the community of practice and community of interest who actively involves in achieving the success of the institution's KM initiative.

4. Conclusion

Within a few years when IFSU has been exercising its innovativeness and creativity in the development of its outcomes-based curriculum, the stakeholders have fully exploited their organizational culture towards KM, however, the implementation of modern KM tools has to be explored in the institution. The KMS architectural framework presented in this study has emphasized informing the stakeholders as to why manage knowledge, what type of knowledge to organize, whose knowledge, and how knowledge is managed. Succeeding studies would include the development and implementation of a KMS prototype based on the framework and an assessment of its impact and acceptability.

References

[1] Philippine Commission on Higher Education 2015 *CMO* No 25 . 2015 (Quezon City)

[2] Limon M R and Vallente J P 2016 Outcomes-Based Education Integration in Home Economics Program: An Evaluative Study *J. of Educational Issues* **2** 1 pp 289-304

[3] Cruz R Z and Cruz R O 2017 Educators' Attitude towards Outcome-Based Information Technology Education in the Philippines *I-manager's J. of Educational Technol.* **13** 4 pp 14-21

[4] Alias R and Aris B 2016 *A Guide to New Academia Learning Innovation* 1st ed (Malaysia: UTM Academic Leadership)

[5] Khan M and Law L 2015 An Integrative Approach to Curriculum Development in Higher Education in the USA: A Theoretical Framework *Int. Education Studies* **8** 3 pp 66-76

[6] Gulikers J, Biemans H, Wesselink R and van der Wel M 2013 Aligning formative and summative assessments: A collaborative action research challenging teacher conceptions

Studies in Educational Evaluation **39** 2 pp 116-24

[7] Biggs J 2014 Constructive alignment in university teaching *HERDSA Review of Higher Education* **1** pp 5-22

[8] Haren R 2000 The integration ladder: a tool for curriculum planning and evaluation *Medical Education* **34** 7 pp 551-7

[9] De Lara M G 2017 Outcome-based program quality assurance accreditation survey instrument: its development and validation *The Normal Lights* **11** 2 pp 1-30

[10] Macayan J 2017 Implementing Outcome-Based Education (OBE) Framework: Implications for Assessment of Students' Performance *Educational Measurement and Evaluation Review* **8** 1 pp 1-10

[11] Telsang M 2015 Outcome Based Education - Design Delivery and Assessment of Product Design and Development Course at Undergraduate Engineering Program *J. of Engineering Education Transformations* special issue p 145

[12] Weistroffer H R and Gasen J 2010 Assessment of Information Systems Programs: A Life Cycle Approach *J. of Education for Business* **70** 5 pp 258-62

[13] Barrett A 2016 Measuring Learning Outcomes and Education for Sustainable Development: The new Education Development Goal *Oxford Studies in Comparative Education* **25** 1 (United Kingdom: Symposium Books)

[14] Castillo R 2014 A Paradigm Shift to Outcomes-Based Higher Education: Policies, Principles and Preparations *Int. J. of Sciences: Basic and Applied Research* **14** 1 pp 174-86

[15] Aziz A, Noor M, Ali A and Jaafar M 2005 A Malaysian outcome-based engineering education model *Int. J. of Engineering and Technol.* **2** 1 pp 14-21

[16] Agrawal S, Sharma P and Kumar M 2008 Knowledge Management Framework for Improving Curriculum Development Processes in Technical Education *2008 Third Int. Conf. on Convergence and Hybrid Information Technol. IEEE*

[17] Wu Y, Lin Y, Wen M, Perng Y and Hsu I 2016 Design, Analysis and User Acceptance of Architectural Design Education in Learning System Based on Knowledge Management Theory *EURASIA J. of Mathematics, Science and Technol. Education* **12** 11 pp 2835-49

[18] Al-Busaidi K, Olfman L, Ryan T and Leroy G 2010 Sharing Knowledge to A Knowledge Management System: Examining the motivators and the benefits in an Omani organization *J. of Organizational Knowledge Management* **2010** pp 928-35

[19] Omona W, van der Weide T and Lubega J 2010 Using ICT to enhance Knowledge Management in higher education: A conceptual framework and research agenda *Int. J. of Education and Development using Information and Communication Technology* **6** 4 pp 83-101

[20] Wang S and Noe R 2010 Knowledge sharing: A review and directions for future research *Human Resource Management Review* **20** 2 pp 115-31

[21] von Krogh G 2012 How does social software change knowledge management? Toward a strategic research agenda *The J. of Strategic Information Systems* **21** 2 pp 154-64

[22] Mushonga C 2014 *Social Networking for Knowledge Management: Group Features as Personal Knowledge Management Tools* thesis (Stellenbosch: Stellenbosch University)

[23] Gunjal B, Gaitanou P and Yasin S 2012 Social Networks and Knowledge Management: An Explorative Study in Library Systems *Knowledge Management 2.0* pp 64-83

[24] Cerchione R and Esposito E 2017 Using knowledge management systems: A taxonomy of SME strategies *Int. J. of Information Management* **37** 2017 pp 1551-62

[25] Liao S 2003 Knowledge management technologies and applications—literature review from 1995 to 2002 *Expert Systems with Applications* **25** 2003 pp 155-64

[26] Pinto M 2014 Knowledge management in higher education institutions: a framework to improve collaboration *9th Iberian Conf. on Information Systems and Technologies* pp 87-90

[27] Butler T, Heavin C and O'Donovan F 2007 A Theoretical Model and Framework for Understanding Knowledge Management System Implementation *J. of Organizational and End User Computing* **19** 4 pp 1-21

[28] Shannak R, Masa'deh R and Akour M 2012 Knowledge management strategy building: literature review *European Scientific J.* **8** 15 pp 143-68
[29] Oluikpe P 2012 Developing a Corporate Knowledge Management Strategy *J. of Knowledge Management* **16** 6 pp 862-78
[30] Smuts H, van der Merwe A, Loock M and Kotzé P 2009 A Framework and Methodology for Knowledge Management System Implementation *Proc. of the 2009 Annual Research Conf. of the South African Institute of Computer Scientists and Information Technologists* pp 70-9
[31] Baldridge Excellence Builder 2019 *National Institute os Standards and Technology* Retrieved from www.nist.gov/baldrige
[32] Abdullah R, Selamat M, Sahibudin S and Alias R 2005 A Framework For Knowledge Management System Implementation In Collaborative Environment For Higher Learning Institution *Institution J. of Knowledge Management Practice* **6** 1 pp 1-8
[33] Anupan A, Nolsook P and Wannapiroon P 2015 **1** 2 pp 146-9
[34] Jia Z, Shi Y, Jia Y and Li D 2012 A framework of knowledge management systems for tourism crisis management *Procedia Engineering* **29** 2012 pp 138-43
[35] Cil I, Alpturk O and Yazgan H 2005 A new collaborative system framework based on a multiple perspective approach: InteliTeam *Decision Support Systems* **39** 2005 pp 619-41
[36] He W and Abdous M 2013 An Online Knowledge-Centered Framework for Faculty Support and Service Innovation *VINE* **43** 1 pp 96-110
[37] Silwattananusarn T and Tuamsuk K 2012 Data Mining and Its Applications for Knowledge Management: A Literature Review from 2007 to 2012 *Int. J. of Data Mining & Knowledge Management Process* **2** 5 pp 13-24
[38] Obaide A 2004 *A model for a successful implementation of knowledge management in engineering organizations* Thesis (Manchester: University of Salford)
[39] Panahi S 2014 *Social Media and Tacit Knowledge Sharing: Physicians' Perspectives and Experiences* Thesis (Brisbane: Queensland University of Technology)
[40] Spector M and Edmonds G 2002 *Knowledge Management in Instructional Design* Retrieved from https://www.ericdigests.org/2003-1/design.htm

Design and Development of Data Warehouse Framework of Highland Vegetable Crops for Benguet

A R Quitaleg and M G Ortiz

University of the Cordilleras, Baguio City, Philippines

oliangelanne@yahoo.com, mariegraceortiz@gmail.com

Abstract. The data warehousing technology has been widely used in organizations providing management, data integration and helps do the decision-making. Data warehouse offers an effective way for analysis and statistics to the big data in agriculture. Agriculture is one of the major issues in the Philippines and needs technological advancement. There is an increasing need for farmers and decisions makers in processing agricultural data. Agricultural datasets are diversified, non- standard and very large. It captures information about crops, farmers, buyers and other factors and requirements pertaining to the agricultural production. Designing a data warehouse for agricultural data is a great challenge for the researchers.
The data warehousing technology has been widely used in organizations providing management, data integration and helps do the decision-making. Data warehouse offers an effective way for analysis and statistics to the big data in agriculture. Agriculture is one of the major issues in the Philippines and needs technological advancement. There is an increasing need for farmers and decisions makers in processing agricultural data. Agricultural datasets are diversified, non-standard and very large. It captures information about crops, farmers, buyers and other factors and requirements pertaining to the agricultural production. Designing a data warehouse for agricultural data is a great challenge for the researchers.

1. Introduction

The world population grows rapidly but the resources for crop production continuously diminish. There is a big shortfall between the amount of food we produce today and the amount needed to feed everyone in 2050 [1]. There will be nearly 10 billion people on Earth by 2050.

In recent years the growth rates of world agricultural production and crop yields have slowed. This has raised fears that the world may not be able to grow enough food and other commodities to ensure that future populations are adequately fed. In the Philippines, according to Philippine Statistics Authority (PSA), agricultural production grew at only 0.56% in 2018, greatly missing the target of 4% for the year. This was a sharp drop from 2017's record of 6.69% [2].

The core question is whether today's agriculture and food systems are capable of meeting the needs of a global population. The international community has recognized these challenges. In particular, the 2030 Agenda for Sustainable Development, provides a compelling, but challenging, vision on how multiple objectives can be combined to define new sustainable development pathways. The second Sustainable Development Goal explicitly aims at ending hunger, achieving food security and improved nutrition, and promoting sustainable agriculture, simultaneously by 2030 [3].

In addition, the World Health Organization (WHO) reported fruit and vegetables are important components of a healthy diet, and their sufficient daily consumption could help prevent major diseases. Approximately 16.0 million (1.0%) disability adjusted life years and 1.7 million (2.8%) of deaths worldwide are attributable to low fruit and vegetable consumption. Thus, WHO recommends 400 grams per day of fruit and vegetables per day for the prevention of chronic diseases and alleviation of several micronutrient deficiencies, especially in less developed countries.

High Value Crops Development Program (HVCDP) has identified top priority vegetables and these are the lowland crops, highland crops, and export crops. The study will focus on the highland vegetable crops. Benguet is popularly known as the "Salad Bowl" of the Philippines because virtually every type of vegetable used in salads is produced here.

To satisfy the increasing demand for food particularly highland vegetables, the use of new farming techniques must significantly increase crop yield, thus increasing agricultural income. Considering that vegetable farming provide employment for the province and its neighboring provinces.

With technological advances in the area of information and communication technology (ICT), the agricultural data processing is an increasing need for farmers and other decision-makers. Agricultural data is varied, non-standard and there are islands of information. Decision-makers including farmers, cooperatives and other agencies need to collect, consolidate and summarize data from various sources to fulfill their needs and meet the increasing demand of vegetables. Due to these circumstances, data warehouse (DW) is an essential need for agriculture. Data from each cooperatives, organizations, farmers, buyers, and agencies must be stored in a common database and used for comprehensive data analysis. This data is used to inform important business decisions. In order to make good decisions, relevant data should be taken into consideration and the best source for that data is a well-designed data warehouse.

This study focuses on the design and development of data warehouse framework of highland vegetable crops for Benguet. Building data storage will be used in the agricultural data analysis process that can be used to facilitate decision-making on farming and marketing. It aims to increase the volume of commodities, better farm management, and better marketing system.

2. Literature Review

Big data analytics can help stakeholders design and implement predictive models based on crop type and algorithm for farm optimization. Big Data is defined [4] as high-volume, high-velocity and/or high-variety information assets that demand cost-effective, innovative forms of information processing that enable enhanced insight, decision-making, and process automation.

The paper will focus on the predictive analytics, which aim to something about "what might happen next?" It involves extrapolating trends and patterns in the future. By considering historical data, making smarter decision about the future in present is called predictive analysis model. According to the study conducted by [5], this model encompasses different patterns from historical facts to analyze the future predictions. This is a type of data mining technique that derives results from previous year data and predicting the behavior using history of data. This includes analyzing what event had occurred in the past, when it is occurred, and monitoring current terminologies and facts about data, then making smarter decision using machine-learning methods. In prediction, previous year data are referred as training set and the data's are classified based on training set [6].

One of the most significant bodies of work for big data is the project named SatSure founded by the 33-year-old Abhishek Raju works on using Big Data and its allied technologies like data science and IoT to better the lives of farmers. Abhishek shares the fact that he was deeply moved by the rate of farmer suicides and the lack of application of science and newer technologies in the oldest Indian industry – agriculture. His solution to this is SatSure [7].

Another notable contribution is The Yield, as stated by [8] in her article entitled "Transforming the agriculture industry using IoT and predictive analytics". The initiative uses data, sensing technologies, and analytics to solve problems for Tasmania's agriculture ecosystem.

The above-mentioned projects mentioned did not consider building a data warehouse for the organization to support their decision-making.

A paper entitled "An Efficient Data Warehouse for Crop Yield Production" [9] was proposed for precision agriculture. It presented system architecture and a database schema for design and implementing a DW. The authors used a constellation schema or galaxy schema because it focuses on various areas in farming such as treatment, operation, trading, and yield. However, in the paper, it did not address a specific issue of a particular organization.

Also in the paper [10] "Indian Agriculture Data Warehouse Design employed the star schema data model, however, it concerns only on time and location dimensions. The DW system provides spatial analysis of data with the help of Geographic Information System (GIS). However, the DW concerns livestock farming, not for agricultural crop production.

The need for integrated information is the biggest driver for investing in data warehouse systems, because it represents both operational and strategic business needs to support day-to-day business operations. With these, the researchers would like to propose a DW design to increase the agricultural production of highland vegetable crops in Benguet.

3. Design and Development of Data Warehouse Framework

The country's biggest agricultural trading center is in La Trinidad, Benguet. Benguet Agri-Pinoy Trading Center (BAPTC) offers research and extension and instruction expertise to further develop the vegetable and root crop industry of Benguet. The increase in the trading operations of the BAPTC has driven more farmers, buyers, market facilitators, packers and porters from different wholesale market sites to transact at the facility. To maximize its potential as catalyst of improving the lives of highland vegetable farmers in the Cordilleras, the center needs good management to efficiently handle both the prospects and challenges of the project. With these, BAPTC, and its accredited farmer groups and individuals need to document all transactions. However, they have not yet utilized technology as a solution to these problems but they are willing to adopt a new technological solution.

One of the most important assets of the center is its information. The goals of this study are to 1) maintain historical information, 2) facilitate reporting and analysis, 3) be the foundation for better decisions making, and 4) be an adaptive source of information. This study is also intended to offer solutions to help BAPTC achieve its mission and vision, "Modernizing farmers' access to global markets" and "Building avenues for farmers and consumers to have better market access through participatory, capability building and provision of innovative technologies and quality facilities."

Data used by BAPTC comes from various sources such as accredited farmers group and individuals, other trading posts which has many forms and formats, structured and unstructured, and they wanted to use this data and convert it into useful information by integrating and storing it in a single location. Data warehouse design is the process of building a solution to integrate data from multiple sources that support analytical reporting and data analysis [11]. The researchers would like to propose a data warehouse framework for the BAPTC so that all information about the center's transactions and activities are consolidated in one place.

3.1. Data Warehouse Architecture

The researchers considered the system architecture of the various technical components and data architecture of the information stored in the DW. Data architecture refers to how data is organized within the warehouse.

The three-tier architecture [7] was used in this study composed of bottom tier, middle tier, and top tier. In the bottom tier, data from multiple sources are cleansed, extracted, transformed, and loaded into the database of the data warehouse serves. The middle tier is an application tier that presents an abstracted view of the database using OLAP. Top tier is the front-end client layer that uses query, reporting, analysis, and data mining tools to extract data from data warehouse. Figure 1 shows the data warehouse architecture for BAPTC.

The end-users of the data warehouse are the BAPTC employees and cooperative's staff. In the data presentation layer, users can interact and access data for business insights for ad-hoc analysis, reports and dashboards. The top-down approach is where the data warehouse is designed first and then data marts are built on top of data warehouse.

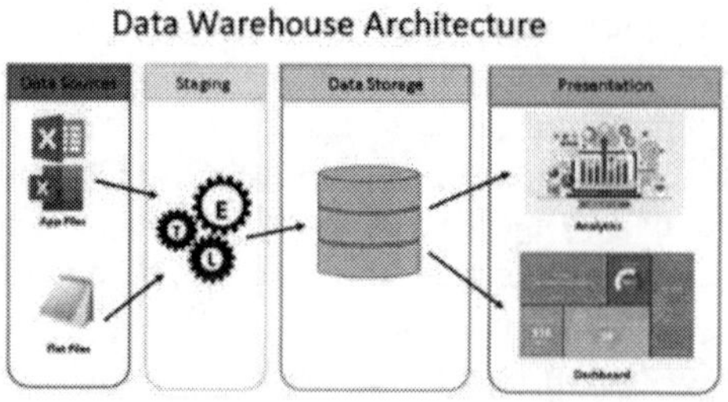

Figure 1. Data warehouse architecture for BAPTC

3.2. Business Process

Understanding the business requirements with an understanding of the available source of data and identifying what business process to model is the first step in the design. In BAPTC, the center wants to better understand management monitor the volume of crops produced by a farmer at a particular field and its selling/buying price to wet market buyers and institutional buyers.

These information is critical to the center to meets its objectives namely 1) A multi-stakeholder approach to sustainable linkage of farmers to market, and 2) to increase the farmers' incomes and protect them from the effects of market price fluctuations.

BAPTC identifies farmers and buyers who will compose the market. It provides processing and storage facility to farmers and area accessible to traders. The center determines the commodity and volume requirement to match the projected of the market with the production capacity of farmers. The logistical cost of moving the commodities from farm to the center and from the center to the market outlets and consumers are also monitored. BAPTC gathers historical and actual production data. The center also identifies potential crops.

3.3. Grain

After the business process has been identified which include monitoring volume and prices of commodities, monitoring buyers and sellers, supervising on crops produced by farmers, and other business processes in the BAPTC, the researchers face a serious decision about granularity. Declaring the grain is central and important step in a dimensional design. The grain establishes exactly what a single fact table row represents. The grain must be declared before choosing dimensions. Atomic data is highly dimensional [12]. In BAPTC, the most granular data is the daily production of commodities of each farmer in each crop on a particular season.

3.4. Dimensions

After the grain of the fact table has been chosen, the choice of dimensions is straightforward. Dimension tables are sometimes called the "soul" of the data warehouse because they contain the entry points and descriptive labels that enable the DW system to be leveraged for business analysis. Dimensions can be attributed to the production of commodities measurements, such as the farmer who produced the commodity, the type of crop produced, the facility used by the farmer, the, the name of the buyer/seller who purchased the commodities, the season when the crop is produced, the BAPTC staff who assisted in the transaction, the coop or group of the farmer, and the place where the crop is harvested. The following descriptive dimensions apply to BAPTC and its descriptive attributes of each dimension are described in Table 1.

3.5. Facts

The fourth and final step in the design is to make a careful determination of which fact will appear in the fact table. The term fact represents a business measure [12]. Facts are determined by answering the question, "What is the process measuring?" The grain declaration helps in the process.

In this case, the daily production of commodities of each crop will be the fact that will include the volume of the commodity produced and purchased, the unit price of the commodity, and the total amount of the commodity produced and purchased. The fact table in this case is presented in Table 2.

Table 1. Dimension Tables

No	Dimension Table	Descriptive Attributes
1	Crop	cropID, crop type, crop name
2	Farmer	farmerID, lastName, firstName
3	Buyer/Seller	buyerID, name, address, contactNo
4	Facility	facilityID, facility name, facility type, facility rental fee
5	Season	seasonID, season name, season month start, season month end
6	Staff	staffID, staff lastName, staff firstName
7	Date	dateID, date, quarter, weekly indicator
8	Location	locationID, location barangay, location municipality
9	Field	fieldID, field name, locationID
10	Truck	truckID, truck type, plate no.
11	Coop	coopID, coop name, address, contact no.

Table 2. Production Fact table

Production Fact
dateID (FK)
cropID (FK)
farmerID (FK)
fieldID (FK)
buyerID (FK)
staffID (FK)
facilityID (FK)
volume or weight of commodity
price per unit
total amount

3.6. Snowflake Schema

After defining requirements, the next step is to determine how data structures will be available, combined, processed and stored in the data warehouse. This process is known as data modeling [12]. DW design and data sources as defined in this step.

The proposed DW design employs snowflake schema. A snowflake schema is a variation on the star schema, in which very large dimension tables are normalized into multiple tables. Figure 2 shows the data warehouse snowflake schema and how data from the data warehouse will be used for BAPTC. Product Fact is connected to multiple dimensions represents the snowflake schema. Dimensions such as Crop, Field, Buyer, and Farmer are normalized into multiple related tables. In the figure, field dimension is elaborated into location dimension to understand on what Barangay and Municipality a crop is produced. The crop dimension is elaborated into season dimension to determine on what season a crop is best produced. The farmer dimension is extended into two (2) relationships to understand what truck farmer in the trading areas and on what cooperative used or organization the farmer belongs to.

A data warehouse contains data that is used to support business decisions which can be used in various ways such as with data visualization tool to provide reports and dashboards; with advanced data analytics such as forecasting, data mining, sentiment analysis; and can be used as a source of data to other systems such as supply chain management and marketing.

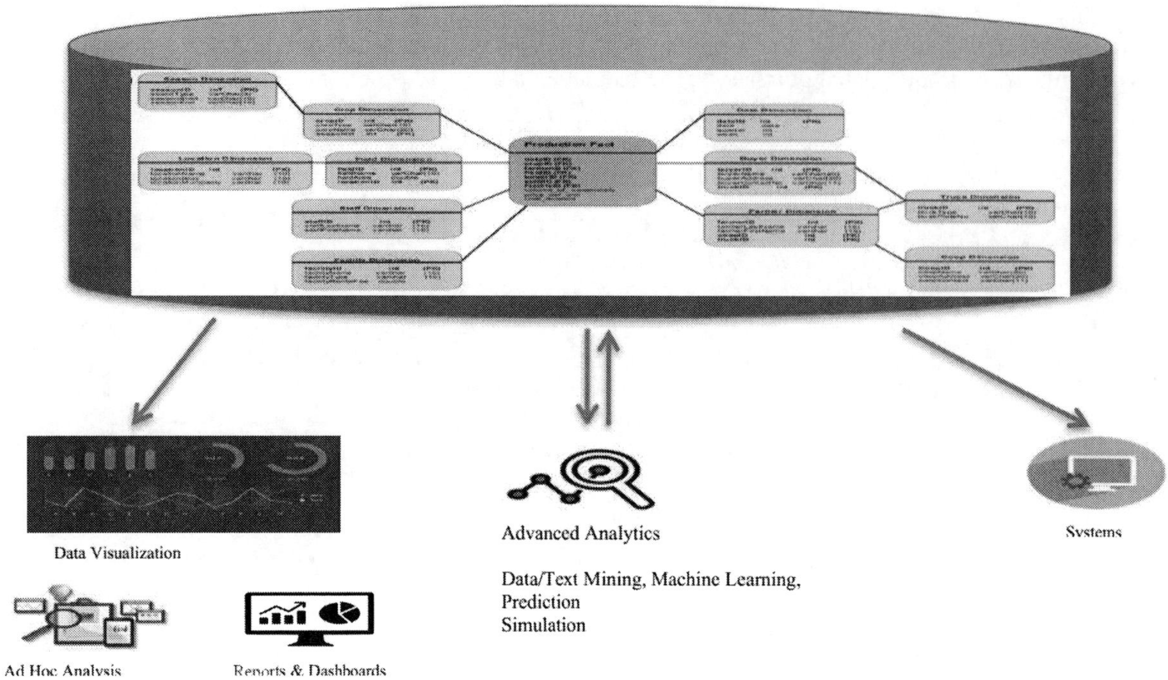

Figure 2. DW Snowflake Schema for BAPTC

4. Conclusion and Recommendations

A data warehouse design was presented as a solution to agricultural data issues. For BAPTC to operate efficiently and achieve their goals, from decisions-makes, farmers to individual contributors need efficient access to data and analytics that provide insights into how the center is operating, areas of concern, and opportunities for competitive advantage. By using snowflake approach in the DW design will save space. Data from DW can be utilized for data mining, forecasting, predictive analysis, accurate reports, and informative dashboard. This will help management resolve various complex queries as per their requirement. Further, this research could be extended to develop predictive models to increase the yield of highland vegetable crops in Benguet.

References

[1] Ranganathan J, Waite R, Searchinger T and Hanson C 2018 *How to sustainably feed 10 billion people by 2050 in 21 charts* (Washington DC: World Resources Institute)
[2] Rivas R 2019 *Agriculture grows a meager 0.56% in 2018, misses target* Retrieved from: https://www.rappler.com/business/221717-agriculture-growth-philippines-2018
[3] United Nations *Sustainable Development Goals*
[4] Bayer M and Laney D 2012 *The Importance of 'Big Data':A Definition* (Gartner, Inc)
[5] Kumar P, Sarangi A, Singh DK, Panhar S S and Sahoo R N 2015 Simulation of salt dynamics in the root zone and yield of wheat crop under irrigated saline regimes using SWAP model *Agricultural Water Management* vol **148** C pp 72-83 DOI: 10.1016/j.agwat.2014.09.014
[6] Gupta S 2018 *Data Science in the Indian Agriculture Industry* May 28th 2018
[7] Guru 2019 *Data Warehouse Architecture, Concepts and Components* Retrieved from https://www.guru99.com/data-warehouse-architecture.html
[8] Bindi T 2017 *Harnessing IoT in the Enterprise*: *Transforming the agriculture industry using IoT and predictive analytics* February 1st 2017 Retrieved from https://www.zdnet.com/article/transforming-the-agriculture-industry-using-iot-and-predictive-analytics/

[9] Ngo V M, Le-Khac N and Kehchadi M 2018 An Efficient Data Warehouse for Crop Yield Prediction *Proc. of the 14th Int. Conf. on Precision Agriculture*

[10] Rai A, Nilakanta S and Scheibe K 2007 Indian Agriculture Data Warehouse Design *Managing Worldwide Operations and Communications with Information technology* pp 535-8

[11] Bui A 2017 *Data Warehouse Design: The Bad, The Good, The Ugly* Retrieved from https://blog.panoply.io/data-warehouse-design-the-good-the-bad-the-ugly

[12] Kimball R and Ross M 2013 *The Data Warehouse Toolkit: The Definitive Guide to Dimensional Model* 3rd edition (New Jersey: John Wiley & Sons, Inc)

Design and Implementation of Retail Product Assortment System with Private Label Considerations and Weighted Apriori

Auliasa Enlina[1] and Jaka Sembiring[2]

School of Electro and Informatics Engineering, Institut Teknologi Bandung, Bandung, Indonesia

[1]auliasaenlina@gmail.com [2]jaka@stei.itb.ac.id

Abstract. Increasing product variety raise the needs for analytical model to determine an efficient and optimal assortment based on consumer preference considerations. However, most of retailers have not considered the private label factor that can generate profit differentiation while building store loyalty. Apriori algorithm, which is often used to find consumer shopping habit patterns in retail, still has many drawbacks that can increase the burden of input, output, and processing time. Thus, this study proposes a specific procedure to establishing basic product based on private label, improving association rules by reducing transaction through Size of Transaction (SoT) attribute, and adopting maximum weighted value concept in apriori algorithm, therefore the result of association rules will be reflecting user considerations. This research produces a model to recommend assortment by considering private label and user reference, and building a system to integrate the whole process. Based on evaluation result, this study was able to reduce number of scanning databases, number of candidate and frequent itemsets generated, as well as strengthen the store image (qualitative) and increase company profit (quantitative).

Keywords — product assortment, private label, apriori algorithm, maximum weighted value, size of Transaction (SoT)

1. Introduction

Increasing product variety and limited shelf space, raises an urgency to determine an efficient and optimal assortment. However, previous assortment method had not considered the private label factor of a retailer. In fact, retailer from strong private label brands can generate profit differentiation while building store loyalty [2]. In addition, assortment decisions were made without useful analytic models, such as considering consumer preferences [1].

Product assortment is all kind of products that available at the retail stores. Optimal assortment should meet the following two criteria, which are consistent with the store image (qualitative) and generate profit (quantitative). Therefore, retailers often distinguish basic product and added product. Basic product aims to build the store characteristics and should be chosen at the assortment stage [6]. Strong private label brands can generate profit differentiation while building store loyalty [2]. So it can be concluded that private label related to the store image [8]. While added product is chosen to increasing product variation and maximizing cross-selling potential with basic products [6]. Therefore, to determine the added product can also be done by knowing purchases affinity with basic products.

Apriori Algorithm [3] is a part of association rule mining. The role is to find consumer shopping habit patterns in retail for assortment needs [4] [5], through shopping basket analysis or transaction data [3]. Eventhough it is the most popular association rule mining method, apriori algorithm still has some drawbacks such as time-consume database scanning and generating too many candidate sets.

Determining support value becomes a challenge when user facing a huge database size. To eliminate useless rules, user must specify a minimum support value. But without specific knowledge, it's difficult to adjust support value to get the desire results. Therefore we need an algorithm that can produce a minimum support and confidence value which is adjusted to dataset needs in database [10].

Most of association rule mining implementation does not consider item weight. Association rules that consider weight factors will attract consumers and profitable. In addition, weight on association rule mining aims to improve traditional apriori algorithms [9].

In previous research, product assortment is determined by using association rule mining to get frequent itemsets, and proposed the PROFSET (PROFitability per frequent SET) model to determine optimal product selection based on frequent itemsets in order to maximizing profit [6]. In that research, the assortment only choose frequent itemset products, so any products that reflected store image but have low selling power, will be automatically removed from the assortment; basic and added products are not specified yet; basic product was not established at the beginning of study, so the selected products did not strengthen the store image; and there is no consideration of private label products in determining the assortment.

Another previous research made some improvements in apriori algorithm by reducing number of comparisons, number of scanning databases, candidate and frequent itemset generated, and transactions that contain product variety less than Lk [7]. However, to access the database when carrying out transaction elimination procedures will be consumes time. In addition, the minimum value of support is chosen randomly, and has not yet applied the weight concept that can reflecting user references.

Thus, in this study several improvements will be proposed as follows. Raises the possibility of cross selling effects based on purchases between basic products and added products, by determining basic products at the beginning of the study. To strengthen the store image, basic products are chosen based on private label products from a retailer [8]. In addition, the basic product automatically becomes part of the assortment, because it is the core form of the retail store formula [6]. There is a specific procedure to determining added products, which is based on the purchase affinity with basic products [6] or based on frequent itemset products (forced by model). Adding apriori weight factor, so the result rules are able to reflect the character of the algorithm based on user considerations, by adopting calculation of the maximum weighted value. The proposed weight value is $0<wj\leq1$. Minimum support value is set by 2, because the dataset has varied items and a huge transaction volume, the selected support value is as minimum as possible, in order to produce optimal association rules. Adds the SoT (Size of Transaction) attribute to the transaction database which is need to be updated in each iteration. SoT is data that stores TID and number of transactions, it proposed to reduce database scanning time. Validate the association rules using confidence weight (Wconf), to produce set of strong association rules (strong rules).

2. Analysis and Model Development

The proposed research describes the development of previous research, and it is divided into two parts. First, the product assortment stage which includes: determine basic product; mining association rule to get frequent itemsets; determine added product; calculate gross margin (per itemset and item); calculate profit maximization with PROFset; determine assortment. Second, apriori weighted improvement stage, which includes: mining association rules by applying the weighting concept and SoT attributes; and validating frequent itemset by calculating confidence weight (Wconf).

2.1. Product Assortment Model

Previous study only consider frequent itemsets and thus infrequent itemsets will not added to the assortments, because it is highly probable that infrequent itemsets exist because of random purchase behaviour. Moreover, basic product and added product are not specified well. In previous study, basic product is only assumed as all the products that generating the highest profit for retail. So, any products that reflected store image but have low selling power, will be automatically removed from the assortment. Whereas, basic product aims to build retail characteristics, so it needs to be chosen in assortment, and basic product is the core materialisation of the retailer's store formula [6] and must be identified by the retailer as a reflection of the store image. Added products are selected to maximise cross-sales potential with basic product, in order to confirm the store image even more [6]. But there is no specific way to determine the differences of those characteristic of products. The previous model can be seen in Figure 1.

In this study, private label is defined as a basic product, with the consideration products that reflected the store image but have low selling power, will remain as assortment. Then added product is determined based on the affinity with basic product, but if there is no association found with basic product, then added product is determined based on frequent itemset results. The proposed model can be seen in Figures 2, 3, and 4.

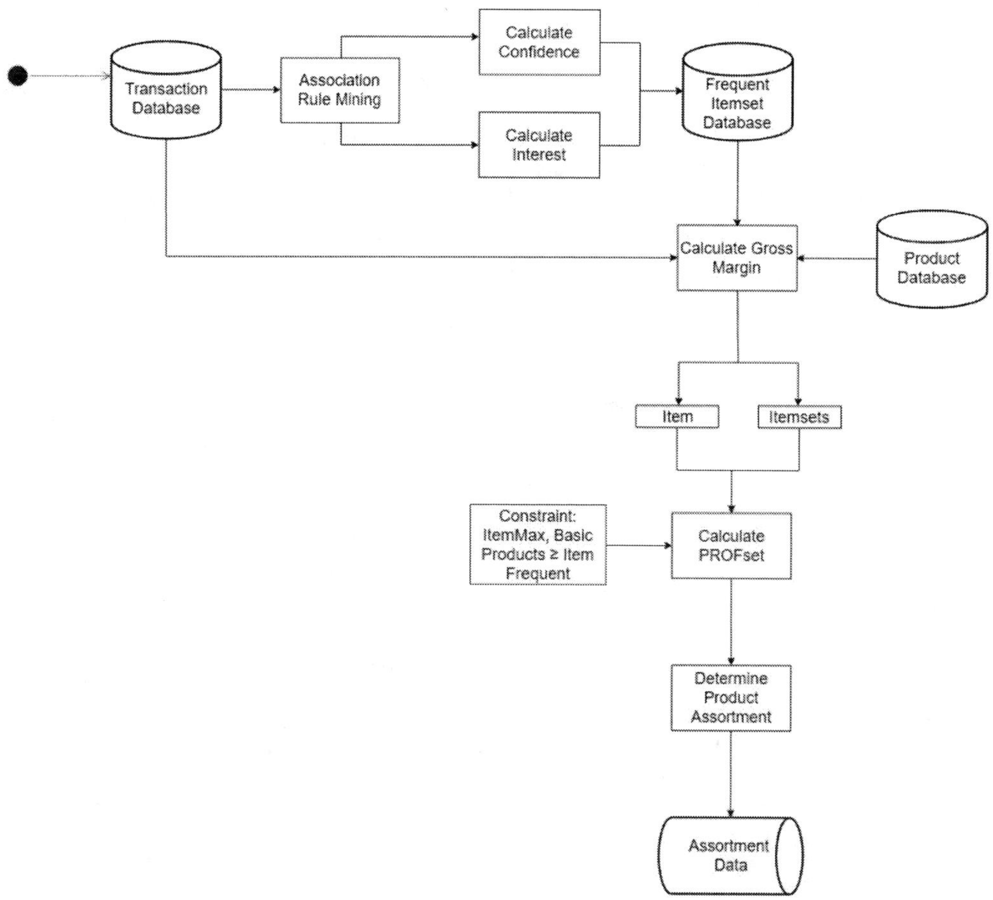

Figure 1. Previous Product Assortment Model

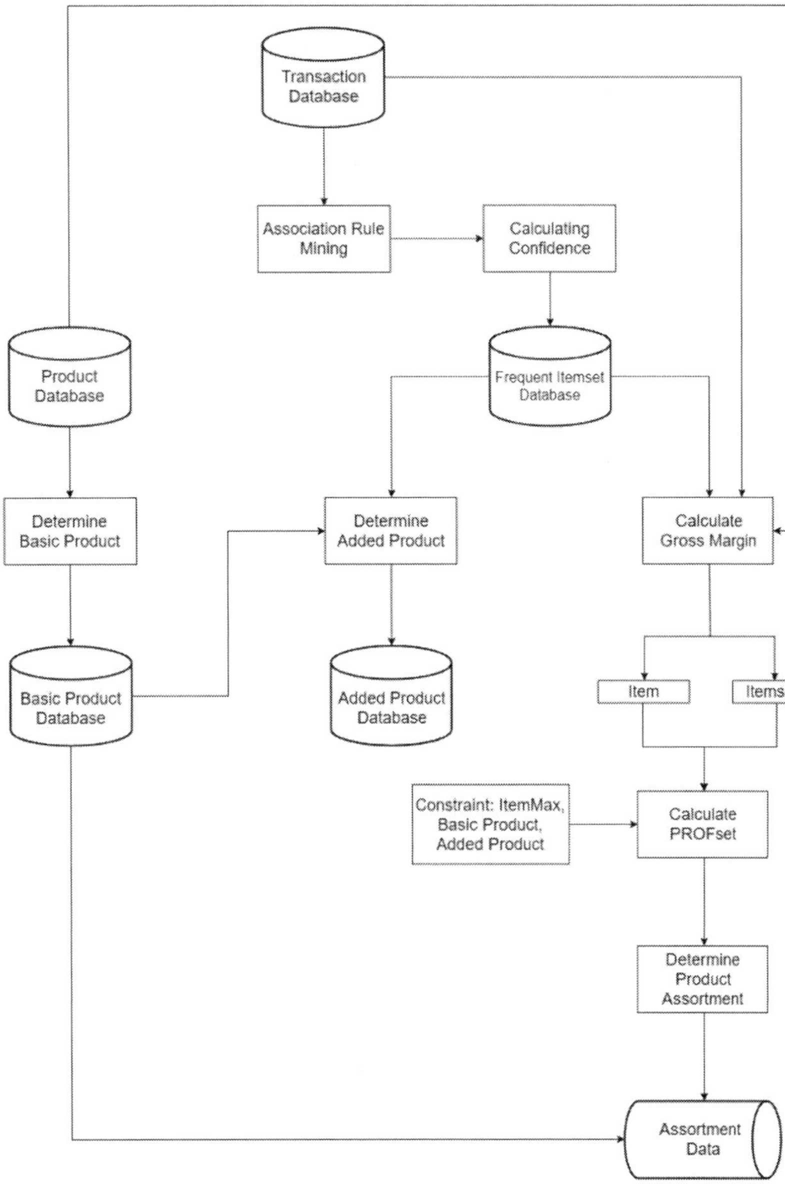

Figure 2. Propose Product Assortment Model

2.2. Weighted Apriori Model

In previous study, apriori is done without consideration of minimum support value, weight factor, and Size of Transaction (SoT). The differences between traditional apriori contained in k + 1 iteration, as follows: transactions whose size is less than the iteration value will be eliminated; database scanning process is minimized based on item with a total minimum support on each itemset; those itemsets that are not included in the frequent itemsets will be eliminated and will not generated in the next iteration. The previous model can be seen in Figure 5.

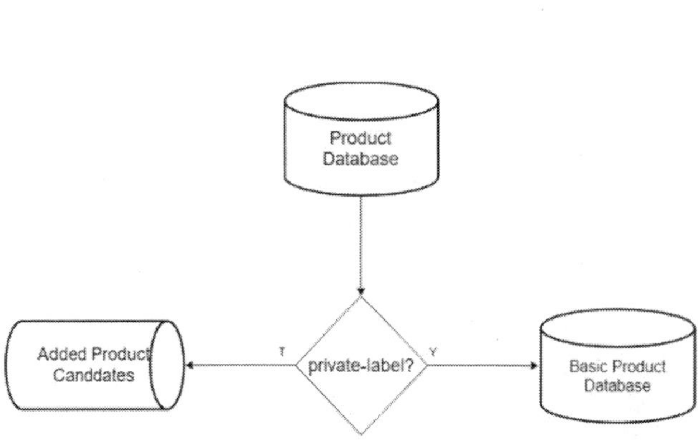

Figure 3. Propose Basic Product Model

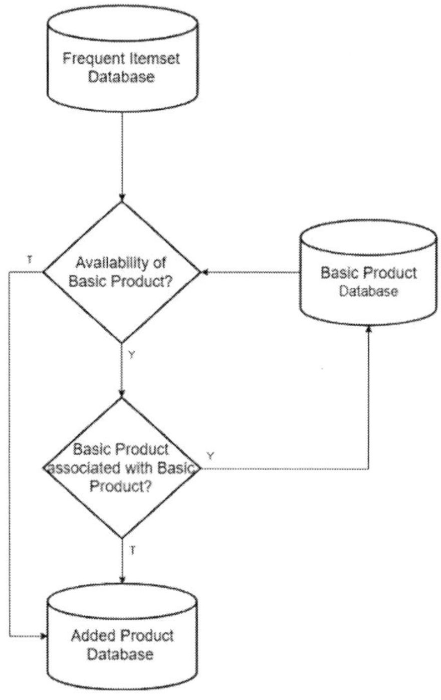

Figure 4. Propose Added Product Model

In this study, a weight factor will be added to association rules procedure. The minimum support value is 2, not 1 or 0. If we choose 0, then retail items that are not sold at all, can be part of frequent itemsets. Meanwhile, retailer is profit-based organization, and tend to choose good selling items. If we choose 1, then the system is considered unsuccessful in capturing the pattern of consumer habits. Because an event is considered to have a pattern if it occurs repeatedly (more than once). So if there are two or more different variants of products that are purchased simultaneously at once, then the event is considered as a random purchase (does not contain a pattern). In addition, the Size of Transaction (SoT) attribute is added to store TID and transaction amount. Then the strong rules are carried out by calculating the confidence weight (Wconf). The proposed model can be seen in Figure 6.

IOP Conf. Series: Materials Science and Engineering **803** (2020) 012036 doi:10.1088/1757-899X/803/1/012036

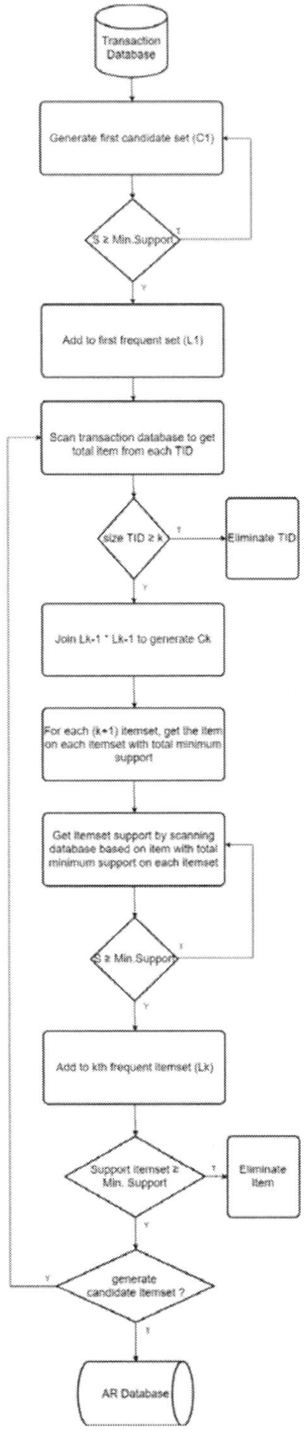

Figure 5. Previous Association Rule Mining Model

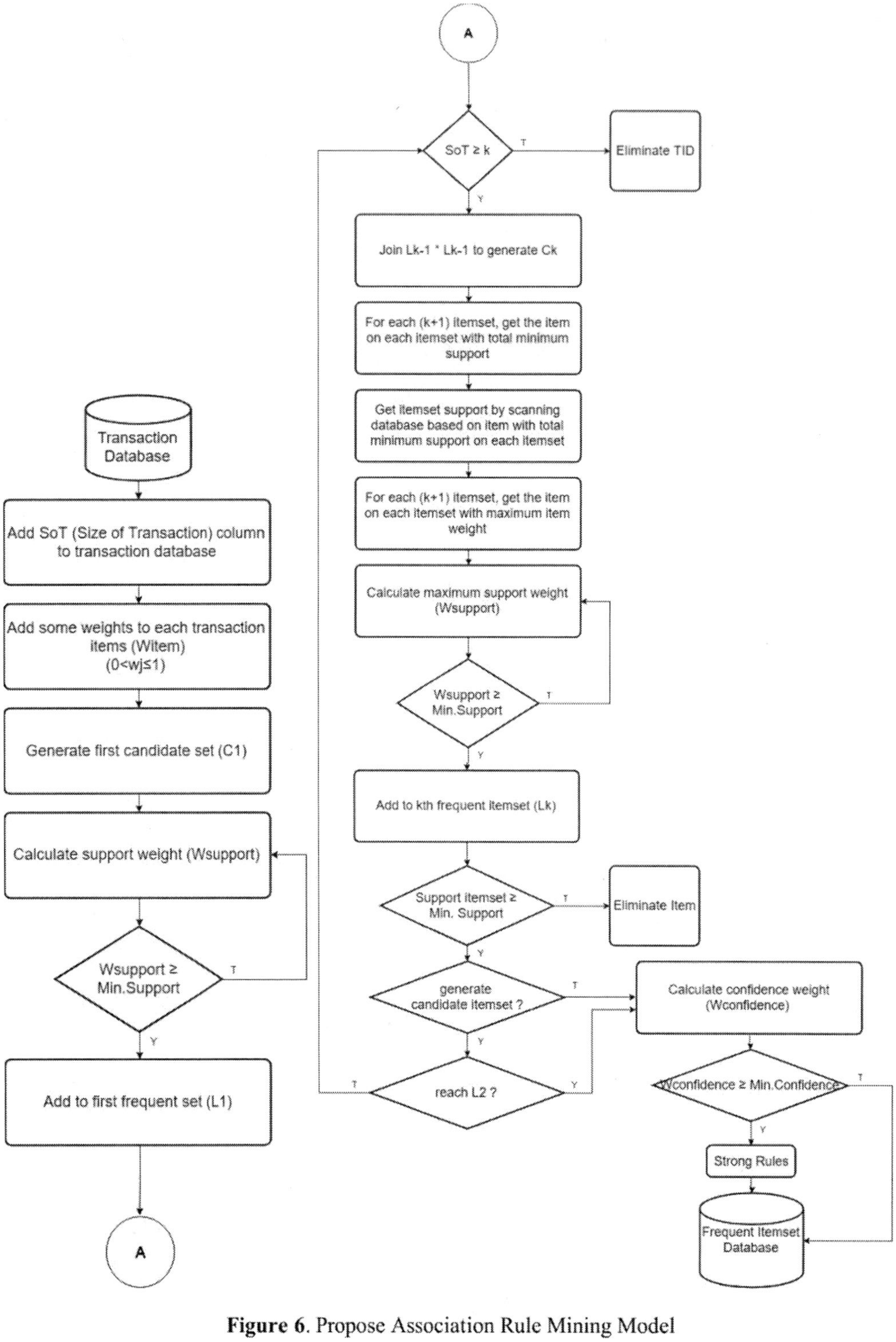

Figure 6. Propose Association Rule Mining Model

2.3. System Design
A web-based retail product assortment system is built with consideration of private label and weighted apriori. Functional system requirement is to assist store managers in determining product assortment, based on private label and the profitability level of product variants, by tracing frequent itemsets with apriori weighted algorithm and PROFSET (PROFitability per SET) method. The system architecture can be seen in Figure 7.

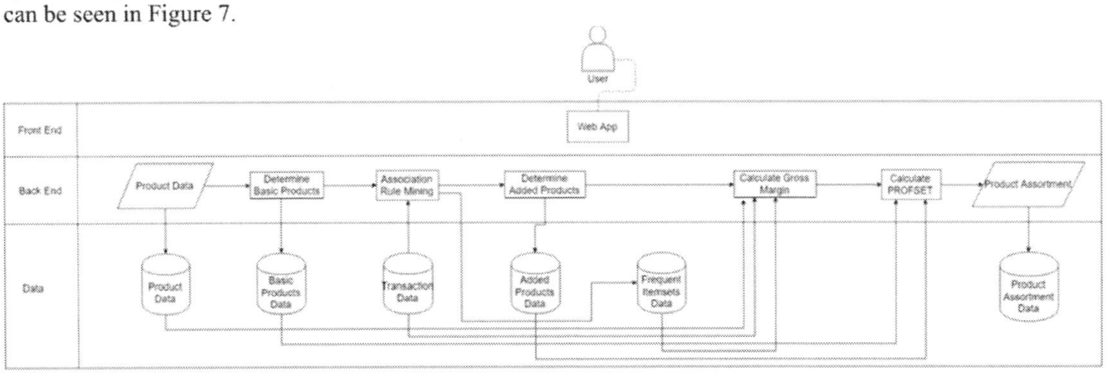

Figure 7. Retail Product Assortment Architecture System

3. Case Study
The proposed model is implemented with a case study at a retail bookstore, located in Bandung. Case studies are implemented to prove that the design of the model (Figures 2 – 4 and 6 – 7) is an appropriate solution to the research problem.

3.1. Determine Basic Product
In this stage, basic product selection will be made from all product variants provided by retailers. Basic products are chosen based on private label products from a retailer. So it produced some basic products by product variant (by title) and private label of the product (publisher) (Tabel 1).

Table 1. Basic Product

Item	Item Name	Vendor	Selling Price	Purchase Price
200378576	Doraemon 01 (Terbit Ulang)	Komputindo, Pt	25000	17500
208017364	Koran Kontan April 2019	Kompas (Kep Koran)	6000	4200
550000712	Bicara Itu Ada Seninya	Gramedia Pustaka, Pt	67000	46900
571840004	Sebuah Seni Untuk Bersikap Bodo Amat	Gramedia Widiasarana Indonesia	67000	46900
581812170	Perang Eropa Jilid 1 Edisi Revisi	Penerbit Buku Kompas, Pt	149000	104300
618143001	Lima Sekawan: Di Pulau Harta	Gramedia Pustaka, Pt	43000	30100
718011170	Detektif Conan 94	Elex Media Komputindo, Pt	25000	17500

3.2. Association Rule Mining
In this stage, product association rule mining will be conducted based on occured transactions. This procedure aims to obtain frequent itemsets, so product contribution in assortment will be obtained, through the information on a set of products that are often purchased simultaneously by consumers (Table 2).

3.3. Determine Added Product
In this stage, the determination of added products will be made. In this study, added products are determined based on the rules of product associations that have been carried out in the previous stage (affinity with basic products) or frequent itemset products (forced by models) (Table 3).

Table 2. Association Rules

L2								
Rule	Itemset	Item with minimum support	Scan Number	Support	Transaction	Item with Max Witem	Wsupport	Confidence
1	207961915 x 207996408	207961915	2	2	10129-011099658, 10129-021189624	207961915	2	1
2	718011953 x 719010224	718011953	2	2	10129-011099649, 10129-011099681	718011953	2	1
3	207993786 x 208001579	207993786	3	3	10129-021189624, 10129-021189826, 10129-041143662	207993786	3	1

Table 3. Added Product

Added product
207961915
207993786
207996408
208001579

3.4. Calculate Gross Margin Frequent Itemset

In this stage, a frequent itemset gross margin calculation will be performed to determine profit generated by the frequent itemset product in each transaction that occurs. From the results, two profit categories will be generated, namely profit per item and profit per itemset (Table 4 and 5). Profit per item is profit contained in a transaction that generated from each items of frequent itemset, while profit per itemset is generated from items combination of frequent itemset.

Table 4. Profit Per Item

TID	ID Item	Item Name	Quantity	Item Profit
10129-011099649	718011953	Boruto – Naruto Next Generation Vol. 1	1	12000
10129-011099649	719010224	Boruto – Naruto Next Generation Vol. 2	1	12000
10129-011099658	207961915	Si Juki The Movie : Panitia Hari Akhir	1	19500
10129-011099658	207996408	Si Juki : Seri Keroyokan X	1	21450
10129-011099677	718011953	Boruto – Naruto Next Generation Vol. 1	1	12000
10129-011099677	719010224	Boruto – Naruto Next Generation Vol. 2	1	12000
10129-021189624	207961915	Si Juki The Movie : Panitia Hari Akhir	1	19500
10129-021189624	207993786	Komik Tahilalats New Edition	1	25050
10129-021189624	207996408	Si Juki : Seri Keroyokan X	1	21450
10129-021189624	208001579	Si Nopal Komik Strip	1	21450
10129-021189716	719010224	Boruto – Naruto Next Generation Vol. 2	1	12000
10129-021189826	207993786	Komik Tahilalats New Edition	1	25050
10129-021189826	208001579	Si Nopal Komik Strip	1	21450
10129-041143662	207993786	Komik Tahilalats New Edition	1	25050
10129-041143662	208001579	Si Nopal Komik Strip	1	21450

3.5. Calculate Profset (Profit Maximize with Integer Programming)

In this stage, profit maximization will be calculated using integer programming, or as known as PROFSET (PROFitability per SET) method, based on the total profit from all frequent itemset transactions, with constraints of ItemMax (maximum number of product variations that determined by retailer), basic product, and added product. The results of PROFSET calculation will be used as a reference to determine the product variants to get the optimum profit contribution for retailers (Table 6 and 7).

Table 5. Profit Per Itemset

Item 1	Item 2	Itemset Profit
718011953	719010224	24000
207961915	207996408	40950
718011953	719010224	24000
207961915	207996408	40950
207993786	208001579	46500
207993786	208001579	46500
207993786	208001579	46500

Table 6. Profset Per Item

ID Item	Item Name	Profit
207961915	Si Juki The Movie : Panitia Hari Akhir	39000
207993786	Komik Tahilalats New Edition	75150
207996408	Si Juki : Seri Keroyokan X	42900
208001579	Si Nopal Komik Strip	64350
ID Item	Item Name	Profit
718011953	Boruto - Naruto Next Generation Vol. 1	24000
719010224	Boruto - Naruto Next Generation Vol. 2	36000

Table 7. Profset Per Itemset

Rule	Itemset	Profit
1	207961915 x 207996408	81900
2	718011953 x 719010224	48000
3	207993786 x 208001579	139500

3.6. Implementation Results
The implementation results of the design and implementation of retail product assortment system with consideration of the private label and weighted apriori can be seen in Table 8. To determine product assortment, then profset per item and profset per itemset data is going to be analyzed. First of all, basic product is automatically becomes part of the assortment. In the next step, we need to compare itemset profit and item profit, with the following conditions:

1. If profit generated from itemset AB is greater than item A and item B, then choose itemset AB.
2. If profit generated from itemset AB is greater then item B, but less then item A, vice versa, then choose itemset AB. So that items that have a minimum profit will include in the hitlist.
3. If profit generated from itemset AB is less then item A and item B, then calculate the confidence weight of each items. If confidence weight is less then minimum confidence weight, then choose both items. However, if confidence weight is greater or same as minimum confidence weight, then choose itemset AB.

Table 8. Assortment Data

Rank	Item or Itemset	Profit
1	207961915 x 207996408	81900
2	718011953 x 719010224	48000
3	207993786 x 208001579	139500

3.7. Testing
Primary data for testing is contains with 30 transaction data and 65 item variants. Testing is done by comparing the proposed apriori algorithm, with the apriori algorithm improvement in previous studies, to find out how much effort to carry out apriori procedures from both studies.

In this test, there are 5 test scenarios. The first scenario is testing previous research with minimum support = 2; the second scenario is testing previous research with minimum support = 3; third scenario is testing previous research with minimum support = 1; the fourth scenario is testing the proposed research with minimum support = 3; and the fifth scenario is testing the proposed research with minimum support = 1. Result of the performance comparison, is in Figures 8-10.

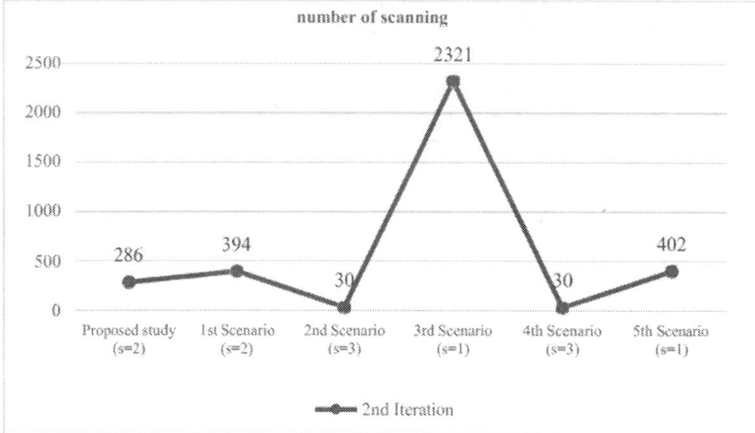

Figure 8. Performance graph of the scanning databases number to generate candidate sets in the 2nd

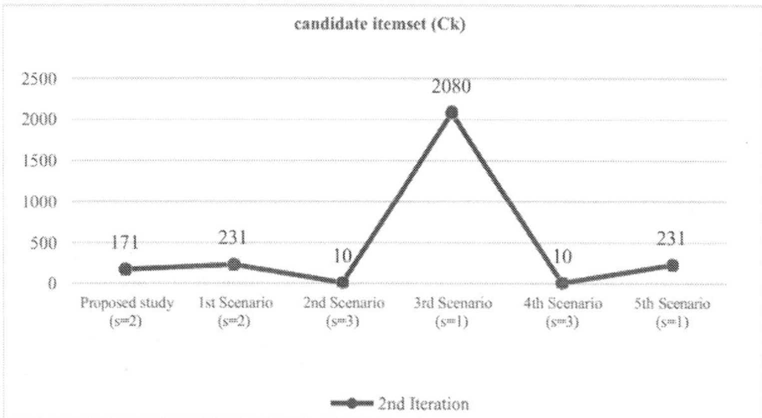

iteration

Figure 9. Performance graph of the generated candidate itemsets number in the 2nd iteration

Based on those graphs above, then.

1. Compared with testing in scenarios 1, 3, and 5, the number of candidate and frequent sets generated in this study tends to be less. Then it can be concluded, this study reduces the number of scans in generating itemset. In scenario 3, purpose to find out which items are frequently bought by consumers, is missing, because the combination of association rules that are generated, includes all variants of items from the transaction data analyzed.
2. Compared with testing in scenarios 2 and 4, although the number of candidate and frequent sets generated in this study is far greater, the association rules generated in scenarios 2 and 4 cannot capture interesting patterns in the database optimally. This is caused by the selection of minimum

threshold support value that is not in accordance with the dataset used. High minimum support value is applied to dataset with very diverse item variations and very large transaction volumes will result a low number generated association rules. Or at the worst case, frequent itemset cannot be traced because the minimum support value is too high. When it happens, then efforts to find interesting patterns from a database cannot be achieved. If something like this happens in a retail environment, then the company is considered unable to capture the pattern of consumer habits when shopping at supermarkets. Of course it will harm the company, if product variants that have good potential to increase sales, are lost from the arrangement (assortment).

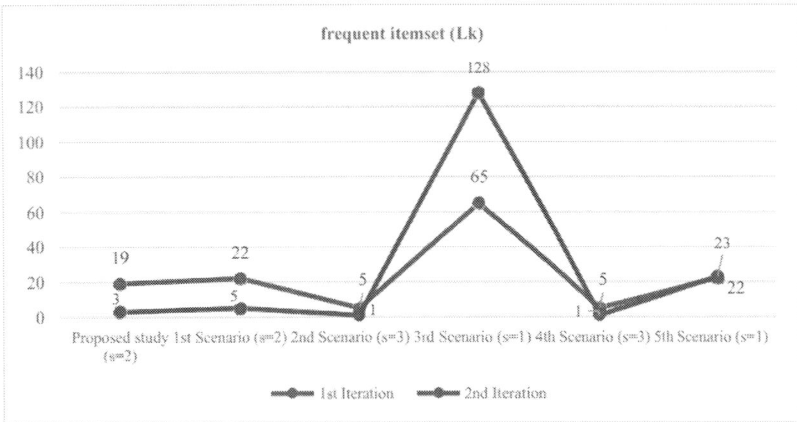

Figure 10. Performance graph of the generated frequent itemsets number in the 1st and 2nd iterations

4. Conclusion

Designing a product assortment system with consideration of private label and mining the weighted association rules, help user to make decisions regarding assortment efficiently. Improvement of an apriori procedure based on adjusting the minimum threshold support to the dataset, applying the concept of weight, and SoT (Size of Transaction) attribute, are able to produce strong association rules, and reducing: number scans of databases, number of candidate and frequent itemsets generated. As well as reduce the burden of input, output, and processing time.

Based on system implementation result, product assortment can strengthen the store image (qualitative), also increase the company profit (quantitative).

Future works are expected to handle unresolved problems in this study, such as: a specific measurement in determining weight value in association rules, generic algorithms that can produce ideal support values, and reducing the scans number on the first iteration.

References

[1] Agrawal N and Smith S A 2003 Optimal retail assortments for substitutable items purchased in sets *Naval Research Logistics (NRL)* **50** pp 793 - 822 DOI:10.1002/nav.10090

[2] Corstjens M and Lal R 2000 Building Store Loyalty Through Store Brands *J. of Marketing Research - J MARKET RES-CHICAGO* **37** pp 281-91 DOI:10.1509/jmkr.37.3.281.18781

[3] Agrawal R and Srikant R 1994 Fast algorithms for mining association rules *Proc. of the 20th VLDB Conf.* pp 487-99

[4] Fageeri S, Ahmad R and Baharudin B 2014 A semi-apriori algorithm for discovering the frequent itemsets *2014 Int. Conf. on Computer and Information Sciences (ICCOINS)* pp 1-5 DOI:10.1109/ICCOINS.2014.6868358

[5] Charanjeet K 2013 Association rule mining using Apriori Algorithm: A Survey *Int. J. of Advanced Research in Computer Eng. & Technol. (IJARCET)* Vol **2** pp 2081-4

[6] Brijs T, Swinnen G, Vanhoof K and Wets G 2004 Building an Association Rules Framework to Improve Product Assortment Decisions *Data Min. Knowl. Discov.* **8** pp 7-23 DOI:10.1023/B:DAMI.0000005256.79013.69

[7] Dutta A and Ghosh A 2016 Improvement of Apriori Algorithm by Reducing Number of Database Scans and Generation of Candidate Keys *Int. J. of Advanced Research in Computer Science* **7** pp 35-41

[8] Ailawadi K and Keller K 2004 Understanding Retail Branding: Conceptual Insights and Research Priorities *J. of Retailing* **80** pp 331-42 DOI:10.1016/j.jretai.2004.10.008

[9] Zhang X 2015 Study of an improved Apriori algorithm for data mining of association rules *2015 Int. Conf. on Applied Science and Engineering Innovation* DOI:10.2991/asei-15.2015.238

[10] Dongre J, Prajapati G and Tokekar S 2014 The role of Apriori algorithm for finding the association rules in Data mining *2014 Int. Conf. on Issues and Challenges in Intelligent Computing Techniques (ICICT)* pp 657-60 DOI:10.1109/ICICICT.2014.6781357

Usability Analysis with Webuse Model in Information System Design in Monitoring Child Growth and Development

Kurniawan Teguh Martono[1,2], Oky Dwi Nurhayati[1], Eko Didik Widianto[1]

[1]Computer Engineering Department, Diponegoro University, Indonesia
[2]Center for Biomechanics, Biomaterial, Biomechatronics and Biosignal Processing, , Diponegoro University, Indonesia

E-mail : k.teguh.m@live.undip.ac.id

Abstract. Information system today becomes a core need for any organizations as it can help in data management for resulting in the structured data. Also, the use of information system can assist the organization in minimizing the errors in making a policy. One of organizations implementing such system is POSYANDU or Integrated Service Post that is predominantly in charge of monitoring the child growth and development. In this research, a process of testing the information system was conducted using the WEBUSE (Web Usability Evaluation) model aimed to observe the level of usability of the system. Based on the results of the test, it has been found that the average values of the system were 0.69 indicating that the system was at the Good level of usability. Thus, the system can work in accordance to the needs of the users.

Keywords: Usability, WEBUSE, Information, Growth and Development

1. Introduction

For being a next generation for both family and state, children are considered as one of the important assets for family and state for playing a role in a very strategic position in supporting a success. This very strategic role then needs a serious concern and management in terms of the child growth and development. Growth and development are two different events, Growth means a phase in which the physical size and body structure are facing a change; while development refers to a change of ability or the functions of body is going to be more complex. The factor influencing the child growth and development is divided into two: internal factor and external factor in which both factors can cause the difference in the pattern of growth and development for each child.

To find out the usefulness of this application, we need a test that can be used as a reference to the level of subjective satisfaction of application users. The method used in this test is to use the Web Usability Evaluation Tool or commonly called the WEBUSE. Usability testing is intended to determine the level of comfort of the information system for monitoring the growth and development of children from the user side of the application.

2. Literature Review

The process of monitoring the child growth and development is deemed necessary for each state. Indonesia is one of the states implementing the monitoring process for the child growth and development using the monitoring model until the rural phase. It is done by the group of community independently.

This group of community has been established by the community itself called as POSYANDU or Integrated Service Post. Here, everyone participated in this group has been given a training by the health department, in this case, represented by PUSKEMAS (Centre for Public Health). POSYANDU is an independent institution established by the community in which it is responsible for providing the health basic service including the measurement and monitoring the child growth and development. In addition, POSYANDU provides a health basic service for mothers, family planning and nutritious food and immunisation [1],[2]. Health data plays a crucial role as it can function as a consideration in making a policy. An error in the process of reading or inputting the data can make an error in decision making. Recording the data about the child growth and development conducted in POSYANDU needs to be accurately done to prevent any errors in determining the status of growth and development of an area. Hence, it needs a system that can help the cadres in POSYANDU in facilitating the process of reading the results of measurement and recording process.

Usability is a software testing model that aims to determine the level of ease in learning the system, the ease of using the system as a device that can help positively solve problems [3]. Usability testing process can be carried out with various models, namely by using a special laboratory or by testing in public places. Usability testing is carried out using three aspects, namely effectiveness, efficiency and satisfaction. The effectiveness aspect is a part that is related to how the user can complete the task precisely and accurately. The efficiency aspect is a part that is related to the time needed to complete a task. The satisfaction aspect is related to the feeling of the user after using the application [4], [5], [6]. Usability Testing in general uses 4 stages in its implementation, namely determining the questionnaire model that will be used, determining the participants or participants who will do the testing, determining the number of participants or participants and finally the data analysis.

3. Research Method

This research was conducted to observe the level of usability of information system application to monitoring the growth and development of children. The aim of this test was to help the application developer in determining how to improve a design of application. In this research, the test of usability was conducted by obtaining the respondents from the cadres of POSYANDU. Then, the cadres were asked to do their task in accordance with the test scenario. The instrument used in this research was WEBUSE (Web Usability Evaluation tool) [7]. The criteria for usability evaluation using WEBUSE are Content, Organization, and Readability, Navigation and Links, User Interface Design, Performance and Effectiveness [8]. The steps in this research are presented as follows.

1. Determining the system to be tested
2. Asking the respondents to answer all questions in the questionnaires
3. Accumulating the scores used based on the answers from the users for each question in each category of usability.
4. The point of usability category is the mean value of each category
5. Point of usability from the website is the mean value from each category.
6. Determining the level of usability based on the point of usability.

The flow chart of this research is shown in Figure 3.1

At this stage of testing, using four variables, namely the level of ease of reading, information content and information structure. The next variable is the use of existing menus in the system which includes buttons and navigation system. The next variable is the interface design of the system, and the last variable is the performance of the system. Data is collected by using a questionnaire that will be distributed to respondents. The choice of the answers in the questionnaires would be divided into 5 answers: Very Disagree, Disagree, Neutral, Agree and Very Agree. Table 3.1 shows the choices of answer with the weight of score from each answer.

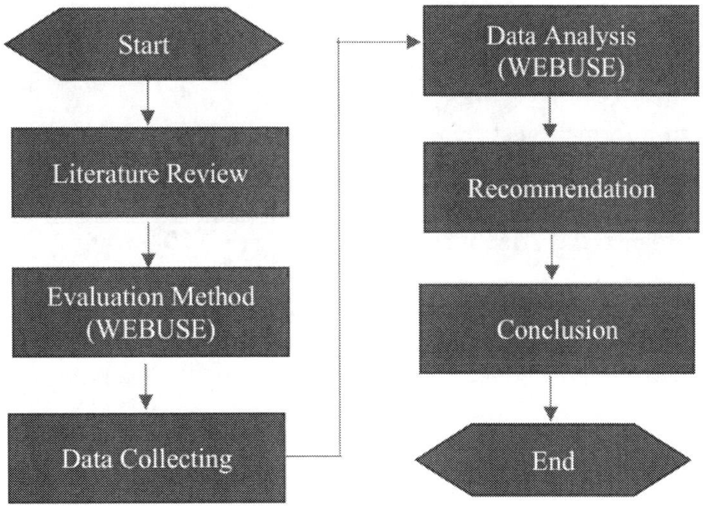

Figure 3.1 Flowchart Diagram of Research

Table 3.1 Weight and Choice of Answer [7]

Choice	Very Disagree	Disagree	Neutral	Agree	Very Agree
Weight	0.00	0.25	0.50	0.75	1.00

To determine the point of usability in the category was by labelling it with the notification x. Equation 1 shows the value of the point of usability.

$$x = \frac{[\Sigma(Weight\ of\ each\ question\ form\ each\ category)]}{number\ of\ questions\ J} \quad \ldots(1)$$

Table 3.2 shows the relationship between usability point and usability level with an explanation of the results obtained.

Table 3.2 shows the relation between usability points and usability levels

Usability Point	0 <= x <= 0.20	0.20< x < =0.40	0.40<x<=0.60	0.60<x<=0.80	0.8<x<=1.0
Level	Bad	Poor	Moderate	Good	Excellent

4. Results

The test involved POSYANDU cadres in Gajahmungkur sub-district, Semarang, Indonesia. POSYANDU cadres involved in this test were 30 people. Testing is done with the initial stage is to test the system that has been developed. This test includes:

1. Ease of reading information, appropriate information content and information structure
2. The use of the menu on each page of the website in accordance with needs and easy to use
3. Design the user interface in accordance with the rules in Human and Computer Interaction
4. Performance and effectiveness of the application so as to facilitate cadres in the process of filling and reading data.

The results obtained based on the equation 1 in WEBUSE method are shown in Table 4.1.

Table 4.1 The results of the test using WEBUSE

No	Test Variables	Point of usability (x)	Usability Level*
X_1	Facilities in reading, information contents, and information structure	0.66	**Good**
X_2	The use of menu	0.62	**Good**
X_3	Design of user interface in application	0.79	**Good**
X_4	Work performance and effectiveness of application	0.69	**Good**
	Total Average (x)	**0.69**	**Good**

* 1. x is greater than 0, and x is less than 0.20, then usability level is Bad
2. x is greater than 0.20, and x is less than 0.40, then usability level is Poor
3. x is greater than 0.40, and x is less than 0.60, then usability level is Moderate
4. x is greater than 0.60, and x is less than 0.80, then usability level is Good
5. x is greater than 0.80, and x is less than 1.00, then usability level is Excellent

Based on the data obtained as shown in Table 4.1, the result on average towards 4 variables of test was 0.69 indicating that it was at the Good level. Thus, the system or application developed has met the needs of the users in this case the cadres of POSYANDU. For each test variable, it obtained the results of the usability point in the range of 0.62 to 0.79 meaning that it is at the **Good** level.

5. Conclusion

Based on the analysis from the test results using WEBUSE, it can be concluded that the application of monitoring the growth and development of children had the Good level of usability. Thus, it can be concluded that the system can be accepted by the users. The acceptance includes the process in terms of learnability, efficiency, Memorability, Error and satisfaction with the average score of 0.69.

6. Acknowledgement

"This research has been funded by Directorate of Research and Community Service, Ministry of Research, Technology and Higher Education with the Scheme of PTUPT, Higher Education in the Budget Year of 2019"

References

[1] Nirwana M D, Utami I H and Utami H N 2015 The Cadre of Integrated Health Service Post (Posyandu) as an Agent in the Socialization of Cervical Cancer Prevention in Malang Regency, Indonesia: A Cultural Approach *Procedia - Soc. Behav. Sci.* vol **211** pp 681–7

[2] Giles J and Satriawan E 2015 Protecting child nutritional status in the aftermath of a financial crisis: Evidence from Indonesia *J. Dev. Econ.* vol **114** pp 97–106

[3] Miller K *et al* 2018 The design of decisions: Matching clinical decision support recommendations to Nielsen's design heuristics *Int. J. Med. Inform.* vol **117** January pp 19–25

[4] Jibb L A *et al* 2017 Development of a mHealth Real-Time Pain Self-Management App for Adolescents With Cancer: An Iterative Usability Testing Study *J. Pediatr. Oncol. Nurs.* vol **34** no 4 pp 283–94

[5] Ratwani R M, Benda N C and Hettinger A Z 2015 Electronic Health Record Vendor Adherence to Usability Certification Requirements and Testing Standards *JAMA* vol **314** no 10 pp 1070–1

[6] Sahfitri V and Ulfa M 2014 Analisis Usability Sistem E-Learning Menggunakan Use Questionnaire *Seminar Nasional Penelitian dan Pengabdian Masyarakat* pp 373–80

[7] Chiew T K and Salim S S 2003 Webuse: Website usability evaluation tool *Malaysian J. Comput. Sci.* vol **16** no 1 pp 47–57

[8] Dewi I K, Mursityo Y T and Mardi R R P 2018 Analisis Usability Aplikasi Mobile Pemesanan

Layanan Taksi Perdana Menggunakan Metode Webuse dan Heuristic Evaluation *J. Pengemb. Teknol. Inf. dan Ilmu Komput. Univ. Brawijaya* vol **2** no 8 pp 2909–18

IOP Conf. Series: Materials Science and Engineering **803** (2020) 012038 doi:10.1088/1757-899X/803/1/012038

Usability Evaluation for User Interface Design of Student Support System: Mixed Method Study

Ahmed AL-sa'di[1], Rasmus Bugge Skammelsen[2], Qiang Ji3, Yufeng Xiang[3], Bhogadi Aakarsh[3] and Sathishkumar Kuppusamy[3]

[1]Department of Information system, UUNZ Institute of Business, Auckland, New Zealand
[2]Department of Architecture, Design and Media Technology, Aalborg University, Aalborg, Denmark
[3]Department of Computer Science, Auckland University of Technology, Auckland, New Zealand

E-mail: ahmed@uunz.ac.nz

Abstract. The efficiency of a website depends on how usable the UI user interface, (UI) is and how well it is structured. The aim of this research was to try and understand the issues with the Arion System. Arion is the student information system used by Auckland University of Technology. We conducted three experimental studies to examine the usability and to improve the user experience of student who use Arion. Three main methods were used namely; survey, open card sorting and situated co-inquiry. The survey was used to inform the study and collect more information about users. The open card sorting used to redesign the navigation menu for the system. Then the Situated co-inquiry used to discover the user's current unexpected or unforeseen ideas, which may be helpful for future research.

1. Introduction

The user interface (UI) provides two main interaction points between user and system (the input and the output), which indicate the effects of manipulation by the user [1]. Thus, interface design can be regarded as the "front-end" product, which allows users to interact, communicate, and converse with the machine. Meanwhile, code and data represent the "back-end" product [2].

Usability is one of the main features of an interface assessment. If the UI is difficult to use, it will likely be abandoned by users. Usability can help creating a pleasant and useful UI and is therefore essential throughout the software industry. A user-oriented application design method benefits not only the users, who enjoy a high-quality application, but also the company that produces and maintains the digital products [3].

The target of UI design is a system with high usability [4]. The International Organization for Standardization (9241-11, 1998) defined usability as "the extent to which a product can be used by specified users to achieve specified goals with effectiveness, efficiency, and satisfaction in a specified context of use" [5].

More simply, Nielsen in [6] defined usability as the quality characteristic that assesses how easy a user interface is to use. The term usability can also refer to the method by which ease-of-use is improved throughout the entire design process [2]. Nielsen in [6] also defines five components of usability: efficiency, learnability, memorability, minimal errors, and satisfaction. By adopting usability concepts, companies can improve their productivity, enhance the quality of their work, increase user satisfaction, and reduce their support and training costs (ISO/IEC, 2011). 0shows the attributes of usability, based on the views of several researchers.

Table 1: Usability as viewed by several researchers [5-8]

ISO9241-1 1998	Nielsen 1993	Shneiderman 2005	Shackel 2009
Efficiency	Efficiency	Speed of performance	Effectiveness
Effectiveness	Learnability	Time to learn	Learnability
Satisfaction	Memorability	Retention over time	Flexibility
	Errors	Rate of user error	Attitude
	Satisfaction	Subjective satisfaction	

As pointed out by [3], human factors (including efficiency and safe, effective interaction between user and tasks) must be considered when measuring and analysing the usability of any system. According to this description, usability is essential to the development of any system.

Usability testing is a routine test of many web sites to ensure that the site has a friendly interface that makes it easy for users to find what they want [9]. The standards of the International Organization for Standardization for usability testing are effectiveness, efficiency and satisfaction [10].

Arion is a student service and support tool used by Auckland University of Technology (AUT) to Manage correspondence between the organisation and students them and students. Students can use Arion to read correspondence from Arion, pay fees, lookup papers and grade, and look up their timetable as shown in figure 1. Informal discussion with some students revealed they struggle of using the Arion system. Also, four of the researcher in this study doing a postgraduate and using the Arion and face some issue.

The academic website often does not do justify to the user experiences who navigate the website[11]. User depends on the website menu (hyperlink) as an overall map structure for the website[12]. Therefore, they need to be labelled in such an easy way to understand and organise in an ideal way to follow user mental model[11]. Therefore this study is to identify problems with the information architecture and the information discoverability of Arion.

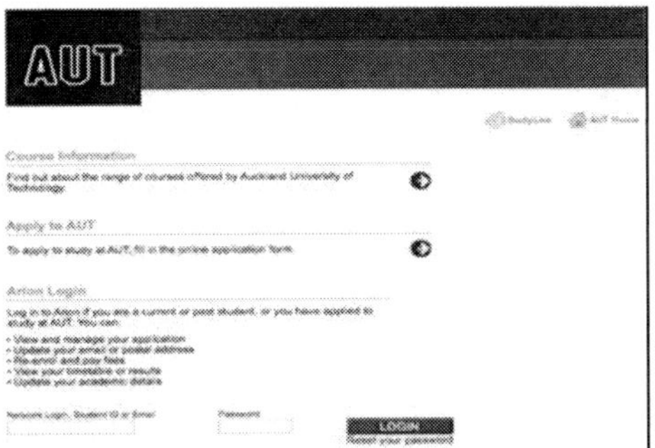

Figure 1: Arion system.

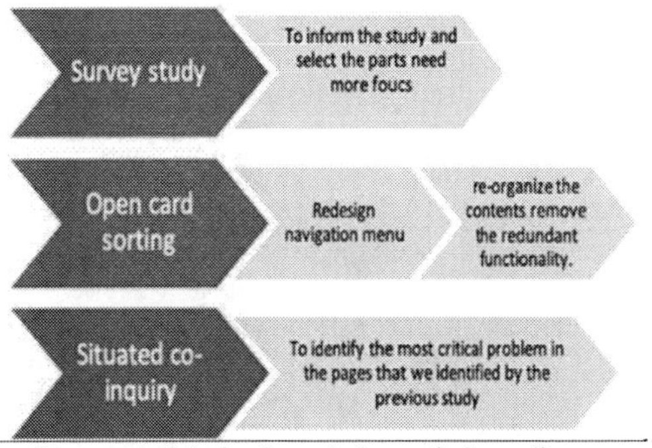

Figure 2: Methods and aims of all studies.

In the next section, we will present the study methods of our research. Each of our studies is then presented and the results of each are discussed.

2. Study Methods

To find relevant user tasks for the usability test, we wanted to identify the areas of Arion that were the primary use for most people and the areas that had the most problems. This way, we would be able to formulate recommendations that not only addresses actual issues, but that would also improve the user experience for most people. To do this, we created an enquiry.

Three main methods were used in this study to enhance the depth of the system evaluation and support the methodological triangulation to increase the validity and reliability of the finding as shown in figure 2. Mixed method approach allowed to a more comprehensive understanding of participants experience and enable the identification of usability issues [13].

The first method used was the survey. We conducted a survey using Google Survey. In order to understand more about the problems and issues that the users face while using Arion. Google Forms was used as it is free and easy to use the tool. 0shows our study aims; the survey study was used to help to know which areas of Arion require our attention. In addition, this also provided us with useful information about the end users.

Following, we used open card sorting to redesign the navigation menu and re-organize the contents by removing the redundant functionality. The card sorting was used before the situated co-inquiry to prevent the user from knowing or using the original navigation menu while they are searching for timetables or running a paper search, as these might create a short-term memory for users and affect the outcome of the card sorting.

In total, 146 participants responded to the survey. The participants who took the survey were aged between 18 and 37, with a mean age of 24.6. 37% of the participants were men, while 60.9% identified as women. Many of the participants had only used Arion for less than a year (69.6%).

Additionally, 78.26% of the participants had not used a similar system in the past, so the target group were relatively inexperienced. People who had used similar systems pointed to systems such as Moodle, Massey Portal, and Gateway - most of which were referred to as being more user-friendly than Arion. The majority of users also did not log into Arion that often, 58.7% only logged in once a week and another 37% logged in a couple of times a week. Their reason for logging in was often to check their timetable or their grades.

We have formulated a primary goal of the usability testing order to fix the problem with navigating the website and for finding the information we have set up an open card sorting test to answer the following question 'how can the navigation of the Arion website be improved to increase discoverability?'

Additionally, we have formulated a secondary goal that we hope to address using the situated co-inquiry and the prepared user tasks in order to understand how can information be presented to the user to make Arion a more enjoyable experience?

2.1. Survey Study

The survey consisted of two main sections; the purpose of the first section was to gather demographic data about our participants. We asked the participants about their age, gender, and for how long they have used Arion, and additionally, if they had used similar systems before. This gave us an understanding of who our user group is. Due to us being temporarily unable to determine the direct relationship between the user experience and these factors we will retain this data for reference later in the design of usability tasks and for analysis purposes.

The purpose of the second section of the survey was to find the most used areas of the system and the areas that people found to be less user friendly. To do this, we used both Likert scales and qualitative text fields. We were concerned about how people would remember their experience, depending on whether their experience of using the system was successful, as people tend to remember either the peaks of or the last part of an experience, as described by the peak-end [14]. We were not really concerned with whether participants were successful, as the whole process of completing an action is more critical. Achieving a goal after five seconds with no problems is vastly different from doing so after five minutes with a lot of frustration. In order to counter this problem, this section of the questionnaire/survey begun with a disclaimer about how to answer the following questions. The disclaimer can be seen below.

"Please answer the questions as accurately and fully as possible. Please do not base your answers on whether you were [successful], but how the experience was overall (was it annoying, troublesome, tiring, pleasant, easy, etc.)."

The results demonstrated that users have an issue with the timetable and paper search functionalities that were existing in the Arion website.

2.2. Open card sorting Study

The second study was open card sorting. Six users participated in this study all of the students at AUT aged between 18-35. In open card sorting study, we had the option of either allowing the participants to work individually or in pairs, as each presented us with advantages and disadvantages. If we used individual participants, we would have more data to compare and analyse, but we would potentially have had to interrupt participants in order to encourage them to think aloud. Allowing the participants to work in pairs helped the discussion to develop naturally, and the think-aloud technique is a natural part of talking and deliberating with a partner.

Before the participants entered the testing room, the cards were arranged in random order on the table in front of two chairs, so as not to create any bias or connection between cards from the get-go. When the participants entered the test was introduced and explained to them. Participants were told to sort the cards into natural and intuitive categories and give these categories names. Additionally, they were told that nothing they did would be considered wrong and that we were testing Arion and not them.
The testing was passive, meaning that we did not take an active role in the test.

The data was analysed using the method adopted by [15] in which we created an Excel document with the categories on the first row and the cards on the first column. The classes consisted of the most common categories created by users, as well as standard terms for similar categories that the participants had named differently; for example, two pairs of users had put same cards into groups called "Interaction" and "Help", so these were counted into one category called "Help". After that, we created a heat map of the data to see in which category cards most often appeared and then we grouped the most common occurrences together.

There were a number of submenu item cards that all three pairs of participants categorised differently, making it difficult to sort these cards into categories based on the user feedback. There could be several reasons for this.

Firstly, the number of participants for this test were very few and the number of data points even lower. According to Wood and Wood in [16] the general consensus is the more participants involved, the better. However, 25-30 participants are considered to be enough if these participants are representative of the group, as this is likely to yield the same results as several hundred participants.

Secondly, it is possible that the cards were named too ambiguously for the participants to be able to sort them into appropriate categories. This happened with the tags "Secondary education" and "Tertiary education", where some participants upon seeing the tag name associated this with signing up for different kinds of educational courses at these levels. These participants stated that they did not believe this was suitable for a university website. In actual fact, a tag name was related to the user's profile and their previous education. This was also utilised with the cards "Course by [...]" which initially just said "By faculty", "By specialisation", etc. It was deemed that these were too contextual for the top menu item they were in, and the participants expressed the same view after the test, where they said that they would have sorted differently if they knew. In any case, these results can also be used to improve the naming of some of the submenus.

Certain cards were consistently put into the same categories, and this can be seen in the table above. The tickets that were not sorted systematically because of the reasons stated above, should still have been sorted by us into appropriate categories. For example, the find papers "By [...]" classes should be in course information, even though our participants placed them all over the place.

Some cards were grouped into subcategories of higher-level architecture. For example, one pair of participants created a finances category which they put into an "Account" category. They thought that these finances should be connected to the user account in some way, but that they were also two distinct categories. It was also a common theme for many of the "Redundant"- category cards, that participant did not know where to put such a menu item. In this category, some participants placed the find papers "By [...]"-cards, the "Graduation application", "Downloadable forms" and more. These cards could be sub-sections of other cards instead of being classed separately, cluttering the information architecture and making the search less intuitive.

2.3. Situated Co-Inquiry Study

Situated Co-Inquiry is described by Carter in [17] as a participatory talk aloud approach at the centre of a user's or participant's experience in a way that is harmonious with their cognitive and affective consciousness at play at that moment.

To ensure the diversity of data collected, we used the survey data above to identify the ideal test users to make sure they are representative of one group of users. Also, the characteristics of different user groups are distributed as shown in table 2. Please note that we use pseudonyms for our testers here to protect their privacy.

Table 2: Participant's information.

ID	Age	Ethnicity	Years of use of Arion	Other systems they have used
User 1	25	European	2 months	Moodle, Lectio
User 2	27	European	2 months	Moodle
User 3	30	Chinese	1 year	N/a
User 4	23	Chinese	2 years	N/a
User 5	24	Indian	8 months	N/a
User 6	35	Indian	9 months	N/a

The researchers started the session by introducing themselves, and all participants were fully informed about the process and purpose of each session. There was no deceit or intimidation involved and participants could withdraw from the study at any time before their data were analysed.

As such, we encouraged the participants to talk-aloud while performing formulated user tasks. The participants would be reminded to do this throughout the test in case they forgot to do it. There were three user tasks as shown in Table 3.

Table 3: Testing tasks

Section	Task#	Task Description
Timetable	1	a. You have been enjoying your semester break, but now it is soon over. You need to know when you start having classes again. Find the first class after the semester break.
	2	b. To make sure you don't miss anything after having much time off, you want to know when all your September classes are. Show us all your courses in September.
Paper information	3	You are a postgraduate student looking for papers about writing journalism to enrol in Semester 2, 2019. Find an article that is relevant to you.

During the testing session, the facilitator would ask the participant to pause in order to clarify certain points. In relation to this, the facilitator and participant discussed various problems or communications during the situated co-inquiry in order to find common ground or even solutions to problems encountered.

The experiments session were placed at AUT campus was used so the participant will be familiar with the room. Two cameras were recording during the test to ensure that the test results and data were saved. One camera was used to record the user's expressions and gestures, and the other to document the process of operating the Arion interface, we also used Windows screen recording for backup. The audio recorder allows us to analyse the user's think-aloud responses repeatedly. By playing back the video, we were also able to record the time required for each user to complete individual steps during the test. From doing this we were also able to analyse how efficient the test was.

Each person participating in the trial was equipped with a logger that could observe the tester's behaviour and what was said. These findings were then recorded in the logbook. Moreover, at the end of the survey to express our gratitude to the participants, we provided them with each with a cookie.

All users completed the task, even though we observed many users with nervous expressions. There were also a number of complaints which we logged in the logbook.

Through the sessions, we identified several problems that we would like to address. For example during the test, we noticed that the users were mostly accustomed to browsing the screen from top to bottom. The navigation should always be at the top of the screen so that it is more accessible to the user. This is in line with what Williams et al, in [18] recommends. That is, the longer the distance and the smaller the size of the target, the longer the operation takes. As we see users complain that the font in the user interface is too small, we recommend increasing the size of the buttons and fonts in the navigation menu and positioning the navigation at the top. The time it takes for the user to move from the edge of the screen to the navigation menu is minimised.

The user is usually not guided during the process of using the system, and the user does not follow a set of procedures to learn and become familiar with the system. Conversely, a pleasant environment can be established by supporting exploratory interaction, and users will learn by themselves [19]. For example, we found that in using the graphical timetable, User 4 found that he/she could display consecutive days of courses by dragging and selecting.

What is surprising is that users with experience using other similar systems, spend more time in the first timetable task, and can quickly find a course on a specific date through text timetable.

The graphical timetable uses interactive data visualisation, which in theory should help users increase decision making speed [20]. Unfortunately, in our test results, it shows that most users do not tend to use a graphical timetable.

Interestingly, the user believes that the corresponding results can be displayed through some clickable content. However, if the clickable content cannot be clicked or does not show any changes, the user will no longer trust the design of the interface. For example, user 3 clicks on the "June" label multiple times and wants to display the entire courses through June, but the corresponding timetable has not changed. The ambiguity of a title or repetition of categories in a different heading tab of a menu can cause confusion for the user, for example, "course info" and "programme info". We see that when user 1 browses through the navigation and finds paper search, the user's misunderstanding of the title prompts the user to spend more time searching. When design deviates from standards and practices, users often take longer to think about or get auxiliary information from other factors. This will cause user emotionalisation to be very detrimental to the user experience [21]. This is a common problem in human-computer interaction. The best way to solve the challenge is to unify the standard so that the user only needs less time to react.

3. Discussion and Conclusion

The study aimed to improve the UI design for the Arion system with the help of three main usability testing methods to see how these could improve the user experience result.

The first method was the survey which helps to indemnify the issue, collect more data about end users in addition to more understanding the problems.

The second method was Open card sorting. The aim of using this was to try and improve the navigation structure of the website. Through the Open card sorting Study, we found that all three groups of users were confused about how labels are classified. Moreover, it was found that each of the user groups had different thoughts and opinions about the existing navigation menu of the Arion system. These results are in line with those of previous studies by Wang & Yen in [22] which found the web structure needs to be simple, clear and make the features easy to find.

The third method was the Situated Co-Inquiry which has been used to find the most critical problem in the pages, in addition, to help us to give the suggestion for the designers such as Appropriately add image information to the page, and the appropriate image can attract the user's attention Schmidt, Liu, & Sridharan [23], and can increase the aesthetics of the page [24].

This study has identified several issues with the Arion system as we mentioned in the previous sections. The purpose of our study is not only to point out the issues with Arion UI but to also provide some solutions. Therefore, the usability report was handed to the design team at AUT in order to improve the user experience for the Arion system.

From the analysis of the results, we discovered that although all users have completed the tasks we set, the less experienced users are less efficient in the way that they operate the system. The current interface is not conducive in allowing users to explore and self-learn, and it is a challenge for students to complete to use. Through the students' reaction, we can see that most people have various opinions on the system; that is, they are dissatisfied with the discoverability of the system architecture and interface. Users feel that there are too many options available in the Arion website and some of them are redundant. Overall, we concluded that Arion could be redesigned in order to create a system that is more efficient and user-friendly.

References

[1] El-Bakry H M *et al* 2010 Adaptive User Interface for Web Applications *Recent Advances in Business Administration: Proc. of the 4th WSEAS Int. Conf. on Business Administration (ICBA'10)* pp 20-2
[2] Galitz W O 2007 *The essential guide to user interface design: an introduction to GUI design principles and techniques* (New Jersey: John Wiley & Sons, Inc)
[3] Sharp H, Rogers Y and Preece J 2007 *Interaction design: beyond human-computer interaction* (Chichester: Wiley)
[4] Rubin J and Chisnell D 2008 *Handbook of usability testing: howto plan, design, and conduct effective tests* (New Jersey: John Wiley & Sons, Inc)
[5] ISO9241-11 1998 *Ergonomic requirements for office work with visual display terminals (VDTs) - Part 11 : Guidance on usability* Available: http://www.it.uu.se/edu/course/homepage/acsd/vt09/ISO9241part11.pdf
[6] Nielsen J 1993 *Usability engineering* (Amsterdam: Elsevier)
[7] Shneiderman B and Plaisant C 2005 *Designing the user interface* 4th edition (USA: Pearson Addison Wesley)
[8] Shackel B 2009 Usability–Context, framework, definition, design and evaluation *Interacting with Computers* vol **21** no 5-6 pp 339-46
[9] Battleson B, Booth A and Weintrop J 2001 Usability testing of an academic library web site: a case study *The J. of academic librarianship* vol **27** no 3 pp 188-98
[10] Kortum P and Sorber M 2015 Measuring the usability of mobile applications for phones and tablets *Int. J. of Human-Computer Interaction* vol **31** no 8 pp 518-29
[11] Silvis I M, Bothma T J and de Beer K J 2018 Evaluating the usability of the information architecture of academic library websites *Library Hi Tech*
[12] Sundt A and Eastman T 2019 Informing Website Navigation Design with Team-Based Card Sorting *J. of Web Librarianship* vol **13** no 1 pp 37-60
[13] Abowitz D A and Toole T M 2009 Mixed method research: Fundamental issues of design, validity, and reliability in construction research *J. of Construction Engineering and Management* vol **136** no 1 pp 108-16
[14] Chajut E, Caspi A, Chen R, Hod M and Ariely D 2014 In pain thou shalt bring forth children: The peak-and-end rule in recall of labor pain *Psychological science* vol **25** no 12 pp 2266-71
[15] Ma S 2010 *Dancing with the Cards: Quick-and-Dirty Analysis of Card-Sorting Data* Available: https://www.uxmatters.com/mt/archives/2010/09/dancing-with-the-cards-quick-and-dirty-analysis-of-card-sorting-data.php
[16] Wood J R and Wood L E 2008 Card sorting: current practices and beyond *J. of Usability Studies* vol **4** no 1 pp 1-6
[17] Carter P 2007 Liberating usability testing *Interactions* vol **14** no 2 pp 18-22
[18] Williams K, Clarke T, Gardiner S, Zimmerman J and Tomasic A 2019 Find and Seek: Assessing the Impact of Table Navigation on Information Look-up with a Screen Reader *ACM Transactions on Accessible Computing (TACCESS)* vol **12** no 3 p 11
[19] Sharp H, Preece J and Rogers Y 2019 *Interaction design: beyond human-computer interaction* (New Jersey: John Wiley & Sons, Inc)
[20] Few S 2013 Data visualization for human perception *The Encyclopedia of Human-Computer Interaction* 2nd Edition (Aarhus: Interaction design foundation)
[21] Wong E 2019 *Principle of Consistency and Standards in User Interface Design* Retrieved from https://www.interaction-design.org/literature/article/principle-of-consistency-and-standards-in-user-interface-design

[22] Wang M and Yen B 2007 Web structure reorganization to improve web navigation efficiency *PACIS 2007 Proceedings* p 46

[23] Schmidt K E, Liu Y and Sridharan S 2009 Webpage aesthetics, performance and usability: Design variables and their effects *Ergonomics* vol **52** no 6 pp 631-43

[24] Stone R W, Baker-Eveleth L J and Eveleth D M 2018 The Role of Aesthetics Influencing User Satisfaction with Recruitment Websites *Int. J. of Humanities and Social Sciences* vol **10** no 1 pp 76-92

IOP Conf. Series: Materials Science and Engineering **803** (2020) 012039 doi:10.1088/1757-899X/803/1/012039

Disaster Sentiment Analysis: Addressing the Challenges of Decision-Makers in Visualizing Netizen Tweets

R A Baro[1] and T D Palaoag[2]

[1]Don Mariano Marcos Memorial State University, Bacnotan, La Union 2515, Philippines
[2]University of the Cordilleras, Gov. Pack Road, Baguio City 2600, Philippines

E-mail: [1]rbaro@dmmmsu.edu.ph, [2]tpalaoag@gmail.com

Abstract. Among other crucial aspects of disaster-related initiatives is decision-making. The capacity to carry out efficient holistic management of measures and programs rely greatly upon the decision-makers and the religious participation of all stakeholders. Adopting the sentiments can help government leaders in their decision-making responsibilities towards disaster management when meaningful patterns are easily visualized. The purpose of this research is to analyze disaster-related sentiments from Twitter and presents the results using data visualization easily understood by decision-makers. The study underscores the use of Python programming with NLP techniques to learn from Twitter data, Sentiment Analysis using TextBlob to identify polarity and subjectivity of tweets and Plotly for interactive data visualization. The implementation of the study contributes to the reduction of injury, damage to infrastructures, properties and especially the loss of life. The challenges of both decision-makers and data scientists working in the discipline are highlighted in the study.

1. Introduction

The common focus of disaster risk management and climate change adaptation is associated with saving lives, reducing losses and supporting both individuals and communities to bounce back from disasters quickly and effectively [1]. The Humanitarian Assistance and Disaster Relief (HADR) National Mission Initiative (NMI) recognizes the importance of software solutions that uses artificial intelligence and machine learning in decision-making to formulate an effective disaster response strategy. At times of disasters caused by natural and anthropogenic hazards, people use social media platforms to share information that can potentially be useful for disaster response [2].

Emergency responses need to be swift and timely. Government leaders are usually blamed when the response to disasters is not met particularly when such events claim lives and damage to properties. United Nations Office for Disaster Risk Reduction (UNISDR) proposes the use and research of new scientific and technological advancements in disaster management [3]. This is where machine learning and data analytics on big data from social media can play a significant role.

Social media has sustained its growth over the last decade and presents tremendous breakthroughs for various ideas such as decision-making [4]. With the aid of social media, which can be accessed by computers or by smartphones, it can have huge potential to facilitate interaction, collaboration, and the sharing of content like disaster data [5].

Twitter became a great data source and an utmost interest for researchers and commercial organizations [6]. Because of the Big Data technologies, it has become very easy to collect millions of tweets and implement data analysis on the data [7].

There are also numerous studies on sentiment analysis using social media [8] [9] as data resource but providing near real-time graphical data presentation of sentiments for decision-makers is scarce – particularly and especially during disasters or emergencies. The goal of sentiment analysis is to have an automated process of understanding a review or opinion which leads to actionable results. One of the challenges of decision managers towards disaster responsiveness is acquiring timely and credible status reports during emergencies [10]. Disaster awareness is everybody's concern. Institutionalizing policies that use systems focused on Twitter as one of the communication channels during disasters can forge unity between government leaders and the citizens. Mining the sentiments on social media during disasters and visualizing these sentiments can empower the netizens and all the stakeholders on disaster management. While many scientists have dealt with numerous studies on the context of disaster management, a lot of challenges are yet to be addressed [11]. Over the years, studies are persistent in bridging the technical gap between data scientists and decision leaders [4]. The gap between data scientists and decision-makers can be bridged by presenting results of studies in a simpler way such as graphs, charts, and maps while other relevant data are also available.

This study utilizes Twitter API to stream data from the social media platform. The sentiment analysis will be programmed in Python using the TextBlob NLP library and Plotly for interactive data visualization.

This study aims to classify and visualize disaster-related sentiments providing decision-makers a near real-time online graphical presentation of data.

2. Related Literature

2.1. Sentiment Analysis

Sentiment analysis is the process of determining whether a piece of writing is positive, negative or neutral. A sentiment analysis system for text analysis combines natural language processing (NLP) and machine learning techniques to assign weighted sentiment scores to the entities, topics, themes, and categories within a sentence or phrase [12]. Sentiment analysis has gained numerous usage for the past years as internet usage and exchange of public opinion increases and becomes the driving force of this discipline [13].

2.2. Social Media

Social media refers to applications such as Facebook, WhatsApp, Twitter, YouTube, LinkedIn, Pinterest, and Instagram. Social media has become an integral part of people's lives. Studies show that government agencies and non-profit organizations use social media and that scholars are focused on communication and response in times of emergencies. Tweets from Twitter proved its usefulness in coordinating and planning the disaster relief operations serving as a back-channel communication system during natural disasters [14]. Analysis of user-generated media provides a deeper understanding of the extent to which media influences decision-making [15].

2.3. Data Visualization

Visualization techniques can help us efficiently analyze and understand how and why users interact on Twitter [16]. Visualization has a persuasive and seductive rhetorical force that transforms a complex idea into something concrete and easier to understand [17].

3. Methodology

Figure 1 is the system architecture of the proposed system. It is composed of four different stages: Data Streaming & Storage, Classification of Tweet Data, Analysis and Interpretation of Data, and Data Visualization.

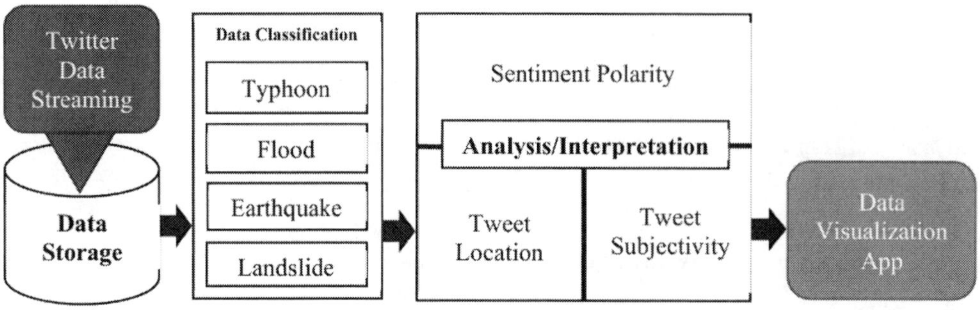

Figure 1. System architecture for the analysis and data visualization of disaster tweets.

3.1. Data streaming & storage
The system uses Twitter Streaming API which captures live tweets from the platform and parses them as string objects. These string objects are processed and stored in MySQL database for further manipulation. Stored data includes the tweet message, profile name, time-stamp of the tweet and location of the tweet based on the user's profile. Retweets are also captured and stored.

3.2. Data classification
Using a predefined set of weighed keywords, data tweets are classified into the four most common disasters in the Philippines: Typhoon, Flood, Earthquake, and Landslide. If a tweet matches forty percent of the listed keywords, a tweet will belong to that particular category of disaster. A tweet has a greater relevance factor if it has higher matching weighted keywords. The tweets are saved for analysis based on their disaster category including its relevance factor. All tweets which do not belong to the predefined categories are automatically removed.

Table 1. Sample Relevance Factor for Typhoon.

Tweet	Matched Keywords	% of Relevance
A moment of silence at International Stadium Yokohama as a symbol of the global rugby family's solidarity with Japan in the wake of Typhoon Hagibis, which claimed more than 70 lives.	Typhoon, solidarity, wake, claimed, lives	95%
Cranes got to work lifting up toppled poles of a golf range in Ichihara, Chiba Prefecture. The poles had been blown over during Typhoon Faxai last month, damaging 22 houses. Nobody was injured.	Typhoon, toppled, blown, damaging, houses, injured	100%
If this gets 400 tweets I'll share an interview with an IAF MiG-21 pilot, including an account of him getting the better of an RAF Typhoon in a training	Typhoon, interview	42%

3.3. Analysis and interpretation of data
The third stage of the system architecture is focused on the analysis and interpretation of the classified tweet messages. It identifies the polarity of the sentiment, tweet location, and tweet subjectivity. Thru the NLP library, the polarity and subjectivity of the sentiment are determined. One of the salient points

in disaster sentiment analysis on tweets is identifying the location of the emergency. Geo-tagging of tweets based on the profile of the user will be applied to track the origin of the tweet since not all tweet message specifies the location of the sentiment. It is understood, however, that the location based on the user profile may not be true.

3.4. Data visualization

The primary contribution of this study is the Data Visualization App. Data analysis results are presented using graphs and charts in an online web dashboard where decision-makers have a glimpse of the status of a disaster based on netizen sentiments via Twitter. A time interval of data refresh can be set in the system.

Guided by the above system architecture, the extraction and transformation of tweets into near real-time online data visualization relevant to decision-making in times of disasters are illustrated in figure 2.

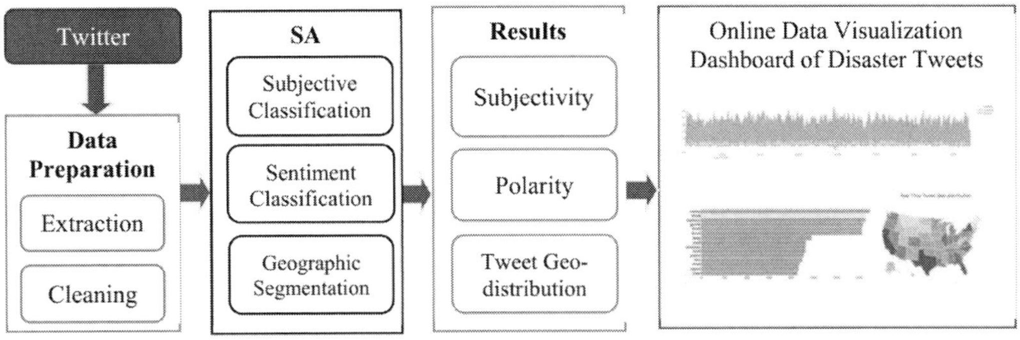

Figure 2. Proposed online sentiments data visualization framework.

Data are extracted from the Twitter microblogging platform using a Python library called Tweepy. Cleaning the data includes the removal of emoji characters by stripping all non-ASCII characters, removing URLs, 'RT' and '&' symbols. Conversion to lowercases of all the cleaned tweet texts will be applied. Using a tokenizer, the text of tweets in the dataset is tokenized to separate text into a list of sentences. The scope of the sentiment analysis will work on sentence-level and non-English words are removed. Natural Language Processing using Regular Expression (RE) & Natural Language Toolkit (NLTK) are utilized to track the most frequent keywords. Commonly used words are removed by Stop Words and 10 most common words are extracted from the frequency distribution of all words. To identify the polarity and subjectivity of the sentiments, TextBlob (NLP library) is employed. The polarities of tweets are a) Positive (1), b) Neutral (0), and c) Negative (-1). Geographic segmentation recognition with text processing is used to identify the location of tweets. The geolocation of sentiments is based on the user profile's location. The result of the tokenization can be presented in a bar chart and bubble chart of commonly used words. Sentiment analysis of the polarity of tweets can be visualized through a line chart. And the location of tweets can be viewed on a map using an interactive graphing library.

4. Findings

At the onset of disasters and emergencies, national and local governments are under the law tasked and expected to respond and rescue. At the onslaught of an emergency, government leaders need to know the crisis/disaster and location of the situation, severity and geographical measure of the impact and which sectors (infrastructure, economic, environmental or social) are affected. Coordination relies primarily on traditional communications and emergency reports usually come after lives had been claimed and damages are done. While datasets of most studies were derived from actual disaster data, current emergencies require live information and actionable reports for decision-makers. Further, studies show that researches are not focused on decision-making using social media. The technical gap also arises between data scientists and decision-makers and results of studies may have not reached entities benefiting from such studies.

The study emphasizes that a rule-based sentiment analysis algorithm using NLP techniques has useful applications in gaining the best out of massive data from Twitter to learn themes and topics relating to disasters and emergency crises. The Tweepy Python library is used to capture live streaming Twitter data. The filter method used Twitter's free Standard Stream which allows 400 keywords and 5000 userids. Using disaster tracking words, the system captures live tweets and stored these sample data in a MySQL database file. This dataset reveals topics, themes and the overall sentiments of the netizens which are useful for decision-making using TextBlob sentiment analysis tool and Plotly's interactive data visualization.

5. Conclusions and recommendations

The results of the study reveal that the challenges of government leaders in disaster management remain an issue that requires sustained attention. Likewise, data scientists need to be persistent in trying to focus researches on disaster management tools and applications particularly utilizing social media. And that results of studies should reach the hands of decision-makers and be in a form understandable by non-technical viewers of data. The main problem of government leaders in addressing crises or emergencies lies in the availability of actionable reports during such situations. To address this problem, a development framework based on IT solutions with the aid of Twitter, Python programming equipped with NLP techniques, Sentiment Analysis, and an interactive, web-based application, data visualization tool is proposed. It is highly recommended that the proposed system be developed and applied to address this problem.

6. Future works

One of the main purposes of this study is to provide an IT model as a solution to the main problem stated above. An application to use Twitter API for further testing and analysis is under review. While the system is under development, the framework presents a new technique in addressing disaster management. Addendum to the results of the sentiment classifications and using other NLP techniques may be done by other researchers to more fully address the problem. As a result, the devastation effect of disaster occurrences may be reduced.

References

[1] IFRC 2016 *World Disaster Report 2016 Resilience: saving lives today, investing for tomorrow* Retrieved from https://www.preventionweb.net/publications/view/50615

[2] Imran M, Castillo C, Diaz F and Vieweg S 2015 Processing social media messages in mass emergency: A survey *ACM Computing Surveys* vol **47** no 4

[3] United Nations Office for Disaster Risk Reduction 2015 *Global Assessment Report on Disaster Risk Reduction*

[4] Gilani E, Salimi D, Jouyandeh M, Tavasoli K and Wong W 2019 A trend study on the impact of social media in decision making *Int. J. of Data and Network Science* vol **3** pp 201-22

[5] Regalado R V, Dela Cruz K M H, Garcia J P, Kalaw K M D and Lu V 2015 FILIET: An information extraction system for Filipino disaster-related tweets *DLSU Research Congress 2015* vol **3**

[6] Flekova L, Ruppert E and Preotiuc-Pietro D 2015 Analysing domain suitability of a sentiment lexicon by identifying distributionally bipolar words *Proc. of the 6th Workshop on Computational Approaches to Subjectivity, Sentiment and Social Media Analysis (WASSA 2015)* pp 77-84

[7] Wan Y and Gao Q 2015 An ensemble sentiment classification system of Twitter data for airline services analysis *2015 IEEE 15th Int. Conf. on Data Mining Workshops* pp 1318-25

[8] Morente-Molinera, J A, Kou G, Samuylov K, Ureña R and Herrera-Viedma E 2018 Carrying out consensual group decision making processes under social networks using sentiment analysis over comparative expressions *Knowledge-Based Systems* vol **165** pp 335-45
[9] Peacock D C and Khan H U 2019 Effectiveness of social media sentiment analysis tools with the support of emoticon/emoji *Advances in Intelligent Systems and Computing* vol **800** pp 491–4
[10] Seifert L, Kunz N and Gold S 2018 Humanitarian supply chain management responding to refugees: a literature review *J. of Humanitarian Logistics and Supply Chain Management* vol **8** no 3 pp 398-426
[11] Wang Z and Ye X 2018 Social media analytics for natural disaster management *Int. J. of Geographical Information Science* vol **32** no 1 pp 49-72
[12] Lexalytics 2019 *What is Sentiment Analysis?* Retrieved September 25th 2019 from https://www.lexalytics.com/technology/sentiment-analysis
[13] Fang X and Zhan J 2015 Sentiment analysis using product review data *J. of Big Data* vol **2** no 5
[14] Kapoor K K, Tamilmani K, Rana N P, Patil P, Dwivedi K and Nerur S 2018 Advances in social media research: past, present and future *Information Systems Frontier* vol **20** no 3 pp 531-58
[15] Greenwood B and Gopal A 2015 Tigerblood: Newspapers, blogs, and the founding of information technology firms *Information Systems Research* vol **26** no 4 pp 812-28
[16] Kumar S, Morstatter F and Liu H 2014 *Twitter data analytics* (New York: Springer-Verlag)
[17] Gold M and Klein L 2019 *Debates in the Digital Humanities* (Minnesota: University of Minnesota Press)

Design and Implementation Network Administrators Account Management System Based on Authentication, Authorization, and Accounting Based on TACACS and LDAP

Ardelia Puri Paramitha, Adian F. Rochim, Adnan Fauzi

Departement of Computer Engineering, Faculty of Engineering
Diponegoro University, Tembalang, Semarang, 50275, Indonesia

Email: apparamitha@student.ce.undip.ac.id, adian@ce.undip.ac.id, adnan@live.undip.ac.id

Abstract. Authentication is required for to authenticate user administrators for accessing network devices. In large-scale network devices management, firstly administrators need centralized user management to modify their account for logging network devices. Secondly, they must have log monitoring data who modified their account. The goal of this research is to develop and implement Authorization, Authentication, Accounting (AAA) system and making Log Monitoring System (LMS). Objectives of this research are implementing a system that can centralize and manage network administrators account. Design and topology of the network use simple star network. Software design uses Rapid Application Development (RAD) method. Research began with analyzing the research problem, next step is creating a system design, then build the system. After the system has been built, a demo is done whether the system is running well. If the system has not functioning as desired yet, refine the system. If already functioning as desired, a testing is done. If there are no obstacles during testing, the system is ready to be implemented by users who will use the system. The result shown that system is able to centralized authentication and authorization using Terminal Access Controller Access-Control System (TACACS), and processing accounting data into information in graphical form using ELK Stack. The result of the research is developing monitor user login and log review. The test result shown that system can manage network administrators account according to the Authentication, Authorization, Accounting (AAA) principle.

Keyword: Authentication; Authorization; Accounting; TACACS

1. Introduction

Information technology advancement nowadays is very rapid along with the increasingly high information needs. Many technologies produced both hardware and software that can simplify human work. The use of information technology cannot be separated from computer for data management. Computers cannot be separated from the role of computer networks [1].

In managing computer networks, there is the role of the network administrators who are responsible for designing and configuring network devices. Authentication is needed to validate the administrators account to access network devices. In large-scale network devices management, administrators will find it difficult to do repeated authentication with different account. Judging from the approach, a solution is needed to manage network devices only with one account. Single Sign On (SSO) which can simplify the number of accounts and support the AAA protocol (Authentication, Authorization, Accounting) [2].

The family of AAA protocols, which stands for Authentication, Authorization and Accounting, were originally designed as remote access control mechanisms and network service providers through modem and dial-in services, but they continue to be presently implemented in multiple architectures [3].

Terminal Access Controller Access-Control System as known as TACACS is a remote authentication protocol that separates out the authorization functionality, so it enables additional flexibility and granular access controls on who can run which commands on specified devices [4]. These protocols works on client server logic and uses messages/ packets to be exchanged between them to enable AAA service [5].

A. Riyanto et al., in 2014, describe the implementation of SSO using LDAP for information system and hotspot access at Universitas Muhammadiyah Surakarta Campus [6]. The implementation used for hotspot access authentication. The results of their research are to ease students to access information system using one account only. SSO also make it easy for admin to manage user data who is accessing the system. With using LDAP, every data will be centralized in LDAP server.

Sikawar et al., in 2017, describe the study of AAA model with special references to TACACS+ and RADIUS [7]. This research aims to describe the AAA system comparison between TACACS+ and Radius. The result of their research are both the protocols help to deploy the AAA model of security. For deploying the secure and effective security, the TACACS+ is much effective then the RADIUS protocol. Both the protocols are effective and acceptable for different applications and their utilities.

Through this research, the author intends to implement network administrators account management system based on TACACS to centralize network devices authentication in order to facilitate network administrators to manage network devices.

First section describes the background of the research. Second section describe the research methodology. Next section describes the result of research. And finally, the last section explains the conclusions of the research.

2. Research Methodology

Rapid Application Development (RAD) methodology, are used to design the implementation of the system. RAD methodology process consists of analysis and quick design, build, demonstrate, and refine system, testing, and implementation [8].

The first step is analyzing system needs and system design. This step analyzes the user needs and system that will be built. The next step known as prototype cycles. The beginning of the cycle is building the system by determining the specification of the requirements that match the system definition. Furthermore, refines are made to the system in accordance with the design. The third step is system testing. Testing is done by testing the results of authentication and authorization in the accounting. The final step is implementation. If there is no obstacle in the testing phase, the system is ready to be implemented.

2.1. Functional Requirements

Functional requirements of the network administrators account management system are the system can authenticate users account that access network devices. The system can authorize users account to grant access rights according to the privilege level based on the group. The last functional requirement is the system can visualize system logs and users account management into dashboard that is easy to use by network administrators.

2.2. Hardware Requirements

Hardware is used as a system support so the system can run. The system is implemented in a computer. The computer used has the specifications in Table 1.

One Cisco Catalyst WS-C2960-24LC-S switch and 3 Cisco 1841 routers as sample network devices that were tested in the research. One Asus A442U laptop to create the application.

Table 1. Computer Specification

Component	Server
Processor	Intel® Core™ i3-5005U CPU @ 2.00GHz
RAM	8 GB
HDD	1 TB
NIC	1 Fast Ethernet

2.3. Software Requirements

The network administrators account management system requires software to build servers and other supporting components. Network administrators account management server use Windows Server 2019 as the operating system. The application for user account management uses Active Directory. Centralized authentication and authorize accounts using TACACS. Accounting visualization using Kibana with data that collected by Logstash and saved in Elasticsearch [9].

Network devices configuration is done remotely using MobaXterm application. Google Chrome application as a web browser that runs on the Windows 10 x64 operating system and used to test the running of the application.

2.4. Topology Design

The topology design of the network administrators account management system consists of various entities involved in system development. The entity is connected using a star topology with a switch as the center [10]. Figure 1 shows the physical topology of the log management system.

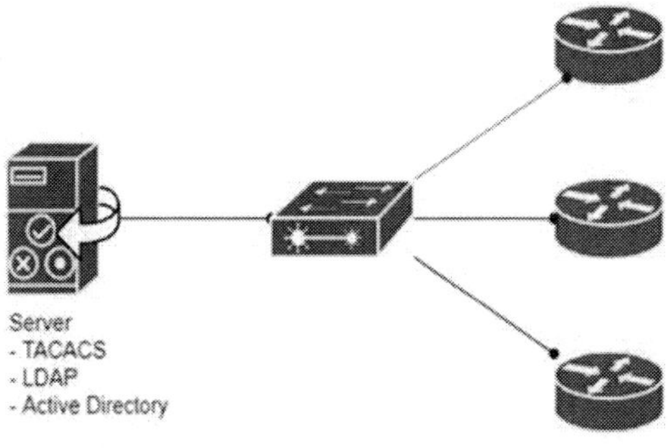

Figure 1. Design and Network Topology System

The server where the system is located will be connected to other network devices and connected to the switch. All devices connected to the server will be authenticated on the TACACS server.

2.5. System Design

System design provides an overview or part of a system. Figure 2 shows the server system design.

2.6. Work Process Design

The design is done on the TACACS configuration including Authentication, Authorization, Clients, and Tacplus. Authentication contains account details that are used to access network devices stored in LDAP. The account will be stored in the Users container and is a member of the group according to the access rights. Authorization contains the distribution of privilege level and commands that can be accessed in each group. Clients contain the IP address of the devices that connected to the server. Tacplus contains the configuration of the server including the IP address and the port used, also contains settings from the TACACS logs.

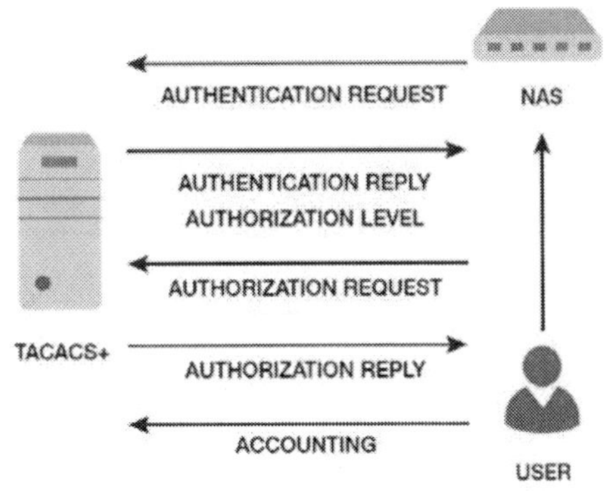

Figure 2. Server System Design [4]

2.7. AAA System Flow Process

The flow chart diagram explains the process of authentication and authorization. Information from both processes will be stored in the accounting files and the information is used to monitor the account who is logging in to the network devices. The work process of the system is shown in Figure 3.

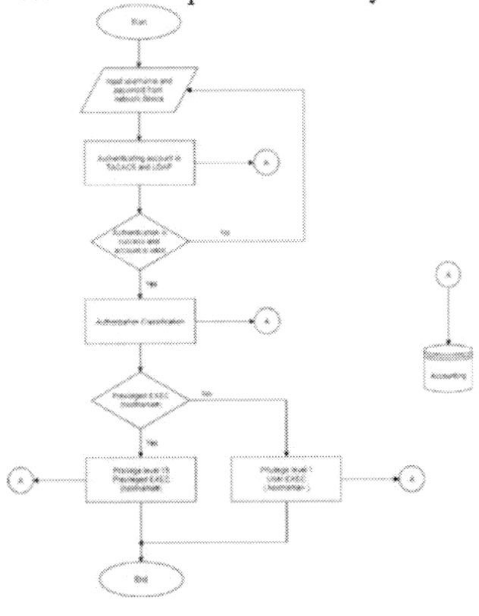

Figure 3. AAA System Flowchart

3. Result and Discussion

System evaluation is done to check the performance of the system being implemented. The main purpose of system evaluation is to ensure that the components of the system are functioning as needed.

3.1. Authentication Evaluation

Based on functional requirements, the network administrators account management must be able to authenticate user account who is accessing network devices. The intended authentication is verifying username and password which is entered by the user when accessing network devices based on Active Directory database.

IOP Conf. Series: Materials Science and Engineering **803** (2020) 012040 doi:10.1088/1757-899X/803/1/012040

Testing is done by logging in to network devices using user admin and operator. The data log results example from authentication testing user admin on router 1 shown below.

Table 2. admin1 Authentication Result

No	Code	Information
1	`<87>`	ID
2	`9/23/2019`	Access date
3	`11:39:34`	Access time
4	`[10.42.12.185:14224]`	IP destination and port used
5	`Authentication for user admin1 passed against group admin_tacacs - Passed`	Information of authentication result for admin1 against group admin_tacacs is success.

The data log results example from authentication testing user operator on router 1 shown below.

Table 3. op1 Authentication Result

No	Code	Information
1	`<87>`	ID
2	`9/23/2019`	Access date
3	`11:43:37`	Access time
4	`[10.42.12.185:36929]`	IP destination and port used
5	`Authentication for user op1 passed against group op_tacacs - Passed`	Information of authentication result for op1 against group op_tacacs is success

Based on authentication data log results above, shown that authentication process is success.

3.2. Authorization Evaluation

Based on the functional requirements of the network administrator account management system is providing authorization to user in accordance with privilege level. In authorization, user who are members of security group admin_tacacs get access rights as administrator which have privilege level 15. Privilege level 15 is the highest privilege to run all commands on network devices. For user who are members of security group op_tacacs get access rights as operator which only have privilege level 1. Privilege level 1 is the lowest privilege and only can run limited commands.

Testing is done by accessing the network devices using user admin and operator. The data log results example from authorization testing user admin on router 1 shown below.

Table 4. admin1 Authorization Result

No	Code	Information
1	`<87>`	ID
2	`9/23/2019`	Access date
3	`11:39:34`	Access time
4	`[10.42.12.185:14328]`	IP destination and port used
5	`User admin1 belong to group admin_tacacs - from cache`	User admin1 is member of group admin_tacacs.
6	`priv-lvl=15`	Have privilege level 15

The data log results example from authorization testing user operator on router 1 shown below.

Table 5. op1 Authorization Result

No	Code	Information
1	<87>	ID
2	9/23/2019	Access date
3	11:39:34	Access time
4	[10.42.12.185:14328]	IP destination and port used
5	User op1 belong to group op_tacacs - from cache	User op1 is member of group op_tacacs.
6	priv-lvl=1	Have privilege level 1

Based on authorization data log results above, shown that authentication process is success.

3.3. Accounting Evaluation

The functional requirements of the network administrator account management system are visualizing TACACS log. The log is stored in TACACS Accounting files. The accounting files content are shown below.

Table 6. Accounting Files Content

No	Code	Information
1	<102>	ID
2	2019-09-23	Access date
3	13:05:31	Access time
4	[10.42.12.187:51565]	IP destination and port used
5	NAS_IP=10.42.12.187	IP destination
6	User=op2	User that accessing the network devices
7	Flags=TAC_PLUS_ACCT_FLAG_START	Log Message

Table 6 shows the contents of accounting files which can be managed by using Elasticsearch also identifying logs which contain AA (Authentication, Authorization) special message. Testing is done starting from September 2nd, 2019 until September 9th, 2019. The testing generates 24.557 log which was collected in server. Processed log is not only from accounting files, it also contains debug and system files. Testing is done by observing Discover page on Kibana as user interface. If there is log detected, the data will show on Kibana's Discover page. Figure 4 shows the results form log receiving.

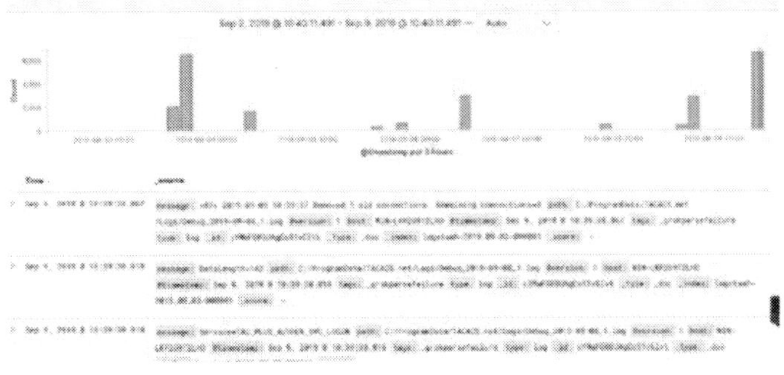

Figure 4. Log Receiving Test Results

3.4. Application Evaluation

Application evaluation explains the results and appearance of the application that has been created. The initial appearance of the application is a login page that serves to authenticate users who will access the system.

After authenticated, the user will be directed to the main page that contains the main dashboard. The main dashboard displays general information about numbers of registered admin, user, admin tacacs, and user tacacs. There is also information about user lists and graphical form of which port that users use to access the network devices. There are several navigation menus available in the application, those are Monitoring, Account, TACACS Account, and System Configuration. The main page view is shown in Figure 5.

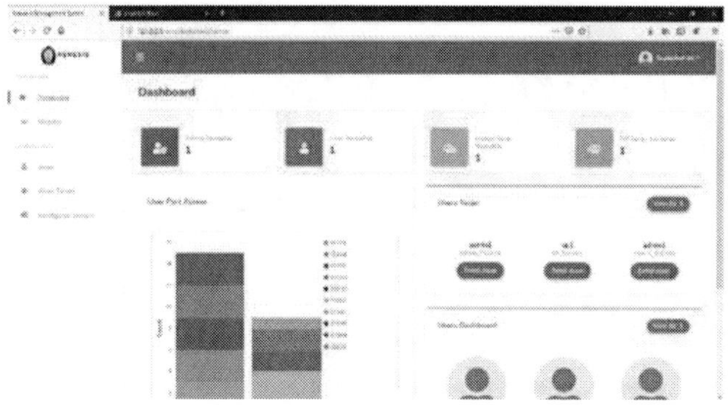

Figure 5. Application Main Page

3.5. Discussion

The result of the research is developing TACACS + LDAP and an application to monitor user login. Secondly, other feature of the application provides early log review.

4. Conclusion

Topology that applied in this research is star topology with Cisco Catalyst WS-C2960-24LC-S Switch as hub. TACACS and server connected using TCP protocols on port number 49. Based on system and application evaluation all functionality run well. Username and password that are not registered either in TACACS or LDAP cannot access the network devices. User operator cannot run configuration commands on network devices because granted privilege level 1. All user's activity when accessing the network devices are recorded and collected as log on TACACS accounting.

User account modification can be logged by LMS system. Priority of the modification filtered by admin to show a password changed, creating user, and failed account logging.

In the future research suggest to build the LMS System to send notification message for admin to notify about the modification account. Suggestions for further research can be developing a function to manage registered network devices on TACACS. Further research can develop the mobile version to ease admin and operator doing the network management efficiently.

Acknowledgement

This research was financially supported by The Faculty of Engineering, Diponegoro University, Semarang, Indonesia through Strategic Research Grant 2019 number: 3161/3/ UNT7.3.3/PG/2019.

References

[1] Mustafa, Hamzah A and Rachmawati Y 2018 Rancangan Infrastruktur Jaringan Backbone Hybrid Di Tiga Kampus IST AKPRIND Yogyakarta *J. JARKOM* vol **5** no 2 pp 122–9

[2] Kerta J M, Adiprabowo P and Kusmiyati E 2011 Penggunaan Single Sign On (SSO) Pada

Jaringan Internet Badan Pengkajian Dan Penerapan Teknologi (BPPT) *ComTech* vol **2** no 2 pp 880–6

[3] López A 2015 *AAA protocols and network access control: Radius* [Online] Available: https://www.incibe-cert.es/en/blog/aaa-radius [Accessed: 10-Feb-2019]

[4] TACACS 2011 *The Advantages of TACACS +* [Online] Available: http://www.tacacs.net/docs/TACACS_Advantages.pdf [Accessed: 19-Mar-2019]

[5] Ravi V, Sunitha N R, Pradeep R and S Verma 2017 Formal methods to verify authentication in TACACS + protocol *2017 2nd Int. Conf. Emerg. Comput. Inf. Technol.* pp 1–4

[6] Riyanto A and Ulinuha A 2014 *Design And Implementation Of SSO (Single Sign On) Using LDAP Authentication For Information System And Hotspot Access At Pesma KH Mas Mansur UMS* thesis (Solo: Universitas Muhammadiyah Surakarta)

[7] Pratap A and Saxena P 2017 An Analytical and Experimental Study of AAA Model with Special Reference to RADIUS and TACACS+ *Int. J. Comput. Appl.* vol **169** no 9 pp 6–10

[8] Rukmana A and Desiyani I D 2017 *Rapid Application Development* pp 2–4

[9] Rochim A F, Fauzi A and Aziz M A 2019 Design Log Management System of Computer Network Devices Infrastructures Based on ELK Stack *ICECOS 2019*

[10] Shekhar A 2016 *What Is Star Topology? Advantages And Disadvantages Of A Star Topology* [Online] Available: https://fossbytes.com/star-topology-advantages-disadvantages-star-topology/ [Accessed: 10-Feb-2019]

IOP Conf. Series: Materials Science and Engineering **803** (2020) 012041 doi:10.1088/1757-899X/803/1/012041

Implementing an ETL-driven Data Integration Framework for State Universities and Colleges

Lorna T. Soriano[1] and Josephine S. Dela Cruz[2]

[1]Bicol State College of Applied Sciences and Technology, Naga City 4400 Philippines,
[2]University of the Cordilleras, Baguio City Philippines

E-mail: [1]ltsoriano@biscast.edu.ph, [2]delacruzpen@gmail.com

Abstract. State Universities and Colleges collect enormous volumes of data scattered across multiple disparate transactional systems. Integration of these data is therefore vital in order to eliminate problems in consolidating and interpreting such data. This paper aimed to develop and implement a data integration framework that employs an ETL-driven data warehousing technique for State Universities and Colleges (SUCs). Moreover, the framework was implemented and tested using heterogeneous data of the Bicol State College of Applied Sciences and Technology (BISCAST), one of the state colleges in Region 5, Philippines. Consequently, the integration framework has been found to be effective in streamlining the process of consolidating and interpreting SUC's data thus enabling the organization to produce reliable reports and eventually can cope with submission deadlines. A reporting tool was also developed for end-users to generate and view the outcome of the integration process. Furthermore, the level of the system's usability was determined. Based on the result, it is effective in accomplishing specific tasks, highly efficient and commendable to adopt.

1. Introduction

Data integration is the process of combining data from different sources and bringing them together to ultimately provide a unified view [1]. Government agencies nowadays are often driven by data. This is evident in the Open Data Philippines (OPD) Program of the government which mandates that government decision-making should be based on available and sound data. Consequently, government institutions, such as State Universities and Colleges (SUC), are tasked to prepare and submit accomplishment and performance reports based on the agreed targets and indicators set by the different agencies. These required reports are actually enormous volumes of data and information from disparate sources of SUCs that need to be integrated to obtain a single comprehensible view of data.

The Bicol State College of Applied Sciences and Technology (BISCAST), one of the SUCs located in Southern Luzon, Philippines, maintains multiple transactional systems with independent databases due to the varying tasks performed by different offices. The transactional systems were designed in different platforms and in diverse database applications without considering integration in the future. In addition, flat files such as electronic spreadsheets are also utilized as record-keeping tools. This scenario has created data silos within the institution. It further led to difficulties in generating accurate reports and problems in consolidating and interpreting data correctly.

The ever-increasing data silos in SUCs have brought problems to their data management particularly in the requirement to report across multiple information systems. This paper aimed to design and implement a data integration framework that employs data warehousing technique using an ETL-driven approach for extracting and integrating data from multiple sources of SUCs. It was implemented and tested in BISCAST specifically in preparing and generating mandatory reportorial requirements where data come from disparate sources. Moreover, the level of the system's usability was also determined.

2. Literature Review

Data warehouse is a set of materialized view across the various operational information sources of an organization designed to provide support for decision making and data analysis [2]. According to [3], data warehouse development is an effective method to extract significant information from disparate source systems into a centralized storage and provide data history. It is further explained by [4] that data warehouses differ from operational databases in the sense that they are integrated, subject-oriented, summarized, time variant, nonvolatile, not normalized, and can perform online analytical processing (OLAP).

In building a data warehouse, one must run an ETL (Extract-Transform-Load) tool to perform the following tasks: extraction of data from the sources, transformation and conformation of data so that separate sources can be used together, and finally loading them into the warehouse. ETL system is the groundwork of the data warehouse. It specifically eliminates mistakes and corrects missing data; offers documented data confidence measures; captures the flow of operational data for custodial purposes; conforms data from multiple sources for collective use; and structures data to make it available to end-user tools [5].

Recent studies related to data integration were noted. The work of [3] investigated on data warehouse that handle big data in educational system. In addition, the authors presented a framework that utilizes Hadoop as data analytic tool for data ingestion. Another is the study of [6] which investigates on the current implementation methods used to solve problems of data integration in multi-sources information systems as well as the current trends such as data warehousing, descriptive logic, and ontologies. A combination of ETL technology and a wrapper layer which allows for the integration of any data sources was described in the work of [7]. Moreover, the study of [8] described the importance of data integration and ETL in business intelligence framework as well as its significance in higher education management. Finally, the paper of [9] discussed the implementation of ETL processes using an open-source tool in integrating heterogeneous databases in academic environment.

3. The Integration Framework

The integration framework was presented in the previous paper and is based on the previous works involving data warehouse. It consists of data sources, ETL processes, data warehouse, and end user layer as shown in Figure 1.

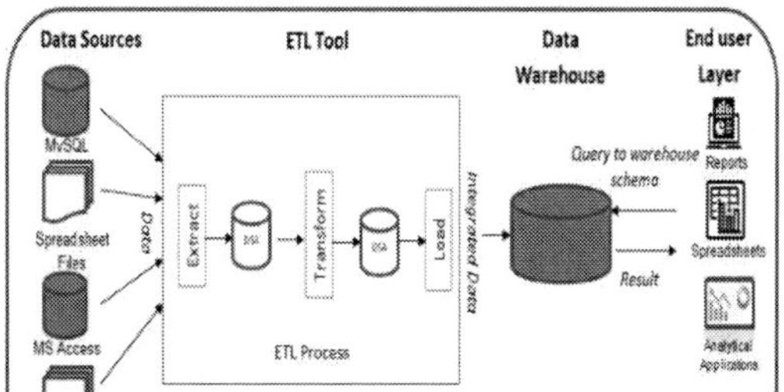

Figure 1. Components of the data integration framework

Data can come from various sources such as relational databases, XML files, flat files, and the like. These heterogeneous sources have unique set of characteristics that should be administered so as to effectively retrieve data for the ETL process. Extract-Transform-Load (ETL) is a data integration process that extracts data from various sources into the staging area where they are transformed and conformed before being loaded into the data warehouse. To perform such process, an open-source tool which provides flexible management can be used. When extracting data from disparate sources, one should be aware of the following: the drivers that are used to connect to database sources; its data structures; and handling sources with different nature. After identified data were extracted, the next step is to change the form of the extracted data to fit into the structure of the target data warehouse. Transformation such as validation, cleansing, decoding, renaming, aggregation, and key generation and management are some of the activities performed into the data. Lastly, the transformed data are stored in the data warehouse to be accessed by the end-users and applications at the end user layer. Various tools can be used by the end user for working with the data such as reports, dashboards, spreadsheets, and published documents.

4. Implementation and Evaluation

The integration framework has been implemented in the Bicol State College of Applied Sciences and Technology (BISCAST). Data sources of BISCAST are often heterogeneous as shown in Table 1. This is due to autonomous acquisition of software by the different units within the institution. It consists of transactional systems designed for specific functions that were purchased, developed in-house and donated. Different database applications were utilized such as MySQL and MariaDB. Spreadsheet file is also used to store and manage research and extension information and the like.

Table 1. List of BISCAST data sources.

Name of Information System / File	Development Strategy	Computing Scheme	Owner/User	Name of Database/file	DBMS
Student Information and Accounting System (SIAS)	Outsourced	Client-server	Registrar, Colleges, Accounting, Cashiering	sias_db	MariaDB
Class Scheduling System	In-house	Client-server	Colleges	scheduler_db	MySQL
Procurement Management System (PMS)	In-house	Client-server	All units	biscast_db	MySQL
Libro System (LS)	Donated	Client-server	Library	rdadb	MySQL
Research Information File (RIF)	In-house	Stand-alone	Research unit	research_file	MS Excel
Extension Information File (EIF)	In-house	Stand-alone	Extension unit	extension_file	MS Excel
Human Resource Information System (HRIS)	In-house	Stand-alone	Human Resource Management Office	bhrms_db	MySQL

Table 2. BAR 1 report contents.

Document Code	Particulars	BISCAST Data Source
BAR 1	% of first- time licensure exam- takers that pass the licensure exams	sias_db, spreadsheet
	% of graduates (2 yrs prior) that are employed	sias_db, tracer_file
	% of undergrad stud population enrolled in CHED-identified priority programs	sias_db, spreadsheet
	% of graduate school faculty pursuing advance research degree	bhrms_db, sias_db, scheduler_db
	% of grad stud population enrolled in CHED-identified priority programs	sias_db, spreadsheet
	Number of research outputs in the last 3 yrs utilized by the industry or other beneficiaries	research_file
	Number of research completed within the year	research_file
	Number of trainees weighted by the length of training	Spreadsheet
	Number of extension programs organized and supported consistent with the SUCs mandated and priority programs	Extension_file

The State Universities and Colleges are required to submit mandatory reportorial requirements by the different government agencies with specific data and format. One of which is the FY Quarterly Physical Report of Operation. It is commonly called Budget Accountability Report (BAR) required by the Department of Budget and Management (DBM). The data for this report are often found in different data sources of BISCAST. There are also instances that a particular set of data has several versions in multiple sources. Table 2 shows the content of the FY Quarterly Physical Report of Operation. It is important to note that BISCAST sources of the report data varied.

Mapping the source systems to the target data warehouse is useful in identifying the data to be extracted as well as in the whole integration process. Figures 2 and 3 show the data flow at the table level from BISCAST data sources to the dimension and fact tables of the warehouse.

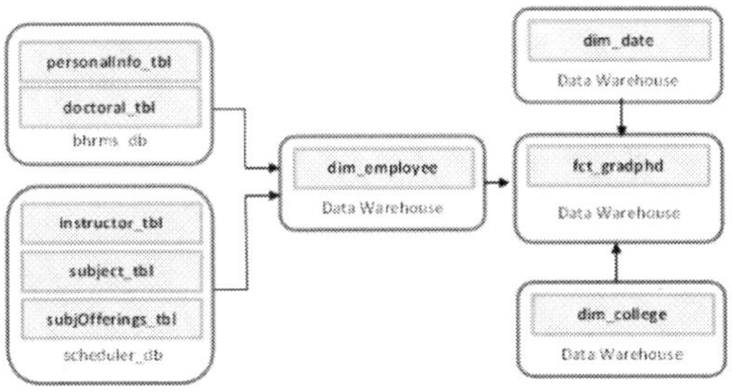

Figure 2. Mapping of bhrms and scheduler databases with target data warehouse.

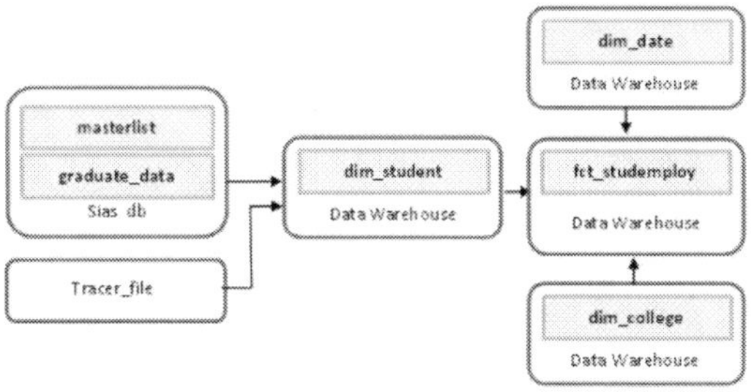

Figure 3. Mapping of sias database and tracer file with target data warehouse.

Table 3. An excerpt of the logical data map.

TARGET				SOURCE			TRANSFORMATION
Table Name	Column Name	Table Type	SCD Type	Dbase Name	Table Name	Column Name	
DIM_STUDENT	STUD_KEY	Dim					Surrogate Key
DIM_STUDENT	STUD_NO	Dim	1	SIAS	MASTERLIST	STUD_NO	Natural key
DIM_STUDENT	STUDNAME	Dim	2	SIAS	MASTERLIST	STUDNAME	initcap(studentname)
DIM_STUDENT	COLLEGEID	Dim	2	SIAS	MASTERLIST	COLLEGEID	initcap(collegeID)
DIM_STUDENT	IS_GRAD	Dim	2	GRADS		IS_GRAD	if exist 1 else 0
DIM_STUDENT	DATEGRAD	Dim	2	GRADS		DATEGRAD	trunc(date_grad)
DIM_STUDENT	IS_EMPLOY	Dim	2	TRACER		STATUS	if exist 1 else 0
DIM_STUDENT	D_APPOINT	Dim	2	TRACER		D_APPOINT	trunc(date_appoint)
FCT_EMPLOYD	EMPL_KEY	Fact	N/A				Surrogate Key
FCT_EMPLOYD	COLL_KEY	Fact	N/A	DW	DIM_COLLEGE	COLL_KEY	dim_college SK
FCT_EMPLOYD	DATE_KEY	Fact	N/A	DW	DIM_DATE	DATE_KEY	dim_date SK
FCT_EMPLOYD	Y_EMPLOY	Fact	N/A	DW, GRADS, TRACER	DIM_STUDENT		year(date_appoint) - year(date_grad)

In addition, a logical data map has been drafted and an excerpt of it is depicted in Table 3. It contains the data definition for the source systems, the target data warehouse data model, and the specific data manipulation required to transform it from its original format to that of its terminal destination. This will be used to map out the physical implementation of the ETL system. It is notable that dimension table DIM_student is composed of three (3) data sources namely *sias_db*, *grads_file*, and *tracer_file*.

The ETL process is carried out using Spoon application, a graphical data integration IDE for creating transformations and jobs. A transformation depicts an ETL task in a narrow sense and is data-oriented while job is procedure and task-oriented which consists of one or more transformations [10].

IOP Conf. Series: Materials Science and Engineering **803** (2020) 012041 doi:10.1088/1757-899X/803/1/012041

Multiple transformation objects were built to integrate BISCAST data. The process starts with the configuration of database connection to hook up to a specified database. Connections to data sources will allow extraction of data whereas connection to the data warehouse enables loading of data. This process prompts for connection details which include the connection name, type and access, as well as setting details such as the host name, database name, port number, username, and password. In extracting data from spreadsheets and databases, Microsoft Excel and table input steps were utilized, respectively. In configuring input for Excel file, the file location, the specific sheet(s), and the header row equivalent to fields ware identified. For databases, on the other hand, table input step was used wherein it prompted for the database connection and the specific query.

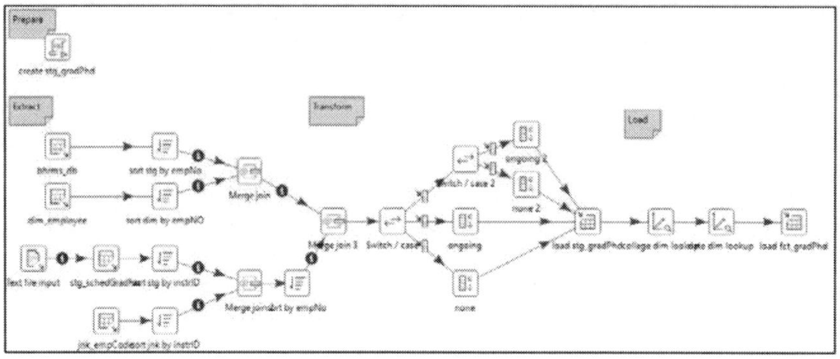

Figure 4. Transformation object for employee data.

To obtain a reliable data and information for the physical report of operation prepared by the Institutional Planning officer, data from heterogeneous sources of BISCAST composed of operational databases and spreadsheet files were integrated. As illustrated in Figure 4, employees' data collected from the Human Resource Info System (HRIS), Spreadsheet-based Class Scheduling System (SBCSS), and Student Information and Accounting System (SIAS) were extracted, performed necessary transformations, and stored into a central storage.

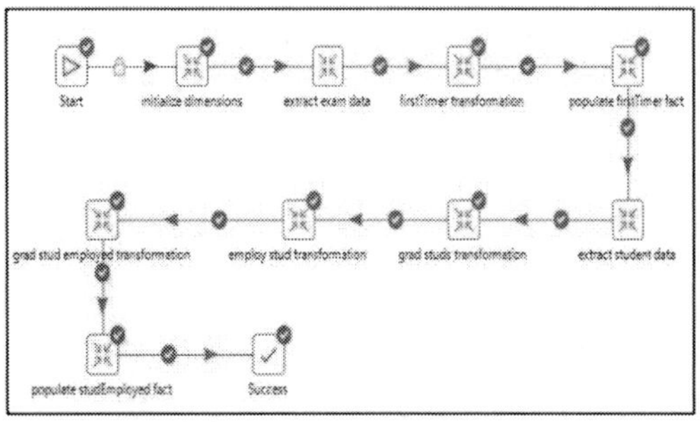

Figure 5. Job to execute multiple transformations.

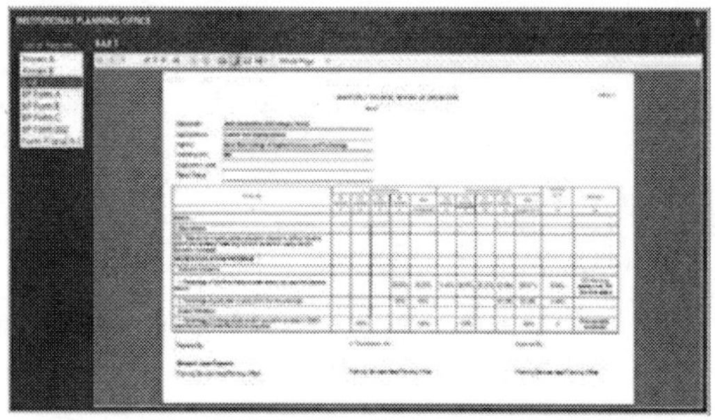

Figure 6. Reporting tool used for accessing the data warehouse.

Various transformation objects created for the specified report should be executed in a proper order. To achieve this, such transformations will be loaded as entries in a job object. Jobs in Spoon application are used to execute all transformation objects in the correct sequence. As shown in Figure 5, transformations such as initializing data warehouse tables, extracting data from sources and loading transformed data into data warehouse were sequentially executed.

Moreover, in order for the end-user to generate and view the result of the integration process, a reporting tool was developed based on the requirements of the institution. A screenshot of the sample generated report utilizing the integrated data is shown in Figure 6.

Furthermore, evaluation of the framework system was done to assess its usability. Questionnaires were distributed among twenty-two key personnel of the concerned offices in BISCAST. The assessed indicators include usefulness, ease of use, and satisfaction. Answers were tallied and the detailed assessment for each area is listed in Table 4.

Table 4. Evaluation summary result for the integration system.

Indicator	1	2	3	4	5	6	7
Usefulness	0	0	0	1	13	6	2
Ease of Use	0	0	0	0	6	15	1
Satisfaction	0	0	0	0	16	6	0

The description of the rating scale is as follows: 1-Worst Imaginable, 2-Awful, 3-Poor, 4-Ok, 5-Good, 6-Excellent, and 7-Best Imaginable.

The rating of the system's usefulness is good based on the answers of 59% of the respondents; the system's ease of use is rated as excellent by 68% of the respondents; and for the system's satisfaction, 73% of them rated it as good.

5. Conclusion

Integrating data from disparate sources of government organizations like State Universities and Colleges is essential in preparing and generating accurate reports. Applying data warehouse technique in BISCAST data has been found to be effective in integrating such heterogeneous data. Evident to which are the system-generated reports that contain all the relevant information from different sources in a less complex process. Implementing an ETL-driven data integration framework in BISCAST streamlined the process to consolidate and interpret its data thus enabling the institution to produce reliable reports eliminating delays in complying with submissions. Based on the evaluation result of system's usability, it is effective in accomplishing specific tasks, highly efficient and commendable to adopt.

References

[1] Sherman R 2015 *Business Intelligence Guidebook: From Data Integration to Analytics* (USA: Elsevier Inc.)

[2] Calvanese D, De Giacomo G, Lenzerini M, Nardi D and Rosati R 2001 Data Integration in Data Warehousing *Int. J. of Cooperative Information System* 10 pp 237-71

[3] Santoso L and Yulia 2017 Data Warehouse with Big Data Technology for Higher Education *Procedia Computer Science* **124** pp 93-9

[4] Ali El-Sappagh S H, Ahmed Hendawi A M and Bastawissy A H 2011 A proposed model for data warehouse ETL processes *J. of King Saud University - Computer and Information Sciences* **23** 2 pp 91-104

[5] Kimball R and Caserta J 2004 *The Data Warehouse ETL Toolkit: Practical Techniques for Extracting, Cleaning, Conforming, and Delivering Data* (Indianapolis: Wiley Publishing, Inc)

[6] Saeed A M 2015 Data Integration in Multi-sources Information Systems *International Journal of Computational Engineering Research (IJCER)* **5** 1 pp 63-9

[7] Macura M 2014 Integration of Data from Heterogeneous Sources using ETL Technology *Computer Science* 15 **2** 109-32

[8] Rodzi N, Othman M and Mi Yusuf L 2015 *Int. Conf. on Science in Information Technology (ICSITech)* pp 181-6

[9] Aziz A 2012 Integration of Heterogeneous Databases in Academic Environment Using Open

Source ETL Tools *The Int. Conf. on Informatics and Applications (ICIA2012)* pp 433-9
[10] Bouman R and Dongen J 2009 *Business Intelligence and Data Warehousing with Pentaho and MySQL* (Indiana: Wiley Publishing, Inc)

Document Tracking and Monitoring of Expenses for State Universities and Colleges Responsibility Units

Joni Neil Capucao[1], Sean Sierra[2], Kennedy Cuya[3], Josephine Dela Cruz[4]

[1]College of Arts and Sciences, Partido State University, Camarines Sur, Philippines
[2]Partido State University, San Jose Campus, Camarines Sur, Philippines
[3]College of Arts and Sciences, Partido State University, Camarines Sur, Philippines
[4]University of the Cordilleras, Benguet, Baguio, Philippines

E-mail: [1]jcapucao265@gmail.com, [2]seanasierra519@gmail.com, [3]kennedy.cuya@parsu.edu.ph, [4]delacruzpen@gmail.com

Abstract. The increasing data volume generated by higher learning institutions, aggregating of expenses, and locating files among tons of others can be a tedious and time-consuming process. Partido State University does not have a document and expense tracking system, making it difficult for the clientele to determine the status of their submitted documents. The development of document tracking and monitoring of expenses system for the University is a responsive, professional committal of the proponents in addressing this gap. The descriptive method explained the existing processes and developmental method in the creation of the system. Semi-structured interview among office personnel revealed the University offices need to adopt online software. In the systems' dry run, the faculty, the staff, and the students gave an overall very good (4.33 and 4.45) ratings for functionality and usability, and also an excellent overall rating of 4.5 for efficiency. The respondents strongly considered the adoption of an online system. The evaluation results manifested by the *Accurateness in Functionality* (4.63, excellent), *Learnability in Usability* (4.63, excellent), and *Time Behaviour in Efficiency* (4.63, excellent) have indicated that the system satisfied all the requirements. After its implementation, if some flaws may surface, adjustments could just be easily adapted.

1. Introduction

The changing technological pressure on management has led to an increasing interest in improving operational processes not merely to enhance performance but as well as staying current. Government organizations in the Philippines, particularly in the academe sector, such as State Universities and Colleges, have experienced various difficulties in implementing efficient processes and process management. These management difficulties are very evident on the following specifics; office transactions are commonly done manually using office applications that frequently lead to a delayed response to inquiries, longer time for dissemination of output, and even reduced service efficiency. If, however, the computerized system has been adopted, they are mostly a standalone application suggesting, therefore, limited access to information. The implementation of online systems, since it has been an intense clamor of interest among practitioners [1], is one of the challenges that need to be addressed. One such justification for this is Republic Act No. 11032, which stressed the need for heightened transparency and guarantees citizen-centric public service and lessen the number of complaints on delayed government transactions with the citizenry allowed to evaluate and give feedback on public services and the bureaucracy [2]. Also, the ISO 9001:2015 standards further specify the requirements for a quality management system [3] for State Universities and Colleges (SUCs).

Indeed, the large and increasing data volume generated by higher learning institutions, aggregating of expenses, and locating files among tons of others can be a tedious and time-consuming process for administrative and academic staff. Previous works have shown evidence that the challenge can be addressed using an online document tracking system. The system may be designed and developed using responsive website design [4] applicable to both mobile devices and desktop computers [5]. The use of barcodes can even be incorporated into the system for ease of entry [6]. Document Tracking System can be useful in administrative and academic institutions [7] for tracking documents and manage the movement of files from desk to desk and different colleges [8]. A mobile application can be used to track expenses incurred [9].

The absence of online systems in the different responsibility units of Partido State University (PSU) such as the Document Control Office, Finance Office, Colleges, and Campuses creates this difficulty. Such shortcoming makes it difficult for the administrators and personnel in monitoring into how well this office is succeeding in meeting its goals. Since most of the execution of processes are being done manually or mostly using application software such as an electronic spreadsheet, often leads to some specific problems; like difficulty in tracking documents; and monitoring of expenses incurred in the different units; consolidation of outputs; and report generation.

The main goal of the present study is for the proponents to provide a solution to efficiently track the document's status and whereabouts, including other features for aggregating of expenses. In addressing this need, the researchers plan to establish an in-depth validation of data on the information processes that can be designed and adapted for tracking administrative and financial documents in terms of a) flow of documents and b) aggregating of expenses. Next is to develop a system that could help enhance the efficiency in tracking administrative and financial documents through the online system. Finally, it will solicit respondents' (faculty, staff, and students) responses for determining the level of effectiveness on the proposed system. As a whole, the development of document tracking and aggregating of expenses system for the Partido State University is but their professional committal in addressing such research gap and by allowing the University in becoming more responsive.

2. Methodology

The research protocol concerning the design of online document tracking and monitoring of expenses system utilized the descriptive and developmental method. Descriptive method is intended to give focus on the information processes describing the current process flow and the processes that need to be enhanced or replaced in the different responsibility units. For the construction of the online system, the developmental method using rapid application development technology is fitting to fast-track the delivery and implementation of the prototype and for the inclusion of the financial requirements in the next University budget cycle.

2.1. Investigating PSU Existing Processes

Conducting a semi-structured interview with the office personnel and collecting sample forms used by different offices such as work instruction, procedure manual, organizational structure, the master list of logbooks, and observing the office procedures in monitoring document whereabouts and existing status may eventually benchmark in identifying the current procedure in monitoring documents as well as in the aggregation of expenditures in the different responsibility units. These may eventually lead to identifying the data requirements and solutions to the problems in tracking documents and aggregation of expenses. The expected information obtained can be made bases in creating the proposed web-based document tracking and monitoring of expenses system.

2.2. Development of the Web-based System

The development of the Document Tracking and Monitoring of Expenses System will focus on handling the various types of documents submitted to the different offices. It is meant to enhance the manner of locating, storing, and distribution of documents. Likewise, it may expedite the processes in monitoring the expenses incurred by the University's stakeholders.

2.2.1. Web Development Framework. Web Development Framework in this endeavour is the appropriate framework in designing the system. The web development framework is a set of resources and tools for software developers to build and manage web applications, web services, and websites [10]. The Laravel Framework being one of the leading web development frameworks that implement the Model-View-Controller (MVC) context, can be the best choice for a PHP framework that aids developers quickly build secure, upgradable web applications [11]. As shown in figure 1, the Model component pertains to all the data-related logic that the user is working. The View component provides all the User Interface (UI) logic of the application. Moreover, the Controllers act as an interface between Model and View components to process all the business logic and incoming requests, manage data using the Model component, and interact with the Views to render the final output [12].

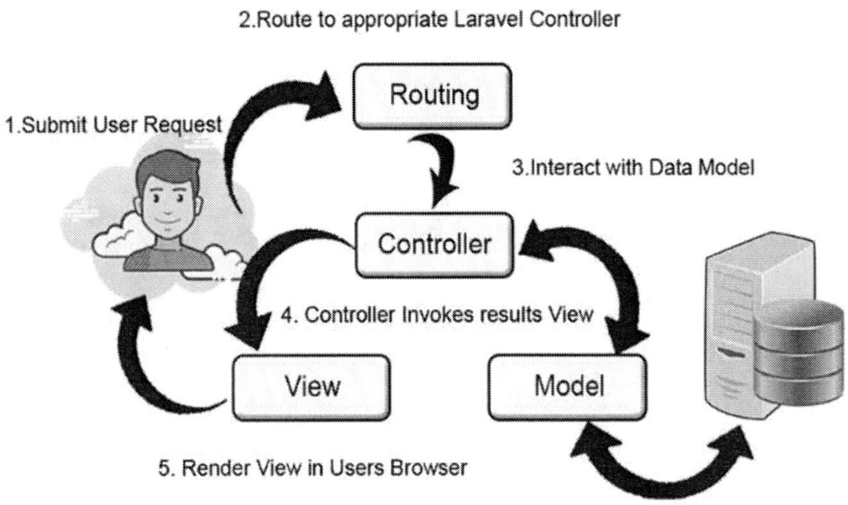

Figure 1. Model, View, and Controller Framework.

2.2.2. Rapid Application Development. For the software development methodology, Rapid Application Development (RAD), shown in figure 2, can be objectively chosen by the researchers, as such it could allow them to get immediate feedback from the end-users to enable necessary adjustments in developing the prototype [13].

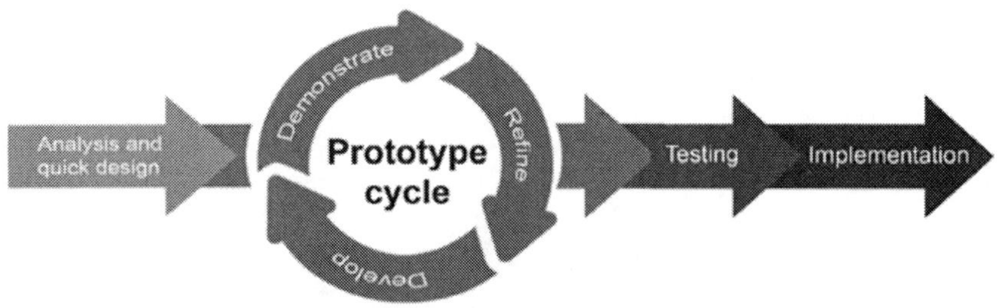

Figure 2. Rapid Application Development (RAD) Methodology.

3. Results and Discussion

3.1. Information Processes Designed and Adapted

This research tried to verify the assumption that information processes designed and Adapted for tracking administrative and financial documents in terms of a) flow of documents and b) aggregation of expenses encountered were essential.

PSU has seven campuses and several administrative and financial offices which, adopts the centralized processing of administrative and financial documents. The primary data source emanates from the Office of the President, Vice-Presidents, Quality Management Office, Document Control Office, Accounting Office, Budget and Finance Office, Cashier Office, College Deans, Campuses, Faculty and Staff, and Students of Partido State University-Main Campus, Goa, Camarines Sur. Additional secondary information collected through grey researches either from the internet, books, related articles, and including theses writings.

It was learned that all of the thirty-two offices of the University were exercising duties and functions as stipulated under the Quality Management Procedures of the ISO 9001:2015 quality management process. These findings affirmed the availability and application of the ISO standards as discovered in the use of Procedure Manuals (PM) and Work Instruction (WI) provided by different auditing offices; General Administration and Support Services (GASS) - 20 offices; Support to Operation (STOs) - 7 offices; and Academics -18 offices. The collected information served basis in determining what documents were processed and the sequence of procedures that were employed. The Master List of Logbooks (508 record books issued) suggested the numerous and complex transactions that transpired from one office to another, although these were all executed through the conventional manual approach, including further the aggregating of financial expenses in the budget and finance office.

3.2. Prototype System Development

The prototype Document Tracking and Monitoring of Expenses System was developed using Rapid Application Development (RAD) methodology and guided by the design of the Use Case Diagram.

3.2.1. Rapid Application Development Methodology. RAD methodology as shown in figure 3 has four phases, namely; 1. Analysis and Quick Design, 2. Prototype Cycle, 3. Testing, and 4. Implementation.

Phase 1: Analysis and quick design. In this stage, the researchers conducted interviews with faculty (n=160), staff (n=77) and students (n=340) of PSU regarding the office flow of documents and the accessibility issues in tracking documents and expense report generation needed in the design of the proposed system.

Phase 2: Prototype cycle. In this phase, the researchers gathered documents such as Procedure Manuals (PM) and Work Instructions (WI) in determining the scope of the system; and identified the needs of the stakeholders in terms of document tracking and aggregating of expenses. These requirements were analyzed, findings of the study were discussed, and the architecture design of the system was made, including the design of the database, components, and modules of the system.

Phase 3: Testing. Under this phase, the researchers analyzed the requirements for the system and designed solutions that were implemented, taking into consideration the requirements, constraints, and all applicable standards and guidelines. The proposed system developed was tested ensuring that it met all the requirements specified.

Phase 4: Implementation. The last part of the RAD methodology was the implementation of the prototype system. The researchers conducted an initial implementation of the prototype delimiting its coverage to those offices with current basic facilities.

3.2.2. Use Case Diagram. The researchers selected the use case modelling, figure 3, as the text-based tool for system analysis and design of the prototype system [14]. The model presents the different roles of the system users. The diagram showed that receiving officers in the university are responsible in encoding details of the documents, generating barcode and issuance of electronic mail to the clientele.

Head of offices can access the documents, check document entry and recommend approval. The president has provisions for checking the user performance logs and approval of documents. The clients are not allowed to access the important features of the system and can only check and track the status of their submitted documents and view other system information.

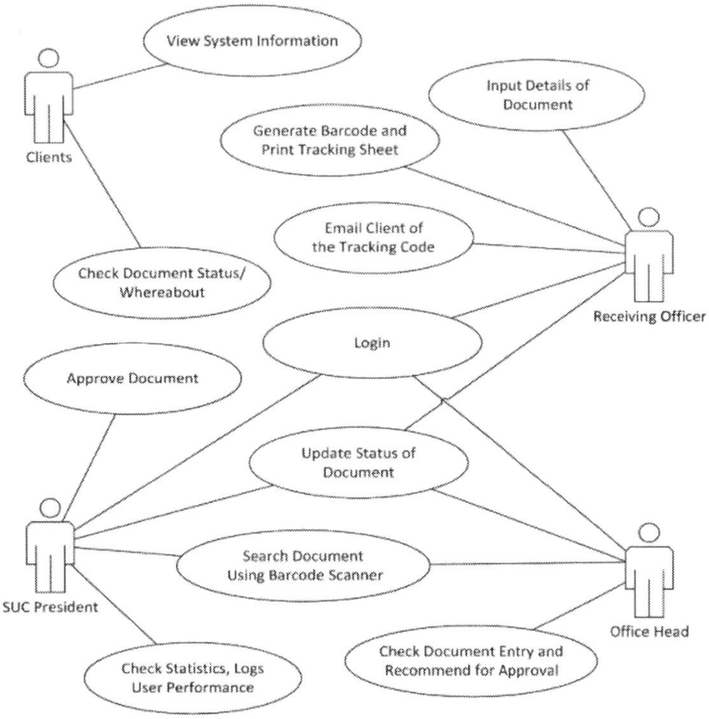

Figure 3. Use Case Diagram.

3.2.3. User Interface. Figure 4 shows sample users' interfaces, indicating among other things that this prototype system can be accessed either through mobile phones or laptop/desktop computers by respective clients and other interested parties.

Figure 4. User Interface for Mobile Phone and Desktop Computer.

3.3. Evaluation of the Proposed System

In assessing the prototype systems' functionality, usability, and efficiency the researchers collected primary data from the total sample size of five hundred seventy-seven (577) represented by faculty members (160), staff (77), and students (340) of Partido State University. This number was within the five percent tolerable margin of sampling error, though purposively (homogenous) selected the intent is to provide higher representative (see table 1).

Table 1. Distribution of Respondents.

Faculty		Staff		Students	
Target Population	Sample Size	Target Population	Sample Size	Target Population	Sample Size
271	160	96	77	2500	340

The evaluation result of the proposed system in terms of: (a) functionality; (b) usability; and (c) efficiency using the ISO 9126 Quality model for External and Internal Quality is shown in table 2. All responses along functionality (4.33, very good), usability (4.45, very good) and efficiency (4.50, excellent) showed no significant difference at alpha .05, suggesting therefore that faculty, staff and students have the same observations favorably indorsing the use of the said prototype system. There were observable low scores in Interoperability as rated by faculty (3.90) and staff (3.90). The respondents manifested that the proposed system provides an avenue to input necessary information about the whereabouts of the documents, expenses incurred, and presents reports generated by the system but is not meant to exchange information with other systems adequately. For Understandability, being new software, the expected interfaces to quickly understand the flow of the system suggests getting more used to it (faculty, 4.10; staff, 4.10). Under Security, the online programming language chosen is one of the general types. Hence, faculty (4.2) and staff (4.00), provided leeway for enhancement.

The adoption of the system may enhance the tracking of documents' whereabouts, and aggregation of expenses with the conversion of unscientific manual processes to online system shows the potential of meeting the main objective of the study as reflected by the excellent (4.50) rating in Efficiency. Respondents' overall rating of very good (4.33, 4.45) for functionality and usability suggests that the prototype meets the expected requirements. Thus, the implementation of such a system by State Universities and Colleges Responsibility Units, under the Philippine context, answers their utmost administrative and finance needs.

Table 2. ISO 9126 Quality model for External and Internal Quality.

Characteristic	Sub-characteristic	Explanation	Weighted Mean				Interpretation
			Faculty	Staff	Students	Total	
Functionality	Suitability	Can software perform the tasks required?	4.30	4.40	4.60	4.43	Very Good
	Accurateness	Is the result as expected?	4.50	4.50	4.90	4.63	Excellent
	Interoperability	Can the system interact with another system?	3.90	3.90	4.30	4.03	Very Good
	Security	Does the software prevent unauthorized access?	4.20	4.00	4.50	4.23	Very Good
Usability	Understandability	Does the user comprehend how to use the system easily?	4.10	4.10	4.30	4.17	Very Good
	Learnability	Can the user learn to use the system easily?	4.50	4.50	4.90	4.63	Excellent
	Operability	Can the user use the system without much effort?	4.50	4.40	4.80	4.57	Excellent
	Attractiveness	Does the interface look good?	4.30	4.40	4.60	4.43	Very Good
Efficiency	Time Behaviour	How quickly does the system respond?	4.60	4.50	4.80	4.63	Excellent
	Resource Utilization	Does the system utilize resources efficiently?	4.30	4.30	4.50	4.37	Very Good

H_o *Significant @.05 **Significant @.01

4. Conclusion

In this study, a prototype document tracking and monitoring of expenses system were presented using an MVC web development framework. The PSU responsibility unit's business processes considered and exhibited using a use case diagram. In the systems' dry run, the faculty, the staff, and the students gave an overall very good (4.33 and 4.45) ratings for functionality and usability, and also an excellent overall rating of 4.5 for efficiency. The respondents strongly recommend the adoption of the online system. The evaluation results manifested by the *Accurateness in Functionality* (4.63, excellent), *Learnability in Usability* (4.63, excellent), and *Time Behaviour in Efficiency* (4.63, excellent) have indicated that the system satisfied all the requirements. After its full implementation, if some minor flaws may surface, adjustments could just be easily adapted.

References

[1] Rhee S, Cho N and Bae H 2008 Increasing the efficiency of business processes using a theory of constraints *Information Systems Frontiers* **12** 4 pp 443–55

[2] GovPh 2018 The implementing rules and regulations of RA.11032 *Official Gazette* Available from: https://www.lawphil.net/statutes/repacts/ ra2018/pdf/ra_11032_2018.pdf

[3] ISO/TC176/SC2 ISO9001:2015 2015 *Quality management systems requirements* Available from: https://www.iso.org/standard/62085.html

[4] Giurgiu L and Gligorea I 2017 Responsive web design techniques *Int. Conf. Knowledge-Based Organization* **23** 10.1515/kbo-2017-0153

[5] Waseem B I and Abdelaziz H I 2016 Responsive web design techniques *Int. J. of Comp. App.* **150** 2 pp 18-27

[6] Gao J Z, Prakash L and Jagatesan R 2007 Understanding 2D-barbode technology and applications in m-commerce design and implementation of a 2D barcode processing solution *31st Annual Int. Computer Software and Applicatioin Conf.*

[7] Ajoge N S and Hamisu I A 2015 Document e-tracker: a panacea for efficient internal document management system in kaduna polytechnic *JOSTE* **3** 2 pp 169-81

[8] Omoregbe N A, Azeta A A and Adewumi A 2014 Development of a file tracking system for tertiary institutions *Proc. of edulearn14 Conf.* pp 3963-70

[9] Thanapal M P, Patel Y, Lokesh R T P and Satheesh K J 2015 *Income and expense tracker* (Tamil Nadu, India: Vellore)

[10] Ryabtsev A *Web frameworks: how to get started* Available from: https://djangostars.com/blog/ what-is-a-web-framework

[11] Armel J *Web application development with laravel PHP framework version 4* Available from: https://www.theseus.fi/bitstream/handle/10024/74052/Author.pdf

[12] Majeed A and Rauf I 2018 MVC architecture: a detailed insight to the modern web applications development *J. of Solar & Photoenergy Systems* **1** pp 1-7

[13] Kralev V S and Kraleva R S 2017 Methods and tools for rapid application development *Proc. of the III Int Scientific and Practical Conf.:Methodology of Modern Research"* pp 21-4

[14] Gemino A and Parker D 2009 Use case diagrams in support of use case modeling: deriving understanding from the picture *J. of Database Management* **20** (1)

Semantic Web in Disaster Management: A Systematic Literature Review

Teduh Dirgahayu[1], Hendrik[2], Hari Setiaji[3]

Department of Informatics, Universitas Islam Indonesia, Yogyakarta, Indonesia

E-mail: {[1]teduh.dirgahayu, [2]hendrik, [3]hari.setiaji}@uii.ac.id

Abstract. Disaster management requires extensive and various information. Semantic web offers a large potential to be utilized in the applications for supporting disaster management with its ability to integrate heterogeneous data from diverse sources. This study aims at investigating the extent to which semantic web is used in disaster management systems. The study conducted as a systematic literature review on 13 papers from academic databases that are leading in the area of computer science, i.e. ACM Digital Library, IEEE Explore and Science Direct. The result shows that the applications in the selected papers are mainly provided to institutions responsible for disaster management. Emergency relief is the phase that is supported by most applications. The study indicates that common geospatial ontology is necessary to facilitate the development of semantic web applications for disaster management. In the selected papers, the use of semantic web for data integration is done via an integration layer. This layer includes semantic processing to ensure that all data from multiple sources have proper semantics and to facilitate the application in querying the integrated information.

Keywords: semantic web, disaster management, ontology, data integration, semantic processing, literature review.

1. Introduction

Disaster management consists of several phases, i.e. disaster risk analysis, disaster preparedness, early warning systems, emergency relief, and reconstruction [1]. To be effective, all relevant parties should be actively involved in disaster management. Those parties can be categorized into three groups, namely (i) institutions that are responsible for disaster management, i.e. governmental agencies or companies whose business has the potential for large-scale accidents e.g. airports and oil mining/refinery, (ii) first responders, i.e. medical staffs, search-and-rescue teams, firemen, and volunteers, and (iii) people hit by disasters or living in the disaster-prone areas [2].

Information technology (IT) has been intensively used in every phase of disaster management. For example, a community-based information system [3] is used in the phase of disaster preparedness to identify the availability of logistics and local volunteers. DERMIS [4] and Sahana [5] are IT applications to support in the phase of emergency relief. Wireless technology is already commonly used to support the coordination between involved parties in emergency relief [6].

Disaster management requires extensive and various information. For example, during emergency relief, first responders do not only need to know the number of the people hit by disasters and their locations, but also the safe evacuation routes and the possibilities that the disaster would hit again.

Some of which have already made publicly available on the Web. When such information is also available in a machine-readable format, IT applications can be developed to make use of that information for supporting the first responders.

Semantic web is intended to extend the information that is available on the Web by giving the information well-defined meaning that can be understood and processed by applications [7]. It allows different applications to share and reuse data between them [8]. In fact, semantic web is mainly used to facilitate the integration of diverse and heterogeneous data [9].

Semantic web relies on ontology for representing conceptualized knowledge from the information. An ontology is a graph database that is composed of vocabularies and their relationships, e.g. in RDF or OWL. Applications may then query the ontology using query languages, e.g. SPARQL or OQL.

With its ability to integrate heterogeneous data from diverse sources, semantic web offers a large potential to be utilized in the applications for supporting disaster management. In this study, we aim at investigating the extent to which semantic web is used in disaster management systems. Hence, we can identify promising research in this area. Also, we can identify what shall be considered when we develop a semantic web application for disaster management.

The paper is further structured as follows. Section 2 describes the method that we use to conduct a systematic literature review. Section 3 presents and discusses the review results. Finally, Section 4 concludes this paper and identifies some future work.

2. Method

We conduct this study by following the guidelines for systematic literature reviews from Kitchenham and Charters [10]. More specifically, we follow a similar study in the same area [2]. The steps are described in the followings subsections.

2.1. Research Questions

We define the following research questions to be answered in this study.

RQ1. In which phase of disaster management is semantic web used?

RQ2. How semantic web are used in systems for disaster management?

2.2. Search Strategy

We define a search string i.e. ("semantic web" AND "disaster"). Instead of using the keyword "disaster management", we use keyword "disaster" to include all papers discussing the use of semantic web in any phases of disaster management. The searching is done in abstracts only.

Also, we define an inclusion criterion to select a list of relevant papers. The criterion is "the paper must describe the use of semantic web in disaster management systems". We conduct the searching on July 19, 2019.

We select academic databases that are leading in the fields of computer science as the data sources of our study. The selected databases are ACM Digital Library (dl.acm.org), IEEE Explore (ieeexplore.ieee.org), and Science Direct (www.sciencedirect.com). We are aware that Google Scholar is a comprehensive index for scholarly papers. However, it has a limitation that makes us exclude it. Using Google Scholars, searching in abstracts is limited to papers published in the last year only. Thus, it does not give us complete demographic information of all available papers. Searching in full text resulted in 10,300 papers that we cannot manage to handle manually.

The searching results in several papers as listed in Table 1. It initially results in 45 papers. In these papers, we apply the inclusion criterion in their abstracts and then in their full texts. We finally get 13 selected papers to review, i.e. [11]-[23]. This number of papers might be considered too few for a systematic literature review, but it can be understood as the topic is quite narrow.

Table 1. Search results.

Database	Initial results	Relevant (abstract)	Relevant (full text)
ACM Digital Library	18	4	3
IEEE Explore	12	12	8
Science Direct	15	7	2
Total	45	23	13

2.3. Data Extraction

The demographic analysis of the selected papers is described as follows. Four papers (30.8%) are published in journals and nine papers (69.2%) are in conference proceedings. About 61.5% (8 papers) are published in IEEE-indexed journals and conference proceedings. They were published between 2015 and 2019. Figure 1 depicts the distribution of the papers based on their publications years. On average, 2.6 papers are published per year. The figure does not indicate any trend in the publication numbers. It should be noted that the number of papers published in 2019 may increase until the end of this year.

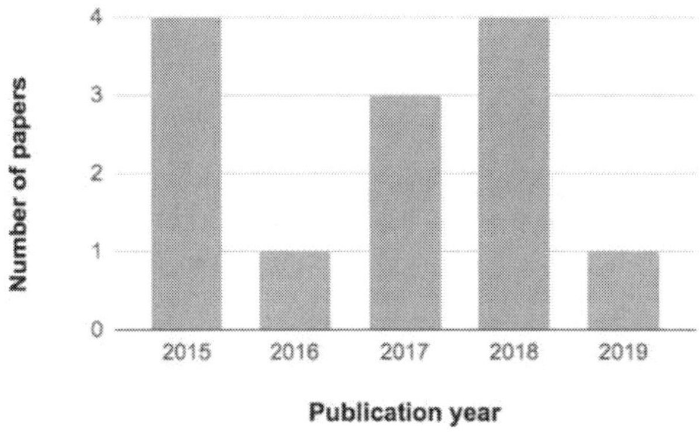

Figure 1. Number of papers vs. publication year.

Based on the affiliations of the selected papers' authors, the papers were recognized from fourteen countries. Note that a paper may have several authors from different countries. Table 2 depicts the distribution of the countries of the authors. The table also indicates the risk index of the corresponding countries. A risk index is a combination of four components, i.e. exposure, susceptibility, coping, and adaption. The index is categorized into *very high* (10.49 – 36.45), *high* (7.36 – 10.48), *medium* (5.63 – 7.35), *low* (3.53 – 5.62), and *very low* (0.09 – 3.52) [1]. Although there are 68 developed and developing countries whose risk indices are *high* and *very high*, only authors from the Netherlands (risk index = 8.41) contributed to the selected papers. Other countries, such as Japan (risk index = 13.47, *very high*) and Indonesia (risk index = 10.49, *very high*) did not contribute to the selected papers. Interestingly, most countries (9 of 14; 64%) have low or very-low risk indices.

Table 2. Countries of the authors of the selected papers.

Country	Number of papers	Number of authors	Risk index	Risk index category
Canada	1	1	3.31	Very low
China	1	6	6.81	Medium
France	1	2	2.73	Very low
Greece	1	3	7.11	Medium
Germany	2	2	3.09	Very low
India	2	5	7.00	Medium
Italy	1	5	4.63	Low
Netherlands	2	4	8.41	High
Slovenia	1	1	3.62	Low
Taiwan	2	8	N/A	N/A
Tunisia	1	2	5.62	Low
Turkey	1	1	5.42	Low
UK	2	4	3.60	Low
US	1	4	3.90	Low

3. Results and Discussion

The answers to our research questions are described in the following subsections.

3.1. *Phases of disaster management (RQ1)*

Applications in the selected papers use semantic web for data integration. Table 3 lists in which phases and by whom the applications are used. Most applications (6 papers) are for supporting emergency relief. Other phases are supported by applications in 2 or 3 papers only. Most applications (7 papers) are available to be used by institutions. Few applications (2 papers) are targeted for people as their users. It should be noted that an application may support several phases and be targeted for many types of users. The application in [14] is used to monitor a disaster management process from risk analysis (diagnosis), preparedness, emergency relief (response) to reconstruction (recovery).

Table 3. Phases and relevant parties.

Phases	Institution	Responder	People	#papers
Risk analysis	[14][16]			2
Preparedness	[14][17]	[23]		3
Early warning	[13][18]		[13]	2
Emergency relief	[14]	[11][15][19][21]	[12]	6
Reconstruction	[14][20][22]			3
#papers	7	5	2	

Different applications in the selected papers address different types of disasters. Disaster can be (i) *natural disasters*, e.g. tsunami, typhoons, and earthquakes, or (ii) *man-made disasters*, e.g. humanitarian crisis, forest fire due to carelessness, and oil spill at sea from drilling. The applications can address *specific disasters*, e.g. disaster at the sea [13], earthquake [17][22][23], tsunami [18], flood [20] or *generic disaster*, i.e. the paper does not mention the type of disaster being addressed. It should be noted that an application may address several types of disasters.

Table 4. Types of disasters addressed.

	Natural	**Man-made**
Specific disaster	[17][18][20][22][23]	[13]
Generic disaster	[11][12][14][15][16] [19][21]	[12]
#papers	12	2

3.2. Use of semantic web (RQ2)

Disasters happen at certain locations. Geospatial information is then very importance in disaster management. Most applications (ten selected papers) [13][14][16]-[23] make use of ontology that related to geospatial information. Applications in other selected papers [11][12][15] might use spatial-related ontology, but the papers do not clearly indicate that. There is no consensus however which geospatial-related ontology shall be used as different applications serve for their purposes. For example, the application in [22] develops a geographic ontology to perform classification and damage assessment in OWL. The ontology consists of three sub-ontologies, i.e. surface area, disaster, and damage. The surface includes geographical concepts including land cover and land use classification. Time is included as disasters happen at a certain time.

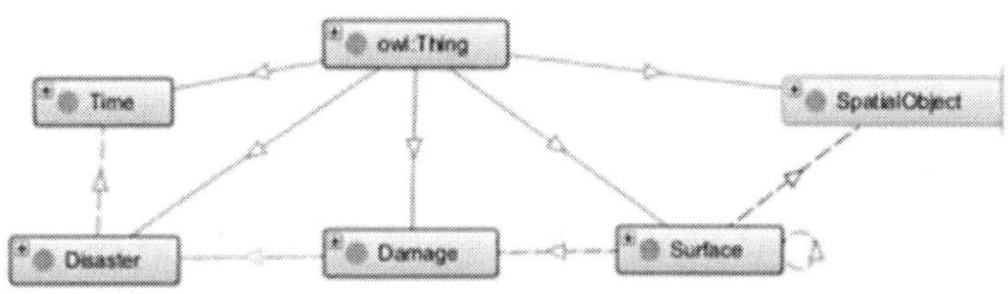

Figure 2. Ontology includes spatial information.

Four selected papers indicate that the latest technology has been used in disaster management, i.e. robotics [11][21], Internet of Things (IoT) [18], and unmanned aerial vehicle (UAV) [19]. This technology relies on sensors to be able to capture the disaster situations. Data observed by the sensors are transmitted to a semantic processing component with predefined ontologies to be interpreted or reasoned into information. For example, Figure 3 depicts an excerpt of W3C Semantic Sensor Network Ontology (SSNO) [24] that defines the relationships between sensors and their observed data that is used in [18][21]. These papers further extend SSNO with other ontologies to satisfy the application requirements.

Semantic web data integration ability should be exploited more in disaster management. Only four selected papers explicitly mention the use of multiple (internal or external) data source in their applications, i.e. Google public alerts [12]; distributed and heterogeneous databases [13][17]; and satellite imagery, weather, and atmospheric condition [16]. When data from diverse sources are integrated, they would enrich the resulted information. The integration, however, should be made transparent to the application. In all those papers, a common approach to providing such transparency is to establish an integration layer with semantic processing between data sources and the application as illustrated in Figure 4. External data sources may include sensors. The semantic processing is necessary for two reasons, i.e. (i) to ensure that all data from different sources have proper semantics and (ii) to facilitate the application in querying the integrated information. For the latter reason, semantic processing can also be found in other selected papers [15][18][20].

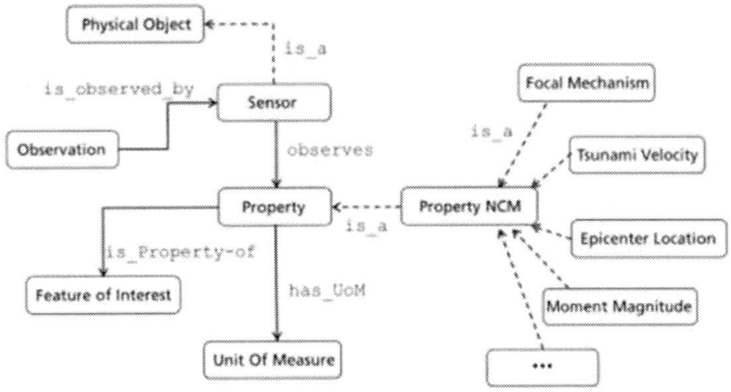

Figure 3. Excerpt from SSNO [18].

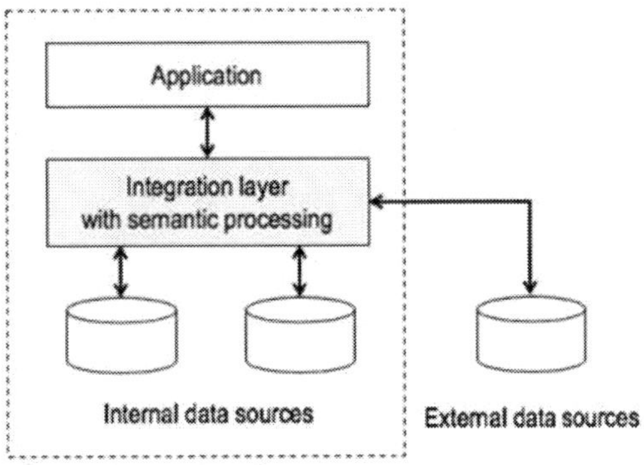

Figure 4. Integration layer with semantic processing.

4. Conclusion and suggestion

The study conducted in this paper has reviewed 13 papers discussing the use of semantic web in disaster management. The papers are selected from ACM Digital Library, IEEE Explore and Science Direct. While the idea of semantic web has arisen since 2001, papers discussing the implementation of semantic web in disaster management were published since 2015 only.

Based on the result of our demographic analysis, we suggest that researchers from developed and developing countries with high and very-high risk indices contribute more to this area. Especially, we may expect more publications in using semantic web for disaster management that are specific to their countries.

The applications in the selected papers are mainly provided to institutions responsible for disaster management. Emergency relief is the dominant phase that is supported by most applications. Future research shall be conducted to address other phases and different parties.

Geospatial ontology is very important in disaster management as disasters happen at certain locations. Each application in the selected papers develops its geospatial ontology. Some work on geospatial ontology has been done, e.g. [25][26]. Future research on a general-purpose geospatial ontology will facilitate the development of a semantic web application for disaster management. This idea has been demonstrated in the case of a semantic sensor network ontology (SSNO).

Disaster management requires extensive and various information. Semantic web allows the integration of heterogeneous data from multiple (internal and external) data sources to enrich information for relevant users. Semantic processing is required during data integration to ensure that all data have proper semantics. We expect more applications on semantic web in disaster management so that we can learn how to use it effectively and efficiently.

References

[1] UNU-EHS 2017 *World Risk Report. Analysis and prospect 2017*

[2] Dirgahayu T, Hendrik and Setiaji S 2019 Context-aware applications for disaster management: A systematic literature review *Proc. 2019 IEEE Int. Conf. Cybernetics and Computational Intelligence (CyberneticsCom 2019)*

[3] Troy DA, Carson A, Vanderbeek J and Hutton A 2008 Enhancing community-based disaster preparedness with information technology *Disasters* vol **32** no 1 pp 149-65

[4] Turoff M, Chumer M, van de Walle B and Yao X 2004 The design of a dynamic emergency response management information system (DERMIS) 2004 *J. Information Technology Theory and Application (JITTA)* vol **5** no 4 p 3

[5] Currion P, de Silva C and van de Walle B 2007 Open source software for disaster management *Communications of the ACM* vol **50** no 3 pp 61-5

[6] Reddick C 2011 Information technology and emergency management: preparedness and planning in US states *Disasters* vol **35** no 1 pp 45-61

[7] Berners-Lee T, Hendler J and Lassila O 2001 The semantic web *Scientific American* vol **284** no 5 pp 28-37

[8] W3C 2011 *W3C Semantic web activity*

[9] Shadbolt N, Berners-Lee T and Hall W 2006 The semantic web revisited *IEEE Intelligent Systems* vol **21** no 3 pp 96-101

[10] Kitchenham B and Charters S 2007 Guidelines for performing systematic literature reviews in software engineering *EBSE Technical Report EBSE-2007-01*

[11] Bagosi T, Hindriks K V and Neerincx M A 2016 Ontological reasoning for human-robot teaming in search and rescue missions *The Eleventh ACM/IEEE Int. Conf. on Human Robot Interaction* pp 595-6

[12] Pandey Y and Bansal S 2017 Safety check: a semantic web application for emergency management *Proc. Int. Workshop on Semantic Big Data* p 7

[13] Santipantakis G, Kotis K I and Vouros G A 2015 Ontology-based data integration for event recognition in the maritime domain *Proc. 5th Int. Conf. on Web Intelligence, Mining and Semantics* p 6

[14] Chen N, Liu Y, Wang C, Xiong C, Chen Z and Xiao C 2018 SWRO-DDPM: A sensor web resource ontology for the dynamic disaster process monitoring *Proc. 26th Int. Conf. on Geoinformatics* pp 1-4

[15] Mobasheri A and Bakillah M 2015 Towards a unified infrastructure for automated management and integration of heterogeneous Geo-datasets in disaster response *2015 IEEE Int. Geoscience and Remote Sensing Symposium (IGARSS)* pp 4570-3

[16] Masmoudi M, Taktak H, Lamine S B A B, Karray M H, Zghal H B, Archimede B and Guegan C G 2018 PREDICAT: a semantic service-oriented platform for data interoperability and linking in earth observation and disaster prediction *Proc. IEEE 11th Conf. on Service-Oriented Computing and Applications (SOCA)* pp 194-201

[17] Liu C H, Bau J F and Lee C R 2015 Integrating various data sources of geospatial information systems with semantics *Proc. 2nd Int. Conf. on Information Science and Control Engineering* pp 274-8

[18] Poslad S, Middleton S E, Chaves F, Tao R, Necmioglu O and Bügel U 2015 A semantic IoT early warning system for natural environment crisis management *IEEE Transactions on*

Emerging Topics in Computing vol **3** no 2 pp 246-57

[19] Cavaliere D, Saggese A, Senatore S, Vento M and Loia V 2018 Empowering UAV scene perception by semantic spatio-temporal features *Proc. 2018 IEEE Int. Conf. on Environmental Engineering (EE)* pp 1-6

[20] Kurte K R, Durbha S S, King R L, Younan N H and Vatsavai R 2016 Semantics-enabled framework for spatial image information mining of linked earth observation data *Proc. IEEE J. of Selected Topics in Applied Earth Observations and Remote Sensing* vol **10** no 1 pp 29-44

[21] Dey S, Bhattacharyya A and Mukherjee A 2017 Semantic data exchange between collaborative robots in fog environment: Can CoAP be a choice? *Proc. 2017 Global Internet of Things Summit (GIoTS)* pp 1-6

[22] Bouyerbou H, Bechkoum K and Lepage R 2019 Geographic ontology for major disasters: Methodology and implementation *Int. J. of disaster risk reduction* no 34 pp 232-42

[23] Chou C C, Jeng A P, Chu C P, Chang C H and Wang R G 2018 Generation and visualization of earthquake drill scripts for first responders using ontology and serious game platforms *Advanced Engineering Informatics* no 38 pp 538-54

[24] W3C 2017 *Semantic Sensor Network Ontology*

[25] Klien E and Probst F 2005 Requirements for geospatial ontology engineering *Proc. 8th Conf. on geographic information science (AGILE 2005)* pp 251-60

[26] Arpinar I B, Sheth A, Ramakrishnan C, Usery E L, Azami M and Kwan M-P 2006 Geospatial ontology development and semantic analytics *Transactions in GIS* vol **10** no 4 pp 551-75

Automation Process to Support an Information System on Extreme Weather Warning

M Tamamadin[1], A Susandi[1], A F Pratama[1], I Faisal[2], A R Wijaya[1], I M Firdaus[1], W S Kuntoro[3]

[1]Department of Meteorology, Faculty of Earth Sciences and Technology, Institut Teknologi Bandung, Jalan Ganesha No. 10, Bandung 40132, Indonesia
[2]Department of Environmental Engineering, Faculty of Civil and Environmental Engineering, Institut Teknologi Bandung, Jalan Ganesha No. 10, Bandung 40132, Indonesia
[3]Department of Oceanography, Faculty of Earth Sciences and Technology, Institut Teknologi Bandung, Jalan Ganesha No. 10, Bandung 40132, Indonesia

Email: mamadtama@meteo.itb.ac.id

Abstract. Hydrometeorological hazard early warning system has been developed in order to support public society in anticipating the extreme weather in Indonesia. Meanwhile, the early warning system in Indonesia still produced locally and need arduous human resources to process the enormous size data to have a good quality early warning. Therefore, in order to increase the effectivity, efficiency, and efficacy of the early warning system to produce the desired result, the automatic process was needed. This study will explain further the mechanism of process automation from the weather forecast to the output of warning displayed in an information system. The automation process combines several scripting in Python and batch script for GIS processing, as well as process scheduling using task scheduler. The output of extreme weather warning is in form of JSON files stored in organized folder directory. The automated process proven to be more useful in increasing the effectivity, efficiency, and efficacy of the early warning system to produce the desired result and later the automated process could be enhanced further to boost the lead time and reducing the arduous task in producing the early warning system.

1. Introduction

Hydrometeorological disasters, such as floods and landslides, are very closely related to the occurrence of extreme weather in the form of rain with high intensity [1]. Extreme weather commonly occurs in Indonesia. Thus, Indonesia is vulnerable to hydrometeorological disasters. Extreme weather in Indonesia is caused by various weather phenomena in the region, which are El-Niño Southern Oscillation (ENSO), Indian Ocean Dipole Mode (IODM), and Madden-Julian Oscillation (MJO) [2,3]. To anticipate and reduce the risk of extreme weather, the public needs to be supported by information technology in form of early warning system.

World Meteorological Organization (WMO) stated that the main objective of a warning system is to enhance individuals and societies in responding timely and appropriately to the hazards so that they are able to reduce the risk of death, injury, property loss and damage [4]. Macherera **et al** [5] also states that community-based early warning systems support collection and analysis of information driven by community; Therefore, warning messages delivered to help a community to respond to a hazard and enable to reduce the loss or harm. Meanwhile several computational systems for disasters early warning have been developed and used operationally by several agencies such as Jakarta-Flood Early Warning System [6], drought early warning system in Pemali-Comal River Basin [7], and other regional early warning system such Hydro-Meteorological Hazard Early Warning System in Indonesia as described in Susandi et al [8].

Previously, the early warning system in Indonesia have been produced locally and need arduous human resources to produce the warning [9]. Furthermore, the warning will be subject to the human error which is unavoidable [9]. In addition due to the enormous size data and the need to process a lot of data to have a good quality early warning, also due the need of daily early warning for National Disaster Management Agency (NDMA)'s operational, the automatic process is needed to increase the effectivity, efficiency, and efficacy of the early warning system to produce the desired result. Meanwhile, another reason for the need of the automatic process was there are a lot of step that need to be executed to reach the desired early warning. Therefore, this paper will be specifically describes the automation process to produce output in the form of markers that provide time- and location-specific information about the potential for extreme weather in Indonesia which has been used to operate the Hydro-Meteorological Hazard Early Warning System in Indonesia as described in Susandi et al [8].

2. Material and Methods

2.1. Material

The data that will be used in to be processed to produce extreme warnings are weather forecast data that have been produced by state of the art of the operational Hydro-Meteorological Hazard Early Warning System as described in Susandi et al [8]. The data is produced daily with the domain coverage throughout Indonesia. The forecast data has a 3-hourly timestep and is updated every day at 00 UTC.

2.2. Methods

The weather forecast data that have been produced by the state of art in Susandi et al [8] is stored a designated directory in form of .dat format. To be able to display the forecast in information system, the data is converted to geotiff (.tif), this study utilized the Geospatial Data Abstraction Library (GDAL) in Python Library. GDAL is a library for translating a raster and vector geospatial data formats released under an X / MIT style Open Source License published by the Open Source Geospatial Foundation [10]. The use of GDAL Library in the conversion of map file format has also been carried out by Wang, et al. [11]. Wang, et al. [11] used the GDAL's library to obtain processing services that enable the bridge between the geo-analysis models and data resources hosted on networks.

Meanwhile, the state of art by Susandi et al [8] displays various weather forecast parameters, which are rainfall, wind speed, humidity, air pressure and air temperature. For wind direction, the forecast result is converted to JavaScript Object Notation (JSON) file, so that it can be displayed in vector form. By using the JavaScript library, the wind direction is displayed in a dynamic animation form on the base map.

Furthermore, weather forecast data in the .dat format, that have been produced by the Susandi et al [8]'s state of the art, is processed to produce extreme weather warning, which are categorized into two categories, namely extreme rainfall and strong winds. Later, to get these desired results, weather forecast data is divided into 2 parts, namely rainfall forecast data and wind speed. These two types of extreme weather forecast have a 3-hourly timestep with a forecast duration of 3 days ahead. The whole process is shown in Figure 1.

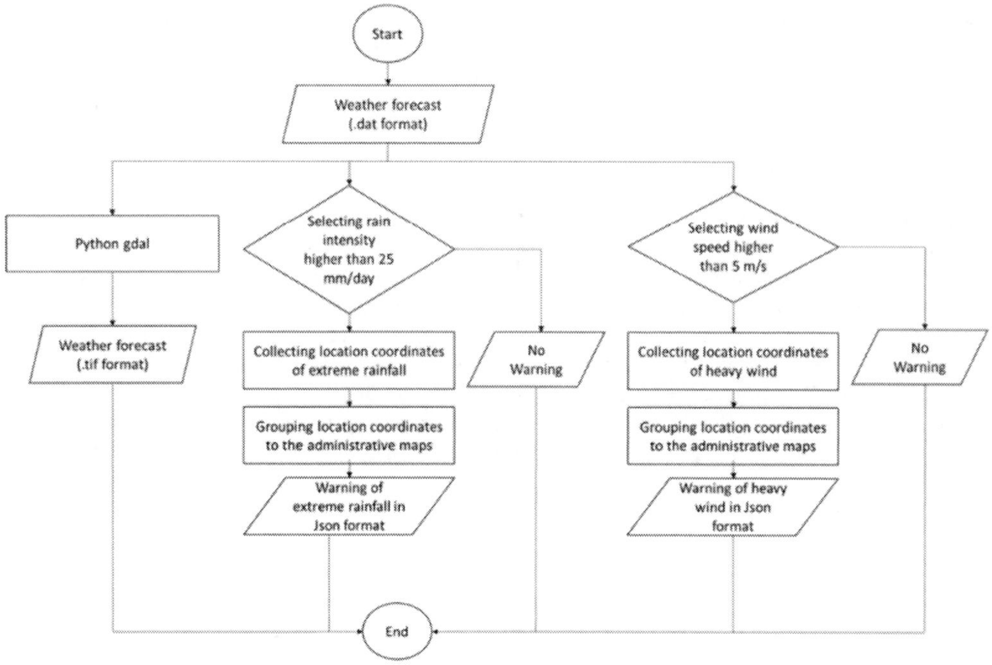

Figure 1. Flowchart of extreme weather processing.

Rainfall forecast data is filtered based on the amount of rainfall threshold. If the rainfall is higher than 25 mm/hour, the information related to the coordinates and the amount of rainfall is taken for further processing. The selected rainfall data will be processed further to be divided based on administrative regions, namely district and village boundaries. Thus, extreme rainfall forecast data will be obtained by district and village. All data is stored in a JSON file format. Each file containing the data is stored in a different directory, namely the district directory and the village directory. The folder directory that store extreme rainfall forecast data are named according to the predicted date.

On the other hand, wind speed forecast data is filtered based on the wind speed to get heavy wind warning. If the wind speed is higher than 10 m/s, the information related to the coordinates and its wind speed is taken for further processing. The selected wind speed data are processed to be divided based on administrative areas, namely district and village boundaries. Thus, the strong wind forecast data will be obtained by district and village. All data is then stored in a JSON file format. The produced file will have same treatment as the extreme rainfall forecast.

3. Result and Discussion

The results of processing automation to produce files containing weather forecast information and extreme weather forecast in Indonesia will be detailed further in this section. Figure 2 shows the directory and the file structures that have been used to display the produced data based on user's requests. By utilizing this structure, the information system will identify the files based on date-time and weather variables chosen by the user. The directory starts from "variable name", then "year", then "month", then "date", and ends with the name of the tif file format with the name "1.tif". To display weather forecasts, .tif format files will be displayed on the base map with the Leaflet library. For this way, Seeger et al [12] wrote about how to create layer controls for base maps and overlays.

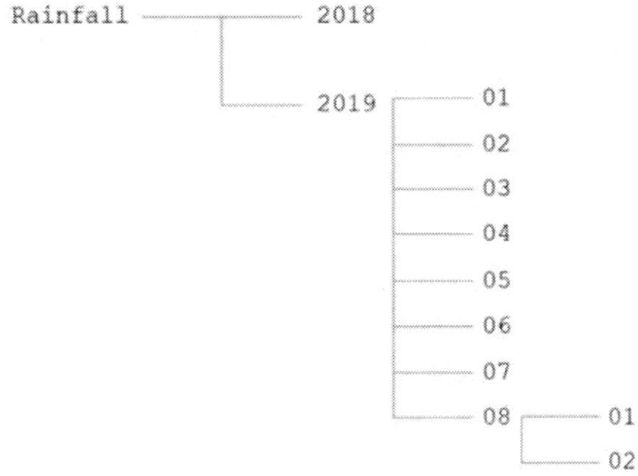

Figure 2. Files and directory tree of weather and extreme weather forecast.

Furthermore, the file used to display the wind direction is using a file with JSON format. Following is an example of the contents of the wind direction file in the JSON format that is displayed in the information system (Figure 3).


```
[ { "header":{ "discipline":0, "disciplineName":"Meteorological products",
"gribEdition":2, "gribLength":121656, "center":0, "centerName":"WMO Secretariat",
"subcenter":0, "refTime":"2019-09-09T03:00:00.000Z", "significanceOfRT":1,
"significanceOfRTName":"Start of forecast", "productStatus":0,
"productStatusName":"Operational products", "productType":1,
"productTypeName":"Forecast products", "productDefinitionTemplate":0,
"productDefinitionTemplateName":"Analysis/forecast at horizontal level/layer at a
point in time", "parameterCategory":2, "parameterCategoryName":"Momentum",
"parameterNumber":2, "parameterNumberName":"U-component_of_wind",
"parameterUnit":"m.s-1", "genProcessType":255, "genProcessTypeName":"Missing",
"forecastTime":3, "surface1Type":100, "surface1TypeName":"Isobaric surface",
"surface1Value":100000.0, "surface2Type":255, "surface2TypeName":"Missing",
"surface2Value":-9.999E-252, "gridDefinitionTemplate":0,
"gridDefinitionTemplateName":"Latitude_Longitude", "numberPoints":125664,
"shape":6, "shapeName":"Earth spherical with radius of 6,371,229.0 m",
"gridUnits":"degrees", "resolution":48, "winds":"true", "scanMode":64, "nx":476,
"ny":264, "basicAngle":0, "lo1":90.5098, "la1":13.8142, "lo2":141.786, "la2":-
14.2903519 , "dx":0.10795, "dy":0.106862 }, "data":[ -6.1976137, -5.8538637, -
5.330426, -4.846051, -4.596051, -4.658551, -5.0257387
```


Figure 3. Example of JSON file produced by Susandi et al [8].

Figure 4 shows the plot results of a weather forecast map consisting of rainfall and wind direction overlaid on the base map. On the base map, there is a shaded contour which indicates the presence of rain in several locations. If there is no shaded contour, then there is no rain at that location. In addition to rainfall, other weather variables that can be displayed on the base map include wind speed, air temperature, air pressure, and humidity.

Furthermore, Figure 5 shows the results of extreme weather forecasts (including extreme rainfall and strong winds) displayed above the base map. The number of extreme weather locations in each province is indicated in the information system. When the user zooms in the province, the number will change to a marker. Each marker will provide an early warning information consisting of the type of disaster, the name of the village, the name of the sub-district, the name of the district and the name of the province. This information is contained in JSON-formatted files that are stored in structured directories where website scripts can identify them precisely.

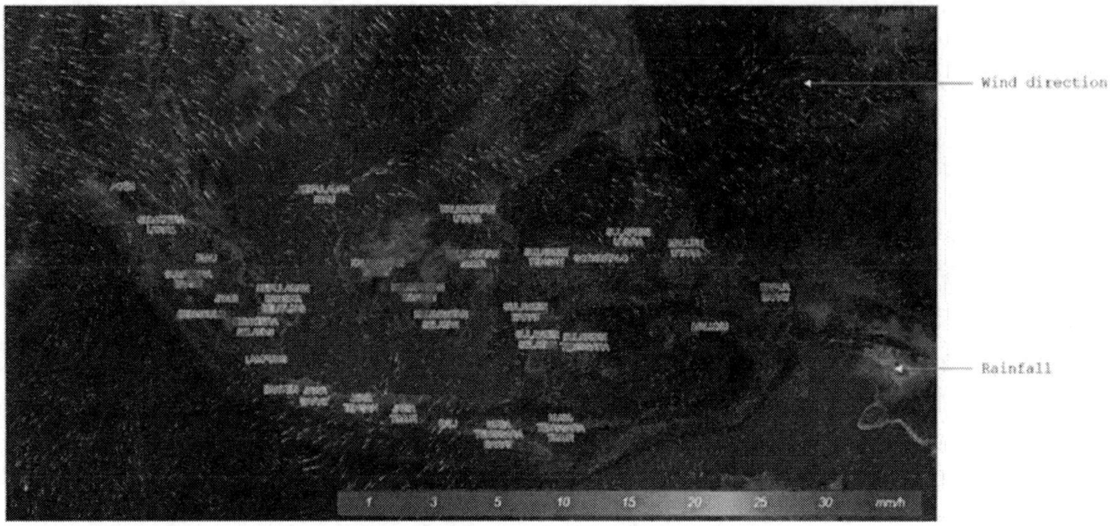

Figure 4. Weather forecast mapping displayed on information system.

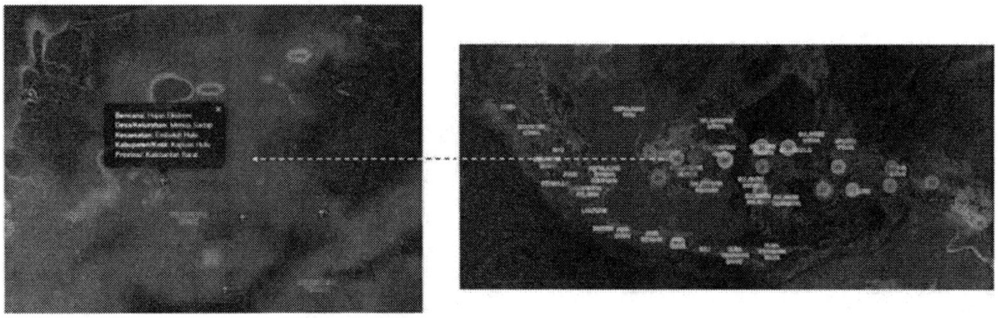

Figure 5. Extreme rainfall warning displayed on information system including warning marker (left) and number of extreme weather event in a province (right).

By using the automated process, this study has tried to reduce the human-error in the information system that previously stated by Rajib et al [9] which is have been done manually in Indonesia transformed into the automated process. The automated process proven to be more useful in increasing the effectivity, efficiency, and efficacy of the early warning system to produce the desired result. Later this automated process could be enhanced further to boost the lead time and reducing the arduous task in producing the early warning system.

4. Conclusion

This paper has explained the automation process to produce an extreme weather early warning system information that have been used in state of art by Susandi et al [8]. This system provides location and time information about the potential for extreme weather with a timestep of 3 hours duration of the next 3 days. The system updates the information once a day, starting from generation of weather forecast until generation of extreme weather early warning information. This system uses weather forecast input with cycle 00 and has been used operationally. By using the automated process, this study has tried to reduce the human-error in the information system have been done manually in Indonesia. The automated process proven to be more useful in increasing the effectivity, efficiency, and efficacy of the early warning system to produce the desired result. Later this automated process could be enhanced further to boost the lead time and reducing the arduous task in producing the early warning system.

Acknowledgment
We appreciate the support from Ministry of Research, Technology & Higher Education through Industrial Technology Development Program. We also thank to the National Disaster Management Authority for his suggestions and support during the development of this Hydro-meteorological Hazard Early Warning System in Indonesia.

References

[1] Ferdiansyah A 2012 *Potensi Parameter Keluaran Raob (Rawinsonde Observation Programs) Sebagai Indikator Kunci Dalam Analisis Curah Hujan* Thesis (Bogor:Institut Pertanian Bogor)
[2] Handayani A S 2010 Analisis Daerah Endemik Bencana Akibat Cuaca Ekstrim Di Sumatera Utara *Jurnal Meteorologi dan Geofisika* vol **11** no 1 – Juli 2010 pp 54–8
[3] Prayuda S 2018 Analisis Karakteristik Lapse Rate Tropopause (LRT) dan Kaitannya dengan Aktivitas MJO di Surabaya *Prosiding Seminar Hari Meteorologi Dunia ke-68 Tahun 2018*
[4] United Nations Development Programme (UNDP) 2016 *UNDP Programme on Climate Information for Resilient Development in Africa. Climate Information & Early Warning Systems Communications Toolkit*
[5] Macherera M and Chimbari M J 2016 A review of studies on community based early warning systems *Jamba* **8(1)** 206 doi:10.4102/jamba.v8i1.206
[6] Ginting S and Adidarma W 2013 Jakarta Flood Early Warning System (J-FEWS) *Workshop on MCCOE Radar Meteorology/Climatology in Indonesia*
[7] Hatmoko W, Raharja B, Tollenaar D, and Vernimmen R 2015 Monitoring and prediction of hydrological drought using a drought early warning system in Pemali-Comal river basin, Indonesia *Procedia Environmental Sciences* **24** pp 56-64
[8] Susandi A, Tamamadin M, Pratama A, Faisal I, Wijaya A R, Pratama A F, Pandini O P and Widiawan D A 2018 Development of Hydro-Meteorological Hazard Early Warning System in Indonesia *Journal of Engineering and Technological Sciences* [Sl] vol **50** no 4 pp 461-78
[9] Rajib M and Fan L 2015 A study on the critical factors of human error in civil aviation: An early warning management perspective in Bangladesh *Management Science Letters* **5(1)** p 21
[10] GDAL/OGR contributors 2019 GDAL/OGR Geospatial Data Abstraction Software Library *Open Source Geospatial Foundation* URL: https://gdal.or
[11] Wang J, Chen M, Lü G, Yue S, Chen K and Wen Y 2018 A study on data processing services for the operation of geo-analysis models in the open web environment *Earth and Space Science* **5** pp 844–62 DOI:10.1029/2018EA000459
[12] Seeger C and Hanson B 2016 *Mapping API's: Leaflet -Adding Google Basemap Layers*

The Potential of Designing a Digital Payment Framework for Philippine HEIs

D I Cendana[1], **T D Palaoag**[2]

[1]PHINMA-University of Pangasinan, Dagupan City, Philippines 2400
[2]University of the Cordilleras, Baguio City, Philippines 2600

dicendana@up.phinma.edu.ph, tpalaoag@gmail.com

Abstract. Cashless payment in educational environment among developed countries have been one of the norms to improving payment services. The main objective of the study was to explore the potential adoption and acceptance of a designed digital payment framework for Higher Educational Institutions in the Philippines to improve the payment services, utilize the use of student ID and solve behavioural issues of students in settling their school fees by converting student ID into Smart ID. A combination of qualitative and quantitative method was applied to present the data and findings. A pre-tested questionnaire was used for data collection and 288 participants among enrolled students were drawn using snowball purposive sampling. Results revealed that 9% among 6596 total enrolled students had remaining balance at the end of the semester. Whereas 66% scholars admitted to having failed in paying their fees on time due to some priorities (WM=3.24). Moreover, 75% among parents working abroad preferred the use of smart card for payment to ensure that money is consumed to tuition fee payment as to where it is intended. Further, USCTAM was used to measure the adoption and acceptability level which has resulted to 91.5% positive perception to using smartID in a university setting.

1. Introduction

Several educational institutions in developing countries such as Singapore, Malaysia and Zimbabwe have adopted smart technology created a norm for cashless payment to be implemented in their school-related transactions. This is evident as impending use of smart card has proliferated the educational environment. In the Philippines, two (2) universities- Lyceum University of the Philippines and De La Salle University attempted to implement the use of digital payment method using smart card. Apparently, no success stories were recorded to rationalize its utilization. After a careful collaboration, it was found out that its implementation has not undergone initial investigation to whether students and educational institution have positive adoption and acceptance regarding the use of smart card technology. Further, no such pre-investigation nor post study was conducted to investigate its usefulness to both the students and the school management. And so, the above-mentioned universities in the Philippines have brought a perspective for the possible use of smart card technology in schools. However, due to unsuccessful implementation, a broader study has to take place to be able to ensure its likelihood to be used in the educational environment. Thus, a new study was conducted to perform another investigation on student, parent and managements' perception to implementing smartcard in educational environment. Thus, this study aims to (1) identify the issues encountered by parents, students and school management in terms of tuition fee payment procedure (2) identify the factors to be considered in conceptualizing the framework (3) design a framework that is suitable in the implementation of smart ID cashless payment among Higher Educational Institutions (HEIs) in the Philippines, and (4) explore the acceptance and adoption of smart card technology in Philippine university setting.

1.1. The adoption of Digital Payment Method in the Philippines

In the Philippines, the Bangko Sentral ng Pilipinas (BSP) is the central bank that oversees, supervises banks and exercises regulatory powers over banks in the Philippines [1]. As part of its vision to adopt technological advancement in business sector, a study [2] confirms that BSP has been trying its best to improve the payment system in the country. As a matter of fact, BSP has developed the National Strategy for Financial Inclusion (NSFI) which focuses on the need to reach out to those who are financially excluded. The motivation to implement NSFI is to provide a facility for people living in urban areas to make online transactions through the use of cards due to poor physical banking access. Furthermore, it aims to encourage more customers to migrate from cash to non-cash payment. Meanwhile, to enable security on transactions, BSP has mandated all banks in the Philippines to adopt and integrate the Europay MasterCard and Visa (EMV) to the debit cards until 2017. Similarly a corroboration was disclosed that BSP adopted the NSFI in 2014 which aims to reach at least 20% increase on cashless transactions in 2020 but still needs to be observe because the success of its goal would still rely on the people's positive adoption and exposure to technology and with that, the merchants are expected to readily implement cashless transactions and be encouraged to go cashless which will enable them to create a surge of cashless payment channels and further improve its inclusion to the payment transactions in the Philippines [3]. It was also reported that the number of cashless transactions in the Philippines had increased to 10% ratio during the first quarter of 2018 as compared for the past 3 years which only gained about 3% increase [4]. Correspondingly, verified that the number of inflow and outflow of e-money transactions as recorded by BSP has evidently increased from 2015-2016 which has even reached almost 1 trillion pesos amounting to PHP 956.1 billion, a 4.3 increase from 2015 transactions [5]. Whereas, an increase to 21.4% and 8.9% was recorded from the number of registered and active e-wallet accounts. Additionally, the issuance of Automated Teller Machine (ATM) debit cards has increased to 24.1%. While prepaid and credit card did not increase, thus, it encourages that there is still a large reason to study the adoption of digital financial transactions in the Philippines [5].

1.2. School-related issues on tuition fee payment and collection

School fees and levies have become a bone of contention when schools open to both parents and the school management, yet alone, to boarding schools. This means that the essential existence of collections based from the tuition fee payment made by parents of students is important. However, several issues have risen as to payment collection and its security to schools. In Zimbabwe, it was discovered that because parents do not honour their financial obligations on time, schools were unable to purchase the necessary materials as soon as the schools open. Because of this, the school management is left with no alternative savings to employ disciplinary measure to compel parents to pay their financial obligations on time [6]. Also, Maryland school [7] has encountered persistent cash shortages due to recurrence of late payment from parents for children's tuition fee which has resulted to cash lying idle in the bank. What the school did was they have used both discount and payment model to address the issue on late cash payment by charging debtors an additional fee to induce parents to make timely payment. Similar issue on fee collection was conducted [8] to whether providing feedback on norms of university tuition fee payment could encourage students to pay their unsettled fees in University College London. To solve the problem, feedback has taken place by utilizing email facility to send reminders to those who have not settled their financial obligations. Accordingly, an investigation on late tuition fee payment collection was conducted in Jamaican

University in a manner that is effective but not sensitive to students' financial obligation [9]. Study has shown that that there was a significant relationship between IT system and students being financially cleared at the release of examination result. As to security issues, in handling payment collections from students, fraud has been prevalent to schools' finance department because staffs have the immediate access and control over the cash that is collected directly from student and it was observed that these staffs were the most vulnerable to fraud when dealing with money. Fraud was defined as a civil crime that provoke a person to do intentional deception to gain or damage another individual [10]. Because of the above-cited issue, this has become their motivation to investigate ways to prevent cash transactions in school.

1.3. Cashless Payment in Educational environment

The impending use of smart card has also proliferated educational institutions. As a matter of fact, schools in India, Nigeria, Singapore and European countries have adopted the Digital Payment Method (DPM) known as cashless transactions [11]. This is evident as to the study [12] which reveals that smart card has been widely used in developed countries not only for products but also in services like public transport system and it was even being usable in food courts, toll system, Enterprise Resource Planning (ERP), parking, supermarkets and even some schools in Singapore where cashless payment has been implemented for school fee related transactions. Similar study [13] has explored the possibility of developing smart card as a university application and investigated its potential to reduce the university business administration process and its positive impact to their service efficiency. Also, a study on the possibility of adopting cashless payment in schools in India where most of college administrators also agreed that the adoption of cashless transactions has no issue because millennial students are becoming excessively tech-savvy. Thus, their positivity to adopting new technology may result to a more convenient process to both parents and management especially when online payment is implemented for paying school fees [14]. Moreover, cashless payment in educational institution can led into easier financial transactions, transparency, lessen fear of theft, efficiency, keep track of expenses, benefits to banks and may even contribute to proper hygiene because students need not to carry cash on hand. With the potential use of cashless payment in school. Similarly, [12] designed a multi-purpose student card system using smart card technology for educational institutions which enhanced the typical student ID. The said multi-purpose student card was mainly intended to replace the magnetic student card and was eventually used in several transactions for retail, identification, transportation, financial transactions and for security measures. Whereas, study has resulted to a positive adoption of smart card among students because of its versatility, practical use, security, efficiency and usefulness. Whilst [15] unveiled that the implementation of cashless payment may require the existence of automated teller machine, electronic purse/wallet, smart card and may even demand banks to offer services such as mobile banking, e-transfer, and mobile money. Moreover, to successfully implement cashless payment in schools, [16] addressed distinct aspects that need to be considered such as economic, social, technological adoption among students, including institutional and law aspects.

1.4. Existing Cashless Payment Frameworks

Studies have been conducted to explore the impending use of smart card in promoting convenience to services and payment transactions. In the study[17], a smartcard framework was presented to securely manage applications on smart card. As a result, a User Centric Card Ownership Model (UCOM) was used to assist cardholders to manage applications on their smart card. The motivation of integrating UCOM was driven by Near Field Communication (NFC), Secure Element (SE) and Service Providers (SPs) which is commonly used in mobile transactions. With the use of aforementioned models, Trusted Service Manager (TSM) could further be employed to establish trust relationship to the industry and the telco. Additionally, using UCOM, cards may be acquired from User-Centric Smart Card (UCSC) suppliers who can also be a smart card manufacturer. By integrating smart card for payment schemes, there may already be an underlying Smart Card Operating System (SCOS).

Correspondingly, [2] revealed the challenges and opportunities of implementing electronic payment system in the Philippines while exploring the usefulness of smart card. Wherein, the researcher's conceptual framework has uncovered participants demographic and the kind of online transactions which was found to influence student's perception on electronic payment. Further, it was found out that the devices used to transact online and the payment scheme may also impact student's acceptance to the use of smart card in schools. While [18] prove that students and parents' initial knowledge and awareness in using smart technology has an impact to the continuous patronage in using smart card technology for payment transactions. Additionally, [19] revealed that students' exposure to smart card using Radio Frequency Identification (RFID) has influenced their adoption of smart ID for school related transactions. Also, the University Smart Card Technology Acceptance Model (USCTAM) was adopted in the study [20] to identify the parameters that can be measured to prove its acceptance in a university setting. The aforementioned study has further enhanced the model by integrating the theory of diffusion to adopt technology.

2. Methodology

A combination of Survey and interview were applied for data collection. The survey instrument was reviewed and approved by the CICTS Research Director from University of the Cordilleras, Baguio City Philippines. The survey form was uploaded in [21] for referencing. Data was collected from PHINMA-University of Pangasinan, Philippines and snowball purposive sampling was applied to draw the 317 students from University of Pangasinan, 8 parents were interviewed during the payment collection period and 2 employees from finance department were interviewed to know their perception to the possible implementation of smartID in the educational environment. Both quantitative and qualitative approach were adopted to assess the validity of findings. Regression analysis was used to analyze the relationship between perceived usefulness and the relative advantage of using smart card. In reference to the conceptual framework, a review of secondary references as conducted to gather data on existing studies relevant to the framework.

The conceptual framework was designed by considering the issues encountered by students, parents and management. Contact Smart Card (SSC) was considered in conceptualizing the framework to be able to manage the data that is stored in a programmable chip. Contact smart card is a type of card that is programmed using a reader that connects to the computer chip on the card. Thus, contact smart card is expected to have its SCOS installed on it and connecting it to Electronic Data Capture (EDC) machine when initializing transactions [12]. Through the programmable and integrated EMV chip, data are more secured and it would be more convenient to customize the data format as student Identification (ID) details should be coming from the institution which has been generated by its Information Technology Services Department (ITSD) during the period of application for students ID. By adopting this type of card, it is possible to manage the data according to (a) students' identification and student type (b) card status and account details (c) transaction details. Also, since bank will provide the smart card for the school, the Application Management Server (AMS) of the bank can be utilized to restrict the use of smart card within the institutional transactions only. AMS is concerned with managing the application processes that is installed into the smart card. This further contributes to ensuring security of transactions because the Service Provider (SP) of the bank will take care of the authentication of card through Application Services Authentication Server (ASAS) [17].

3. Discussion

A response rate of 90.85% (N=288) from PHINMA- University of Pangasinan students were gathered. 57.99% (N=167) of them were Male and 42.01% (N=121) were female. Participants also came from various year level knowing that students who have been in the university for several years would better express their experiences and thoughts regarding the existing payment system in the institution. Statistics has revealed that 46.88% were seniors, 31.60% were juniors, 14.93% were sophomores, 3.47% were 5th year and 3.13% came from freshmen level. Study has revealed that among 288 students, 67.01% of them are supported by parents for their education, 16.32% were supported by their relatives while 10% were supported by siblings and the source of income of the person supporting their education are based from their salary. Noticeably, 35.07% of the salaried employees are employed in their home country while 32.29% of them are Overseas Filipino Workers (OFW). Surprisingly, while parent/ relatives are employed in the homeland, analysis has shown that 31.06% prefers to give their children's allowance and tuition fee payment through bank deposit and remittance rather than giving it to the student's guardian since they are working outside provinces. While 54.51% among parents who are living with children prefers to give their allowance and payment fee in cash.

3.1. Issues Encountered by Parents and Students when Settling Account

Eight (8) parents were interviewed to know the issues they encounter when monitoring children's school record. Interview has revealed that 5 parents had a difficulty when talking to their children about the status of their grade and financial obligation in school because their child cannot directly answer the question of their parent on why their grades doesn't show on the student portal whenever these parents attempted to check their child's grade. They need to go to school and talk to the dean or college clerk to confirm the grade of their child, only to find out that the tuition fee of their child was still unsettled- the reason why grades are not viewable due to unsettled financial obligation. Moreover, when parents were asked why they opt to experience the long queue to pay their child's tuition fee instead of their child doing it alone, 5 parents revealed that trust issue has gave them the reason to do it so. Also, 3 parents have revealed that because students nowadays are more exposed to technology and photo editing, their children even tried imitate examination permit just to let the parents know that they have settled their account but actually not. Moreover, 6 OFW parents revealed that they usually give the budget directly to their children for their tuition fee payment before the examination period but they were unable to pay their fees because of the influence of their friends. Because of these behavioural issues of their children, they are left nothing but to avail of student loan in the university just to settle their tuition fee. However, parents are still left obliged to pay the student loan as indicated in the promissory note. Thus, resulting to another financial obligation for parents.

Two (2) groups of students were asked on issues they encounter when settling their school fees. Analysis has revealed that 45% among participants were scholars while 55% were non-scholars. Despite being a scholar where a Number (N=130), 66% among students still fails to pay their fees on time due to some priorities using a Weighted Mean (WM=3.24). Because of the notion that their allowances are given on hand, the students are tempted to spend the money on other important things rather than paying their school fee. This is evident as 70.8% (N=92) among them agreed that they still fail to pay the remaining fees they need to pay during exam period. The reason why 3.3% of 6596 enrolled tertiary students applied for student loan at the end of semester, knowing that there is a expected fund to be received for the next semester. The result also coincides with data from finance department that 166 scholars applied for promissory note ending the 2nd semester of the academic year. While 59.7% (N=77) were unable to get their exam permit despite scholarship funds have been forwarded to the institution. As a result, scholars were unable to get permit automatically and with that, they were unable to take the examination. Expectedly, 48% (N=61) scholars had no issue on long queue because they need not to pay on the teller as their names have been forwarded to the institution by the scholarship sponsors and so, payments are collected automatically by the institution. Non-scholar students (N=158) had comparable issue with scholars wherein 61% (N=89) were unable to pay their fees on time because of some personal expenditures once they have received their allowance. Thus, regardless of the type of student, students were enticed to spend the money once they have it hand but is supposed to be intended for tuition fee payment. Also, 63.3% (N=100) were unable to pay their tuition fee on time because they were unable to manage their money vs. personal expenses. Additionally, 60.9% (N=50) were unable to get their exam permit for the 3-exam period even fees are fully settled before the class semester starts. Hence, they need to wait for the announcement of per term exam period for the teller to release the exam permit. As to the convenience, 58.3% (N=91) students were unable to pay during payment period because the procedure is hassle and time consuming. This can be attributed to 57.7% (N=90) students who prefer not to pay on time because of the long queue. Thus, 58.3% (N=91) non-scholars prefers to pay at the end of semester instead. Remarkably, 65.2% (N=103) students agreed that they were unable to pay tuition fee on time because it's not them who manage their own budget and sometimes it is spent to something else rather than to their tuition fee payment.

3.2. Factors Influencing Students' Perception in Embracing Smart ID

Two hundred eighty-eight (288) respondents were asked about perception to embrace smart ID in school. Survey has revealed that 95% has positive response to adapt smart ID while 5% are not. This can be attributed to their answers which influences their perception to embracing the use of smart ID. Fig. 1 shows students feedback on the factors influencing their ability to embrace smart card technology.

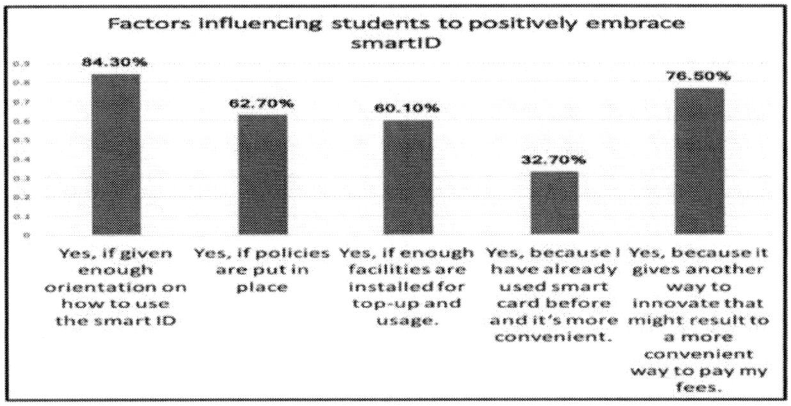

Figure 1. Factors influencing students' perception to embrace smart ID.

274 respondents conformed their positive perception to using smart id. 84.3% said YES if there is an appropriate orientation administered by the institution on how to use the smart ID; 62.7% would like that policy should be put in place before its implementation and; Consequent to the study of [15] as to the availability of facilities, 60.10% practically considered the availability of installed facilities in the campus such as top-up machines, card reader, smart ID production and collection of information relevant to the need of bank for opening a bank account for students. This is to ensure sustainable usage and implementation of smart card technology. Subsequently, 32.7% were already exposed to using smartcard and since they have already experience the convenience and security of using it, they are keen and positive to the potential implementation of smart ID for ease, convenience and innovative way of payment system in the university. Relative to the need of students for the orientation, survey also revealed that among 5% (N=14) students who have expressed their doubt to embracing smart ID 80% (N-11) have fear in using smart card. Analysis has found out that these students never experienced using smartcard for any transactions. Also, 30% (N=4) confirmed that they have no knowledge in using smart card and these students have never experience going to the bank for any transactions. Thus, they have this perception that smart ID might be impractical and insignificant for them. Further, students with positive perception in using smart ID have expressed their expectations whenever student ID will be converted to smart ID. Surprisingly, students nowadays have positive response to using technology because with WM=4.3 (N=280) they believe that using smart card can promote innovation, greater responsibility and awareness to students' behaviour in spending their money for payment transactions as comparable to the study undertaken by [12]; thus, using card for payment transactions can be more practical because transactions history can be easily monitored rather than paying in cash (WM=4.1, N=283).Wherein through transactions history, it might already help the student in prioritizing and managing their important expenses (WM=4.1, N=283) which might led to a more convenient and efficient payment procedures. Whilst, 64.7% (WM=3.9) feels that they might be

more secured if they only have the card with them instead of cash in their pocket while on their way to school. Table1 shows the actual data on student's openness to implement smart ID in HEIs.

Table 1. Students' openness to implement smart ID in HEIs

	STATEMENT	SA	A	N	D	SDA	N	WM	STD
a.	Using smart card for payment might promote greater responsibility, awareness and innovation.	38.9%	48.6%	12.1%	0.4%	0.0%	280	4.3	0.7
b.	I think using smart card in payment transaction is more convenient than cash	37.9%	42.1%	19.3%	0.7%	0.7%	285	4.1	0.8
c.	Using smart card for payment transaction might help me in prioritizing and managing my important expenses.	31.8%	45.9%	21.9%	0.4%	0.0%	283	4.1	0.7
d.	I think using card for payment transactions is more practical because I can easily monitor my expenses through transaction history rather than paying in cash.	28.3%	49.5%	21.6%	0.7%	0.0%	283	4.1	0.7
e.	I feel more secured on my way to school if I use card for payment rather than cash.	25.8%	38.9%	32.2%	2.5%	0.7%	283	3.9	0.9
f.	I have already used smart card before (e.i atm card, credit card, gcash, smartmoney, etc.)	29.0%	32.2%	24.4%	9.5%	4.9%	283	3.7	1.1
g.	Using cash for payment is more convenient than using card.	10.7%	31.8%	50.0%	4.3%	3.2%	280	3.4	0.9

3.3. Digital Payment Design Framework for HEIs

Factors on the above-mentioned results were taken into consideration to design a conceptual framework for the digital payment method that might be applicable in HEIs. Factors influencing students to embrace smart ID such as orientation, policies and, availability of facilities were discussed with the management to ensure sustainable implementation of the new technology. A positive response from the management was obtained to implement the new technology. Hence, active cooperation from Marketing, Centre for Students Development and Learning (CSDL) and Finance department is considered for policy making and orientation purposes. Also, a diplomatic partnership with bank was observed to properly negotiate the procurement of facilities essential to its implementation. This is to ensure that facilities are provided by a partner bank and data security is managed by the bank. Facilities to be provided by the bank includes production of smart card, acquisition of card readers, cellular phone for data connection and efficient transactions. The IT Services Department (ITSD) of the institution takes care of the production of smart ID which already integrate students picture, student number, card number and other details.

The digital payment framework works by acquiring the smart card by bank supplier. The number of cards to be produced is requested by the institution through the marketing department based from the projected number of freshmen students to enrol in the institution. After which, the bank sends request to their supplier. In this process, the bank may adopt UCSC framework when acquiring smart card from supplier where SCOS and EMV chip are already expected to have been integrated into the card. Thus, the card already contains the EMV chip for security and the card number. Once the bank has delivered the smart cards to the institution, ITSD may start producing the smart ID by collecting information from students. Prior to printing, ITSD should record the card number tagged to student number with their respective student details. A report is then forwarded to bank for monitoring. Once report is generated by ITSD, the bank may already assign a bank account number to students' account based from the forwarded student details with card number, student number, contact number of parents/ student and the address. Then a letter is sent by bank to the student/parent of student for them to receive their account number and temporary pin. Once received, the recipient is required to activate the card and change their password for security purposes. Parent then may already top-up the smart ID to any bank affiliated to bancnet. To pay school related fees, student may go directly to the teller section, selects the type of transactions, present his card for payment and the teller inserts the card to the Electronic Data Capture (EDC) machine. Once connection is established, the teller requests the student to enter his password on the EDC machine. For successful transactions, the machines will print two (2) copies of receipt. The first copy is given to student and the 2nd copy is left to the teller. Finally, the teller issues examination permit to student and electronic copy is also sent to students/parents' email address for immediate claim of the examination permit upon payment. SMS alert shall also be received by parents for notification purposes. The advantage of the smart ID is that it does not support withdrawal using the Automated Teller Machine (ATM) which are located outside the campus. The smart ID is purposely designed to solve student's behavioural issue in handling the money sent by parents that is intended for school related fees. Thus, the smart ID can only be used within the university campus and within the teller section of the university. Fig. 2 shows the conceptual design for card acquisition and digital payment framework.

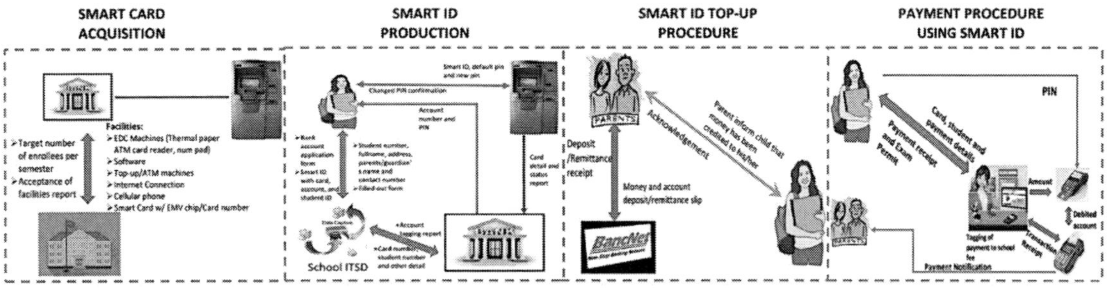

Figure 2. Digital payment framework for HEIs using smart

Additionally, since the contact details of parents were gathered by the institution, it is possible that the institution may come to an agreement with the bank to facilitate the auto messaging that may remind the parents of the upcoming tuition fee payment to be settled. Moreover, an auto debit agreement can be made possibly if the institution will make its effort to communicate the process to the parents and scholarship sponsors to ensure collection of payment at the end of the semester. However, appropriate guidelines and policy should also take place to ensure that parents and sponsors are made aware of the clearances.

3.4. Acceptability and adoptability of smart card in a university setting

A simulation on the use of smart card was conducted to measure its acceptability and student's adoption to using it in a university setting. Analysis has revealed that 67.8% had over 6-year exposure to using computing devices which found to have a significant difference with student's preferences on the indicator of images and information within the card resulting to $p=0.002$ where $p<0.005$. To measure the acceptability of smart card in a university setting, USCTAM was used which has 10 parameters namely (a) relative advantage (b) compatibility (c) triability (d) ease of use (e) demonstrability (f) visibility (g) indicators of image (f) indicators of information (g) perceived usefulness and (h) indicator of adoption. Using likert type of questions, participants answer was rated from 1(lowest)-5 (highest). Fig 3 shows the computed WM for each parameter. Additionally, students were asked on their perception on the need for smart card in the university and 91.5% agreed that smart card technology will really help in the manifestation of efficient payment transactions and solve behavioural issues of students in managing their money to prioritize school-related fees.

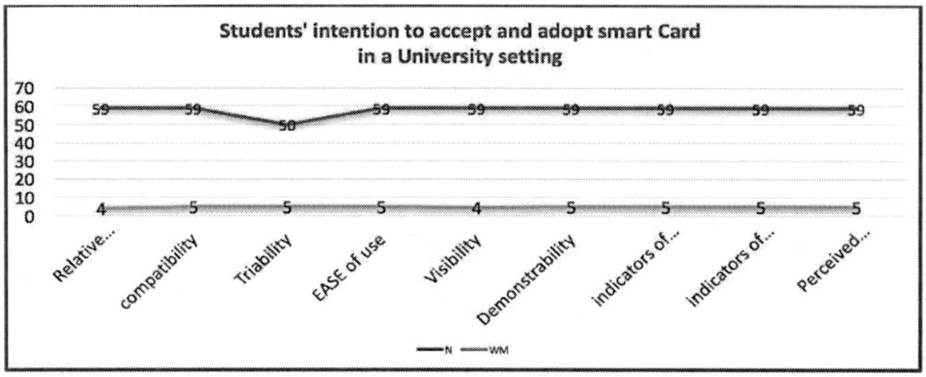

Figure 3. Parameters in assessing the acceptability of smart card

4. Conclusion

The study has provided relevant and adequate information for the HEI's to potentially embrace the use of Smart ID in a university setting. However, the conversion of student ID into smart ID requires the involvement of people that will comprise the successful implementation of the framework. The adaptability of both parents/scholarship sponsors and students to using smart card is needed to ensure a sustainable culture of embracing technology in the institution. The ease of use and convenience can be assured only when parents were able to transact through banks and students can afford to continuously use smart card instead of cash. Thus, the introduction of smart ID in HEI's may positively result to a potential adaptation among students, the school management and the banks. Study further concludes that the implementation of smart ID may possibly contribute to educational environment giving them the assurance that school fees are paid on time and promotes sense of responsibility for students to prioritize their financial obligations. While this study concludes 95% positive perception to embrace smart ID and 91.5% perceived need to use smart ID in a university setting, the parent's perception through wide data collection should have been considered to paramount the impact and sustainable use of smart ID.

Acknowledgments

This project study would have not been possible without the guidance and help of several individuals who in one way or another contributed and extended their valuable assistance in the preparation and completion of this study. Also, the researcher would like to acknowledge Mr. Norberto Padiernos and Arjay Perreras for sharing their ideas to construct the conceptual framework. The researcher's immeasurable gratitude is extended to the Commission on Higher Education for the sponsorship and PHINMA-University of Pangasinan for the support throughout the development of this research. Above all, the researcher would like to thank God for always giving her the courage, perseverance, the faith and determination to finish this research despite all the hardships encountered.

References

[1] Bangko Sentral ng Pilipinas 2015 *Overview of Functions and Operations* Retrieved from http://www.bsp.gov.ph/about/functions.asp

[2] Ching M R D 2017 Challenges and Opportunities of Electronic Payment Systems in the Philippines *DLSU Research Congress* Retrieved from https://www.dlsu.edu.ph/wp-content/uploads/pdf/conferences/research-congress-proceedings/2017/HCT/HCT-I-006.pdf

[3] Gamboa R 2018 Before going cashless *Philippine Star* Retrieved on 11 Feb 2019 from https://www.philstar.com/business/2018/12/18/1877827/going-cashles

[4] Manila Times 2018 Cashless transactions, e-payments soar in PH *The manila times* Retrieved on 11 February 2019 from https://www.manilatimes.net/cashless-transactions-e-payments-soar-in-ph/416936/

[5] Quimba F M A and Calizo Jr S 2018 *Going Digital: From Innovation to Inclusive Growth in the Philippines (No. DP 2018-19)* Retrieved from https://pidswebs.pids.gov.ph/CDN/PUBLICATIONS/pidsdps1819.pdf
[6] Ngwenya V C 2016 The Best Way of Collecting Fees without Infringing on the Liberties of Learners in Zimbabwean Primary Schools *Int. J. of Research in Business Technologies* **8(3)** pp 974-81
[7] Morrison H A 2018 *Cash management tool for Maryland Montessori Learning Centre, Agona Swedru* Retrieved on 18 February 2019 from https://air.ashesi.edu.gh/handle/20.500.11988/396
[8] Silva A and John P 2017 Social norms don't always work: An experiment to encourage more efficient fees collection for students *Plos one* **12(5)** p e0177354
[9] Chevers D, Archie J, Kerr-Gordon L and Hazel K 2018 The impact of information technology on the efficacy of tuition fee collection: a case in Jamaica *J. of Further and Higher Education* **42(2)** pp 237-47
[10] Wango G and Gatere A 2016 Integrity and Financial Accountability in Schools: Role of Principals' of Schools in Kenya *Int. J. of Education and Research* **4(4)** pp 1-14
[11] Bustillo N V, Cendana D I and Palaoag T D 2017 E-purse transit pass: The potential of public transport smart card system in the Philippines *2017 3rd IEEE Int. Conf. on Computer and Communications (ICCC)* pp 2786-90
[12] Omar S and Djuhari H 2004 Multi-purpose student card system using smart card technology *Proc. of the Fifth Int. Conf. on Information Technology Based Higher Education and Training 2004 ITHET 2004* pp 527-32
[13] Lee C H M, Cheng Y W and Depickere A 2003 Comparing smart card adoption in Singapore and Australian universities *Int. J. of Human-Computer Studies* **58(3)** pp 307-25
[14] Jain P 2017 Cashless System of Colleges in India *Int. J. of Social Sciences and Humanities (IJSSH)* **1(3)** pp 1-7
[15] Kumari N and Khanna J 2017 Cashless Payment: A Behaviourial Change To Economic Growth *Qualitative and Quantitative Research Review* **2(2)**
[16] Kabir M A, Saidin S Z and Ahmi A 2015 Adoption of e-payment systems: a review of literature *Int. Conf. on E-Commerce, Kuching, Sarawak* pp 112-20
[17] Akram R N, Markantonakis K and Mayes K 2009 Application management framework in user centric smart card ownership model *Int. Workshop on Information Security Applications* (Berlin, Heidelberg: Springer) pp 20-35
[18] Taherdoost H, Zamani M and Namayandeh M 2009 Study of smart card technology and probe user awareness about it: A case study of Middle Eastern students *2nd IEEE Int. Conf. on Computer Science and Information Technology 2009* pp 334-8
[19] Mirza A A and Alghathbar K 2009 Acceptance and Applications of Smart Cards Technology in University Settings *Eighth IEEE Int. Conf. on Dependable, Autonomic and Secure Computing 2009* pp 746-8
[20] Zahedi A 2008 *Intention to adopt smart cards: case of application in universities* disertation Retrived from http://urn.kb.se/resolve?urn=urn:nbn:se:ltu:diva-57701
[21] Survey Questionnaire Form 2019 *Designing a digital payment framework with an integration of SmartID for HEIs* Uploaded in http://bit.ly/SmartIDsqform dated 3 July 2019

IOP Conf. Series: Materials Science and Engineering **803** (2020) 012046 doi:10.1088/1757-899X/803/1/012046

Designing a Mobile Application Framework as an Innovative IT Solution for Waste Recycling

Ma. Melanie Ablaza- Cruz

Institute of Engineering and Applied Technology, Bulacan Agricultural State College, Bulacan, Philippines

E-mail: melanieablazacruz@gmail.com

Abstract. In today's modern society, we are facing the greatest environmental challenge, wastes management. With the growth of mobile application development, opportunities in enhancing the community's waste recycling practices have become easier. The study focuses on designing a mobile application as an IT intervention in helping to re-shape individuals in waste recycling habits. The purpose of this paper is to support everyone in strengthening their waste recycling practices within the community. This study focusses on designing a mobile application by integrating the hybrid architectural framework. Also, the researcher used a co-design methodology to define the content features, system requirements and to identify the user of the system. This framework and design consideration are recommended for utilization and implementation to aid every member of the community in their waste management practices through innovative solutions. The result of the study could raise awareness of how emerging technology be a solution to climate change. It attests that there is a great opportunity in improving the recycling habits of every individual through IT innovation, which different stakeholders could take advantage of. Somehow, this study helps individuals to recognize the importance of mobile technology to contribute to the solution to global concerns associated with climate change.

1. Introduction

Every individual is vital for valuable waste recycling; even the smallest alteration in recycling practices could influence the environmental and efficient sustainability of a city. Doing an active commitment to realizing righteous behaviors is a real challenge task since individuals living in the city are varied, have diverse social and cultural upbringings, and nationalities [1]. Despite several initiatives directed to work out the issues related to wastes, waste production as well as accumulation has persisted to increase every year [2]. Increasing population level, urbanization and growing living standards have enhanced solid waste generation in developing countries [1] [3] [4]. Improper disposal formed severe ecological problems such as destroyed water bodies, damaged the habitat of the wildlife, air pollution and it decreases the aesthetic value of the environment, etc. that have a direct impact on the society and the country's progress. Several studies prove that municipal solid waste has a greater percentage in having the impacts of the environment especially global warming [4] [5] [6]. Additionally, the Environmental Management Bureau specified that the waste in residential level constitutes the majority (56.7%) of waste within a community, this includes papers, plastic and glass bottles, kitchen scraps, backyard waste, foils, soiled tissues and diapers, and special wastes such as containers of household cleaning agents batteries and waste electrical and electronic equipment (WEEE) [4] [5]. Likewise, it is identified that the different establishments contribute 27.1% of which, in some provinces, about two- thirds of commercial wastes come from the latter. Approximately

12.1% accumulated from Institutional sources, while the remaining 4.1% is waste coming from the industrial or manufacturing sector [6] [7]. Similarly, the World Bank organization reported that by the year 2050 the world is expected to generate 3.4 billion tons of municipal solid waste annually [6].

In the same scenario, the Philippines problem in solid waste is highly crucial in urban areas, it remains the most pressing environmental challenge in the country. Currently estimated solid waste generation will increase by 0.5 kg/day to 0.9 kg/day in 2025, daily generation by 2025 is equal to 104,536 tons or about 38,155,640 tons annually [5] [8] [9]. However, the Municipality of Baliwag is a highly urban municipality in the province of Bulacan with a population of 153, 762 inhabitants in 36,091 households and considered as the economic hub in the westernmost part of the province. Additionally, an average of 85 tons of solid waste are generated every day, this has led to the volume of solid waste generated to go beyond what the available facilities can accommodate. According to the head of the Municipal Environmental Natural Resource Office (MENRO), one of the major factors that have contributed to poor waste collection and management in this municipality is limited community involvement in waste management. Waste management has been a great challenge, especially in highly populated cities around the world. One of the United Nations goals is to make cities inclusive, safe, resilient and sustainable, for an instance is an increase in the municipal waste collection [10].

Nowadays, mobile development and digital evolution are carving new era that affects human behavior and global governance. A mobile application can become valuable tools for every individual by integrating IT innovation in the battle for environmental protection, waste recycling and waste prevention [11]. Consequently, the explosive growth and usability of different gadgets such as tablets and smartphones unlock the possibility for several uses of technology, where end-users create applications as pragmatic tools for their interest and work [12]. It's time for each individual to embrace the advantages this technology in applying to waste management to contribute to a better and more responsible citizen within the community, to have direct contact between people who are willing to participate in the zero-waste goal of the country. The "trash tag challenge", where citizens are taking up photos on before and after cleaning specific area that is heavily littered, then sharing it to the community through social media to challenge other citizens to do the same [13]. Unilever Philippines also has its initiatives in helping the country to achieve the zero-waste goal within the community. They have this advertisement in social media to encourage citizens to refill shampoo/conditioner in their empty used bottles [13]. People among communities taking advantage of the usefulness of this emerging technology nowadays.

Given the fact that different efforts are being made by the government and non-government organizations related to waste problems, yet there are still major gaps to be filled in this area [14]. Consequently, despite the several already available mobile applications and studies regarding on the improving waste management and recycling activities has been presented [1] [15] [16], evidently there is still needs more study that would emphasize the individual level through the use of mobile application [17]. However, various research conducted in this area, a lack of user-centered design in recycling habit support tools can be one of the causes [1] [17]. For such reason, designing a framework for a mobile application to support waste recycling practice of individual become the core of this study. This study focuses on defining an ICT solution in helping re-shaping every individual in waste recycling practices. The co-design methodology approach has been applied in the study to identify a design consideration in terms of its content features, system requirements and the user of the system in developing a mobile application framework, that will support the solid waste management within the community, since this methodology has been proven as a good approach to be able to study the common features of the existing mobile application and the behavior of its user [11]. The result of the study could help the individual realize that emerging technology today can be part of a solution to climate change. It will enhance the involvement of every citizen who regards waste management. It would also attest that there is a huge room for the adoption of recycling habits and waste segregation that is facilitated through mobile applications, which different stakeholders could take advantage of. This study reinforces the citizens within the community to have a smart way of waste recycling. Through this application, the simple household has more opportunity to be a more responsible citizen and be part of a solution in environmental and natural resources security issues in the country. Finally, this study draws readers' attention to both the content and the importance of mobile technology.

2. Materials and Methods

Before gathering the data, approval from the Municipal Mayor was sought. Likewise, a courtesy call was made to the barangay officials and the Head of the Municipal Environmental and National Resources Office (MENRO). The qualitative research approach has been utilized by the researcher to attain its objectives. The Likert scale-based survey questionnaire was also distributed as a data-gathering tool. The sample size was also determined. This study specifically aims to determine the design consideration in re-shaping both household and community waste recycling habits in terms of a) architectural framework b) content and features c) system requirements and d) determine the users of the system. The initial stage of the methodology has been started using guided questions in conducting an in-depth interview with various respondents to be able to identify the different users and desired content and features of the mobile application. The acceptability of the application was measured using the Likert Scale. The evaluation criteria used in the study were based on the International Organization for Standard (ISO) 9241-210 [18] [19]. Also, the generally accepted definition of usability given by ISO in 92411-11 [20], which states that usability is the "extent to which a system, product or service can be used by specified users to achieve specified goals with effectiveness, efficiency, and satisfaction in a specified context of use" [20] [21]. This definition has been adopted in the majority of studies on the subject of mobile application usability. The co-design methodology was also used in this study [22]. Literature review from diverse published articles and journals from different databases similarly being done; to support the existing practices in addressing the needs of the study. An in-depth analysis of different documents was also done to design the proposed architectural framework of the mobile application.

3. Results and Discussion

3.1. The proposed architectural framework for the mobile application

The architectural design framework that would appropriate to address the need of the community regarding waste management through IT innovation is being shown in Figure 1. The proposed framework composed of the Web and Mobile platform that supports the requirements in mobile application development.

Fig 1. Architectural Framework for the Mobile Application

Determining the device, considering the bandwidth, navigating the methods, and defining the user interface are equally important factors to be considered in the study. Three different layers are included. Namely: *a. Presentation Layer:* This layer comprises UI Components and UI Process Components (Views and Controllers). In this layer, the user interface of the proposed mobile application has been considered. The different features and the location of every button and menu have been identified. However, concurrently, additional attributes like the design, background colors, font size, theme, etc. have been considered. *b. Business layer:* This layer focuses on the way business will be displayed for end-users. This comprises business components, workflows, and entities under the Service and Domain model layer. The service layer emphases on setting common application functions that will be an available user of the proposed application. While the domain model layer signifies knowledge linked to the problem domain. This layer focuses on enhancing the feature of the application. How the different functions of the mobile application are being implemented, how waste recycling activity is presented was also considered. The data synchronization also is considered as there is a web application support that enables every individual within the community access to the entire application. *c. Data layer:* In this layer, data related factors are being considered. The Data access components, data helpers/utilities, and service agents are included It also includes access to data, which can be stored in a database or backend, the latter is responsible for networking calls.

3.2. Content and features of the proposed mobile application

Bonino et al [1] emphasized that based on the several studies conducted that lack a user-centered design approach can be one of the reasons for the low adoption rate of the recycling application, a co-design methodology has been selected to design and develop a recycling support tool. This gives an idea in this study to adopt the co-design methodology to collaborate with the community to mutually define the needs of every individual about waste recycling application.

The involvement of the community was strategic and motivated by a specific purpose. The researchers consider the results of the data-gathering phase to finalize the content and features of the system in consonance with the needs of the users. The features were considered helpful for the users are the following:

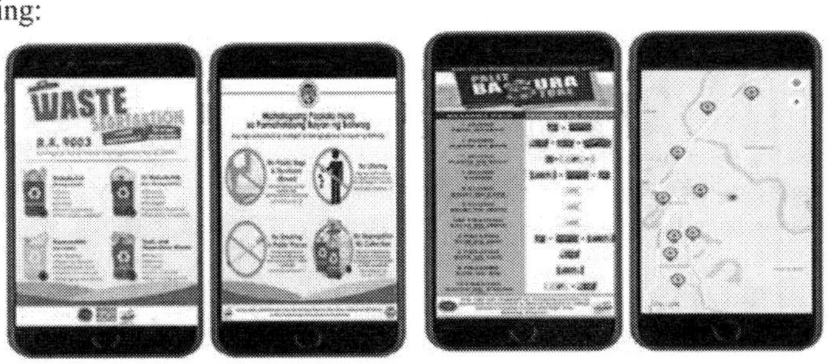

Fig 2. Interface design for the Mobile Application

1. Garbage Collection Schedule: This feature of the system will provide the list of the scheduled days for the trash collection and pickup services of every barangay. The Administrator can set-up the scheduled days of garbage collection in every barangay. This enables the MENRO head to easily monitor their staff who was assigned in a certain area through the remark's menu. This is where the staff has to fill-up whether a certain task has done or not. Households within the community will personally be updated for garbage collection days in their area so they can prepare it in advance.
2. Location Map: This feature helps the household owner to locate a participating store nearby where they can trade their waste recycled items into grocery items. Moreover, this feature of the system will also provide the household to find possible junkshop and remotely communicate with the owner to trade their recycled items into money. For the junkshop owner, this system features to aid them to easily locate the address of their customer.
3. Notification Features: These features help the MENRO office to be easily connected to the community. Through this mobile application, the MENRO Head can update their constituents about the different activities in waste management. Likewise, the junk shop owner will get informed once a household owner chooses them to collect their items. For the household owner, they will also be notified if the junkshop owner will accept their request.

4. Statistical Reports: This feature of the system is requested by the MENRO personnel for them to easily monitor the most traded waste recycled items from both junkshop and participating stores. The report generated from the application will help them in decision and policymaking.

3.3. Systems requirements of the proposed mobile application
Users shall use android smartphones to be able to access the different features of the application. Consequently, the mobile application will use cloud-based databases that are hosted by the android-based API to support the implementation of the system development.

3.4. Users of the system
This will be a great help for the community to reinforce each household to strengthen their recycling habits. Innovative learning strategy and practice can be adopted for the policymaker within the municipality. This Mobile application will fetch more junkshop owner to participate and could probably develop their active involvement, likewise, enhance their business. After the conduct of the interview and data gathering techniques, this study is thought significant for the following:

1. To the Household Owner.
   The application will cater to everyone to enhance their recycling habits and be able to update on the different policies of the municipality regarding waste management. Through the help of the mobile application, which in a way allow them to sell the recycled items directly to the junkshop owner to earn money and be part of the solution to climate change. They can also have easy access to the store nearby to exchange their recycled items into grocery products.
2. To the Junkshop Owner.
   Through the mobile application, they could have more opportunities to have more clients who will probably be their regular customer. This will allow them to be more accessible to their target clientele. By utilizing the application in their business process, it will surely lessen their time in looking for possible sellers.
3. To the participating store owner.
   This will be a great avenue to have more customers, and perhaps to get more involved in helping the community to lessen its waste disposal.
4. To the Municipality (MENRO Office).
   The application will be intended to enhance citizens' participation in waste management. Through this application, citizens can be updated on the different policies and announcements on the different programs of the municipality regarding waste management. It helps the management to be updated in the performance of the assigned garbage picker. Moreover, an avenue to track the registered junkshop owner within the community. It will also help the MENRO office to monitor the bulk of solid waste produces among households within the community for reference. However, the different participating stores in their "Palit-Basura Project" will also be monitored. Following the above considerations would be able to target the specific consumers. They can be able to engage themselves in using the mobile application in mitigating climate change.

3.5. Software quality factors evaluation for the mobile application
Quantitative Analytic was used in the study to measure the data gathered to satisfy the questions of the user's evaluation based on its usability.

Table 1. Assessment of the functionality of the mobile application

Characteristics	Items	AWM	Verbal Interpretation
Suitability	The mobile application has suitable and has an appropriate set of functions by its objectives.	3.79	Very Acceptable
Accuracy	The application delivers accurate results.	4.08	Very Acceptable
Compliance	The application addresses a certain set of needs.	4.09	Very Acceptable
Overall Weighted Mean		**3.98**	**Very Acceptable**

Table 1 shows the result of the assessment of the mobile application. Based on the given result, the compliance of the systems got the highest mean value with the average mean of 4.09 and it is interpreted as very acceptable. This means that the application addresses the needs of the users. However, the suitability got the lowest value with the average mean of 3.79. The overall weighted mean of the system based on the functionality of the system was 3.98 which is interpreted as very acceptable. Based on the given result in Table 2, fault tolerance has the highest mean, it was rated as 4.09. It means that the mobile application can maintain a specified level of performance in case of software faults or intrusion of its particular interface. Considering recommendations presented in the ISO 25010, functionality is not filled by the mobile limitations.

Table 2. Assessment of the Reliability of the mobile application

Characteristics	Items	AWM	Verbal Interpretation
Fault tolerance	The application can be easily sustained in case of software faults on its specified interface.	4.09	Very Acceptable
Recoverability	It can re-establish its level of performance and recover the data directly affected in case of failure	4.05	Very Acceptable
Overall Weighted Mean		**4.07**	**Very Acceptable**

The assessment in the reliability of the application has an overall weighted mean value of the 4.04, which is interpreted as very acceptable. Table 3 shows the result of the user's assessment of the usability of the mobile application. Based on the result, the understandability of the system has the highest mean value with 4.09 averages which is interpreted as very acceptable. It only shows that the interface design and the organization of the application can be easily understood by the user. The operability has the lowest mean value of 4.05 but still, it is interpreted as very acceptable. It only means that the application can be easily used. The overall weighted mean of the usability assessment has an average mean of 4.07 and it is interpreted as very acceptable. The result shows that the application was easy to operate. During the evaluation of this characteristic, the limitations of the mobile application was supported using the ISO 9126 measures and by providing other ones specific to mobile environments [23].

Table 3. Assessment on the Usability of the mobile application

Characteristics	Items	AWM	Verbal Interpretation
Understandability	It is easy for users to recognize its logical concept and applicability through its interface design.	4.09	Very Acceptable
Learnability	It is easy for the users to learn to use the application	4.07	Very Acceptable
Operability	The mobile application is easy to operate.	4.05	Very Acceptable
Overall Weighted Mean		**4.07**	**Very Acceptable**

Table 4. Assessment of the Efficiency of the mobile application

Characteristics	Items	AWM	Verbal Interpretation
Time behavior	The application has an acceptable response and processing time.	4.07	Very Acceptable
Resource behavior	It consistently uses mobile memory space	3.79	Very Acceptable
Overall Weighted Mean		**3.93**	**Very Acceptable**

The mobile application was evaluated by the users as very acceptable in terms of its efficiency as shown in Table 4. The highest mean value of the system was 4.07 in terms of its time behavior. The lowest rated mean was the resource behavior with the average weighted mean of 3.79 but still interpreted as very acceptable. The overall weighted mean value of the system was rated as 3.93, which is interpreted as very acceptable. According on the ISO 1998, the efficiency was measured by the resources expended in relation to the accuracy and completeness of accomplished goals.

From the result given in Table 5, based on the overall weighted mean result, the system can be easily maintained. Overall, the users evaluated the mobile application as very acceptable with a mean rating of 4.05. This implied that a mobile application can be easily sustained.

Table 5. Assessment on the Maintainability of the mobile application

Characteristics	Items	AWM	Verbal Interpretation
Stability	It is deemed stable when modified.	4.03	Very Acceptable
Testability	It is easy to validate any modification made.	4.07	Very Acceptable
Overall Weighted Mean		**4.05**	**Very Acceptable**

Table 6. Summary of Mean Performance Evaluation of the mobile application

Items	Mean	Interpretation
Functionality	3.98	Very Acceptable
Reliability	4.07	Very Acceptable
Efficiency	4.08	Very Acceptable
Usability	3.93	Very Acceptable
Maintainability	4.05	Very Acceptable
Overall Weighted Mean	**4.02**	**Very Acceptable**

Based on the summary evaluation result, the efficiency got the same highest rating of 4.08, which means that the end-users found out that the mobile application is effective to attain its objectives in terms of its purpose. While the usability indicator got the lowest rating of 3.93 which only means that some users need enough training in terms of using the application. The overall performance of the mobile application was based on the ISO 9241-11:1998, ISO 25062:2006 standards evaluation.

4. Conclusion

Improper waste disposal of every mankind is one of the issues that have afflicted for so long in different countries. Solid Waste Management (SWM) in the Philippines is an issue that needs serious attention. Though different programs have been implemented by the local government unit in every municipality, the participation of individuals is highly encouraged. In this paper, Innovation of technology has been studied and used in conceptually designing a system solution to help every individual to their waste recycling habits, which is an important step in waste management. Consequently, as mobile technology is becoming a norm for society, even simple household, garbage pickers, and junk shops cannot be left behind. This application could be very helpful in mitigating the climate change issues in the countries, through this mobile application, the communication, data collection, and processing, within the community will be easier. The study gives the potential of improving the waste recycling habit of an individual through IT innovation. This also increases the awareness of waste recycling and proper dissemination of information regarding IT innovation, this would be beneficial to both individuals in every municipality. Hence, support from the municipal office would have to be further considered. Finally, for further studies, researchers like to assess the mobile application in terms of its impact on the community.

References

[1] Bamodu 2013 Application of Mobile Technology in Waste Collection *Int. J. of Applied Sciences (IJAS)*

[2] Aljaradin P 2012 Comparison of Different Waste Management Technologies and Climate Change Effect-Jordan *American J. of Climate Change* pp 57-63

[3] Warunasinghe W Y 2016 A survey on Household Solid Waste Management(SWM) with Special Reference to Peri-urban Area (Kottawa) in Colombo *Procedia Food Science*

[4] Putthakasem 2018 Scenarios of Municipal Solid Waste Mangement for Mitigating Greenhouse Gas

Emission *7th Int. Conf. on Informatics, Environment, Energy and Applications*

[5] Environmental Management Bureau 2018 *Solid Waste- Environmental Management Bureau* [Online] Available: https://emb.gov.ph/wp-content/uploads/2018/09/3-Solid-Waste-1.8.pdf

[6] World Bank org 2018 Global Waste to Grow by 2050 *Unless Urgent Action is Taken: World Bank Report, Washinton September 2018* Press Release No: 2018/037/S [Online] [Accessed April 2019]

[7] U Philippines 2019 *Beat Plastic Pollution Phipippines* 16 March 2019 [Online] Available: https://www.facebook.com/BeatPlasticPollutionPH/posts/540435906363837

[8] Grace S 2018 Solid Waste Management and Sustainable Development: The Case of Philippine Cities *Int. Conf. in Urban and Regional Planning*

[9] Williams I and K J 2003 Green waste collections and public recycling behavior in the borough of wyre resources,Conservation and Recycling *Elsevier Applied Science Publisher* vol **38** pp 139-59

[10] U NATIONS 2019 *Sustainable Development Goals 11: Make cities inclusive, safe, resilient and sustainable* [Online] Available: https://www.un.org/sustainabledevelopment/cities/

[11] Bonino D, Pastrone C, Delgado M T and Spirito M 2016 WasteApp: Smarter Waste Recycling for Smarter Citizens *Splitech 2016*

[12] Tran 2014 *Int. J. of Pervasive Computing and Communications* vol **10** no 2

[13] Unilever Philippines 2019 *Beat Plastic Sollution Philippines* 16 march 2019 [Online] Available: https://www.facebook.com/BeatPlasticPollutionPH/posts/540435906363837

[14] Aljaradin M and Persson K 2012 Comparison of Different Waste Management Technologies and Climate Change Effect-Jordan *American J. of CLimate Change*

[15] Mavropoulos A, Tsakona M and Anthouli A 2015 Urban Waste Management and the mobile challenge *Waste Management and Research*

[16] Mavropoulos A 2009 Recycling behavior: the present focus brain and a framework to understand personal differences in recycling *ISWA World Conference 2009 12-15 October 2009*

[17] Mavopoulos A, Anthouli A and Tsakona M 2019 *D-WASTE White Paper* D-Waste Team [Online] Available: http://wwwd-waste.com

[18] Bevan N, Carter J and Harker S 2015 What Have We Learnt About Usability Since 1998? *Computer Science* vol **9169** (Cham: Springer) pp 143-51

[19] Dillon A 2016 The evaluation of Software Usability *Usability evaluation and redesign recommendations for MyNet portal thesis guidelines* November 21

[20] Weichbroth P 2019 A mixed-methods measurement and evaluation methodology for mobile application usability studies *Communication Papers of the Federated Conf. on Computer Science and Information Systems* vol **20** ISSN 2300-5963 ACSIS pp 101–6

[21] ISO 2018 *Ergonomics of human-system interaction Part 11: Usability: Definitions and concepts* ISO 9241-11:2018(en) E

[22] Lucero A *et al* 2012 *The dialogue-labs method: Process, space and materials as structuring elements to spark dialogue in co-design events* (Abingdon: Taylor & Francis, Informa Ltd) Registered in England and Wales Registered Number march 2012

[23] Moumane K, Idri A and Abran A 2016 Usability evaluation of mobile applications using ISO 9241 and ISO 25062 standards *SpringerPlus*

[24] DA-BAR 2018 *Climate Change* [Online] Available: https://www.bar.gov.ph/index.php/programs/major-programs/climate-change [Accessed 27 June 2018]

[25] World RiskReport 2014 World Risk Report 2014 *Alliance Development Works 2014*

[26] ITU 2012 *Measuring the information society: executive summary CH-1211 Geneva, Switzerland* (2012) [Online]

[27] Bonino D 2017 *The Internet of Things* p 173

IOP Conf. Series: Materials Science and Engineering **803** (2020) 012047 doi:10.1088/1757-899X/803/1/012047

Design considerations of a collaborative and knowledge sharing tool for a sustainable outcomes-based education

D H Galeon[1], T D Palaoag2

[1] College of Computing Sciences, Ifugao State University, Lamut Ifugao, 3605, Philippines
[2] College of Information Technology and Computer Science, University of the Cordilleras, Baguio City, 2600, Philippines

Email: galeon.dh@gmail.com, tpalaoag@gmail.com

Abstract. Through a directive, the Philippine higher education institutions (HEIs) are instructed to continuously perform the plan-do-check-act cycle in developing an outcomes-based, progressively brought together and less departmentally based educational plans as they advanced into a mature institution which requires the stakeholders to communicate and collaborate among each other and coordinate their activities. The paper aims to unfold the design parameters of a collaborative and knowledge sharing tool (CKT) considering the most critical factors affecting its implementation and shall serve as a guide to its development. Using a descriptive-analytical method, it shows that the institution has a strong knowledge management (KM) culture but needs to improve and retool their KM initiative and that the stakeholders are engaged more on socializing and control activities yet a lesser sum on networking, coordinating, and collaborating activities. Decomposed from the problem, design parameters of the said tool includes the plan and usage of a messaging module used for communicating and disseminating knowledge to a network; authoring, assessment and expert profiling modules used for a collaborative acquisition, development and dissemination of knowledge; and a digital library and an academic profiling module used for storing, retrieving and mining the institutional outcomes-based education (OBE) knowledge.

1. Introduction

1.1. Definition, background, and application

While a few universities find in sustainability another method for sorting out and profiling themselves [1], educational sustainability in numerous institutions is at beginning periods [2] where sustainable development (SD) endeavors are not comprehensively incorporated all through the HEI framework [3]. The Ifugao State University (IFSU) can gain from the experience of Malaysian universities where they had developed meaningful and collaborative learning activities, environments, materials, systems in support of their OBE implementation [4]. While Malaysian educational institutions implement OBE at all levels of education especially at HEIs [5], it is now a policy of the Philippine Commission on Higher Education (CHED) to require, support and monitor the formation of an OBE in all its HEIs offering computing programs. In compliance with the policy, HEIs shall develop a complete set of program outcomes and their performance indicators together with its assessment system; a curriculum; an outcomes-based syllabus; a system of program assessment and evaluation; and a system of program continuous quality improvement (CQI) as inputs to the institutional quality assessment (ISA) promoted by CHED [6]. Studies show that faculty members in some Philippine HEIs need more training in OBE implementation [7] including outcomes-based syllabus preparation, use of student-focused methodologies and learning path sequencing [8]. While the activity of curriculum development ought to be profoundly comprehensive, participative and democratic [9], curriculum mapping requires a ton of contemplations including but not limited to collaboration [10], institutional support [11], leadership, communication in exchanging information and coordination to come up with unified and less departmentally based educational plans [12]. While learning outcomes assessment requires considerably more complex and tedious, as it demands more time [13], it ought to be a progressing, iterative and coordinated process [14] that should keep on being a part of worldwide monitoring of quality education [15]. This will finally result in decisions and activities with respect to CQI [16], which is believed to be the driving force to keep institutions headed straight toward curriculum excellence and in consonance with the ISO 9001 quality administration framework [17]. While IFSU is advancing its maturity in its OBE implementation, there is a need for the institution to design and implement a collaborative and knowledge-sharing tool (CKT) for stakeholder support and service innovation. While KM initiatives rely upon knowledge sharing (KS) [18], it is smarter to assess how individuals share information normally and then fabricate a framework to help those activities. The paper focuses on unfolding the design parameters of a CKT considering the most critical factors affecting its implementation. The design considerations shall serve as a guide to the development of a CKT.

1.2. Related studies

KM is a socio-technical procedure [19] where social association, correspondence, and cooperation are in its center [20]. In fact, the present pattern is toward KM by social technologies (ST) [21] such as social networks, social media, and KMS since they are instrumental and have provided an effective channel for the KM processes [22]. It is also a fact that KM and STs equally require each other in order to succeed [23]. It is believed that the group feature of these technologies is a total KM package [22]. These tools are developed using communication technologies that allow users to access needed knowledge and to communicate with each other; collaboration technologies that provide the means to perform group work; and storage and retrieval technologies that are using a database management system to store and manage knowledge [24]. In many cases, these tools should be custom-built when used as support to KM to be capable of addressing the special KM development needs of the organization [25]. With the fast expansion of technological opportunities to improve interaction and increment coordinated effort in organizations, HEIs seeking after KM will bound to succeed if they complement innovative parts of knowledge management system (KMS) developments with the collaborative systems [26] as it increases productivity [27]. Collaborative systems are, per magnificence, tools that advance formal and casual interaction, vital to knowledge sharing and discovery. They bolster communication, coordination, and cooperation both regarding managerial administrations and strategic administrations, at the level of educating, learning and research activities [28]. STs should be designed so as to be easy to use, build with web technologies and guarantee that security concerns are offset with the requirement for transparency [29] and created with better search strategies [30]. SMS-based tools shall be highly accessible by both knowledge experts and knowledge end-users, comply with standards and authority regulations, practical and sustainable [31-32]. It shall allow expressive creativity by enabling the creative authorship of own ideas, creative conversation by supporting learning and fruitful endeavors, personal mastery by assisting in maintaining and developing one's intellectual capital, encouraging empowerment by offering the means for assisting and mentoring others, and for the common good while it eases the wider sharing and faster diffusion of ideas [33]. It shall also provide tools such as creating online meetings [34] and faculty contact tracking system [2]. Typical elements of gamification like badges, leader boards, ratings, or levels can also be integrated into the design since it motivates individuals to participate in such platforms [35]. Overall, the collaboration function of these tools is intended to associate people with people through communities of practices (CoPs), to preserve the discourses, and to invigorate cooperation by incorporating the knowledge repositories and collaboration applications such as work process [36].

2. Methodology

The study is part of the project of building a KMS as a tool to support the quality OBE processes in IFSU. It includes the following phases of KMS design, development, implementation and evaluation. The studies of design considerations and architectural frameworks are the main deliverables in the design phase. Other deliverables of the project include a system prototype in the development phase, a study on the acceptability level of the developed system in the implementation phase, and a study on the impact of the KMS implementation in the evaluation phase. The result of investigating the KM practices of IFSU serves as the primary source of data needed in the KMS design phase. A descriptive-analytical and survey method was utilized in this paper to understand the KM practices of IFSU. It focuses and investigated the level of the commitment and awareness, organizational culture, strategy, organizational structure, people, and technology factors in facilitating KM and knowledge acquisition, development, distribution, measurement and review processes[37]; the level of stakeholders' personal knowledge sharing along the areas of socializing, networking, practicing, storytelling, encountering [38], communicating, collaborating, coordinating and controlling [39]. The KM factors identified in this paper are based on various studies. The commitment and awareness, organizational culture, strategy, organizational structure, people, and technology factors were formed based on the study of Obaide (2004). The personal knowledge sharing practices of the stakeholders along with the areas of socializing, networking, practicing, storytelling, encountering were based on the study of Panahi (2014). The KMS functional domains of communicating, collaborating, coordinating and controlling were based on the study of Spector et al. (2002). The stakeholder's perception of the outcome of their knowledge sharing activities was based on the study of Galeon et al. (2019). Asrar-ul-Haq and Anwar (2016) conducted a systematic review of KM and KS on trends issues and challenges. Their study evidently shows that trust emerged as the most significant factor. Among other factors are culture, organizational structure, information technology, top management support, leadership, organizational commitment, social relations, communication, and an individual's willingness and enthusiasm to impart knowledge [40]. Thirty-four participants composed of the directors from various service departments, the information technology education (ITE) faculty members and the senior students from the college are selected using purposive judgment non-probability expert sampling and are able to provide data for analysis. The participants composing of administrative officials (35%), faculty members (33%), students (32%); below 25 years old (33%), 26-40 years old (35%), 41-55 years old (26%), above 55 years old (6%); high school graduate (29%), baccalaureate (3%), masters (47%), doctorate (21%) are selected since they are observant, of interest, reflective member of the institution, who know much about the institution's KM practices and are both able and willing to share their knowledge. Research analysis was performed by using statistics, tables, and charts.

3. Findings and Discussion

3.1. Level of institutional KM practices

A baseline value of 51% is taken to portray the majority opinion against the KM practices. Any practices reported strongly agree and agree for its implementation and critical and important for its importance by less than 51% of the respondents are not accounted for here. From questionnaire findings, it remarkably shows that the majority of the participants are fully aware of the importance of investing resources in the field of KM (see Table 1 and Table 2), however, they are still exploring the opportunities of a variety of KM practices and KM tools.

From this study, it became evident that the stakeholders' positive behaviors towards KM were fully exploited in the institution while they lack a commitment to a program of KM improvement and managing it to ensure benefits. To improve their strategy and to invest in their KM tools with modern technologies, detail shows that there shall be a senior level ongoing review of the effectiveness of KM in the institution; there shall be defined responsibilities and a budget set for the KM initiative; OBE intellectual assets shall be inventoried or recognized and some measure of value shall be attached to each; key performance indicators for KM shall be in place; modeling systems, decision support systems and artificial intelligence shall be implemented; directories of OBE experts shall be available; search for OBE information shall be easier; and OBE procedures and lessons learned from experiences shall be documented and stored in a database. Such improvements shall increase their performances with their KM practices along with the areas of acquisition and development, and measurement and review.

Table 1. Overall view of questionnaire responses for the KM factors

KM Factors	Implemented (%)	Important (%)
Culture	100	100
Structure	80	100
Awareness and Commitment	80	100
People	78	100
Technology	55	100
Strategy	50	100

Table 2. Overall view of questionnaire responses for the KM lifecycle phases

KM Lifecycle Phases	Implemented (%)	Important (%)
Distribution	75	100
Acquisition and Development	57	100
Measurement and Review	40	100

3.2. Level of the stakeholders' personal knowledge sharing practices

With a baseline value of 51% taken to portray the majority opinion against the KS practices, the study evidently shows that they communicate and exchange experiences, beliefs, and ideas via informal dialogues and conversations (80%) and they ensure the integrity of these information (100%) to a great extent while it is striking that locating eminent experts to track and follow their experiences or opinions while developing collaborative relationship (0%), coordinating their activities (25%), and collaborating among each other to create, modify and disseminate OBE knowledge (40%) are practiced to the less extent (see Table 3 and Table 4). The details show that the stakeholders needed more support while they engage on interacting, participating and networking to other individuals to create and share knowledge products and discuss OBE practices; locating eminent OBE experts in the field who are willing to help; documenting and storing professional OBE experiences, lessons learned, or some important information found in the literature; broadcasting or publishing OBE materials or share particular skills, tips, ideas, and thoughts; searching articulated tacit knowledge; and reporting unusual OBE practices and developing discussions about them.

Table 3. Overall view of questionnaire responses for the KS domains

KS Domains	Implemented (%)	Important (%)
Socializing	80	100
Practicing	67	100
Storytelling	67	100
Encountering	50	100
Networking	0	100

Table 4. Overall view of questionnaire responses for the KMS functional domains

KMS Functional Domains	Implemented (%)	Important (%)
Control	100	100
Communicating	67	100
Collaborating	40	100
Coordinating	25	100

Result also shows that the stakeholders will more likely to continuously share knowledge since the majority of the respondents confidently perceived that they have gained positive experience in it.

There is strong support for KS among the stakeholders (82%), they understand the importance of sharing knowledge effectively (79%), they know how they benefit from sharing knowledge (77%), and they share knowledge easily (56%). Likewise, the majority of the respondents have a positive response to the antecedents of KS including affective commitment (82%), trust (79%), organizational commitment (79%), helping colleagues (77%), and reward and sufficient motivation (56%).

3.3. Design Considerations of a Collaborative and Knowledge Sharing Tool for a Sustainable OBE

The axiom design theory (ADT) was applied to produce the descriptive design considerations of the CKT. Relative to the IFSU's vision of upholding excellence and its quality policy, at the top level, the system has the primary design intent of "Sustaining quality and excellence in OBE by socio-technical system" (see Figure 1). This requirement has tended to position the institution from complying with the CHED OBE policy and it can be accomplished by the design and implementation of a CKT. From the primary design intent, three functional requirements are described including establishing a knowledge network, facilitating collaboration among the stakeholders and building institutional memory. These requirements appear to be the immediate need of the institution based on the result of the investigation on their KM practices. While IFSU is exploring KM tools and KM practices, the Universiti Teknologi Malaysia (UTM) has perceived the strategic value of KM and has been intensively exploiting a variety of KM tools and KM practices. UTM have invested through their new academic learning innovation (NALI) initiative the advancement of their institutional memory for their digital learning resources in the forms of UTM e-Learning, open courseware (OCW), UTM MOOC, BLOSSOMS, student-to-student edutainment, and video of exemplary professionals (VoEP) [4]. The design intent of establishing a knowledge network can be achieved by the implementation of a social tool (ST). Its implementation intends to facilitate communication among the stakeholders allowing them to interact, participate and network to other individuals to share and discuss their personal experiences and lessons learned. It also allows them to synchronize and integrate their activities, responsibilities, and command and control. The ST, specifically a chat and discussion forum system, shall facilitate the preservation of discourses and communication among the stakeholders while acquiring knowledge, externalizing experiential knowledge, coordinating, and attending discussions and knowledge dissemination activities via online meetings and online chat rooms. The OCW and VoEP used by UTM are exemplary instruments in knowledge sharing. It empowers the users from around the world to share learning materials created by instructors and recordings displaying knowledge and aptitudes by experts from UTM. Moreover, the e-Learning utilized by UTM has support for real-time communication. The functional requirement of "facilitating collaboration among the stakeholders" can be attained by designing an interactive learning and co-production environment specifically a knowledge desk (KD). The KD has three components including designing an authoring tool, knowledge assessment tool and an expert profiler. The KD generally provides tools to capture, configure, reuse and control institutional OBE knowledge and manage users, experts' profiles and permissions. Knowledge objects being captured include curriculum maps; course syllabus; assessment plans; documented procedures, lessons-learned, experiences, and knowledge in the form of articles; perceived knowledge conceptual and operational justification in the form of ratings; and high-impact educational programs contents, performance, and feedback. To have a one-stop place for OBE analysis for all instructors UTM had implemented an online OBE analysis system (OAS) to mechanize the entire procedure of OBE analysis which turns the OBE investigation for all students and programs into a basic task and has demonstrated to give incredible help to the staff during the accreditation practice [4]. The KD controls knowledge by enabling combination, reconfiguration, updating, organizing, and validating knowledge. The digital library (DL) as another requirement provides tools for storing and retrieving explicit OBE knowledge objects such as OBE resources, articles, manuals, policies and the likes. The academic profiling (AP) provides a tool for the stakeholders to view summaries of their OBE knowledge and practices for decision-making. A relationship formed between the functional requirement (FR) of "Facilitating collaboration among stakeholders" and the design parameter (DP) of "Design a knowledge disseminating and synthesising tool" in the first level decomposition of the design leads to a triangular matrix that makes the design a decoupled. Other levels of decompositions for the FRs are uncoupled, which is an ideal design. The full design matrix table of the design is decoupled, which is good, and an acceptable design in ADT. It is the best feasible design since it satisfies the independent axiom.

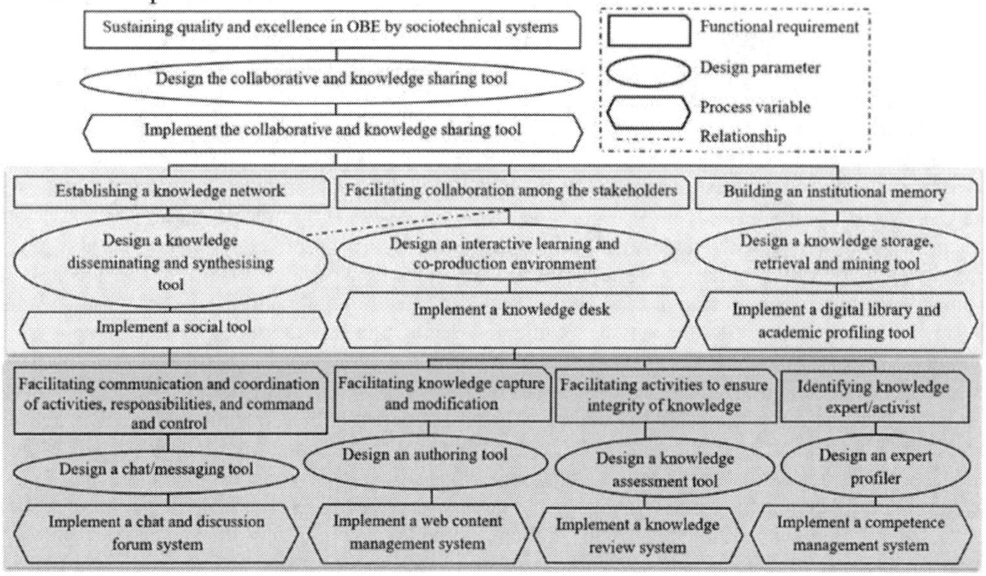

Figure 1. Design Considerations for a CKT

4. Conclusion

With almost four years when the institution has shifted to OBE while exercising their innovativeness and creativity in the development of their outcomes-based curriculum, the stakeholders have a positive behavior to the CHED's OBE mandates and they need strategic and sociotechnical solutions to sustain in the competitive world. The identified design considerations presented in this study serves as a CKT development and implementation guide for the institution that could eventually produce sharable OBE knowledge objects and taxonomies which can eliminate the education divide and will help keep the institution sustain curriculum excellence. It serves as a benchmark to researchers on KM concept and design. It helps future researchers in coming up with a framework directed to create a global KM standard for HEIs considering the different levels of analysis and supporting scientific design theories. Succeeding studies include the design of a CKT framework; development of a CKT prototype; and an assessment on the significant input of the CKT implementation to the institution as well as its acceptability by the users.

References

[1] Wals A 2013 *J. of Cleaner Production* pp 1-8
[2] Abdous M and He W 2013 An online knowledge-centred framework for faculty support and service innovation *VINE* **43** 1 pp 96-110
[3] Lozano R, Ceulemans K, Alonso M, Huisingh D, Lozano F, Waas T, Lambrechts W, Lukman R and Hugé J 2015 A review of commitment and implementation of sustainable development in higher education: results from a worldwide survey *J. of Cleaner Production* **108** pp 1-18
[4] Alias R and Aris B 2016 *A Guide to New Academia Learning Innovation 1st* (Malaysia: UTM Academic Leadership)
[5] Hassan S, Admodisastro N, Kamaruddin A, Baharom S and Pa N 2016 Developing a Learning Outcome-Based Question Examination Paper Tool for Universiti Putra Malaysia *Int.*

Education Studies **9** 2 p 132
[6] Philippine Commission on Higher Education 2015 *CMO* No. 25 s. 2015 Quezon City
[7] Pepito T 2019 Perspectives on outcome-based education among faculty members teaching business courses at a Philippine university *The Palawan Scientist* **11**
[8] De Guzman M F, Edaño D and Umayan Z 2017 Understanding the Essence of the Outcomes-Based Education (OBE) and Knowledge of its Implementation in a Technological University in the Philippines *Asia Pacific J. of Multidisciplinary Research* **5** 4 pp 64-71
[9] Khan M and Law L 2015 An Integrative Approach to Curriculum Development in Higher Education in the USA: A Theoretical Framework *Int. Education Studies* **8** 3 pp 66-76
[10] Gulikers J, Biemans H, Wesselink R and van der Wel M 2013 Aligning formative and summative assessments: A collaborative action research challenging teacher conceptions *Studies in Educational Evaluation* **39** 2 pp 116-124
[11] Biggs J 2014 Constructive alignment in university teaching *HERDSA Review of Higher Education* **1** pp 5-22
[12] Harden R M 2000 The integration ladder: a tool for curriculum planning and evaluation *Medical Education* **34** 7 pp 551-557
[13] Telsang M 2015 Outcome Based Education - Design Delivery and Assessment of Product Design and Development Course at Undergraduate Engineering Program *J. of Engineering Education Transformations* special issue p 145-9
[14] Weistroffer H R and Gasen J 2010 Assessment of Information Systems Programs: A Life Cycle Approach *J. of Education for Business* **70** 5 pp 258-62
[15] Barrett A 2016 Measuring Learning Outcomes and Education for Sustainable Development: The new Education Development Goal *Oxford Studies in Comparative Education* ed W C Smith (United Kingdom: Symposium Books) **25** 1
[16] Castillo R 2014 A Paradigm Shift to Outcomes-Based Higher Education: Policies, Principles and Preparations *Int. J. of Sciences: Basic and Applied Research* **14** 1 pp 174-86
[17] Aziz A, Noor M, Ali A and Jaafar M 2005 A Malaysian outcome-based engineering education model *Int. J. of Engineering and Technology* **2** 1 pp 14-21
[18] Wang S and Noe R 2010 Knowledge Sharing: A Review and Directions for Future Research *Human Resource Management Review* **20** 2 pp 115-31
[19] Al-Busaidi K, Olfman L, Ryan T and Leroy G 2010 Sharing Knowledge to A Knowledge Management System: Examining the motivators and the benefits in an Omani organization *J. of Organizational Knowledge Management* 2010
[20] Omona W, van der Weide T and Lubega J 2010 Using ICT to enhance Knowledge Management in higher education: A conceptual framework and research agenda *Int. J. of Education and Development using Information and Communication Technology* **6** 4 pp 83-101
[21] von Krogh G 2012 How does social software change knowledge management? Toward a strategic research agenda *The J. of Strategic Information Systems* **21** 2 pp 154-64
[22] Mushonga C 2014 *Social Networking for Knowledge Management: Group Features as Personal Knowledge Management Tools* Thesis (Stellenbosch: Stellenbosch University)
[23] Gunjal B, Gaitanou P and Yasin S 2012 Social Networks and Knowledge Management: An Explorative Study in Library Systems *Knowledge Management* **2** pp 64-83
[24] Halawi L, McCarthy R and Aronson J 2017 Success stories in knowledge management systems *Issues in Information Systems* **18** 1
[25] Dehghani R and Ramsin R 2015 Methodologies for developing knowledge management systems: an evaluation framework *J. of Knowledge Management* **19** 4 pp 682-710
[26] Abdullah R, Selamat M, Sahibudin S and Alias R 2005 A Framework For Knowledge Management System Implementation In Collaborative Environment For Higher Learning Institution *J. of Knowledge Management Practice*
[27] Nunamaker J Jr, Romano N Jr and Briggs R 2001 A Framework for Collaboration and Knowledge Management *Proc. of the 34th Hawaii Int. Conf. on System Sciences*

[28] Pinto M 2014 Knowledge management in higher education institutions: A framework to improve collaboration *9th Iberian Conf. on Information Systems and Technologies*

[29] Butler T, Heavin C and O'Donovan F 2007 A theoretical model and framework for understanding knowledge management system implementation *J. of Organizational and End User Computing* **19** 4 pp 1-21

[30] Ward V 2017 Why, whose, what and how? A framework for knowledge mobilisers *Evidence & Policy* **13** 3 pp 477–97

[31] Garcia P Jr, Galeon D and Palaoag T 2018 TEXTnology: A Managed SMS Knowledge Sharing for Farmers *Proc. of the 2018 7th Int. Conf. on Software and Computer Applications* pp 279-84

[32] Galeon D, Garcia P Jr and Palaoag T 2019 SMS-Based ICT Tool for Knowledge Sharing in Agriculture *Int. J. on Advanced Science, Engineering and Information Technology* **9** 1 p 342-9

[33] Schmitt U 2014 Making Sense of e-Skills at the Dawn of a New Personal Knowledge Management Paradigm *Proc. of the e-Skills for Knowledge Production and Innovation Conf.* pp 417- 47

[34] Chen M, Liou Y, Wang C, Fan Y and Chi Y 2007 TeamSpirit: Design, implementation, and evaluation of a Web-based group decision support system *Decision Support Systems* **43** 4 pp 1186-202

[35] Schacht S and Maedche A 2015 Project Knowledge Management While Simply Playing! Gaming Mechanics in Project Knowledge Management Systems *Gamification in Education and Business* ed T Reiners (Berlin: Springer) Chapter 30 pp 593-614

[36] Bose R 2003 Knowledge management-enabled health care management systems: Capabilities, infrastructure, and decision-support *Expert Systems with Applications* **24** 1 pp 59-71

[37] Obaide A 2004 *A model for a successful implementation of knowledge management in engineering organizations* (Salford: University of Salford)

[38] Panahi S 2014 *Social Media and Tacit Knowledge Sharing: Physicians' Perspectives and Experiences* (Brisbane City: Queensland University of Technology)

[39] Spector J, Michael E and Gerald S 2002 Knowledge Management in Instructional Design *ERIC Digest* pp 1-9

[40] Asrar-ul-Haq M and Anwar S 2016 A systematic review of knowledge management and knowledge sharing: Trends, issues, and challenges *Cogent Business & Management* **3** 1 DOI:10.1080/23311975.2015.1127744

Customizing Academic Knowledge Management System Architecture Framework based on Enterprise Architecture

D E Maligat Jr.[1], J O Torio[2], R T Bigueras[3], M C Arispe[4] and T D Palaoag[5]

[1,2,3]Camarines Norte State College, Daet, Camarines Norte, Philippines 4600
[4]Bicol University Polangui Campus; [5]University of Cordilleras

Abstract. This work is centered on making sure that it fosters out the purpose of knowledge management systems framework for the sake of making use understand how best it is used in terms of capturing, utilizing and transferring knowledge from one entity to another. The literature review as well as the document analysis were both used in combination for the purpose of making sure that all the interviews that were used were better expounded in different platforms such as stakeholder, faculty and administrator. Further, the academic knowledge management system Architectural framework had to be utterly customized on matters to do with the intelligence concept, Debrose and the Enterprise Architecture. It is through the use of this framework that then functionality of the organization is categorized in to three namely; the people, technology as well as the application that is required for use. Further, it is true that there are some benefits that are linked with the use of this kind of framework that are easily accessed on the available knowledge and helps in decision making process for the entire administration and the knowledge repository creational matters.

1. Introduction

The There are various techniques as well as approaches linked with the Knowledge Management System. The KMS architecture is one of techniques that are used based on the four layers and it requires functions such as the internet, extranet, intranet for the sake of making or setting up a client server [1] [2]. The second step entails making sure that the organization framework is taken into considerations required, this process will be very essential as it will help to ensure that there is prosper observation that is done on the survey to ensure that the system periphery is well cared for. It is by taking into consideration all the steps that are required in KMS that a person will be in a better position of interpreting the information as well as the decision and the information that is used and analyze how one came to such a conclusion. On the same note, we realize how useful it is for one to embrace the use of two path association as it is very essential in terms of coordinating any learning disclosure as well as sharing and even creating of any form of information [3].

KMS is very profound network that is used for the purpose of improving knowledge, and any other kind of information that is useful to the entire organization [4]. KMS is also useful in terms of making sure that it offers great support that is useful with the information that is provided in the organization. In addition, such a process link with the knowledge creation which is a very useful process for use in the organization. Apparently, any form of organization will tend to make use of knowledge to ensure that its progress are working out in the right manner [5]. Therefore, knowledge creation through KM is of greater use to the company success. Importantly, it is worth noting that KM is used for the sake of reaching and bring out positive outcome that linked well with the aspects of knowledge creation and information that is required for use by the entire organization. On the same not, KM is used by different entities as they all look forwards to behind different knowledge aspects in the organization. Again, it is then responsibility of KM to ensure that supplies and outlines all that areas that require more attention in any given organization. In simples terms it helps the management identify areas of weakness and seek for means to better such areas.

Further, KM is useful for the purpose of conveying various dynamics that ensure that the right models of business operations are well chosen for the organization to document its activities in a more beneficial manner. KM help in making sure that the dynamic information is utterly and best conveyed through a well presented reporting mechanisms [6].

Studying and analyzing Knowledge Management (KM) is of greater essence as it helps in making sure that human thinking and the ways humans tend to view different activities across the globe is well analyze. On the same note, there is a need to work out better plans and strategies that will to ensure that KM prove more beneficial to the entire organizational activities that are taking place wither inside or externally. This implies that KM is linked with new ideas and new through that are used by the organizational management to bring advancement in the technological world of the entire organization. KM help manager to employ the use of better strategies that are of greater use and of benefit to the organization. In addition it is worth noting that KM is of greater use to the organizational activities as it helps with the development of dynamics of the organization [7].

Apparently, there are various challenges that link with KM. These challenges include; the IT infrastructure that need some new reformations and urgency of getting restored within the shortest time possible. The IT of any organization ought to be on the stand still to ensure that the functionality and activities taking place are of greater essence [8]. KM is of greater use in one knowledge orientation [9]. However, it face the challenge of changes that are taking place in the knowledge world as there is the issue linked with many infrastructures that have to be to be us3d for the purpose of forming new information. In addition, there is the issue that link with the development and deploying of knowledge that proof to be a very tough process [10].

Consequently, there are other challenges that are evident when Knowledge management is applied in any higher Institution of learning. At the present age, there are no links to measure the kind of knowledge that is exist in a given institution. In the higher institutions of learning, there is the growing need for KM s to ensure that is contributes to other educational services that can be render to entire learning community. However, there are many challenges linked with the success of education that is administered in these institutions. In addition, there is a need to recognize and acknowledge the fact that there management efficient techniques that have to be employed for the sake of perfecting the educational services in these institutions. Therefore, this simplifies the fact that KMS can be of greater use in terms offering quality education services and in ensuring that there is increased managerial efficiency in every learning institution. Under increased efficiency there will be institutional competency and fiscal autonomy.

At present there are no HEI's in the Bicol Region that is equipped with a KMS deployed in their school. Therefore, it is clear that this kind of program is of greater use as it helps with the study of the present KM s framework that is useful in the technology advancement of the region. This kind of knowledge is used for the sake of capturing, utilizing and sharing as well as transferring all the required knowledge in the required manner that will help in improving the academic performance of the students in the identified region. In addition, the framework is very useful as it shows how best KMS implementation should be done for the sake of capturing any explicit and implicit knowledge useful in higher education learning institutions. The framework shows how KMS implementation should be done in order to capture explicit and implicit knowledge for the HEIs.

The main objective of this study is to present a KMS framework that can served as guide of academic institutions to capture, utilize, share and transfer knowledge.

2. Literature Review

To ensure that the contribution that were made in the study were utterly improved, there was a need to conduct a literature review that would be much essential and necessary in deducing the contributions of both theories and concepts in the study success.

2.1. Enterprise Architecture (EA)

Form critical analysis, we can deduce the Enterprise Architecture plays a sensitive role in any given organization as it helped to control and coordinate all the underpinning by working to predict the present state in link with the present day that can be of greater use to any organizational planning matters. Also, we can explain Enterprise Architecture as a guide that coordinate various strategies and procedures that underpins all the undertaking of any given organization [11]. Apparently, any organization is comprised of diverse personnel who are working from various division of work and they all have different views in regard to the organization. Also, there are different methodologies that are endorsed by person working in different executives as they tend to endorse various aggressive methodologies. Notably, on the side of enterprise architecture [present a single system, that is coherent in structure and is much characterized with various activities as it links with different parts and section of the organization. Moreover, it is the work of EA to ensure that it delivers the need to ensure that it oversees as well as compose various parts of the organization [12].

It is the work of EA to ensure that it administers all the required practice that are useful in supporting the digitalization of learning to help ensure that it improve the exhibition of the entire organization [13]. Whenever EA tend to join the objectives, methodologies and various strategies that are useful in coordinating and advocating for all organizational automations. It is true that the architectural objectives are all centered on the view that the organization is accorded in terms of its performance and creativity. Figure 1 shows the pyramid that contain the organizational architectural layers namely; people, applications, the technology employed for use and the entire business at large [14].

Figure 1. (Tucker and Debrosse, 2003) Enterprise Architecture

2.2. Business Intelligence (BI)

It is true that there are major's ways that are used for the purpose of designing a program. To some extend the individual are much more centered on thinking in terms of info structure and infrastructure [15]. It is true that knowledge is very essential and is useful to working things out and make sure that there is a significant standard that is accepted [16]. In other words, knowledge helps to give better results that will be important in a real scientific discipline [17]. Moreover, we realize that the common KADS are much more important for use in working all activities out as linked to all articulated programs in the British or the European programs [18].

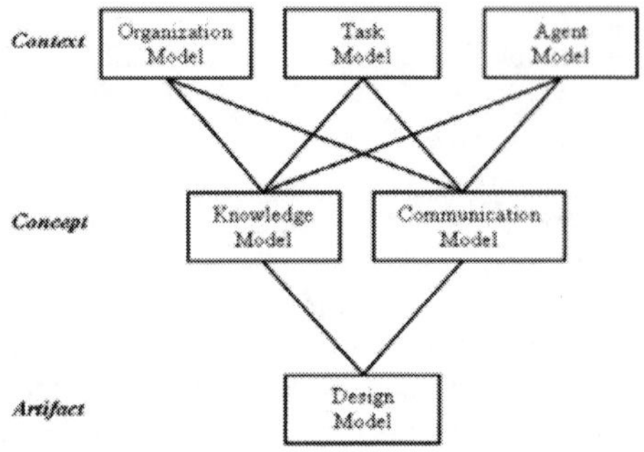

Figure 2. CommonKADS model suite

The major goals of KMS include; promoting different methods for use include storage and communication factors. In addition, the three major goals that help with the promotion of new the flow of encouraging ideas taking place between the entire organization and the knowledge being deposited there in nutshell.

The services used include; storage and communication knowledge services. All these form of services that were used were mostly geared on presenting as well as analyzing the knowledge required for use in perfecting the entire situation [19].

Figure 2 shows evidenced that the KMS design was used in various business groups. Three major goals are to promote the method of producing new information, encourage the flow of ideas between organization members and guarantee the ease of access to knowledge repository. Lastly, presentation services involved with developing the interface among user and the information/knowledge source

3. Methodology

There were several interviews and consultations that were conducted with other academic stakeholders such as administrators, various campus directors, students and the entire faculty. In addition, the IT section were also interviewed for the purpose of acquiring data that is essential for the use in different academic institutions. In addition, there was the inclusion of interviews and other document analysis process for the sake of perfecting the proposed academic architecture framework.

4. Results and Discussions

The proposed academic Knowledge Management System framework is centered on business intelligence concepts and both the Debrose and Tucker. In simple terms the proposed framework modifies the organizational architecture layers into three categories namely; technology, application and the people. All the three organizational architecture layers comprise of People, technology and the entire application process.

Figure 3 shows the three (3) organizational architecture layers as: people, technology and application, and process.

4.1. People

In this architecture, the people are viewed as being the primary source of data in KMS. These are then articulated as different data source in an academic institution. It comprises of faculty, the non-teaching and well as the personnel program and schools heads alongside all the primary uses of the KMS.

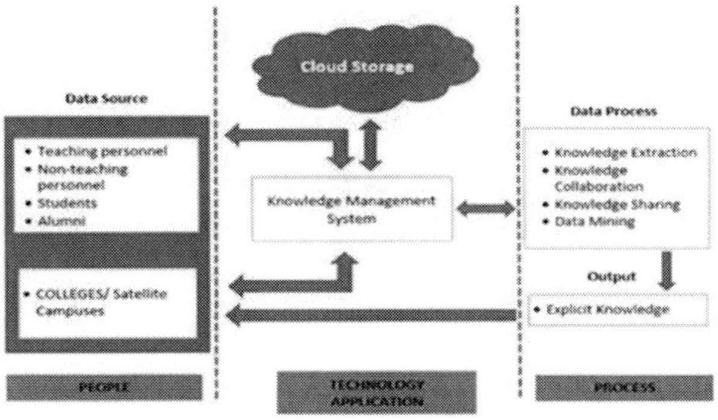

Figure 3. Proposed Academic KMS Architecture Framework

4.2. Technology and Application

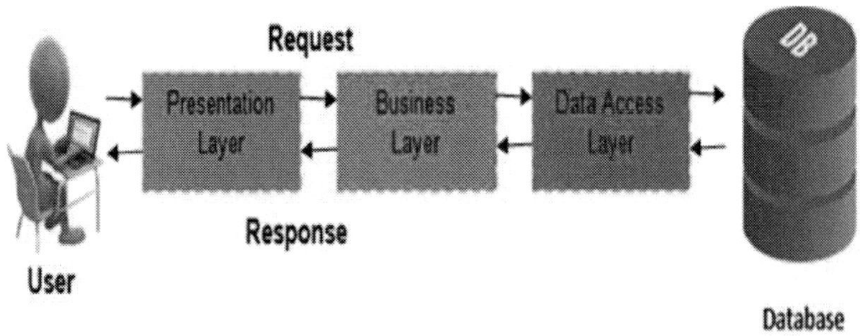

Figure 4. Simplified Three-Tier Architecture

In this architecture, Evident the technology and application process that requires both internet and the application technology. Further, the propose KMS is bound to be developed in Three-tier architecture as figure 4 tries to show us in general. Notably, the hierarchal architecture is consisted of three distinct and even in depended tiers of layers of use. The user pushed a request over presentation layer after that it sends directly the request to business layer. The business layer afterwards manages the request by supplementing business and data rules, also business processes and so on, and forward the enhanced request through data access layer. Data access layer opens database for the reply and send it back to the business layer and in turn sends the response back to the presentation layer, which shows the information to the user on a user interface (UI) [20].

4.3. Process

It is true that knowledge sharing as well as the aspect of data mining are the major essential processes that are evident in the proposed academic architecture. The figure below evidenced how KM can be of greater use in terms of getting implemented on an academic setting. Again knowledge sharing is more technological bound rather than just addressing the future sections

In figure 5, the model served as guide on how KM process should be implemented on the academic settings, moreover, model displays which of the three groups are more people oriented and which are more technology focused. Knowledge sharing should be mainly technology focused is certainly arguable and it is something that I will address in future sections. However, this is largely how organizations manage to contact the issue i.e. as a technological relatively as organizational and social challenge [21].

IOP Conf. Series: Materials Science and Engineering **803** (2020) 012048 doi:10.1088/1757-899X/803/1/012048

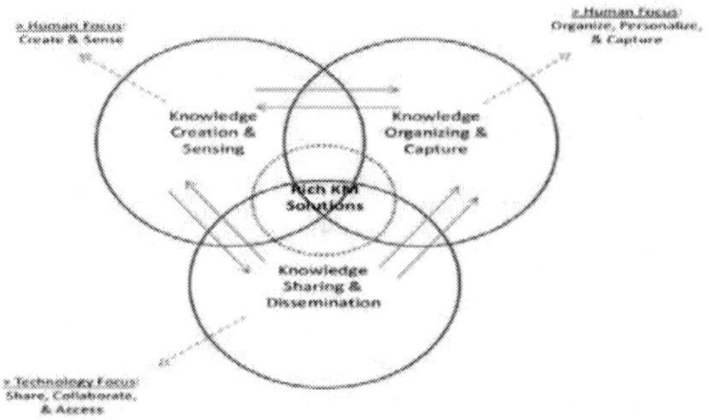

Figure 5. The Knowledge Management Process Model by Botha et al (2008)

4.4. Output

It is through the use of explicit knowledge that a better decision making process is attained. In addition, the knowledge is essential as it bolsters the exchange of information that is attained through workers. [22].

5. Conclusion

The study was highly centered on customization of knowledge management system that is of higher essence architectural framework that is highly essential with the development of the KMS. Evidently, there are many knowledge frameworks that exists and are highly centered on business/enterprise knowledge Management architectural framework. Assertively, the proposed architectural framework based on academic needs as well as processes that help with the facilitation of the transfer, help with sharing and the collaboration of knowledge linked with various academic stakeholders. Moreover, the academic framework that was created was highly customized on business alongside enterprise KM framework that helped to suit all the needs of the academic institution at large. Some of benefits linked with the use of this framework are the creation of available knowledge and availability 24/7.

References

[1] Abdullah R, Selamat M H, Sahibudin S and Alias R A 2005 A framework for knowledge management system implementation in collaborative environment for higher learning institution *J. of Knowledge Management Practice* Mar **6(1)** pp 1-8

[2] Bourdon I and Kimble C 2005 An analysis of key factors for the success of the communal management of knowledge *Proc. of 13th UKAIS Conf. Bournemouth* **x** Apr 4th

[3] Dube R 2017 *An analysis of the knowledge management systems in an organization with a center-periphery knowledge management landscape* dissertation (Stellenbosch: Stellenbosch University)

[4] Jennex M E and Olfman L 2002 Organizational memory/knowledge effects on productivity, a longitudinal study *Proc. of the 35th Annual Hawaii Int. Conf. on System Sciences* Jan 10th pp 1029-38

[5] Schreiber A T, Schreiber G, Akkermans H, Anjewierden A, Shadbolt N, de Hoog R, Van de Velde W, Shadbolt N R and Wielinga B 2000 *Knowledge engineering and management: the CommonKADS methodology* (Cambridge: MIT Press)

[6] Grover T H V 2001 General perspectives on knowledge management: Fostering a research agenda *J. of management information systems* May 31st **18(1)** pp 5-21

[7] Holsapple C W and Joshi K D 2003 A knowledge management ontology *Handbook on knowledge management: Knowledge matters* ed C W Holsapple (Berlin/Heidelberg: Springer-Verlag)

[8] Marhraoui M A and Manouar A E 2017 Towards a new Framework linking knowledge management systems and organizational agility: an empirical study *arXiv preprint*

arXiv:1707.08182 Jul 24th

[9] Campbell H M 2006 The role of organizational knowledge management strategies in the quest for business intelligence *2006 IEEE Int. Engineering Management Conf.* Sep 17th pp 231-6

[10] Edwards J 2010 A process view of knowledge management: it ain't want you do, it's the way that you do it *11th European Conf. on Knowledge Management*

[11] FEA Program Management Office 2007 *FEA Practice Guidance* November Retrieved from https://goo.gl/QIqI1V.Ffff

[12] Nonaka I 1991 *The knowledge creating company Harvard Business Review,* (November-December) pp 96-104

[13] De Vries M and Van Rensburg A C 2008 Enterprise Architecture-New business value perspectives *South African J. of Industrial Engineering* **19(1)** pp 1-6

[14] Tucker R and Debrosse D 2003 *Enterprise Architecture Roadmap for Modernization* (Virginia: Mitre Corporation)

[15] Gartner Inc 2013 *Gartner Business intelligence* Retrieved November 9th 2015 from http://goo.gl/LmJRG3

[16] Bichsel and Jacqueline 2012 Analytics in Higher Education *Rep. EDUCAUSE Center for Applied Research* Retrieved from https://library.educause.edu/resources/2012/6/2012-ecar-study-of-analyticsin-higher-education

[17] Soares L, Steele P and Wayt L 2016 Evolving Higher Education Business Models: Leading with Data to Deliver Results *American Council on Education* Retrieved from http://www.acenet.edu/newsroom/Pages/Evolving-Higher-Education-Business-Models-Leading-with-Data-to-Deliver-Results.aspx

[18] CommonKADs *What is CommonKADS?* Retrieved from https://commonkads.org/

[19] Tiwana A 2000 *The knowledge management toolkit: practical techniques for building a knowledge management system* (New Jersey: Prentice Hall)

[20] Techopedia *A Detailed Look at 3-Tier Software Architecture* Retrieved from https://www.techopedia.com/2/32100/software/a-detailed-look-at-3-tier-software-architecture

[21] Knowledge Management Tools 2012 *Three KM Models* Retrieved from http://www.knowledge-management-tools.net/three-km-models.html

[22] Moscoso-Zea O, Luján-Mora S, Cáceres C E and Schweimanns N 2016 Knowledge Management Framework using Enterprise Architecture and Business Intelligence *ICEIS* **1** pp 244-9

e-LAHOK: An e-Participatory Platform for Disaster Risk Reduction and Management

Jennifer L Llovido[1, 3] and Thelma D Palaoag[2, 4]

[1] Bicol University, Legazpi City, Philippines
[2] University of the Cordilleras, Baguio City, Philippines

E-mail: [3] jllovido@bicol-u.edu.ph and [4] tpalaoag@gmail.com

Abstract. The study aimed to design, develop and determine the extent of user acceptability of e-LAHOK (Legazpi Alisto at Handang magbigay ng Opinyon para sa Kaligtasan), an electronic and participatory Disaster Risk Reduction and Management Platform for the City of Legazpi that fosters civic engagement and open participatory governance by soliciting Disaster Risk Reduction (DRR) strategies and incident reports from the local populations. An integrated Design Thinking (DT) and Extreme Programming (XP) framework (DT@XP) was utilized in the design and development of e-LAHOK whereas, a user's evaluation based on ISO 9126 software quality metrics was conducted to determine the extent of user acceptability. Significant findings showed that e-LAHOK facilitated orderly solicitation and submission of DRR strategies and incident reports from the local populations by allowing the creation, management, consolidation, analysis and visualization of survey responses as well as submission, verification and acceptance of submitted incident reports for analysis. These results could serve as basis to concerned agencies in the development and implementation of DRR plans and policies to ensure preparedness and effective response by the barangays and its citizens in the event of a disaster. Further, evaluation results show that the system was rated highly acceptable in terms of functionality and usability. This concluded that the features of e-LAHOK provided an organized way of gathering inputs from the local communities of Legazpi City, fostering civic engagement and open, participatory governance. Thus, utilization of the proposed system is recommended.

1. Introduction

The world has witnessed an alarming increase in the frequency and severity of disasters resulting to loss of lives or injuries, damage to properties and economic losses. Disaster Risk Reduction (DRR) is the concept and practice of reducing disaster risks through systematic efforts to analyze and reduce the causal factors of disasters [1].

The province of Albay is one of these places in the country whose vulnerability from natural disasters is relatively high. Currently, the Provincial Government of Albay has successfully minimized and avoided loss of lives as well as damages to properties by actively implementing the zero-casualty program, however, during the tropical depression Usman last December 30, 2018, Albay had 18 casualties with Legazpi City, its capital having three deaths [2]. Since Legazpi City has to deal with disaster risks and vulnerabilities, there is a need to be prepared for such.

Maceda et al. designed and developed a disaster risk damage assessment system to facilitate and enhance the organization of submission of disaster related reports of each coordinating agency, and the generation of damage assessment reports to effectively manage the process in assessing the damages for Filipinos [3]. Secillano and Local Association of DRRMOs of Albay [4] recommended for a more collaborative approach to learning from experiences towards improving practices and policies, hence, the need to further improve the disaster management practices and policies of Albay.

Community-based disaster management (CBDM) is anchored in the disaster risk reduction framework that covers a broad range of interventions, measures, activities, projects and programs to reduce disaster risks, which are primarily designed by people in at-risk localities and are based on their urgent needs and capacities [5], [6]. Hence, vulnerable groups and communities can be transformed to disaster resilient communities. Concepts of community participation is observed in this study, specifically with the use of Information and Communications Technologies (ICTs).

E-Participation is about fostering civic engagement and open, participatory governance through ICTs. Its objective is to improve access to information and public services as well as to promote participation in policy-making, both for the empowerment of individual citizens and the benefit of society as a whole [7]. It is from this perspective that a participatory online platform can be used as an instrument that could better harness the potential of ICT within citizen participation initiatives. Hence, citizens and government can become partners with shared responsibility for policy formulation and decision-making.

Many existing Philippine government-initiated practices take a top-down approach to knowledge sharing, treating local communities as passive clients instead of active partners. At present, there is no organized way of gathering inputs from the local communities of Legazpi City on how the barangays can help them be better prepared in times of disasters. Thus, the main purpose of this study is to design, develop and determine the extent of user acceptability of e-LAHOK, an e-participatory platform for Disaster Risk Reduction and Management (DRRM) that will solicit DRR strategies and incident reports from the local populations specifically the children, youth, elderly, persons with disability as well as the pregnant and lactating mothers who are the most vulnerable to disasters. Such collected responses are subjected to computer-assisted analysis and the results of which may be considered for a more inclusive disaster risk management approach in the development and implementation of DRR plans such as the Barangay Disaster Risk and Reduction Management Plan (BDRRMP) of barangays in Legazpi City.

2. Methods

In the design and development of e-LAHOK, the integrated Design Thinking and Extreme Programming (DT@XP) framework as proposed by Sohaib et al. was utilized. Various DT practices (empathy, define, persona, DT user stories) are adapted into XP exploration phase, while, prototyping and usability evaluation into XP planning phase [8]. This type of methodology is flexible and agile which has the capability to adapt to change according to the needs of the researchers.

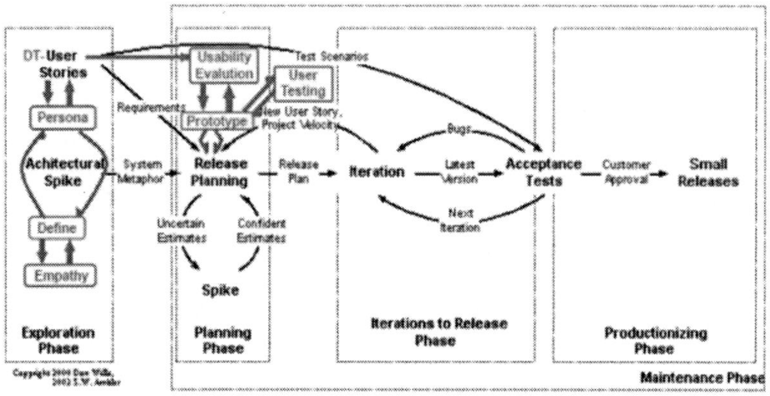

Figure 1. Integrated Framework of DT@XP

IOP Conf. Series: Materials Science and Engineering **803** (2020) 012049 doi:10.1088/1757-899X/803/1/012049

2.1. Exploration Phase

The initial phase of XP is exploration. User stories are created and released in this phase for the planning phase. To make a realistic customer expectation, empathy and define practices of DT along with user persona to create a DT-user story was performed. DT-user stories were built by personas and kept on evolving along with 'Empathy-Define-Personas' cycle approach to produce user feedback in the development of a workable prototype.

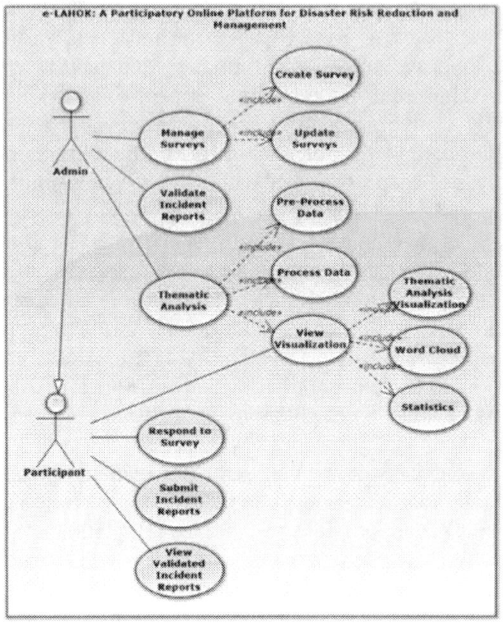

Figure 2. Use Case Diagram

This is where requirements were identified by the clients and researcher in the form of DT-user stories to determine how the system should function. Initial consultation was conducted by preparing letter requests, preliminary consultative meetings and key informant interviews with key personnel of Local Association of DRRM Officers of Albay (LADA) and Albay Public Safety and Emergency Management Office (APSEMO), as the technical and administrative head of Provincial Disaster Risk Reduction Management Council (PDRRMC).

More in-depth interviews were likewise conducted among residents of Legazpi City which include members of the vulnerable groups as well as other stakeholders to focus on describing individual experiences and beliefs in relation to disaster preparedness in their areas. These formed part of the DT-user stories as well as data gathered resulting to identification of the initial requirements up-front presented in the use case diagram in Figure 2 which is a representation of a user's interaction with the system. It is evident that there are two types of user's for e-LAHOK namely the admin and the participant which could either be registered or unregistered users. Creation and management of surveys, validation of submitted incident reports as well as performing thematic analysis are the functionalities provided to the administrator account. On the other hand participants may respond to surveys, submit incident reports and view visualizations of the analysis of results processed by the admin.

2.2. Planning Phase

In the planning phase, further development of the e-LAHOK platform is decided. This begins with the sorting and prioritization of the user stories and ends with a release plan. To create a release plan, the development tasks were broken into iterations. The release plan defines each iteration plan, which drives the development for that iteration.

The usability evaluation is likewise performed on DT-user stories and prototypes. Finally, actual users tested the prototype so that the feedback from the users were collected and changes were made accordingly. The final prototype generated for user stories were passed to the release planning phase.

2.3. Iterations to Release Phase

This is the phase where major development efforts occur. In the development of the participatory platform, the iterations observed include designing, programming, testing, listening, debugging and integration. Important aspects of development such as development time, resources and project schedule are taken into account. The user stories were converted to iterations that cover a small part of the functionality or features required. At the end of an iteration, users perform acceptance tests against the user stories. Identification and fixing of bugs become a step in the next iteration. A combination of iterations provided the customer with the final fully functional product.

2.4. Productionizing Phase

This phase encompasses the acceptance tests leading to customer approval then small releases which eventually provides the customer with the final fully functional system. The extent of user's acceptability in terms of functionality and usability was performed by utilizing an evaluation tool based on the ISO 9126 Software Quality Metrics in terms of functionality and usability.

2.5. Maintenance Phase

This phase encompasses the Planning, Iterations to Release, and Productionizing phases from the second release until the last release of the system. However, this phase is already outside the scope of the study.

3. Results and Evaluation

In this section, the results of the study conducted are presented. This includes the features of the e-LAHOK and the level of technical and user acceptance of the participatory platform in terms of functionality and usability.

3.1. e-LAHOK Features

The participatory platform provided an organized way of gathering inputs from the local communities of Legazpi City by facilitating orderly solicitation and submission of DRR strategies from the local populations with the use of data capture forms shown in Figure 3. Maintenance is supported with the survey management feature which include creation and administration of surveys as shown in Figure 4.

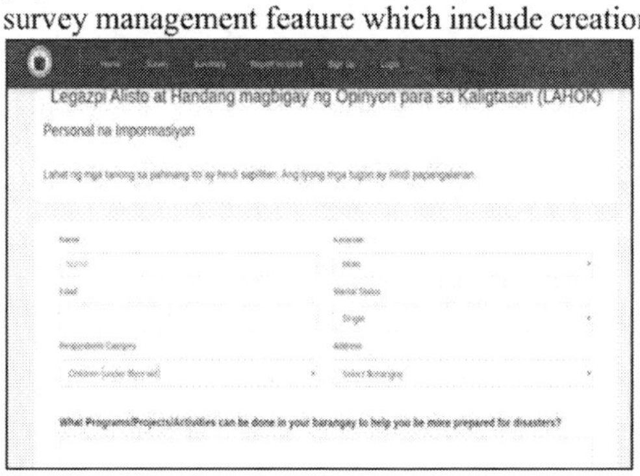

Figure 3. Data Capture Form

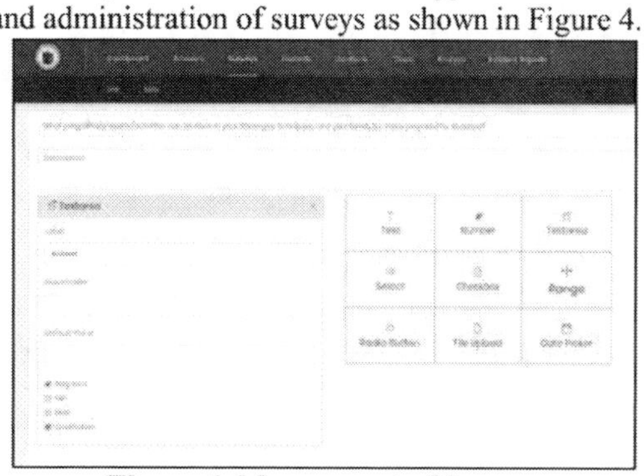

Figure 4. Management of Surveys

e-LAHOK likewise provided a platform for the local populations to submit incident reports, shown in Figure 5 which will be made available to the general public only upon validation by the administrators as shown in Figure 6.

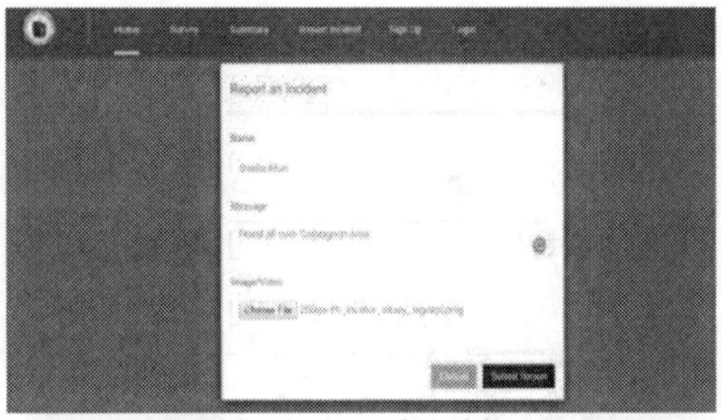

Figure 5. Submission of Incident Reports

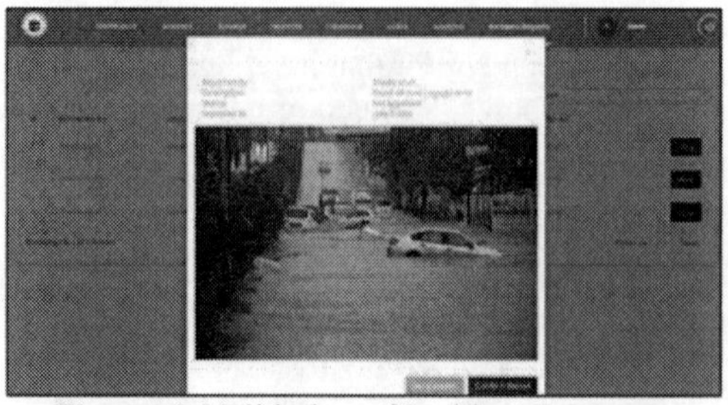

Figure 6. Validation of Incident Reports

The administrator account of e-LAHOK is provided with a platform to perform thematic analysis on the collected responses to identify patterns in the said corpus using topic modelling and word cloud visualization. A wizard guides the user through a series of well-defined steps as provided in Figure 7 to perform thematic analysis. Further, an interactive topic model visualization as provided in Figure 8 is made available using pyLDAvis, a Python library designed to help users interpret the topics in a topic model that has been fit to a corpus of text data [9].

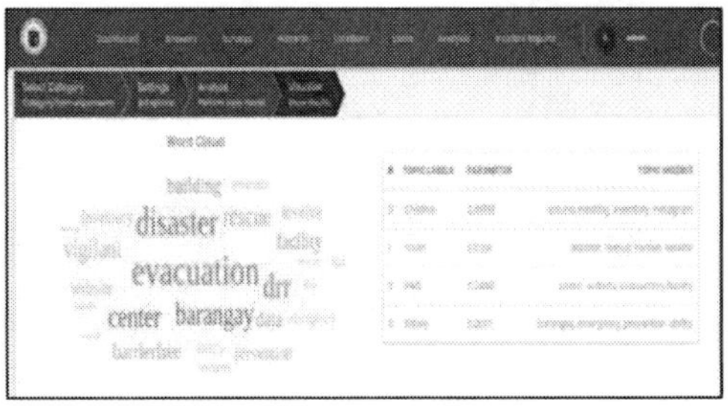

Figure 7. Thematic Analysis and Word Cloud Visualization

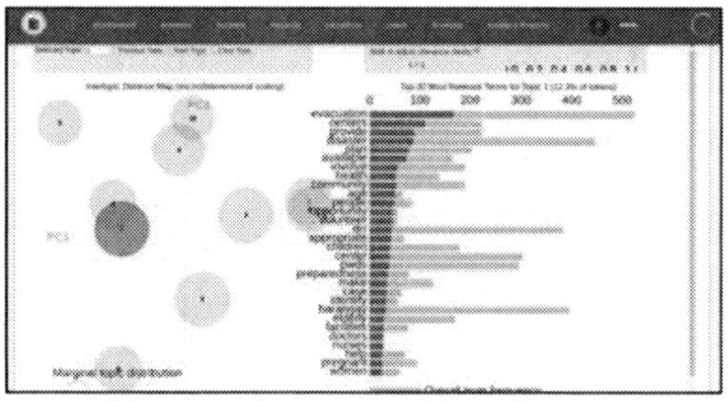

Figure 8. Interactive Visualization

3.2. Extent of User's Acceptability in Terms of Functionality and Usability

Functionality and usability are important aspects of software products. In order to determine the extent of acceptability of the participatory platform, respondents were identified to test and assess the functionalities and usability of the system using random sampling. A total of 40 respondents evaluated e-LAHOK of which 10 are children, 10 are from the youth sector, 5 are elderly, 5 PWDs, 5 pregnant & lactating mothers and 5 are DRR experts. The evaluation was based on the ISO 9126 Software Quality Metrics in terms of Functionality and Usability. Table 1 provides the consolidated data of all the ratings and indicators by metrics. It presents the level of user acceptance of the e-LAHOK based on ISO 9126 in terms of functionality and usability.

3.2.1 Functionality. Functionality means that the system meets all its functional requirements. The average weighted mean is 4.53 or Strongly Agree in terms of functionality. The respondents have strongly agreed that e-LAHOK performed the tasks required with a weighted mean of 4.55; provided creation, management, consolidation, analysis and visualization of survey responses for analysis with a weighted mean of 4.53 and provided submission, verification, and acceptance of incident reports, with a weighted mean of 4.55. Further, the respondents agreed that e-LAHOK provides accurate results, with a weighted mean of 4.48.

Table 1. Evaluation of e-LAHOK

Functionality			Usability		
Indicators	**Weighted Mean**	**Interpre-tation**	**Indicators**	**Weighted Mean**	**Interpre-tation**
1. The system performed the tasks required.	4.55	Strongly Agree	1. The user can easily learn how to use the system.	4.53	Strongly Agree
2. The system provided creation, management, consolidation, analysis and visualization of survey responses for analysis to authorized users.	4.53	Strongly Agree	2. The user interface's organization is clear, logical, effective and appealing to the user which makes it easy to understand the process flow.	4.58	Strongly Agree
3. The system provided submission, verification, and acceptance of incident reports to authorized users.	4.55	Strongly Agree	3. The user can easily operate and control the system.	4.38	Agree
4. The system provides accurate results.	4.48	Agree	4. The user could use the system without much effort.	4.55	Strongly Agree
Average Weighted Mean	**4.53**	**Strongly Agree**	**Average Weighted Mean**	**4.51**	**Strongly Agree**
Overall Weighted Mean	**4.52**	**Strongly Agree**			

5 – Strongly Agree (Highly Acceptable) 4 – Agree (Acceptable) 3 – Moderately Agree (Moderately Acceptable) 2 – Disagree (Unacceptable) 1 – Strongly Disagree (Highly Unacceptable)

3.2.2 Usability. Usability refers to the ease of access or use of the participatory platform. With an average weighted mean of 4.51 or Strongly Agree in terms of the usability of the system, the respondents have strongly agreed that learning to use the system is easy with a weighted mean of 4.53. They also strongly agreed that the user interface's organization is clear, logical, effective and appealing to the user which makes the process flow easy to understand with a weighted mean of 4.58 and that the platform could easily be operated and controlled resulting to a weighted mean of 4.38. Finally, the respondents agreed that e-LAHOK could be used without much effort with a weighted mean of 4.55.

4. Conclusions

Results show that the features of e-LAHOK provided an organized way of gathering inputs from the local communities of Legazpi City by facilitating orderly solicitation and submission of DRR strategies on how the barangays can help them be better prepared in times of disasters as well as incident reports from the local populations by allowing creation, management, consolidation, analysis and visualization of survey responses as well as submission, verification, and acceptance of submitted incident reports for analysis. These results could serve as basis to concerned agencies in the development and implementation of DRR plans and policies to ensure preparedness and effective response by the barangays and its citizens in the event of a disaster.This concluded that the features of e-LAHOK can effectively provide an organized way of gathering inputs from the local communities of Legazpi City, fostering civic engagement and open, participatory governance. As evaluated, the over-all weighted mean of 4.52 or Strongly Agree implied that e-LAHOK is highly acceptable as it provides the necessary requirements along with the necessary features. Thus, utilization of the proposed system is recommended for the benefits of Legazpi City.

References

[1] UNISDR *What is Disaster Risk Reduction?* [Internet] Available from: https://www.unisdr.org/who-we-are/what-is-drr

[2] Ostria R A 2019 Bicol 'Usman' death toll hits 105,20 Still Missing *Inquirer* [Internet] Accessed on 2019 Jan 3rd from: https://newsinfo.inquirer.net/1069020/bicol-death-toll-now-at-105-23-still-missing

[3] Maceda L L, Llovido J L and Serrano J E 2018 System Design for Disaster Risk Damage Assessment *2018 Int. Symp. on Computer, Consumer and Control (IS3C)* pp 246–9

[4] Secillano I J S and LADA 2018 From Experience to Change: Lessons Learned from 2018 Mayon

Volcanic Eruption *2018 South East Asia Disaster Risk Governance Academic Seminar* pp 150–8

[5] UNISDR 2012 *Community-based disaster risk management* [Internet] Available from: https://www.unisdr.org/we/inform/events/28132

[6] Rahman H A 2018 *Community Based Approach Towards Disaster Management in Malaysia Asian J. of Environment, History and Heritage* **2** 2 pp 55-66

[7] United Nations Department of Economic and Social Affairs 2019 *Citizen-Engagement* Available from: https://publicadministration.un.org/en/eparticipation

[8] Sohaib O, Solanki H, Dhaliwa N, Hussain W and Asif M 2019 Integrating design thinking into extreme programming *J. Ambient Intelligence and Humanized Computing* **10(6)** pp 2485–92

[9] Mabey B 2015 *pyLDAvis* [Internet] Available from: https://pyldavis.readthedocs.io/en/latest/

AUTHOR INDEX